STARFISH

STARFISH

Biology and Ecology of the Asteroidea

Edited by John M. Lawrence

THE JOHNS HOPKINS UNIVERSITY PRESS | BALTIMORE

The Johns Hopkins University Press
2715 North Charles Street
Baltimore, Maryland 21218-4363
www.press.jhu.edu

Library of Congress Cataloging-in-Publication Data

Starfish : biology and ecology of the Asteroidea / edited by
John M. Lawrence.
      p. cm.
  Includes bibliographical references and index.
  ISBN 978-1-4214-0787-6 (hdbk. : alk. paper) –
ISBN 1-4214-0787-6 (hdbk. : alk. paper)
1. Starfishes. I. Lawrence, John M.
  QL384.A8S73 2013
  593.9'3–dc23   2012020461

A catalog record for this book is available from the
British Library.

*Special discounts are available for bulk purchases of this book. For
more information, please contact Special Sales at 410-516-6936
or specialsales@press.jhu.edu.*

Contents

Contributors

Charles D. Amsler
Department of Biology
University of Alabama at
    Birmingham
Birmingham, Alabama, USA

Bill J. Baker
Department of Chemistry and
    Center for Drug Discovery
    and Innovation
University of South Florida
Tampa, Florida, USA

Mario Barahona
Departamento de Ecología and
    Centro Mileno de Conserva-
    ción Marina, Estación Costera
    de Investigaciones Marinas,
Las Cruces, Chile
Facultad de Ciencias Biológicas
Universidad Católica de Chile
Santiago, Chile

Michael F. Barker
Department of Marine Science
    and Portobello Marine
    Laboratory
University of Otago, Dunedin,
    New Zealand

Maria Byrne
Schools of Medical and
    Biological Sciences
University of Sydney, Sydney,
    New South Wales,
    Australia

Juan Carlos Castilla
Departamento de Ecología and
    Centro Mileno de Conserva-
    ción Marina, Estación Costera
    de Investigaciones Marinas,
Las Cruces, Chile
Facultad de Ciencias Biológicas.
    Universidad Católica de Chile
Santiago, Chile

Katharina Fabricius
Australian Institute of Marine
    Science
Townsville, Queensland,
    Australia

Patrick Flammang
Université de Mons
Laboratoire de Biologie Marine
Académie Universitaire
    Wallonie-Bruxelles
Mons, Belgium

Andrew S. Gale
School of Earth and
    Environmental Sciences
University of Portsmouth
Portsmouth, United Kingdom

Carlos F. Gaymer
Departamento de Biologia
    Marina, Centro de Estudios
    Avanzados en
    Zonas Áridas
Universidad Católica del Norte
Coquimbo, Chile

Jean-François Hamel
Society for the Exploration and
    Valuing of the Environment
Portugal Cove, St. Philips,
    Newfoundland and Labrador,
    Canada

Elise Hennebert
Université de Mons
Laboratoire de Biologie Marine
Académie Universitaire
    Wallonie-Bruxelles
Mons, Belgium

John H. Himmelman
Départment de Biologie,
    Université Laval
Quebec City, Quebec,
    Canada

Michel Jangoux
Laboratoire de Biologie Marine
Université de Mons
Académie Universitaire
    Wallonie-Bruxelles
Mons, Belgium
Laboratoire de Biologie Marine
Université Libre de Bruxelles
Académie Universitaire
    Wallonie-Bruxelles
Brussels, Belgium

John M. Lawrence
Department of Integrative Biology
University of South Florida
Tampa, Florida, USA

Tatiana Manzur
Departamento de Ecología and
  Centro Mileno de Conserva-
  ción Marina, Estación Cos-
  tera de Investigaciones
  Marinas
Las Cruces, Chile
Facultad de Ciencias Biológicas
Universidad Católica de Chile
Santiago, Chile

James B. McClintock
Department of Biology
University of Alabama at
  Birmingham
Birmingham, Alabama, USA

Bruce A. Menge
Department of Zoology
Oregon State University
Corvallis, Oregon, USA

Annie Mercier
Ocean Sciences Centre
Memorial University
St. John's, Newfoundland and
  Labrador, Canada

Anna Metaxas
Department of Oceanography
Dalhousie University
Halifax, Nova Scotia,
  Canada

Sergio A. Navarette
Departamento de Ecología and
  Centro Mileno de Conserva-
  ción Marina, Estación Costera
  de Investigaciones Marinas
Las Cruces, Chile
Facultad de Ciencias Biológicas
Universidad Católica de Chile
Santiago, Chile

Timothy D. O'Hara
Museum Victoria
Melbourne, Victoria, Australia

John S. Pearse
Long Marine Laboratory
University of California,
  Santa Cruz
Santa Cruz, California, USA

Carlos Robles
Department of Biology
California State University
  at Los Angeles
Los Angeles, California, USA

Eric Sanford
Department of Evolution and
  Ecology
University of California
Davis, California, USA
Bodega Marine Laboratory
Bodega Bay, California, USA

Robert E. Scheibling
Department of Biology
Dalhousie University
Halifax, Nova Scotia, Canada

Richard L. Turner
Department of Biological
  Sciences
Florida Institute of Technology
Melbourne, Florida, USA

Carlos Renato R. Ventura
Museu Nacional
Departamento de Invertebrados
Universidade Federal do Rio de
  Janeiro
Quinta da Boa Vista, São
  Cristóvão
Rio de Janeiro, Brazil

Kristina M. Wasson
Department of Biology
University of Alabama at
  Birmingham
Birmingham, Alabama, USA

Stephen A. Watts
Department of Biology
University of Alabama at
  Birmingham
Birmingham, Alabama, USA

Preface

Starfish are icons of the sea. But they also have immense biological and ecological importance. They are one of the major classes of Echinodermata, with a distinctive and varied body form. Starfish live in a variety of marine habitats where they are major predators that greatly affect their communities. With a long fossil record, starfish are of tremendous interest to paleontologists and evolutionary biologists.

Although study of starfish biology began in the second half of the nineteenth century, many basic aspects of their physiology, nutrition, reproduction, and even anatomy still are not well known. Study of starfish ecology began in the first half of the twentieth century but has been limited primarily to species whose effect can be conspicuous, including the crown-of-thorns starfish on coral reefs and species that affect shellfish fisheries. Effects of other species are not as well understood because they are not as accessible or conspicuous and they do not have an economic impact.

This book provides a comprehensive understanding of the biology and ecology of starfish. It is divided into two sections: comparative chapters that consider aspects of biology and ecology across the class and integrative chapters that consider aspects of biology and ecology of individual species or genera. These approaches are complementary. Species and genera were selected for the integrative chapters based on the scope of knowledge about them and the availability of experts to be authors.

I am grateful to my friends and colleagues, experts in their fields, who have contributed to this book. It is a pleasure to thank executive editor Vincent Burke, Jennifer Malat, and Michele Callaghan at the Johns Hopkins University Press for their kind assistance in its production.

STARFISH

PART I • COMPARATIVE BIOLOGY AND ECOLOGY

1

Phylogeny of the Asteroidea

Andrew S. Gale

The phylogeny of asteroids has been a controversial topic for more than 100 years. The detailed morphology of spines and pedicellariae formed the basis of an ordinal classification by Perrier (1884, 1894), consisting of the Paxillosida, Spinulosida, Valvatida, and Forcipulatida, which have obtained widespread use among neontologists. Apart from some rearrangement of taxa included variously in the Spinulosida and Valvatida (Blake 1981) and the resurrection of Perrier's Velatida (Blake 1987; Fig. 1.1a), the classification has been remarkably stable. In contrast, the interrelationships of the orders have remained obscure and controversial.

Fossil asteroids have been known since the early nineteenth century. It was clear from early studies (e.g., Forbes 1848) that many Jurassic and later asteroids were surprisingly similar to extant taxa at the level of families or even genera. Concomitantly, it became increasingly clear that Paleozoic asteroids differed significantly from later ones. In spite of this, Spencer and Wright (1966) attempted to integrate Paleozoic asteroids into a modified version of Perrier's orders, thus tracing the orders Spinulosida, Paxillosida, Valvatida, and Forcipulatida back into the Ordovician. They were strongly influenced by H. B. Fell's (1963) interpretation of the living luidiid *Platasterias* as a "living fossil" and by a surviving early Paleozoic somasteroid (a basal asterozoan group; see Shackleton 2005). They also placed the Eocene-Recent family Luidiidae (generally classified in the Paxillosida) in the Paleozoic group Platyasterida. Their tentative phylogeny showed the somasteroids, Platyasterida and Paxillosida, as closely related to each other and basal to the asteroid tree, with Valvatida positioned next to Forcipulatida. *Platasterias* was duly replaced, correctly, in the Luidiidae and the Paxillosida (J. Madsen 1966, Blake 1972, 1982).

Subsequently, two independent studies (Blake 1987; Gale 1987) proposed that the post-Paleozoic asteroids formed a monophyletic group (Neoasteroidea GALE, 1987) which evolved in the Late Paleozoic or earliest Mesozoic. This group was distinguished by numerous shared characters of the musculoskeletal system of the ambulacral groove and mouth. The authors of these publications concluded that Perrier's orders could be traced back only as far as the Jurassic or,

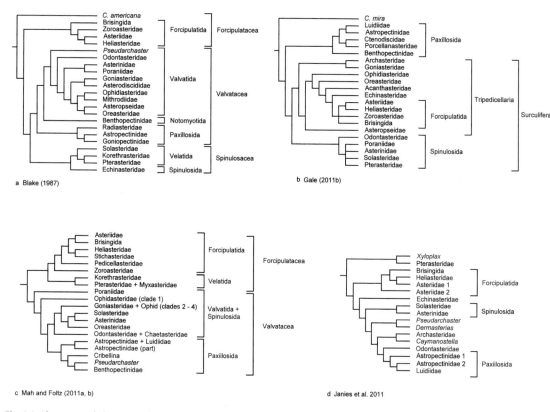

Fig. 1.1. Alternative phylogenies of asteroids, *a* and *b* based on morphological evidence and *c* and *d* on molecular data (12S, 16S, histone H3). Note the considerable differences between these trees in the identification of basal taxa; the only consistent features are the monophyly of the Paxillosida and Forcipulatida and (in all except *a*) the position of Solasteridae and Asterinidae as sister taxa. Note that the basal position of the Forcipulatida in *a, c* is an assumption, and is not based on reference to an outgroup. *a, Blake 1987; b, Gale 2011b; c, Mah and Foltz 2011a,b; d, Janies et al. 2011.*

more controversially, the Triassic (Blake and Hagdorn 2003). However, the phylogenies of Gale and Blake differed in important respects. For Gale (1987), the Paxillosida formed the basal group of neoasteroids, whereas Blake (1987; Fig. 1.1a) considered the Forci-

pulatacea (an extended Forcipulatida, including the Carboniferous *Calliasterella*; see Kesling and Strimple 1966) to be most primitive.

A fundamental problem in the study of asteroid phylogeny is the frequent occurrence of homoplasy—

Fig. 1.2. *(Opposite)* Skeletal construction of ambulacral groove (*a,b,f,g*) and mouth frame (*c,h–m*) in asteroids. Ambulacral groove ossicles in abactinal (*a,d,f*) and lateral-abradial (*b,e,g*) views. Outline of ambulacral ossicles dotted line in *a,d,f. a,b,* Ophidiasterid *Nardoa variolata* (Valvatida). *c,* Abactinal view of mouth frame of goniasterid *Ceramaster granularis*, Recent, North Sea. Note the presence of five unpaired interradial odontophores (od), 10 paired orals and circumorals (co), and radial (riom) and interradial (abiim) muscles that serve to close the mouth frame. The proximal blades of the oral ossicles (pb) articulate. *d,* Poraniid *Porania pulvillus. e,* Solasterid *Crossaster papposus* (both Spinulosida). *f,g,* Luidia sp. (Paxillosida). The articular facet ada3 of paxillosids is restricted to successive adambulacrals (as in Late Paleozoic taxa). In all other neoasteroids, it also articulates with the ambulacral (*a,b,d,e*). In paxillosids, the articular facet ada1 is single and forms a firm articulation, effectively locked with the distal ambulacral (*f,g*); in other neoasteroids, a double surface (ada1a,b) contacts the ambulacral. This permits freer articulation between ambulacrals and adambulacrals, such that the ambulacral surfaces can rotate in the concave structures formed by ada1a/ada2 and ada1b/ada3 (*a,d*). *h,k,* Dissociated oral ossicle of Recent ophidiasterid *Nardoa variolata* from Mauritius, in radial (*h*) and interradial (*k*) views. Two structures articulate with the circumoral (pcoa, dcoa); the first adambulacral ossicle articulates with the oral (orada) and is joined to it by a muscle (oradm). Notches for the ring nerve (nrg) and ring vessel (rvg) are present. Pairs of orals articulate interradially by means

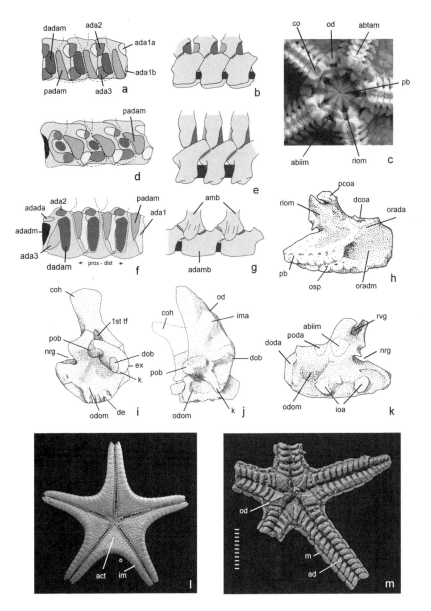

of interoral articular surfaces (ioa). Oral spines have distinctive basal attachments (osp). *i,j,* Interradial view of articulating circumoral-oral and odontophore/axillary in Recent *Luidia* sp. (*i*) and Pennsylvanian (Moscovian) *Calliasterella mira* from Moscow illustrates fundamental differences between Paleozoic asteroids and neoasteroids in mouth frame construction. While landmark structures are homologous, the huge abactinal extension in *C. mira* (od) articulates with the first two inferomarginal ossicles (ima). In *Luidia* and all neoasteroids, the odontophore is entirely separate from the marginals, although it retains a small external face (ex) found in many Paxillosida. The odontophore muscle (odom) runs from the keel (k) of the odontophore/axillary to the inner face of the oral. *l,* Oreasterid *Protoreaster* in actinal view, to show inferomarginals (im) and actinals (act). *m,* Actinal view of *Permaster grandis,* from the Artinskian (Permian) of Wandagee, Western Australia, shows single marginal row (m) that contacts the adambulacrals (ad) in the absence of actinal ossicles. The small external face of the odontophore/ axillary is in contact with the interradial marginals, as in all Late Paleozoic taxa.

Other abbreviations: abtam, abactinal transverse ambulacral muscle; act, actinal ossicles; adada, adambulacral-adambulacral articulation; adadm, adambulacral-adambulacral muscle; adamb, adambulacral; amb, ambulacral; coh, circumoral head; dadam, distal adambulacral-ambulacral muscle; de, dentition; dob, distal odontophore bar; doda, distal odontophore-oral articulation; orada, oral-adambulacral articulation; padam, proximal adamb-amb muscle; pob, proximal bar of odontophore; poda, proximal odontophore-oral articulation; 1st tf, opening for first tube foot.

the independent evolution of strikingly similar body forms and ossicle arrangements. For example, in the Carboniferous (Late Paleozoic), morphologies evolved that in some detail superficially resemble the post-Paleozoic (dominantly Recent) families Luidiidae and Zoroasteridae. *Illusioluidia* (Blake and Guensburg 1989a) superficially resembles an extant luidiid, having parallel-sided arms, with broad, short ossicles framing the ambulacral groove, and an abactinal surface constructed of paxillae as in *Luidia*. In this case, recognition of convergence is straightforward, because the short, broad ossicles are adambulacrals in *Illusioluidia* but inferomarginals in Luidiidae (Blake and Guensburg 1989a).

Downey (1970) noted the close morphological similarities in disc ossicle arrangement and form between the Pennsylvanian *Calliasterella mira* and the extant zoroasterid *Doraster constellatus,* and hinted at a close relationship, which presumably prompted Blake's (1987) argument for the basal position of the Forcipulatida. Detailed examination of the mouth frame and ambulacral groove ossicles of *C. mira* demonstrates the existence of fundamental differences between this taxon and neoasteroids. Thus, the similarities are convergent (Gale 2011a). In fact, the post-Palaeozoic taxa are united by numerous skeletal characters of the ambulacral groove and mouth frame, the presence or absence of which can be readily identified in fossils.

Molecular phylogeny should provide irrefutable evidence of higher level phylogenetic relationships, but studies on asteroids to date have each generated very different trees, none of which are congruent with morphology (Matsubara et al. 2004, Mah and Foltz 2011a,b, Janies et al. 2011; see Fig. 1.1c,d). The only consensus achieved by these studies are that the Paxillosida and Forcipulatida are monophyletic and that the Asterinidae and Solasteridae are sister taxa. Other than this, most of the proposed interrelationships between higher groups are not supported by morphological evidence. In particular, the "Forcipulatacea" (Forcipulatida + Velatida), identified as the basal asteroid group by Mah and Foltz (2011a,b), is not supported by any morphological features and therefore appears to be a highly improbable grouping. The basal position of this group is entirely a supposition. Likewise, the basal position in the asteroid tree of the highly derived taxa Pterasteridae + *Xyloplax* (Janies et al. 2011) lacks any supporting morphological evidence. Recently, Mah and Blake (2012) produced an asteroid phylogeny based on a summary of published molecular studies and Blake's 1987 cladogram. This shows a five-way basal polytomy at the base of the tree (*Xyloplax*, Echinasteridae, Forcipulatida, Velatida, and all other asteroids).

The morphological phylogeny followed here uses the consensus tree generated by Gale (2011a) which has 128 characters and 23 extant species representing 23 families, with the Carboniferous *C. mira* as outgroup. The main features of this phylogeny are as follows (Fig. 1.1b):

- The basal position of a monophyletic Paxillosida.
- All other neoasteroids form a monophyletic group called the Surculifera (Gale 1987), characterized by flat-tipped tube feet and possession of entire digestive systems and brachiolaria larvae.
- A basal division of the Surculifera into two clades, the Spinulosida and Tripedicellaria.
- The Tripedicellaria includes the Forcipulatida, to which the valvatids form a stem group.

Additional important phylogenetic information has been obtained from the Jurassic record of fossil asteroids, which provides evidence of the derivation of the extant families Korethrasteridae and Pterasteridae from close to the asterinid genus *Tremaster*. However, the fossil record provides little if any evidence of the interrelationships between the higher groups (Paxillosida, Spinulosida, Valvatida, Forcipulatida), which all appear in the Early Jurassic. This tree obtains some support from the fossil record, which indicates very early radiation of the neoasteroids, within the Triassic.

Monophyly and Early Radiation of the Neoasteroidea

It is now generally agreed that the post-Paleozoic asteroids are a monophyletic group that likely evolved from a single ancestor surviving the Permian-Triassic boundary (Gale 1987, 2011a). Fundamental, functionally significant changes in mouth frame construction and ambulacral groove articulation and musculature are the most important unifying characters of the clade Neoasteroidea, which includes all known post-Paleozoic asteroids (Gale 1987, 2011a).

Ambulacral Groove

Ambulacral ossicles articulate across the mid-radial line by means of an interlocking dentition, which serves as a fulcrum for the opposing muscle pairs that open and close the groove—the abactinal transverse ambulacral and actinal transverse adambulacral muscles (Fig. 1.2c). Ambulacral and adambulacral ossicles alternate, and articulate with each other by means of specialized surfaces (Fig. 1.2a,b,d–g). A transverse ridge (or two rounded facets) on the proximal part of the base of each ambulacral ossicle articulates with the distal part of an adambulacral (Fig. 1.2a, ada1). Distal articular surfaces on the ambulacral base articulate with corresponding surfaces on the proximal adambulacral (ada 2,3). Muscles attach the proximal wing of the ambulacral to a facet on the distal adambulacral (padam), and the distal wing of the adjacent ambulacral to the proximal adambulacral (dadam). Successive adambulacrals articulate by means of one or two surfaces, and contraction of an adambulacral muscle (Fig. 1.2f, adadm) aids downward flexing of the arm. Adjacent ambulacral heads are linked by articular surfaces and muscles. This muscle/articulation system permits fine adjustment in ray flexure and allows twisting motions to take place.

The musculoskeletal framework of the neoasteroid arm, involving six different serially repeated muscles and seven articular surfaces (i.e., for each ambulacral-adambulacral pair) is one of the most complex found in the Echinodermata and probably enables the extraordinary range of movements observed in asteroid rays, permitting them to perform diverse functions. These movements are aided and amplified by transverse and longitudinal muscles in the body wall of the rays (Heddle 1967). Extreme examples of movements are the burrowing habits observed in the paxillosids *Astropecten* and *Luidia* (Heddle 1967; see Chapters 10 and 11), which also possess an additional ossicle (superambulacral) and an additional two muscles. The powerful A-frame provided by the ambulacral arch and the abactinal transverse muscles in *Asterias* (Eylers 1976, fig. 2.3) permits immense pressure to be applied to the valves of bivalve molluscs by the muscular tube feet.

Some Paleozoic asteroids possessed elements of the ambulacral groove musculoskeletal system; for example, *Calliasterella* has dentition, longitudinal interambulacral and interadambulacral muscles, and ambulacral-adambulacral muscles (Gale 2011a). However, it lacks transverse ambulacral muscles and therefore cannot have functioned in the same way as a neoasteroid. All post-Palaeozoic asteroids known (Neoasteroidea) share all the characters described above, with the single exception of the highly derived *Xyloplax* (Gale 2011a).

Mouth Frame

The asteroid mouth frame consists of ten pairs of oral and circumoral muscles and five unpaired, interradial odontophores (Viguier 1879, Gale 2011a; Fig. 1.2c). The odontophores are positioned between paired oral ossicles with which they articulate by means of proximal and distal processes (Fig. 1.2i). These processes contact smooth vertical articular surfaces on the interradial face of the orals (Fig. 1.2k), and an odontophore muscle (Fig. 1.2i, odom) inserts into the actinal surface of the odontophore and the inner surface of the oral ossicle. The orals of a pair are united by two or more muscles (abactinal interradial interoral muscles, abiim) and adjacent pairs by further muscles (radial interoral muscles, abbreviated as riom, Fig. 1.2h). The oral ossicles articulate with the first adambulacral ossicle distally and abactinally with the circumoral ossicles. The first tube foot emerges through the rounded slot formed by the circumorals and oral ossicles. The mechanics of mouth frame function are not well understood and still based largely on the interpretation of Viguier (1879). Essentially, the five pairs of interradial muscles, five radial pairs, the five transverse actinal muscles on the circumorals, and the five odontophore-oral muscles close the peristome by contracting (Fig. 1.2c). When they relax, the peristome is opened, partly by the resistance provided the extraoral skeleton (e.g., the odontophore is attached to the interradial body wall) perhaps aided to a minor extent by contraction of the adambulacral muscles.

The mouth frame ossicle system is known completely in the Carboniferous *Calliasterella mira* (Gale 2011a). It differs significantly from any neoasteroid mouth frame, specifically in that the homologue of the odontophore, the axillary ossicle, is very large, unpaired interradial ossicle in that is contact with the marginals (Fig. 1.2j). An odontophore muscle is apparently present.

Actinal Ossicles

Almost all adult neoasteroids possess rows of actinal ossicles in the triangular interareas of the disc and arms formed by the adambulacrals, the marginal frame, and the oral ossicles (Fig. 1.2l). These are absent in almost all known Late Paleozoic asteroids (Blake and Elliot 2003), in which the axillary ossicle invariably contacts both marginal and oral ossicles (Gale 2011a; Fig. 1.2j,l). In the Devonian Xenasteridae (Schöndorf 1909) actinals are present, and pairs of marginal ossicles became incorporated interradially in the disc, increasing the minor radius. The dissociation of the axillary from the marginal frame was probably the critical evolutionary event and allowed the development of actinal ossicles in neoasteroids. The development of actinal ossicles permitted expansion of the disc size and created increased space for digestive organs and gonads.

Early Fossil Record of the Neoasteroids

A gap of about 40 million years in the fossil record exists between the latest Paleozoic asteroid fauna known from the Artinskian (Late Permian) of Australia (Kesling 1969) and the first Mesozoic records. The earliest neoasteroids appeared in the Anisian-Ladinian (245-228 Ma, Early Triassic) Muschelkalk of Germany and include distinctive morphologies represented by *Trichasteropsis* and *Migmaster* (Blake and Hagdorn 2003). Although these are clearly neoasteroids (ambulacral groove and mouth frame characters), they have a number of distinctive characters, such as the very broad adambulacrals that are never seen in later taxa. Therefore, assignation of Triassic asteroids to extant orders does not seem to be justified. They are morphologically unique and taxonomically enigmatic, in the opinion of the author. It is apparent that much of the early radiation of the neoasteroids is undocumented from the fossil record as presently known.

By the earliest Jurassic (Hettangian Stage, about 199 Ma), undisputable members of all extant orders are present, although it is not clear how many of these should be assigned to extant families. For example, Hettangian taxa referred to the Asteriidae have unique primitive pedicellariae of a type unknown in other forcipulatids (Gale 2011a). By the Toarcian (180 Ma), Spinulosida (Korethrasteridae, Tropidasteridae, Plumasteridae), Paxillosida (Benthopectinidae, Astropectinidae), and Valvatida (Sphaerasteridae, Staurande-rasteridae, Goniasteridae) were present.

Basal Neoasteroids—the Paxillosida

Two strongly contrasting views on the polarity of neoasteroids have coexisted for more than a century. Researchers have singularly failed to achieve any consensus to date, in spite of the recent application of molecular techniques to the problem. One viewpoint, championed by Mortensen (1921), identified the Paxillosida as basal to the Asteroidea because these asteroids lack brachiolaria larvae. Conversely, MacBride (1921) argued that the loss of the brachiolaria was secondary and a result of specialization to soft substrata on which brachiolar arms had no use. The morphological evidence for the basal position of paxillosids was developed further by Gale (1987), supported by Heddle (1995), and refuted by Blake (1987, 1988). Blake argued that paxillosids evolved from valvatids by the loss of suckered tube feet, much of the digestive system, and the brachiolaria and became soft substrate specialists.

While the paxillosids are undoubtedly highly specialized, dominantly infaunal asteroids that live as predators and deposit feeders on soft substrata, the weight of developmental and morphological evidence supports their basal position (Gale 2011a). A summary of this evidence includes the following:

- Nature of adambulacral and ambulacral contacts with a single distal ambulacral-adambulacral articulation (ada1, Fig. 1.2e,f) as in Paleozoic asteroids.
- The proximal abradial adambulacral articulation surface (ada3, Fig. 1.2e) does not contact an ambulacral, as in Paleozoic asteroids, but simply involves adambulacrals.
- Ambulacral heads are not imbricated, as in Paleozoic taxa.
- The circumoral ossicle morphology is similar to that of Paleozoic asteroids.
- Only elementary pedicellariae are present, similar to those found in the few Paleozoic asteroids known to possess pedicellariae.
- Major differences exist in early development between paxillosids and all other neoasteroids; the madreporite appears in the ring of five first formed ossicles (terminals) in paxillosids (e.g. Kano et al. 1974; Komatsu 1975).
- The madreporite in paxillosids has a marginal positions, more like that in Paleozoic asteroids, than in other asteroids.
- The external face on the odontophore is shared with Paleozoic asteroids.
- The tube feet are always pointed or semi-pointed and nonsuckered.
- A complete digestive system is lacking in most paxillosids, and they are unable to feed extraorally. McClintock et al. (1983) report extrusion of the stomach, apparently to feed on organic-rich particles.
- Brachiolaria larvae are absent.

It is concluded that the paxillosids are a monophyletic group at the base of the neoasteroid radiation, which live as infaunal predators (Astropectinidae, Luidiidae; see Chapters 10, 11) or deposit feeders (Cribellina) on soft substrata. Separating characters that are strictly plesiomorphic from those that are adaptive to specific life habits will always remain controversial. The phylogeny of the paxillosids shows some congruence between morphological and molecular trees (Gale 2011a; Mah and Foltz 2011a), with Benthopectinidae basal to the group and Luidiidae and Astropectinidae as sister taxa.

Paxillosida first appeared in the Carnian (Triassic) of China (personal observation), and have a good fossil record from the Jurassic to the present. The astropectinid *Archastropecten* is a widespread and often abundant element of Mid Jurassic to Early Cretaceous soft substrate shelf faunas, and its detailed skeletal morphology confirms its position as a basal member of the family (Gale 2011b). In the Early Cretaceous, essentially modern forms close to the present day deeper water *Tethyaster* (genera *Coulonia* and *Capellia*) occupied similar habitats. Astropectinids with skeletal features of *Astropecten sensu stricto* (highly asymmetrical proximal and distal ambulacral-ambulacral muscles, transversely arranged) appear in the Eocene, and the worldwide radiation of the genus in dominantly shallow marine habitats took place through the Cenozoic with extensive vicariance speciation related to ocean spreading and closure (Zulliger and Lessios 2010).

Benthopectinids appeared in the Early Jurassic (Pleinsbachian). The best-known genus *Jurapecten* shared most important characters with extant forms, such as the highly specialized ambulacral-adambulacral articulation, alternating supero- and inferomarginals, but entirely lacked the transverse ambulacral ridges that carry the longitudinal arm muscles in extant taxa (Gale 2011a,b). Goniopectinids appeared in the Late Jurassic (Oxfordian) and already possessed the distinctive ambulacral-adambulacral articulation and mouth frame characters of the clade Cribellina (Gale 2005, 2011a,b). However, these apparently lacked the cribriform organs and their lamellar cover spines found in Cretaceous to present-day goniopectinids (Gale 2005). *Luidia*, which was probably derived from an astropectinid ancestor, is present by the Eocene (35 Ma).

In summary, paxillosids appeared later in the fossil record than their inferred basal position to other asteroids would predict. Jurassic paxillosids were common and diverse and include basal astropectinids, benthopectinids, and goniopectinids, which all lack characters found in later representatives of the respective families. Astropectinids show stepwise changes from the primitive Jurassic *Archastropecten*, through an intermediate level of organization (Cretaceous *Tethyaster* grade) to taxa close to *Astropecten* (Eocene to present day). The Radiasteridae have no fossil record (Gale 2011a).

The Surculifera: Stomach Eversion, Brachiolariae, and Arm Flexibility

The Surculifera (Gale 1987) include all non-paxillosid neoasteroids, most obviously characterized by the presence of adhesive, flat-tipped tube feet with or without sucking discs (Vickery and McClintock 2000a; see Chapter 3); the ability to feed extraorally by stomach eversion (Jangoux 1982a); and development including a brachiolaria larval stage (direct and indirect modes of Oguro 1989, Wada et al. 1996). In terms of early skeletal development, all taxa (>20 investigated across diverse Goniasteridae, Oreasteridae, Ophidiasteridae, Asterinidae, Solasteridae, Asteriidae; Kano et al. 1974, Oguro et al. 1976, Komatsu et al. 1979, Chia et al. 1993) studied to date show the Late Madreporic Mode (Gale 2011a) in which the madreporite appears only after all other major ossicles types are fully formed. This developmental mode probably results in a more centrally placed madreporite in Surculifera (Gale 2011a). The odontophore is invariably completely internal. Surculifera are also united by distinctive ambulacral-adambulacral articulation structures that differ significantly from those of Paxillosida and Paleozoic asteroids:

- The distal ambulacral-adambulacral articulation (Fig. 1.2a, ada1a,b) is divided into two discrete surfaces, one adradial and the other abradial (Fig. 1.2a,b,d,e).
- The proximal abradial adambulacral articular surface (Fig. 1.2a, ada3) contact both the adjacent adambulacral and the ambulacral ridge (Fig. 1.2a,b,d,e).
- The proximal bar of the circumoral ossicle is parallel with the body of the ossicle.
- These ambulacral articulations provide a much freer, "three way roller" articulation between ambulacral and adambulacral ossicles (Gale 2011a; Fig. 1.2a,b,d,e). In paxillosids, the firm, broad attachment on the single distal ambulacral-adambulacral articulation (Fig. 1.2a, ada1) limits the flexibility of the articulation (Fig. 1.2f,g).

The larval, developmental soft tissue and skeletal morphological features outlined above demarcate surculiferids sharply from paxillosids, with the exception of the genus *Pseudarchaster* and related taxa. Traditionally assigned to the Goniasteridae, *Pseudarchaster* (and other members of the Pseudarchasterinae, *Paragonaster* and *Gephyreaster*) is actually a paxillosid, as confirmed by the adambulacral-ambulacral morphology and the presence of distinctive intermarginal fascioles. This is confirmed by molecular studies (Mah and Foltz 2011). However, it has a number of unique characters for paxillosids that place it close to the Surculifera, notably the robust, goniasterid-like form of the skeleton, possession of a brachiolaria larva (Wada et al. 1996) and flat non-suckered tube feet (Vickery and McClintock 2000a). *Pseudarchaster* can perhaps be considered as a basal surculiferid, retaining a number of paxillosid characters.

Early in their history, probably within the Early Triassic, the Surculifera divided into two well-demarcated groups, the Spinulosida and Tripedicellaria (Gale 2011a; Fig. 1.1b).

Radiation of the Spinulosida

The Spinulosida (minus the distantly related Echinasteridae) are a highly distinctive group of asteroids, characterized externally by clusters of glassy spines composed of elongated trabeculae and rather lightly developed, often paxilliform abactinal and (reduced) marginal skeletons. True pedicellariae are absent through secondary loss. The ossice morphologies of the ambulacral groove and mouth frame are highly distinctive:

- The ambulacral heads are highly asymmetrical, elongated proximally, and show strong imbrication proximally (Gale 2011a, text fig. 9)
- The interadambulacral articulations and the proximal articular surfaces unite form a single continuous articular surface (Fig. 1.2d)
- Oral ossicles have a rectangular body, and a vertical apophysis, and well-demarcated oral and suboral spines (Gale 2011a, text figs. 17, 18)

Precise phylogenetic relationships of spinulosids are not currently well understood, with the exception of the velatid families Korethrasteridae and Pteroasteridae (see below). The Early Jurassic record also demonstrates the existence of a significant early spinulosid radiation, from which few taxa survived into the Mid Jurassic. This includes diverse taxa of unique morphological forms such as the multiarmed Plumasteridae, with several separate genera (Gale 2011a), the Tropidasteridae (Blake 1996), and the genera *Decacuminaster, Plesiosolaster, Plesiastropecten,* and *Xandarosaster*. In this context, extant spinulosids such as Solasteridae and Asterinidae are seen to be morphologically conservative, but highly successful, survivors of a major Early Jurassic radiation of the group. Refinement of spinulosid phylogeny must await cladistic analysis of both fossil and living forms. However, molecular phylogenies consistently show a close relationship between asterinids and solasterids (Matsubara et al 2005; Janies et al. 2011; Mah & Foltz 2011a), supporting the morphological concept of Spinulosida reintroduced by Gale (2011a).

Origin of the Velatid Clade: Evidence from the Jurassic Fossil Record

The asterinid subfamily Tremasterinae is represented in present-day seas by *Tremaster mirabilis*, which has a global distribution on rocky substrata in deep water (Jangoux 1982c, Clark and Downey 1992). The species has a rounded outline, a domed, concavo-convex form, and, uniquely, calcified internal interradial ducts leading from five openings on the abactinal surface (adjacent to the primary interradial ossicles) to open on the actinal surface close to the oral ossicles. The ducts are used to brood young (Jangoux 1982b) and are floored by paired interradial rows of I- and Y-shaped chevron ossicles (Gale 2011a, text fig. 2). Tremasterines are represented in the Jurassic by a number of species, including *Protremaster uniserialis* SMITH AND TRANTER (1985) from the Sinemurian (Lower Jurassic, 195-190 Ma) of Antarctica. This is in many ways a typical tremasterine (overall body form, calcified ducts), but it shows remarkable similarities with the living *Korethraster hispidus*, a boreal Atlantic species belonging to the velatid family Korethrasteridae (Clark and Downey 1992; Gale 2011a, text fig. 25). These similarities include the following:

- The first adambulacrals are narrow, the others are short and very broad, extending to the lateral margin and carry two to three rows of large spine attachment sites.
- The interradial ducts are opened actinally, to expose a groove formed by the chevron plates
- Actinal and marginal ossicles are lost or fused with the adambulacrals.
- Large, flared, flattened actinolateral spines are present, attached to the adambulacral extensions.
- The abactinal ossicles are flat and imbricate like tiles, and carry a central attachment site for a ring of spines.

However, the interradial chevron ossicles are not exposed abactinally in *Protremaster* but are covered by abactinal ossicles. The Jurassic fossil *P. uniserialis* thus presents a form that is morphologically intermediate between an asterinid and a korethrasterid. This is significant, because korethrasterids are stem group to the highly derived Pterasteridae (Gale 2011a). *Protremaster* therefore appears to be basal to this important modern deep-sea group. However, it is necessary to ask the question: is this a convergent similarity or does it represent a real (albeit unexpected) phylogenetic relationship?

This question can be tested using ossicle morphology. The oral ossicles of asterinids, *Tremaster, Protremaster*, korethrasterids, and pterasterids have a unique morphology, with a rectangular oral body and a tall, vertically positioned apophyse (Gale 2011a, text fig. 17). Oral ossicles of these taxa are morphologically very similar, with an elongated radial vessel groove, a tall proximal flange that carries the radial and interradial muscles and the nature of the circumoral articulation (Gale 2011a, pl. 13, text figs. 17, 18). The radial faces of the oral ossicles of the taxa are also very similar, with a sharply deflected proximal margin, bearing a row of oral spines, and few, centrally placed large suboral spine bases. About 15 shared, derived mouth frame characters unite the tremasterine-korethrasterid-pterasterid clade and establish beyond reasonable doubt the monophyly of these taxa.

The taxa *Tremaster-Mesotremaster-Protremaster-Korethraster-Remaster-Peribolaster* form a stem group to the Pterasteridae and show progressive modification of the abactinal ossicles to form an open trellis-like reticulum as present in pterasterids, opening of the interradial ducts to form grooves floored by the paired chevron ossicles (*Tremaster* to *Korethraster*) and modifications of the adambulacrals (development of an adambulacral extension; Gale 2011a, pl. 9, figs. 9–12, text figs. 11, 12).

Pterasterids are among the most highly derived living asteroids, having developed an elaborate respiratory mechanism with a precise and complex muscular control (Nance and Braithwaite 1972). The muscular canopy that envelops the body is key to respiration but also functions to produce copious defensive mucus and to brood the young in some species. A large number of morphological developments make this behavior possible, including

- Development of a reticulum of abactinal megapaxillae, with tall pedicels each carrying a crown of elongated spines, supporting the canopy
- Imbricated abactinal and interradial chevron ossicles are united by muscles that contract to reduce the coelomic volume and cause the papulae to swell.

- A muscular canopy enveloping the entire body, except the mouth and ambulacral grooves
- Specialized openings between the adambulacrals, each with a guard spine
- A highly muscularized opening in the center of the abactinal disc, surrounded by five highly modified interradial ossicles (the osculum)

An unbroken chain of morphologically intermediate forms, known from both living and fossil taxa, connects the Asterinidae, through successive korethrasterid taxa, with the Pterasteridae. Many of the evolutionary stages at the base of the Pterasteridae are found in the Jurassic fossil record; for example, the basal pterasterid *Savignaster wardi* from the Oxfordian (Late Jurassic, 158 Ma) possessed tall pedicels, enlarged primary interradials, and muscularized abactinal ossicles but lacked an enveloping muscular canopy (see Gale 2011a,b). Other Late Jurassic pterasterids had a fully functioning, powerfully muscularized osculum and presumably also a canopy.

All known Jurassic pterasterids lack one synapomorphy of living *Pteraster*, the presence of an additional muscle on the adambulacral ossicles (ambulacral extension muscle; Gale 2011a), which permits successive adambulacrals to rotate against each other. This muscle, probably used in opening and closing the operculae, is found in all Late Cretaceous (Villier et al. 2004) to present-day pterasterids.

The evidence provided for the origin of the velatid families is interesting for several reasons. First, it demonstrates the importance of the Jurassic fossil record in elucidating evolutionary relationships among higher asteroid groups. This is not possible for most groups, because these originated during the Triassic for which the fossil record is very limited. Second, it provides evidence directly contradicting the molecular phylogeny of Mah and Foltz (2011a,b), which places the Velatida (Korethrasteridae, Pterasteridae, Myxasteridae) with the Forcipulatida in a basal asteroid clade (Forcipulatacea; Fig. 1.1c). It also disagrees with the molecular phylogeny of Janies et al. (2011), in which the Pterasteridae + *Xyloplax* are basal to all other asteroids. In the case of the evolutionary origin of the velatids, it is not currently possible to reconcile molecular and morphological evidence.

Tripedicellaria: Valvatid Origins of the Forcipulatida

Since the late nineteenth century, it has been clear that the Forcipulatida are a sharply demarcated group, characterized particularly by the presence of complex (forciculate) pedicellariae attached by stalks (Jangoux and Lambert 1987). These highly efficient organs, used for removing parasites, deterring predators, and capturing prey (Chia and Amerongen 1975) are present in every species of extant forcipulatids except the zoroasterid *Pholidaster*. However, their closest relatives among other asteroids have never been clear. The highly derived skeletal construction of the Forcipulatida (Gale 2011a) does not support their position as a basal neoasteroid group as proposed by Blake (1987). A number of previously unregarded features of forcipulatid morphology support their derivation from valvatid ancestors, and the cladistic analysis of Gale (2011a) showed the Valvatida as a paraphyletic stem group to the Forcipulatida. There are two main lines of evidence supporting this relationship. The homologies between alveolar and forcipulate pedicellariae (Fig. 1.3) are the first line of evidence. The incorporation of a third ossicle, with which the valves articulate and to which the adductors and abductors are inserted, is seen in both types of pedicellariae. In stauranderasterids, oreasterids, acanthasterids and ophidiasterids a small cup-shaped basal ossicle (cupula) similar to and homologous with the basal piece of straight forcipulate pedicellariae, is present (Gale 2011a; Fig. 1.3c–e). Thus, alveolar ("valvate") pedicellariae represent an intermediate stage between the simple elementary pedicellariae of Paleozoic asteroids and paxillosids and the complex pedicellariae of forcipulatids. The nature of the ambulacral-adambulacral articulation in more derived valvatids and basal forcipulatids is remarkably similar, with symmetrical development of structures on either side of the proximal adambulacral-ambulacral and distal adambulacral- ambulacral contacts contact and a well developed inter-adambulacral articulation.

The second line of evidence is the presence of small superambulacral ossicles which articulate with the tip of an abradial extension on the ambulacral (zoroasterids, ophidiasterids mithrodiids).

Echinasterids share many detailed morphological similarities with the forcipulatids, including the presence of an adoral carina, an actinostome, and an odontophore-circumoral articulation (Gale 2011a), in addition to the comparable body form of a tiny disc and subcylindrical arms. They also have suckered disc-ending tube feet identical to those of the non-brisingid forcipulatids (Vickery and McClintock 2000a). They entirely lack pedicellariae. In the proposed phylogeny (Fig. 1.1b), echinasterids form a clade close to the base of the Forcipulatida.

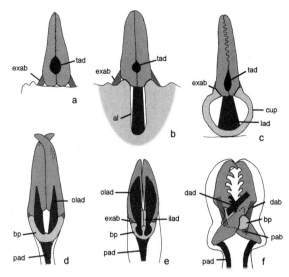

Fig. 1.3. Homologies and evolution of asteroid pedicellariae. *a*, Elementary pedicellaria in *Luidia* sp. *b*, Alveolar pedicellaria in *Archaster* sp. *c*, Alveolar pedicellaria in *Acanthaster*. *d*, Straight complex pedicellaria in Lower Jurassic forcipulatid. *e*, Straight complex pedicellaria in *Marthasterias*. *f*, Crossed complex pedicellaria in *Stylasterias*. Evolution is characterized by a progressive increase of complexity, inclusion of a basal piece as an integral part of the pedicellariae (*b–e*), which eventually becomes a cross-strut between the valves in crossed pedicellariae (*f*). The basal piece of complex pedicellariae (bp) is homologous with the cupula (cup) in the alveolar pedicellariae of valvatids. Dab, distal abductor muscle; dad, distal adductor muscle; exab, external abductor muscle; ilad, inner longitudinal adductor muscle; lad, longitudinal adductor muscle; olad, outer longitudinal adductor muscle; pab, proximal abductor muscle; pad, peduncular adductor muscle; tad, transverse adductor muscle. *Modified after Gale 2011a, text-fig. 3.*

Radiation of the Valvatida

Although the Valvatida are paraphyletic (i.e., they do not form a strictly monophyletic group, because they do not include the common ancestor and all its descendants), they are a distinctive and widely used group. Goniasterids are probably basal to the valvatids and are characterized by large marginal ossicles, a flattened body form, and a flat, tesselate arrangement of small abactinal ossicles. They lack calcified interradial septa and intermarginal ossicles, and papular openings are restricted to the abactinal ossicles. More derived valvatids (Oreasteridae, Ophidiasteridae, Asterodiscididae, Stauranderasteridae) have domed abactinal surfaces that sometimes have enlarged radial and primary ossicles, less conspicuous marginals, and secondary ossicles infilling the reticular spaces between abactinals. Ophidiasterids and oreasterids have a dense coat of granular spines, intermarginal ossicles, and intermarginal and actinal papular openings. The detailed interrelationships of these families are not clear. Do they represent one, several, or many lineages derived from goniasterid ancestors?

The derivation of *Acanthaster* from an oreasterid ancestry was outlined by Blake (1988) and was further supported by evidence from the morphology of the pedicellariae, which possess cupulae similar to those of oreasterids (Gale 2011a).

The presumed close relationships of the valvatid families has been challenged by a new molecular phylogeny (Mah and Foltz 2011a) in which the ophidiasterids form four separate clades, one of which includes the mithrodiids and is basal to all non-forcipulatid asteroids except the poraniids, a second basal to the goniasterids, and the third and fourth are set within the main goniasterid clade (Fig. 1.1c). The pedicellariae of valvatids provide an underused source of phylogenetic information. Many Ophidiasteridae possess highly derived pedicellariae, including forms with oval, boat-shaped basal cupulae; the hourglass-shaped cavity in which the valves rest has a narrow raised rim and the valves are of "sugar tongs" morphology (see Fisher 1906, pl. 30, fig. 4, Clark and Downey 1992, fig. 3c). Essentially identical pedicellariae of this type are found in a least three of the four separate proposed clades of ophidiasterids, and it seems improbable that such complex structures should have evolved independently and repeatedly (see Mah and Foltz 2011a; Fig. 1.1c). However, some "ophidiasterids" that possess large marginal ossicles, such as *Fromia* species (Fisher 1919), are very goniasterid-like, so the story is not a simple one.

The oreasterids, asteropsids, and acanthasterids form a sister group to the combined Goniasteridae and Asterinidae + Solasteridae in the molecular phylogeny of Mah and Foltz 2011a (Fig. 1.1c). This is very surprising for two reasons. The goniasterids and oreasterids are morphologically very close, and historically, genera have been moved backward and forward from one family to the other (compare the classifications of Döderlein 1935, Rowe and Clark 1971, with those of A. M. Clark 1993). In addition, juveniles of some derived oreasterids such as *Culcita novaeguinae* closely resemble goniasterids (Rowe and Clark 1971 pl. 7, figs. 3–5), which certainly suggests a close relationship. Also, types of pedicellariae found in oreasterids and goniasterids are essentially identical, including a distinctive elongated bivalved alveolar form. As in the case of the ophidiasterids, the morphological and molecular data conflict strikingly, and a cautious approach is needed.

The most highly derived valvatids are perhaps the spherical to subspherical Sphaerasterida, including the Jurassic Sphaerasteridae and the Cretaceous to Recent Podosphaerasteridae, both of which superficially resemble echinoids. These evolved from a stauranderasterid ancestor in the Triassic or Early Jurassic, by reduction of the arms, and the undifferentiated abactinal-marginal-actinal ossicles are plate-like and notched for papulae.

The valvatids have an excellent fossil record, largely because the robust skeleton tends to preserve well and forms conspicuous fossils. Valvatids of uncertain affinity appeared in the Carnian (Mid Triassic; Gale 2011a). Goniasterids, sphaerasterids, and stauranderasterids appeared simultaneously in the Pliensbachian (Early Jurassic, 185 Ma), and the latter two groups remained common until the Paleocene. The earliest ophidiasterids appeared in the Albian (Mid Cretaceous; Blake and Reid 1998), and oreasterids appeared in the Eocene (Blake 1988).

Radiation of the Forcipulatida

Diverse lines of evidence support the Zorocallida (Mesozoic terminasterids, and Cenozoic–present-day zoroasterids; Gale 2011a) as the basal forcipulatid group, including the fact that they possess only straight (never crossed) forciculate pedicellariae. This is in agreement with molecular evidence (Mah and Foltz 2011b). Classification of the more derived Forcipulatida (with exception of the clearly monophyletic and highly derived Brisingida) has presented problems for taxonomists. Examination of pedicellariae (Gale and Villier in press) support the following divisions:

- A basal group, which possesses straight and scattered crossed pedicellariae, including five-rayed Pedicellasteridae, Neomorphasteridae, asteriids like *Stichaster,* and the multi-rayed labidiasterid *Labidiaster.*
- A derived group with rosette organs developed around primary abactinal and marginal spines. These comprise numerous crossed pedicellariae set in a pad of connective tissue and muscles (A. Lambert et al. 1984) and include many asteriids and the labidiasterid *Coronaster.*
- Brisingida, characterized by a rounded disc well demarcated from numerous long, narrow arms.

The Brisingida possess a unique type of tube foot (semiflat, non-suckered; Vickery and McClintock 2000a). Evolution of the brisingids involved dramatic modifications of the forcipulatid skeleton with development of firm, fixed articulations between orals, circumorals and odontophores, such that the mouth frame ossicles form a perfect ring around the peristome. In some Brisingidae (e.g., *Brisinga costata*, Gale 2011a), the orals and odontophores are fused, unique for asteroids. Opposing pairs of ambulacral ossicles are locked firmly together by extensive dentition and the internal position of the abactinal transverse ambulacral muscle, and they thus superficially resemble ophiuroid vertebrae. The Brisingids do not appear in the fossil record until the Miocene.

Forcipulatid asteroids with skeletal morphology similar to that of extant Asteriidae are present in the Early and Mid Jurassic (Blake 1990b; Hess 1972). However, Early Jurassic taxa (Hettangian-Pleinsbachian) possess only large, primitive, straight pedicellariae (Gale 2011a, text fig 24; Fig. 1.3d), up to 5 mm in length, quite unlike those found in extant asteriids, which carry a combination of uniformly small, crossed, and straight pedicellariae. The giant pedicellariae have very diverse morphologies—biting terminations of "crossbill" type; tiny, acuminate pincers; elongated triangular blades—and all possess a single pair of large inner adductor muscles. Evidently these Early Jurassic taxa required diverse forms of powerful pedicellariae

to deal with predators and parasites, paralleling the radiation of echinoid pedicellariae during the Mesozoic marine revolution (Coppard et al. 2010). Crossed and straight pedicellariae are present on *Dermaster bohemi* (DE LORIOL) from the Bajocian (Hess 1972), and rosette organs (clusters of crossed pedicellariae set in a muscular structure) are present in the latest Cretaceous *Cretasterias* (Gale and Villier in press). Stem group Zorocallida are represented by the Jurassic-Cretaceous terminasterids *Terminaster* and *Alkaidia* (Gale 2011a), which are sister taxa to the Eocene-Recent Zoroasteridae, and lack characters such as the alternately carinate and non-carinate adambulacrals characteristic of the family (Downey 1970).

All Mesozoic forcipulatids have a similar, simple arm construction, comprising one radial row, two adradials, and two supero- and two inferomarginal rows (Gale and Villier in press). Actinals are absent or rudimentary. Post-Mesozoic forcipulates invariably have multiple actinal rows (Zoroasteridae, Pedicellasteridae) and a broad adradial zone of numerous small ossicles (Asteridae, Pedicellasteridae). Parallel evolution toward the development of broader arms occurred in at least three lineages. The specialized deep-sea Brisingida appeared in the Miocene.

2

The Asteroid Arm

John M. Lawrence

All classes of echinoderms have rays, the radial water canals, and the associated structures (Hyman 1955). The defining character of stellate echinoderms (crinoids, ophiuroids, asteroids) is that their rays extend as appendages from the disc and give them their basic form (Lawrence in press a). Hyman (1955) referred to extensions from the disc as arms or rays. However, the terms *arm* and *ray* are interchangeable only when the rays are distinctly set off from the disc. If the basal parts of the rays are joined, the arm is only a fraction of the ray (H. L. Clark 1907). Asteroids that have a pentagonal or circular body form (e.g., *Culcita schmideliana* and *Tremaster mirabilis*, respectively) have no arms.

Although crinoids, ophiuroids, and asteroids all possess arms, they are distinct in ways that affect their function. This seems related to the way in which the arm is supported (Lawrence in press a). In crinoids and ophiuroids, the arm is supported by an internal skeleton. The support of the arm is external in asteroids, a body wall composed of ossicles and connective tissue, called the *surficial skeleton* by Blake and Guensburg (1988).

This external support of the arm has provided the potential for variation in the body form of asteroids (Fig. 2.1) that can affect these functions. This potential for variation in form does not exist in crinoids and ophiuroids (Lawrence in press a). The variation includes their general shape and number of arms. It does not include branching of the arms (Blake 1989) that is so conspicuous in crinoids and gorgonocephalid ophiuroids. Substantial phylogenetic changes have occurred in the musculoskeleton of the neoasteroid arm (Gale 2011a; see Chapter 1).

The external support of the arm results in an extensive interior coelom that does not exist in crinoids and ophiuroids. The form of the arm affects the volume of this coelom. It may extend the length of the arm (e.g., *Luidia*, *Asterias*) or may be restricted to the proximal part of the ray (e.g., *Styracaster*; Fig. 2.1d). The volume of the coelom is important because it provides space for the development of large gonads and pyloric ceca. No other class of echinoderms has a distinct organ for nutrient reserves. The asteroid arm has additional functions: feeding, locomotion, and respiration (Lawrence 1987).

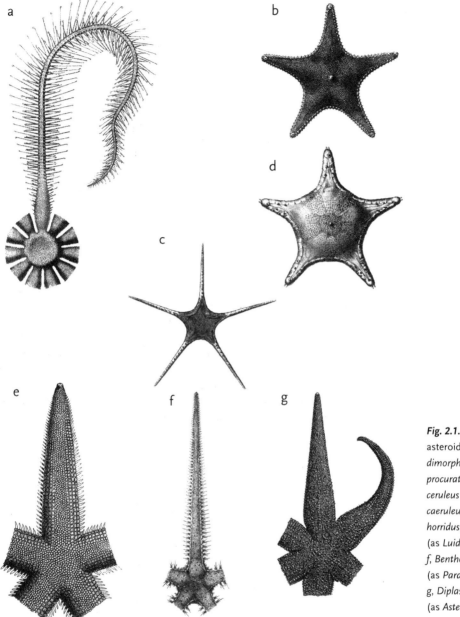

Fig. 2.1. Body form of asteroids. *a, Freyella dimorpha; b, Ctenodiscus procurator; c, Porcellanaster ceruleus* (as *Porcellanaster caeruleus*); *d, Styracaster horridus; e, Luidia hardwicki* (as *Luidia forficifer*); *f, Benthopecten pedicifer* (as *Parachaster pedicifer*); *g, Diplasterias meridionalis* (as *Asterias visculosa*). *Sladen 1889.*

Arm Structure
Anatomy

The body wall of asteroids consists of calcareous pieces (ossicles), connective tissue composed of collagen fibers (dermis), and small bundles of muscle (myocytes; Wilkie 2002). These three components provide the basis for the variation in the arm in asteroids. In some asteroids the ossicles are reduced and may be scattered. In the latter case, the body wall is supported by extensive development of the dermis (Fig. 2.2). The degree of development and interconnection of the ossicles affect the robustness and flexibility of the arm (Blake 1989). Gale has attributed the diversity and success of neoasteroids to the muscularization of the arm associated with the skeleton (see Chapter 1).

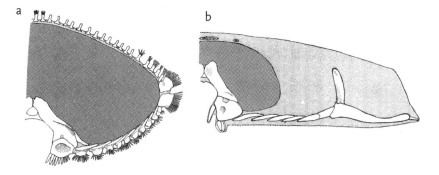

Fig. 2.2. Half arm section of asteroids showing ossicles (*white*), connective tissue (*gray*) and body coelom (*stippled*). *a, Radiaster tizard*; *b, Chondraster grandis. A. M. Clark and Downey 1992.*

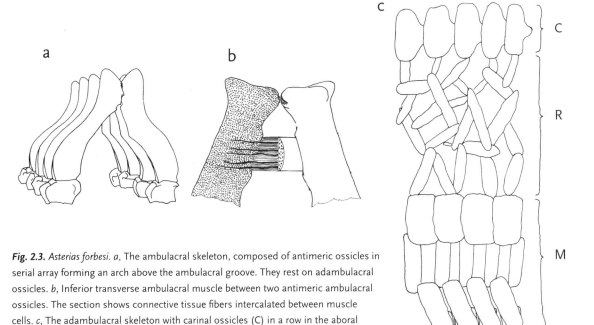

Fig. 2.3. *Asterias forbesi. a,* The ambulacral skeleton, composed of antimeric ossicles in serial array forming an arch above the ambulacral groove. They rest on adambulacral ossicles. *b,* Inferior transverse ambulacral muscle between two antimeric ambulacral ossicles. The section shows connective tissue fibers intercalated between muscle cells. *c,* The adambulacral skeleton with carinal ossicles (C) in a row in the aboral midline and ossicles of the reticulum (R) between it and the marginal ossicles (M) at the base. The arrow points toward the disc. *Eylers 1976.*

Biomechanics

Differences in the biomechanical functioning of the various types of asteroid arms are unknown. The few studies on the biomechanical properties of the asteroid arm have analyzed species in which ossicles are conspicuous. Eylers (1976) examined the skeletal mechanics of the arm of the forcipulatid *Asterias forbesi*. The arm is nearly round in cross-section except for the deep notch on the oral side formed by the ambulacral groove. This notch is formed from antimeric pairs of stout ambulacral ossicles that form an arch (Fig. 2.3). The aboral skeleton is a reticulum of more delicate ossicles that forms a high arc between the adambulacral ossicles (Fig. 2.3). The carinal ossicles in the midline of the aboral surface form a ridge along the length of the arm. Eylers (1976) identified two points of maximum stress in the arm, the aboral ridge of the arm close to the disc and the inferior transverse ambulacral muscle (Fig. 2.3). He concluded the whole body wall acts as a unit sustaining compressive forces across the oral surface and tensile stresses across the aboral surface during contraction of the tube feet. Structural stability results from the position of the aboral ridge in the body and from strong tensile fibers in parallel with the muscle.

Marrs et al. (2000) reported stresses (force / cross-sectional area) and strains (extension / initial length)

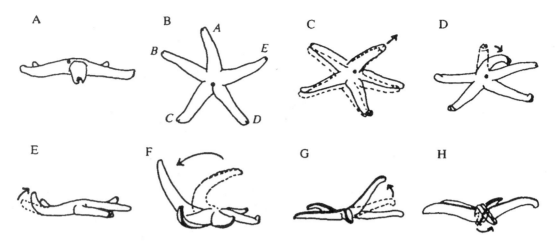

Fig. 2.4. Movements of *Echinaster spinulosus* that load the aboral body wall. *A*, Resting, side view; *B*, Resting, aboral view; *C*, Movement on a horizontal surface; *D*, Lateral bending of the arm; *E*, Arm flexion; *F*, Arm extension; *G*, Arm hyperextension when animal is lying on its aboral surface; *H*, Arm torsion during righting. Arms are designated by Carpenter's system, with A being the arm opposite the madreporite and the other rays being designated B, C, D, and E clockwise when viewed from the oral side. *O'Neill 1989.*

and Young's modulus (stiffness, stress/strain) of the arm increased with size in *Asterias rubens*. They concluded this meant an increased mechanical toughness with an increase in size. This should be associated with a change in the ossicle-connective tissue relationship.

The arm of the spinulosid *E. spinulosus* can undergo great contortion in movement (Fig. 2.4). O'Neill (1989, 1990) studied the biomechanical properties of the body wall of the arm of *E. spinulosus* that affect movement and the effect of physical force. She described the collagen fibers of the body wall as a three-dimensional orthogonal web with ossicles and papulae in the voids. She concluded the orthogonal web gives dimensional stability but allows the shear necessary for arm torsion. The ossicles and fibers interact to load the fibers in tension and the ossicles in compression. She suggested the form of the arm and ossicles differs from that of species such as *Asterias forbesi* studied by Eylers (1976) that generates considerable tension when the tube feet attached to prey contract and *E. spinulosus* that does not.

Mutable Collagenous Tissue

Christensen (1957) concluded from studies on the feeding of *Evasterias troscheli* that there were strong indications that the asteroid is capable of locking its arms into a rigid state before the tube feet attached to a mussel begin to contract. He postulated an internal mechanism was responsible for the necessary revers-

ible condition of the arm, flexible when the asteroid assumes the feeding posture and rigid when the tube feet contract. Motokawa (1982) confirmed the presence of reversible mechanical properties of the connective tissue of *Acanthaster planci*. The stiffness of the arm can vary greatly. The stiffness of the arm of *Linckia laevigata* without stimulation ranges from 2.24 to 51.3 MPa (Motokawa and Wainwright 1991). The variation was interpreted as differences in the state of the mutable collagenous tissue (MCT).

Wilkie (1996, 2002, 2005) has synthesized knowledge about MCT in echinoderms. This type of tissue is found in the extracellular matrix of connective tissue with fibers made up of parallel groups of collagen fibrils. Wilkie (personal communication) has suggested the dermis of all asteroids is MCT or at least that MCT is a primitive feature of the class. Various molecules are attached to them that are the binding sites for molecules responsible for interfibrillar cohesion. MCT can undergo rapid, nervously mediated changes in mechanical properties so that it can be either rigid or flexible (Wilkie 2002, 2005). Wilkie (2002) identified three patterns of change in passive tensile properties: (1) only reversible stiffening and destiffening, (2) reversible stiffening and destiffening or irreversible destiffening that is always associated with autotomy, or (3) only irreversible destiffening that is always associated with autotomy. The change from one state to the other appears to be under neural control.

The first pattern has been demonstrated outside the plane of ray autotomy in the aboral dermis and longitudinal interambulacral ligaments of *Asterias rubens* (Wilkie et al. 1990, 1995). MCT connects ossicles in the arm and provides the basis for the arm being either stiff or flexible. It loosens when the arm flexes. The second has been demonstrated in the aboral dermis and longitudinal interambulacral ligaments within the plane of autotomy of *A. rubens* (Wilkie et al. 1990, 1995). It provides the basis for the arm being either stiff or flexible for autotomy. The third pattern has not been found in asteroids.

Number of Arms

One basic way the asteroid body form can change is by increasing the number of arms. (Lawrence in press a). Cuénot (1948) and Hyman (1955) noted that, although asteroids are fundamentally pentaradial, multiradial species occur. The earliest known multirarmed asteroid, *Lepidaster grayi*, occurred in the Silurian, and several multiarmed species occurred in the Devonian (Blake and Guensburg 1989b). Hotchkiss (2000) concluded multirayed asteroids evolved independently in 14 of the 34 extant families, with 20 being strictly pentaradiate.

Data from A. M. Clark and Downey (1992) indicate variation in number of arms within a family and genus. The number of arms range from eight to 20 in genera of the family Brisingidae. Seven species of the genus *Leptasterias* have five arms, and two have six. Asteroids with multiple arms can be divided into two categories (Lawrence and Komatsu 1990). One includes those asteroids with from six to ca. 16 arms (e.g., *Leptasterias* spp., *Luidia* spp., *Solaster* spp., and *Crossaster*). The frequency of variation from the modal number of arms is low in these species. The second category includes those asteroids with between ca. 15 to 50 arms (e.g., *Heliaster, Acanthaster, Pycnopodia,* and *Labidiaster*). The frequency of variation from the modal number of arms is high in these species.

Multiarmed asteroids also can be divided according to the mode of arm formation. The supernumerary radial canals of all studied species of multiarmed asteroids form either before or after metamorphosis, but not both, in the same species (Lawrence and Komatsu 1990). Supernumerary arms are formed before metamorphosis in some species of *Luidia* and after in others (Komatsu et al. 1991). Species that form the supernumerary radial canals before metamorphosis have direct or non-brachiolarian development (see Chapter 5), while those that form them after

metamorphosis have indirect development (Lawrence and Komatsu 1990).

Hotchkiss (2000) proposed supernumerary rays develop "en bloc" separately from the five primary rays in specific interradii (Fig. 2.5). In none of the species studied do the supernumerary rays appear in the interradius with the madreporite. The location of the numerous supernumerary rays of *Acanthaster, Pycnopodia,* and *Crossaster* is between the same two rays according to Carpenter's designation (see Fig. 2.4). In contrast, the numerous supernumerary rays are located between each of the four interradii on either side of the interradius with the madreporite in *Heliaster*. In the seven-rayed *Luidia ciliaris,* the two supernumerary rays appear in the two interradii on either side of the interradius with the madreporite. Similarly, four supernumerary rays appear at the same time in four interradii in the nine-rayed *Luidia senegalensis* (Lawrence and Avery 2010).

The Interrradius

Cuénot (1948) and Hyman (1955) described five-armed asteroids as variable in form, sometimes with a central disc with more or less long arms, sometimes with a less pronounced disc and short arms, sometimes pentagonal so that arms are not apparent, and sometimes even nearly circular. In these five-armed species, this difference in form results from a decrease in arm length by an extension of the interradius. As a result, the arm widens at its base. This involves addition of actinal ossicles between the rays (Blake and Hotchkiss 2004). Extension of the interradius in this way in five-armed asteroids appeared early. *Doliaster brachyactis* with a near-pentagonal form occurred in the Silurian (Herringshaw et al. 2007a).

An extension of the interradius can also occur by the joining of the proximal part of the rays in multiarmed asteroids. This does not involve the addition of actinal ossicles between the rays. This type of increase in interradius is found in multiarmed species such as *Heliaster* and *Pycnopodia*.

The different degree of extension of the interradius is the basis for the traditional use of the ratio R:r (R = radius of the ray, distance from the center of the disc to the tip of the ray; r = distance from the center of the disc to the edge of the interradius) to indicate body form for asteroid species. A. M. Clark and Downey (1992) described five-armed species with a ratio of less than two as pentagonal or nearly pentagonal. A ratio with a high value occurs in species in which the length of the arm approaches the length of the ray. The area

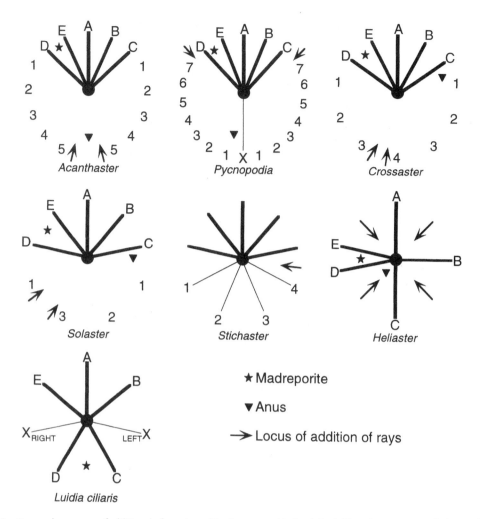

Fig. 2.5. Position and sequence of additional of rays in multiradiate asteroids (oral view). The primary rays are designated by Carpenter's system, with A being the arm opposite the madreporite and the other rays being designated B, C, D, and E clockwise when viewed from the oral side., except for *Stichaster*, because the madreporite is not visible during the period of addition of rays. The supernumerary rays are designated by X or by sequential numerals. All figures represent oral view with mouth at center. Hotchkiss 2000.

of the oral disc of an *Asterias forbesi* with an R of 91 mm and an R:r ratio of 5.7 is 804 mm², while that of a *Poraniomorpha borealis* with an R of 100 mm and an R:r ratio of 1.4 is 20,867 mm² (Lawrence in press a).

Adaptiveness of an Increase in Arm Number and the Interradius

Several advantages of multiple arms have been proposed (Verrill 1914, Lawrence 1988, Blake 1989, Blake and Guensburg 1989b, Herringshaw et al. 2007b). These include an increase in the number of tube feet that could improve the ability to attach, burrow, move, or feed. An increase in the number of gonads could im-

prove fecundity. An increase in the number of pyloric ceca could improve the capacity for digestion and for storage of nutrient reserves. A disadvantage of a greater number of arms is an increase in organic matter in the arm that would require allocation of nutrients for production and maintenance (Lawrence 2010). The body wall of the multiarmed *Acanthaster planci* is the greatest portion of the animal and contains more energy than do the pyloric ceca (Lawrence and Moran 1992).

Suggestions for the advantage of multiple arms in asteroids tend to consider species with 15 or more arms and to ignore those with fewer than 15. The number of supernumerary arms may be few, only one in *Lep-*

tasterias and *Henricia* while luidiid species have five to 15 (A. M. Clark and Downey 1992). The adaptiveness of a small number of supernumerary arms, such as the five-armed *Henricia sanguinolenta* and the six-armed *Henricia hexactis*, has not been demonstrated. This would require comparison of production by individuals that consumed equal amounts of food. Annual production of the gonads and pyloric ceca per unit body weight in the five-armed *Luidia clathrata* is greater than that of the nine-armed *Luidia senegalensis* (S. R. Miller and Lawrence 1999), but comparison is difficult because of the difference in size and possible variability in the quantity and quality of food consumed in the field.

The difference between the effect of attachment of multiple arms to a distinct disc and of joining of the proximal part of adjacent rays in multiarmed asteroids has not been considered. The arms of multiarmed labidiasterids and brisingids are attached to a distinct disc. These asteroids feed primarily in the water column like ophiuroids do. *Labidiaster annulatus* feeds on the substrate and also extends its flexible arms into the water column to capture crustaceans and small fish with its large pedicellariae (Dearborn et al. 1991). Another labidiasterid, *Rathbunaster californicus*, uses its arms to prey on small benthic organisms and on crustaceans in the water column (Lauerman 1998). Similarly, the brisingid *Novodinia antillensis* extends its arms into the water column and captures planktonic crustaceans with its pedicellariae and arm hooks (Emson and Young 1994).

However, rays of asteroids with a large number of arms are usually joined proximally. This results in an increase in the interradius. An increase in interradius also occurs in five-armed asteroids by a greater number of actinal ossicles (Blake and Hotchkiss 2004). H. L. Clark (1907) noted the proximal part of the rays are joined in the five-armed *Pisaster ochraceus* (as *Asterias ochracea*).

An increase in the interradius could be adaptive in several ways. It decreases the length of the exposed arm and could decrease the potential for arm loss from physical damage or predation. In five-armed asteroids, an increase in disc size by extension of the interradius increases the width of the proximal part of the arm and, consequently, increases the space available for gonad and pyloric ceca production. The viscera are restricted to the proximal part of the arm in five-armed species with an increased interradius, such as *Nymphaster arenatus*, in which most of the distal ray width is reduced to the ambulacrum. Blake and Hotchkiss (2004) suggested a large disc increases internal capacity that would be useful for sediment-swallowing Procellansteridae and would provide support on particulate substrata for groups such as the Goniasteridae and Oreasteridae.

Finally, the increase in the interradius, whether by addition of actinal ossicles in five-armed asteroids or by joining the proximal parts of the rays in multiarmed asteroids, has an important consequence: an increase in feeding capacity of extraoral feeders (Lawrence in press a). This is related to the mode of digestion in asteroids that involves direct contact between the stomach and the prey (Feder and Christensen 1966; Jangoux 1982b). An increase in the interradius expands the area over which the stomach can extend in extraoral feeding and should add to the amount of food consumed. *Heliaster helianthus*, which has a very large interradius, can consume multiple macroprey simultaneously (see Chapter 15).

For species that feed on large prey, such as *Crossaster papposus*, a large oral disc area allows enveloping the prey extraorally with the stomach without exposing that organ to damage. This is also true for those species that feed on small prey, small encrusting organisms, and large corals that are preyed upon simply by covering them with the everted stomach. Jangoux (1982b) pointed out the size of the everted stomach is very large in the multiarmed *Acanthaster planci* but restricted to the size of the peristome in *Henricia* and *Echinaster* (see Chapter 20). This is also true for the Ophidiasteridae (e.g., *Linckia*) that have a small disc and four to seven (usually five) long, slender, and cylindrical arms and an R:r ratio ranging from ca. 6 to 10 (A. M. Clark and Downey 1992). *Porania pulvillus*, with an R:r ratio of 1.8 (Clark and Downey 1992), is not only a ciliary feeder but also preys on an octocoral, several species of ascidians, and a brachiopod (Ericsson and Hansson 1973). *Culcita novaguineae*, with an R:r ratio of 1.1 (calculated from A. M. Clark and Rowe 1971) is a surface feeder on corals (Glynn and Krupp 1986).

Arm Autotomy and Regeneration
Arm Autotomy

Although arm loss has been considered characteristic of asteroids (Cuénot 1948, Hyman 1955), Cuénot (1948) noted it is rare in asteroids with a pentagonal body or those with arms with a broad base. Arm loss has not been reported for several pentagonal *Patiriella* species and *Pentaceratus* (as *Oreaster*) *mammillatus* with broad-based arms (Lawrence 1992). It is also rare in species with a sturdy body wall like *Pisaster ochraceus* (O'Donoghue 1926). Frequency of arm loss can vary with population: 1 to 60% in *Acanthaster planci* (Lawrence 1992) and 20 to 60% in *Luidia magellanica* (Viviani 1978). Arm loss has been reported to be the result of autotomy in response to predators and from sloughing the arm off in response to physical damage (Lawrence in press b). Wilkie (2001) referred to this process as opportunistic self-detachment. Asteroids have both structural and chemical defenses that deter predation and decrease arm loss (Lawrence in press b; see Chapter 8).

The plane of autotomy with MCT is restricted to the proximal base of the arm except in the luidiids (Emson and Wilkie 1980), astropectinids (Hotchkiss 2009), and archasterids (Lawrence et al. 2010), where it can occur all along the length of the arm. Autotomy occurs at the base of the ray in *Asterias rubens* (Marrs et al. 2000) and *Coscinasterias muricata* (Mazzone et al. 2003). Autotomy in multiarmed asteroids, such as *Pycnopodia helianthoides* (Wilkie et al. 1995) and *H. helianthus* (Lawrence and Gaymer in press) also occurs at the base of the ray, well below the base of the arm. The pyloric ceca and gonads would be ruptured if autotomy occurred away from the pyloric ducts and gonoducts at the base of the ray. Rupture of the gonads would be limited in those species that have serial gonads in the arms. These include luidiids and several genera of astropectinids, goniasterids, and brisingids (Hyman 1955).

Lawrence (2010) reviewed the costs of arm loss, which would include loss of investment and in functions associated with arms. Lawrence and Larrain (1994) attributed the higher amount of nutrient reserves in the pyloric ceca in intact *Stichaster striatus* than in regenerating individuals to loss of functioning resulting from arm loss. They noted the loss of the pyloric ceca and gonads in the missing arms would result in a decrease in digestive capacity, nutrient reserves, and fecundity. A decrease in food consumption after arm autotomy in the laboratory continued for several months in *Stichaster striatus* and *H. helianthus*, long after an effect due to trauma would be expected, probably was the result of a decrease in the ability to feed (Diaz-Guisado et al. 2006, Barrios et al. 2008, respectively).

Arm Regeneration

Arm regeneration in *Asterias rubens* begins with a combination of morphallaxis (cells are derived from existing tissues by differentiation or migration) and epimorphosis (cells are derived from undifferentiated cells, either stem cells or de-differentiated cells that form a blastema; Bonasoro et al.1998). Subsequent regeneration of the new arm is assumed to be like that of growing tips of intact arms. Regeneration of the pyloric ceca occurs only after much of the arm has been regenerated. Pyloric ceca appeared after about 30% of the arm of *Luidia clathrata* had regenerated (Pomory and Lares 2000).

The cost of regenerating the amount of organic matter in a single arm of *Acanthaster planci* was calculated to be1675 kJ for the body wall and 1220 kJ for the pyloric ceca (Lawrence and Moran 1992). The cost of arm regeneration also includes the amount of energy required for anabolism and deposition of calcium carbonate (Lawrence 2010). These costs have not been measured but can be calculated from the amount of organic matter in intact arms and the cost of anabolism.

Asteroid Arms and Life History Strategies

Life history strategy refers to the coordinated traits of a species that are adaptive to a particular environment (Stearns 1992). Grime (1977) considered stress (defined as conditions leading to decreased acquisition and deposition of energy and production) and disturbance (defined as conditions leading to removal of energy from the organism or loss of biomass) to be environmental factors that limit energy in organisms. He proposed they are the basis for life history strategies. Decreased acquisition of energy results from low quantity or quality of food available or from a low ability to feed. Disturbance results from death or loss of a body part from abiotic or biotic causes. Grime (1977) noted that species in unproductive environments have structural and chemical protection to decrease predation.

Predation is lethal for most animals but is sublethal in asteroids with arm loss. Because the arm is useful and its loss is not lethal, Goss's paradigm (Goss 1969) suggests it would be regenerated. The cost of regeneration of an arm is high (Lawrence in press b). Species with low availability of food or ability to feed would have a low capacity to regenerate the lost part. These species should allocate energy to protection to avoid arm loss. In contrast, species with high availability of food or ability to feed have the capacity to regenerate the arm and should allocate less energy to protection.

Application of Grime's theory to asteroids leads to the prediction of alternative strategies that depend on the availability of food, ability to feed, and the probability of arm loss (Lawrence in press b). This can be seen by contrasting asteroids that ingest their prey to those that evert their stomach and do, or do not, increase the area over which the everted stomach can extend. Paxillosids (see Chapters 10 and 11) would be in the first group. Spinulosids such as echinasterids (see Chapter 20) would be in the second. Valvatids such as *Acanthaster* (see Chapter 13) and forcipulatids such as heliasterids (see Chapter 15) would be in the third. Quality of food must also be considered. Asteroids that are microphagous feeders or feed on detritus, mud, substrate films, encrusting sponges, bryozoans, and ascidians have a lower quality of food than do those that feed on corals or small prey such as pelecypods, branchiopods, and gastropods.

From this, predictions can be made about the life history strategies of asteroids. Blake (1983, 1989) noted the variability in the sturdiness of asteroid arms. He attributed a protective function to armor in the arm. He referred to Yamaguchi's (1975) conclusion that asteroids exposed on the reef are relatively heavily armored to reduce predation, while softer bodied ones are cryptic. Blake (1983) suggested high predation pressure in tropical waters was the basis for extensive development of structural protection in valvatids and echiniasterids and that the microphagous mode of feeding was the consequence of this. Lawrence (1990) suggested the contrary: protection against predation may be the consequence of the microphagous feeding and the advantage of allocating resources to structural defense of the arm.

Blake (1983, 1989) emphasized the conflicting demands of flexibility and protection of arms of asteroids that affect capacity to feed. Flexibility makes the arm more vulnerable to predation, because it cannot be structurally robust. In addition, the area of the oral disc and the size of the mouth affect capacity to feed. This trade-off can be seen by contrasting *Pisaster* and *Asterias* species (Lawrence in press b). *Asterias rubens* and *Asterias forbesi* are active predators and scavengers (Sloan and Aldridge 1981, Ramsay et al. 1997) and should have a high intake of food. In contrast, *Pisaster ochracues* is food limited (Castilla and Paine 1987). The arm of *Asterias forbesi* has a relatively weak body wall but that of *Pisaster ochraceus* is robust (Fisher 1930). The R:r ratio is 5.7 for *A. forbesi* (A. M. Clark and Downey 1992) and 1.4 for *P. ochraceus* (calculated from H. L. Clark 1907). This means the arm of *A. forbesi* is sharply set off from the disc but the arm of *P. ochraceus* is not. In addition, the arm of *P. ochraceus* has a sturdy interbrachial septum supporting this extended interradius (Clark 1907). The frequency of arm loss is high in *Asterias* species and low in *P. ochraceus* (Lawrence 1992).

The characteristics of the arms of brisingids support this interpretation. They are deep-sea species with a small, circular disc sharply set off from multiple long, slender, and weakly attached arms (A. M. Clark and Downey 1992). They are suspension feeders on small plankton (Emson and Young 1994). Because both quantity and quality of available food should be low, why are the arms and their attachment to the disc not robust? Lawrence (in press b) suggested brisingids are ecologically restricted to the deep sea for the same reason Meyer and Macurda (1977) suggested for the restriction of stalked crinoids: the appearance of durophagous teleosts in shallow water during the Mesozoic. The implication is predation pressure is low and frequency of arm loss is not great in the deep-sea brisingids.

Acknowledgments
The author would like to thank A. S. Gale and F. H. C. Hotchkiss for their useful comments.

3

Functional Biology of Asteroid Tube Feet

Elise Hennebert,
Michel Jangoux, and
Patrick Flammang

E chinoderms usually have a rather thick integument in which well-developed calcareous ossicles are embedded, resulting in an armor-like endoskeleton. They also develop soft epidermis-covered coelomic projections (called ambulacral tentacles, tube feet, or podia) through which individuals communicate with their environment. Although these tentacles were used originally as food-collecting organs, in eleutherozoan echinoderms they become involved in locomotion or attachment on or within the substratum (Fig. 3.1). Some of them also develop into strictly sensory appendages, such as the terminal or aboral tentacles of asteroids or echinoids, respectively, or the sensory papillae of holothuroids. Tube feet are the visible part of a tubular coelomic entity—the so-called ambulacral or water-vascular system—that develops from the left mesocoel (or hydrocoel) of the larva and extends radially in or along the integument of the adult echinoderm (Cuénot 1948, Dawydoff 1948, Hyman 1955, Lawrence 1987).

The Water-Vascular System

The water-vascular system of asteroids consists of canals and tentacles filled with a fluid almost identical to seawater, but with a higher potassium concentration and osmolarity and with coelomocytes and small quantities of proteins (Binyon 1976, Prusch and Whoriskey 1976, Prusch 1977).

This system is made up of two main parts, the central and peripheral (Fig. 3.2). The central part is comprised of the ring canal and the stone canal (Nichols 1972). The ring canal encircles the esophagus and has two types of accessory structures: the Polian vesicles and the Tiedemann's bodies. Polian vesicles are muscular sacs whose most plausible function is to act as low pressure reservoirs for water-vascular fluid (Nichols 1966). They occur in astropectinids, oreasterids, and asteriinids and are missing in asteriids (Cuénot 1891). Tiedemann's bodies, in contrast, are always present in asteroids and may have a coelomocyte-forming function (Cuénot 1948, Bargmann and Behrens 1964, Nichols 1966). The stone canal runs vertically from the ring canal through

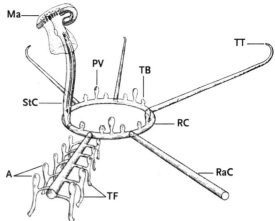

Fig. 3.1. The asteroid *Asterias rubens* lying on its aboral side and using its tube feet (TF) to right itself. *Adapted from Romanes and Ewart 1881.*

Fig. 3.2. Diagrammatic representation of the asteroid water-vascular system. Tube feet are drawn on one arm only. Arrangement of polian vesicles and Tiedemann's bodies is generalized. A, ampullae; Ma, madreporite; PV, polian vesicle; RC, ring canal; RaC, radial canal; StC, stone canal; TB, Tiedemann's body; TF, tube feet; TT, terminal tentacle. *Adapted from Nichols 1966.*

the general body cavity up to a perforated plate of the integument, the madreporite, which opens to the exterior and allows a communication between the water-vascular system and the surrounding seawater. The stone canal is supported by delicate ossicles of calcite embedded in its wall, hence its name (Nichols 1966).

The peripheral part of the water-vascular system is composed of the radial canals, the tube feet, and their accessory structures (Nichols 1972). The radial canals are usually five or multiples of five in number, although some species have odd number of rays (Nichols 1966, Lawrence 1987). They arise from the ring canal and extend along each arm of the asteroid. Lateral canals lead from the radial canals to each tube foot and its ampulla, a small muscular sac. The tube feet are located externally, whereas the ampullae are inside the body cavity. Both are connected through an ambulacral pore that opens between two successive ambulacral ossicles. Each tube foot–ampulla unit is separated from the lateral canal by a muscular valve, allowing it to function independently (Hyman 1955, Nichols 1966, 1972).

Tube feet are hydraulic structures and thus must contain a fluid maintained at a sufficient hydrostatic pressure for their correct functioning. The ring canal and the radial canals deliver that fluid to all tube feet (Nichols 1966, 1972). Hydrostatic pressure is generated by the hyperosmolarity of the fluid that causes water

to come in through the wall of the water-vascular system (Prusch and Whoriskey 1976, Ferguson 1990). The antagonistic action of this pressure and of the muscle systems in both the tube foot and the ampulla results in tube foot mobility, which can be explained by a combination of three basic movements: protraction, flexion, and retraction (J. E. Smith 1947, Nichols 1966). Tube foot protraction is initiated by the contraction of the ampullar muscles, forcing the water-vascular fluid into the tube foot lumen. The valve separating the tube foot from the lateral canal prevents the fluid from escaping into the lateral canal when the ampulla contracts. The elongation process is made possible by the properties of the tube foot connective tissue sheath. Within this sheath, the internal layer prevents the wall of the tube foot from being bent out of shape and maintains its diameter so that the whole hydrostatic pressure is exerted at its distal end (McCurley and Kier 1995). The external layer stretches to allow the tube foot to lengthen gradually (see discussion of the stem below). Retraction is brought about by the contraction of the tube foot retractor muscle, and the fluid is driven back into the ampulla. The retractor muscle is also involved in bending the tube foot in one direction or another. It seems that each sector of the muscular cylinder can contract independently from the others and therefore act on the hydraulic skeleton to produce flexion (Smith 1947, Florey and Cahill 1977).

Tube Foot Histology

Whatever the asteroid species considered, the histological structure of tube feet remains constant. They consist of three tissue layers: a connective tissue layer sandwiched between two epithelia (Nichols 1966; Fig. 3.3). The inner epithelium (i.e., the mesothelium) is continuous with that lining the inner parts of the water-vascular system, whereas the outer epithelium (i.e., the epidermis) is continuous with that covering the rest of the body. A conspicuous basiepidermal (epineural) nerve plexus is present, but there are no reports of a basimesothelial (hyponeural) nerve plexus (J. E. Smith 1945, Perpeet and Jangoux 1973, McCurley and Kier 1995). However, at the base of the tube foot, there is morphological evidence for direct innervation of the mesothelium by a lateral branch of the hyponeural division of the radial nerve (Smith 1945, Cavey 1998).

Mesothelium

The mesothelium (coelomic epithelium) lines the tube foot lumen. It is a pseudostratified myoepithelium consisting of apically situated adluminal cells and subapical myocytes (Fig. 3.3). Both cell types contact the underlying basal lamina to which they attach via hemidesmosomes (Wood and Cavey 1981, Rieger and Lombardi 1987). Adluminal cells are monociliated cells bearing a long vibratile cilium surrounded by a ring of about 10 microvilli. They enclose a bundle of filaments connecting their apical and basal membranes. The adluminal cells are connected apicolaterally by junctional complexes consisting of a distal zonula adherens and a proximal septate desmosome (Wood and Cavey 1981, Rieger and Lombardi 1987). Myocytes contain a bundle of myofilaments associated with numerous mitochondria. These myofibrils are oriented longitudinally. Together they form an extensive longi-

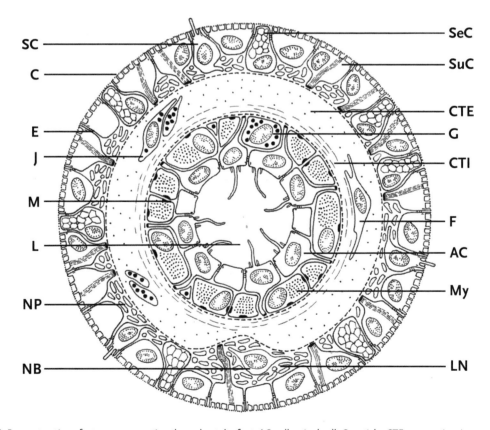

Fig. 3.3. Reconstruction of a transverse section through a tube foot. AC, adluminal cell; C, cuticle; CTE, connective tissue external layer; CTI, connective tissue internal layer; E, epidermis; F, fibrocyte; G, granulocyte; J, juxtaligamental cell; L, water-vascular lumen; LN, longitudinal nerve; M, myomesothelium; My, myocyte; NB, nerve body; NP, nerve plexus; SC, sensory cell; SeC, secretory cell; SuC, support cell. *Adapted from Flammang 1996. Not to scale.*

tudinal muscle layer (i.e., the retractor muscle of the tube foot). Adjacent myocytes are connected by spot desmosomes (Wood and Cavey 1981, Rieger and Lombardi 1987). An additional cell type may be found in the tube foot mesothelium, the granulocytes (Fig. 3.3). However, these are much less common than adluminal cells and myocytes (Wood and Cavey 1981, Rieger and Lombardi 1987). Their function remains obscure.

Connective Tissue Sheath

The connective tissue sheath is made up of an amorphous ground substance that encloses bundles of collagen fibrils (i.e., fibers), elongated cells with an electron-dense cytoplasm that may be fibrocytes, and various types of mesenchymal cells (macrophages, spherulocytes, etc.; Nichols 1966, Santos et al. 2005a). In the proximal part of the tube foot, i.e., the stem, the connective tissue consists of a diffuse external layer and a more compact internal layer (Fig. 3.3). The internal layer comprises a crossed-fiber helical array of collagen fibers, whereas in the external layer, the collagen fibers are oriented longitudinally (Nichols 1966, McCurley and Kier 1995). This layer contains juxtaligamental cells filled with numerous electron-dense granules, typical of mutable collagenous tissues, and a well-developed microfibrillar network surrounding the collagen fibers (Santos et al. 2005a, Hennebert et al. 2010).

Nerve Plexus

The nerve plexus is a cylindrical sheath of epineural nervous tissue (J. L. S. Cobb 1987). It is located just beneath the epidermis, its neurites running between the basal parts of epidermal cells (Fig. 3.3). A single continuous basal lamina lines both the epidermis and the nerve plexus, separating them from the connective tissue (Nichols 1966). The plexus is thickened on one side of the tube foot to form the longitudinal nerve and at the proximal and distal extremities to form two nerve rings (Nichols 1966). It is also well developed at the level of the secretory or sensory areas of the tube foot epidermis. Nerve cell bodies occur in the longitudinal nerve, in the nerve rings, and in areas where the plexus is thickened; the rest of the plexus consists of a criss-cross of nerve processes (Fig. 3.3). The latter contain mitochondria, microtubules, and clear or dense-core vesicles.

Epidermis

The epidermis, which is a monostratified epithelium, covers the tube foot externally (Fig. 3.3). It may enclose different cell types: support, sensory, and secretory (Holland 1984). The last two are particularly abundant in the epidermal adhesive areas of the tube foot (see discussion of the tube foot adhesive area below). All epidermal cells are connected apicolaterally by junctional complexes made up of a distal zonula adherens and a proximal septate desmosome. The epidermis is coated by a well-developed, multi-layered glycocalyx, the so-called cuticle (Holland and Nealson 1978, Ameye et al. 2000). Support cells are the most numerous and form a supportive meshwork in which the other cell types are homogeneously distributed. These cells are traversed by a bundle of intermediate filaments joining their apical and basal membranes (P. Harris and Shaw 1984). The apical cytoplasm always contains "empty" vesicles and vacuoles of various sizes. At their apex, support cells bear numerous branched and unbranched microvilli that are closely associated with the fibrous or granular materials constituting the cuticle. In addition to their supportive function, these cells are also presumably involved in uptake of dissolved organic material and in synthesis of cuticular material (Engster and Brown 1972, Souza Santos and Silva Sasso 1974, Flammang et al. 1998). Sensory cells are usually scattered singly or in small groups. They are neuroepithelial cells having an axonal process and bearing a single short apical cilium arising from a basal body associated to a long ciliary rootlet. It is generally assumed that such cells are either chemoreceptors or mechanoreceptors (J. L. S. Cobb 1987). Secretory cells may be scattered individually or aggregated into small or large groups. Their predominant cytoplasmic inclusions are membrane-bound secretory granules. Each tube foot possesses at least three different types of secretory cells, each having typically only one kind of secretory granule (Souza Santos and Silva Sasso 1968, Holland 1984, Flammang et al. 1994). These different cell types may be characteristic of different areas of the tube foot epidermis. Although some are known to be involved in the adhesive process (see discussion of the tube foot adhesive area below), the function of the others is in general poorly understood.

Diversity and Functions of Tube Feet

Based on their external morphology, asteroid tube feet were originally subdivided into two main categories: those in which the distal part is pointed, the knob-ending tube feet, and those in which the tip is flattened, the disc-ending tube feet (Hyman 1955, Lawrence 1987, A. M. Clark and Downey 1992, Flammang 1996). Close observation of the organization of the different tissue layers of tube feet from a large number of species later led to a revision of this classification (Vickery and McClintock 2000a, Santos et al. 2005b). Three tube foot morphotypes are now described: knob-ending, simple disc-ending, and reinforced disc-ending tube feet (Santos et al. 2005b; Fig. 3.4).

Knob-ending tube feet are made up of a basal cylindrical stem ending distally with a pointed knob (Vickery and McClintock 2000a, Santos et al. 2005b; Fig. 3.4a). These tube feet occur exclusively in paxillosid asteroids (Vickery and McClintock 2000a, Santos et al. 2005b). In these tube feet, the water-vascular lumen extends into the knob where it tappers off to a point. Except for the myomesothelium, all tissue layers are thicker in the knob than in the stem. The knob epidermis is a tall columnar epithelium consisting mostly of granule-filled secretory cells. The nervous tissue consists of a nerve ring on the proximal side of the knob and a thick basiepithelial nerve plexus underlying the knob epidermis. Both layers lie on a thick layer of loose connective tissue (Santos et al. 2005b).

Knob-ending tube feet take part in locomotion and also allow paxillosids to bury themselves by digging vertically into the sediment (Heddle 1967, Engster and Brown 1972, Santos et al. 2005b). They also allow the asteroid to handle food items or particles (McCurley and Kier 1995).

Simple disc-ending tube feet consist of a basal cylindrical stem with an apical extremity that is enlarged and flattened to form the so-called disc (Vickery and McClintock 2000a, Santos et al. 2005b; Fig. 3.4b). They occur in most species of the order Valvatida (Vickery and McClintock 2000a, Santos et al. 2005b). The disc has the same basic structure as the knob of knob-ending tube feet: it encloses the distal extremity of the water-vascular lumen. Its epidermis, nerve plexus, and connective tissue layer are thicker than their equivalents in the stem. In most species, the disc is much larger than the stem, giving the tube foot a flared shape. The lumen extends radially into the disc margin, where it usually presents a scalloped circumference. The disc lumen may even be entirely partitioned

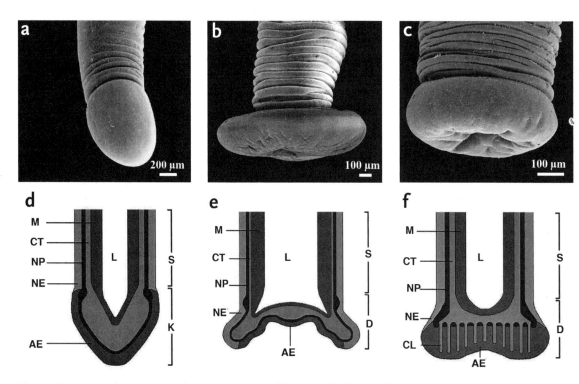

Fig. 3.4. External morphology and histological organization of the three tube foot morphotypes observed in asteroids. *a–c*, SEM views of a tube foot ending in a knob (*a*, *Luidia maculata*, original), of a tube foot ending in a simple disc (*b*, *Mithrodia clavigera*), and of a tube foot ending in a reinforced disc (*c*, *Asterina gibbosa*, original). *d–f*, Schematic representations of longitudinal sections through a tube foot ending in a knob (*d*), a tube foot ending in a simple disc (*e*), and a tube foot ending in a reinforced disc (*f*). AE, adhesive epidermis; CL, connective tissue radial lamellae; CT, connective tissue layer; D, disc; K, knob; L, water-vascular lumen; M, myomesothelium; NE, non-adhesive epidermis; NP, nerve plexus; S, stem. *b*, Hennebert 2010; *d–f*, Santos et al. 2005b.

by radial connective tissue septa, as it is the case in the archasterid *Archaster typicus*. In species of the family Oreasteridae, the large disc is supported by a skeleton made up of numerous calcareous ossicles and spicules located within the connective tissue layer on the proximal side of the disc lumen. Tube feet that end in simple discs are used mainly for locomotion. The tissue organization of the disc provides the tube foot with a large flat distal surface that can be used in locomotion on both soft and hard substrata. These tube feet are also used for attachment to the substratum in a few valvatid species (e.g., *Acanthaster planci*, *Linckia laevigata*) and for burrowing in the sediment in *A. typicus*. In this species, radial division of the disc lumen by collagenous septa would enable modifications of the disc shape and orientation to allow digging (Santos et al. 2005b).

Reinforced disc-ending tube feet have more or less the same external morphology as simple disc-ending tube feet. They are made up of a basal cylindrical stem topped by a flattened disc (Fig. 3.4c). However, the diameter of the disc never greatly exceeds that of the stem (Vickery and McClintock 2000a, Santos et al. 2005b). These tube feet occur in all the investigated species of the orders Velatida, Spinulosida, and Forcipulatida and in a few valvatids (Vickery and McClintock 2000a, Santos et al. 2005b). The histological organization of reinforced discs differs from that of the two other morphotypes. The lumen does not extend into the disc margin. The disc connective tissue layer also sends distally numerous bundles of collagen fibers that insinuate themselves within the adhesive epidermis, reaching up to the cuticle. This tissue organization varies according with species, from the tube feet of velatids, which have few uniformly distributed bundles of collagen fibers, to the tube feet of forcipulatids, which possess a complex array of radial collagenous laminae (Santos et al. 2005b). Tube feet ending in reinforced discs are used in locomotion, attachment to the substratum, and, in some species, in opening bivalves for feeding (Hennebert et al. 2010). They appear to be better designed for strong adhesion than do tube feet that end in simple discs. Indeed, the collagen fibers present in the disc epidermis may function as tension-bearing structures, transferring the stress due to adhesion from the distal cuticle to the connective tissue of the stem (Santos et al. 2005b,c).

The morphological diversity of tube feet in asteroids appears to be a result of adaptation of species to various habitats, but within limits imposed by the evolutionary lineage (Santos et al. 2005b). For instance, most species inhabiting turbulent environments, where strong attachment to the substratum is required, have tube feet ending in reinforced discs regardless of their taxonomic position. In this case, adaptation to habitat overrides belonging to a particular order. However, *A. typicus*, which has the capacity to bury itself completely, is homeomorphic with paxillosid astropectinids *Astropecten* spp. that also bury themselves (see Chapter 10). This homeomorphy, however, does not extend to the tube feet, which end in simple discs in the former and knobs in the latter. In this case, therefore, the taxonomic position of the species seems to supplant habitat constraints (Santos et al. 2005b).

Although tube feet from each morphotype are preferentially associated with specific functions (see above), all tube feet are usually also involved in respiration. This takes advantage of the proximity of the external medium to the fluid of the tube feet. They are also involved in sensory perception, each of them possessing epidermal sensory cells either associated with the epidermal adhesive areas or simply scattered all over the surface of the tube foot (Flammang 1996). In addition, the tip of each arm in all species bears a few slender knob-ending tube feet. These tube feet, in which the knob is rounded, are always in motion and are chemo- or mechanosensory (Sloan and Campbell 1982, Flammang 1996).

Functional Parts of the Tube Foot

Locomotion, fixation to the substratum, and prey handling are activities that require a harmonious functioning of the two parts of the tube feet: the proximal stem and the distal knob or disc (Lawrence 1987, Hennebert et al. 2010). The distal part makes contact with the substratum and allows the tube foot to attach to the surface. The stem, in contrast, acts as a tough tether connecting the knob or the disc to the animal's body. It is also mobile and flexible and thus gives the tube foot the capacity to perform various movements. Despite the diversity in asteroid tube foot forms and functions, only reinforced disc-ending tube feet of forcipulatids have been studied in detail with respect to their attachment strength and mechanical properties.

The Stem

In asteroid species living in the intertidal or in the shallow subtidal, one important role of the tube foot stem is to bear the tensions placed on the animal by hydrodynamic forces (Santos et al. 2005a). This load-bearing function may be critical. Indeed, when asteroids are subjected to a constant pull, a large number of their tube feet sometimes rupture before they are

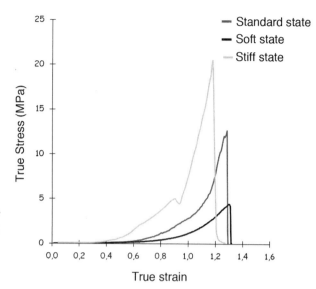

Fig. 3.5. Typical J-shaped stress-strain curves illustrating the mutable character of the stem connective tissue in the tube feet of the asteriid *Marthasterias glacialis*. Mechanical properties were measured in artificial seawater (standard state) and in solutions inducing the soft and stiff states. *Adapted from Santos et al. 2005a.*

detached from the substratum (Berger and Naumov 1996, Flammang and Walker 1997). The tensile strength of the stem can thus limit the capacity of asteroids to be strongly anchored to the substratum. Among stem tissues, only the connective tissue and the retractor muscle contain fibrillar elements (i.e., collagen fibers, microfibrils, myofilaments) oriented in parallel to the tube foot axis in the direction of the tensile stresses (Flammang 1996, Santos et al. 2005a). The comprehension of their respective mechanical properties is therefore important in understanding the functioning of tube feet.

Connective Tissue

Stem mechanical properties have been investigated only in the tube feet ending in a reinforced disc of the asteriids *Asterias rubens* and *Marthasterias glacialis*, by tensile testing (Santos et al. 2005a, Hennebert et al. 2010). When they are pulled, their tube feet present a complex stress-strain curve in which stress increases with strain, first slowly, and then more rapidly until stem rupture (Fig. 3.5). This results in a typical J-shaped stress-strain curve, which is characteristic of shock-absorbing materials such as mammalian skin and arteries (Vincent 1990, Vogel 2003). Four material properties can be calculated from this curve: (1) extensibility, which is the value of strain when the stem fails (breaking strain); (2) strength, which is the maximal value of stress (breaking stress); (3) stiffness, which is calculated as the slope of the last portion of the stress-strain curve; and (4) toughness, which is a measure of the energy required to extend and break the stem and

corresponds to the area under the stress-strain curve (Santos et al. 2005a). The values of these parameters for tube feet bathed in artificial seawater are respectively 1.6 (~500%), 21 MPa, 122 MPa, and 3.5 MJ m^{-3} for *A. rubens* and 1.8 (~600 %), 23 MPa, 170 MPa, and 3.7 MJ m^{-3} for *M. glacialis* (Hennebert et al. 2010). These values are characteristic of an extensible and tough material. It has also been reported that the stem material properties are strain rate dependent; extensibility, strength, stiffness, and toughness all increase as strain rate increases (Hennebert et al. 2010). In the field, this means that tube foot stems present a higher resistance to rapid loads (such as waves) than to slower, self-imposed loads (such as natural extension; Hennebert et al. 2010).

The presence of numerous juxtaligamental cell processes with electron-dense granules in the outer connective tissue sheath of the stem wall (see discussion of histology above) in the tube feet of *A. rubens* and *M. glacialis* suggests this tissue is a mutable collagenous tissue (MCT; Santos et al. 2005a, Hennebert et al. 2010). MCTs, which are characteristic of echinoderms, can undergo rapid changes in their passive mechanical properties under nervous control via the juxtaligamental cells (Wilkie 1996, 2005, Trotter et al. 2000). These cells are neurosecretory and are believed to control the mechanical properties of the connective tissue by secreting stiffening compounds that increase interactions between collagen fibrils (Wilkie 2005). The mutable character of the connective tissue in a particular structure is usually demonstrated by mechanical tests applied to this structure in different solutions, among which the most commonly used are solutions depleted of calcium or solutions causing destabilization of cell membranes. These treatments are known to influence the physiological state of MCTs, the former mimicking its soft state and the latter inducing its stiff state (e.g., Trotter and Koob 1995, Szulgit and Shadwick 2000). As in other known echinoderm MCTs, tube feet from both species become more compliant when placed in a calcium-free solution and stiffen when juxtaligamental cells are completely lysed (Santos et al. 2005a, Hennebert et al. 2010; Fig. 3.5). Having tube feet containing an MCT is obviously adaptive. In its soft state, MCT may facilitate the action of the ampulla muscle in tube foot protraction and of the retractor muscle of the stem in bending and retraction. In its stiff state, MCT could play an energy sparing role in maintenance of position, for example, during strong attachment to the substratum to resist hydrodynamically generated loads (Santos et al. 2005a).

The Retractor Muscle

The retractor muscle allows flexion and retraction of unattached tube feet during their activities. When tube foot discs are attached, contraction of the stem retractor muscle allows asteroids to climb vertical surfaces or right themselves (Lawrence 1987). It also allows them to clamp their body against the substratum, a behavior that may play an important role in the attachment mechanism, because friction generated in this way decreases the risk of dislodgement by shear forces (Kerkut 1953). Finally, some asteroids use retractor muscle contraction to open bivalve mollusks on which they feed (Lawrence 1987). The isometric contraction force of the tube foot retractor muscle has been measured in a few species. It has been measured in a K^+-rich solution (in *A. rubens* and *M. glacialis*; Hennebert et al. 2010), after application of acetylcholine (in the asterinid *Asterina pectinifera*; Saha et al. 2006) and in natural conditions, during surface climbing (in *A. rubens*; Kerkut 1953). Mean contraction values ranged between 3 and 17 mN, with maxima of 40 mN. Several workers reported the total pull exerted by the tube feet of asteroids during prey opening ranged from 7 to 60 N (Feder 1955, Lavoie 1956, Christensen 1957, Norberg and Tedengren 1995). These data indicate that hundreds of tube feet are used cooperatively in this feeding behavior.

To estimate the contribution of the retractor muscle to the tensile strength of the tube foot stem, echinoderm muscle-breaking forces had to be measured.

This has been done on holothurian longitudinal muscles, which contain only a small amount of loose connective tissue. These are considered as a good experimental model to study the mechanical properties of echinoderm muscles (Hill 2004), and the ultrastructure of their cells is very similar to that of the cells from the asteroid tube foot retractor muscle (Wood and Cavey 1981, Hill 1993). The forces needed to break tube foot retractor muscles were then calculated using the data obtained for holothurian muscles and the respective cross-sectional surface areas of both types of muscles and were shown to account for 18 to 25% of the total breaking force of whole tube feet in *A. rubens* and *M. glacialis* (Hennebert et al. 2010). It is apparent therefore that, although the connective tissue layer supports most of the load exerted on the stem, the contribution of the retractor muscle cannot be ignored (Hennebert et al. 2010).

The Disc
Structure and Functioning

Compared with the basic structure of the tube foot wall, a typical reinforced disc shows development of support structures (connective tissue) and differentiation of the musculature. The connective tissue forms a circular plate, the so-called terminal plate, that is continuous on its proximal side with the connective tissue sheath of the stem. The center of this plate, the diaphragm, is much thinner than its margin, which consists of a thick ring (J. E. Smith 1947, Flammang 1996). Numerous arborescent septa of connective tissue radiate from the distal surface of the terminal plate and insinuate themselves between the epidermal cells (Fig. 3.4f). The thinnest, distal branches of these septa attach apically to the support cells of the epidermis (J. E. Smith 1937, 1947, Flammang et al. 1994). The thick adhesive epidermis reinforced by branching connective tissue septa forms a soft (stiffness of 6.0 kPa in *A. rubens*), visco-elastic adhesive pad (Santos et al. 2005c). As for the musculature, the disc possesses a levator muscle that is arranged longitudinally and attaches distally to the center of the diaphragm (J. E. Smith 1947, Flammang 1996).

Since the first scientific descriptions of asteroids (e.g., Forbes 1841, Romanes 1885, Cuénot 1891), tube feet that end in discs have been described as miniature suckers. This idea, imparted by most zoology textbooks, has since pervaded popular knowledge. This assumption presumably originated from the observation of the external shape of the disc, which looks like a cup with a central depression. In his model for tube foot attachment, J. E. Smith (1947) postulated that the suction effect when a tube foot contacts a surface is the consequence of the contraction of the levator muscle which lifts the diaphragm, thus creating a suction cavity. Other authors noted the presence of a well-developed secretory epidermis in the tube foot disc, as well as the presence of a footprint completely filled with material after tube foot detachment (V. L. Paine 1926, Chaet 1965, Souza Santos and Silva Sasso 1968, L. A. Thomas and Hermans 1985, Flammang et al. 1994, 1998). The involvement of suction in tube foot attachment was therefore questioned.

In *A. rubens*, microscopic observations of tube feet rapidly fixed while they were attached to a smooth substratum showed a flat distal surface of the disc, with no suction cavity (Hennebert et al. in press). Moreover, the tube foot attachment strength was independent of the pulling angle, meaning that the introduction of a shear component to the pulling force does not decrease the attachment strength as would be expected for a sucker (Hennebert et al. in press). Taken together with the qualitative observations made by L. A. Thomas and Hermans (1985) that tube feet of the asteriid *Leptasterias hexactis* adhere strongly to a fine-meshed, stainless steel plankton screen (i.e., a perforated surface preventing suction), these results unambiguously demonstrate that, contrary to the generally accepted notion, asteroid tube feet ending in a reinforced disc are not functioning as active suckers.

Attachment Strength

The maximal attachment force of single tube feet has been evaluated in several species and amounts to 0.29 N in *Asterias vulgaris* (V. L. Paine 1926), 0.15 N in *Asterias rubens* (Hennebert et al. 2010), and about 0.3 N in *Marthasterias glacialis* (Preyer 1886, Hennebert et al. 2010). Disc adhesive strength may also be evaluated in terms of tenacity, measured in Pa. Tenacity of single tube feet has been quantified under different conditions and range from 0.17 to 0.43 MPa (V. L. Paine 1926, Flammang and Walker 1997, Santos et al. 2005c, Hennebert et al. 2010). Tube foot tenacity was shown to be dependent on the roughness of the substratum. In *A. rubens* adhesion is stronger on a rough substratum than on its smooth counterpart. This increase in tenacity is explained by the compliance of the adhesive pad of the tube foot disc (see discussion of structure and function of the disc above) and its ability to replicate the substratum profile, increasing the total contact area (Santos et al. 2005c).

The attachment strength of an asteroid depends on the attachment strength of each tube foot (i.e., the adhesive power of the disc) and of the number of tube feet used for attachment (Lawrence 1987). Data are scarce in literature, presumably because of the difficulty to grab and pull a multiarmed, soft-bodied asteroid. The attachment strength of asteroids has only been investigated in two species. Berger and Naumov (1996) measured an attachment force of 5 N for *A. rubens* attached to a glass plate, and Siddon and Witman (2003) measured attachment forces ranging from 23 to 34 N for *A. forbesi* attached to rocks. Neither study, however, reported the number of tube feet adhering to the substratum, rendering valid comparisons impossible.

Adhesiveness of the Tube Foot

Locomotion, attachment, and prey handling require the formation of a typically temporary adhesive bond between the tube feet and the substratum that is the result of secretory activity by the adhesive areas of the tube feet. Indeed, although tube feet can adhere very strongly to the substratum, they are also able to detach easily and voluntarily before reinitiating another attachment-detachment cycle (L. A. Thomas and Hermans 1985, Flammang 1996). Hermans (1983) was the first to propose a model to explain this temporary adhesion. In his hypothesis, tube feet possess a duo-gland adhesive system involving two types of secretory cells: cells releasing an adhesive secretion and cells releasing a de-adhesive secretion. TEM observation of tube feet chemically fixed during and after attachment corroborated this (Flammang et al. 1994, 2005).

Ultrastructure of the Adhesive Areas of the Tube Foot

Adhesive areas are located in functionally important epidermal parts of the tube feet, i.e., the whole knob surface or the distal disc surface (Smith 1937, Engster and Brown 1972, Perpeet and Jangoux 1973, Flammang 1996, Santos et al. 2005b). These epidermal adhesive areas always consist of four cell categories: adhesive, de-adhesive, sensory, and support.

Adhesive cells are flask-shaped. Their enlarged cell bodies are located basally and attach to the basal lamina via hemidesmosomes. Each cell body sends out a long apical process that reaches the surface of the tube foot. The cytoplasm of both the cell body and the apical process is filled with membrane-bound secretory granules. Depending on the species, adhesive cells may be of one or two types that can be distinguished on the basis of the ultrastructure of the secretory granules, recognized by Flammang (1996). Type 1 granules are ellipsoid and consist of a large electron-dense core made up of a parallel arrangement of fibrils or rods, surrounded by a thin layer of less electron-dense material. Type 2 granules resemble the previous group but with a smaller, more homogeneous core. They are spherical to ellipsoid and are always smaller than type 1 granules. Subcategories exist in these groups, reflecting a high variability in granule ultrastructure among different species (Hennebert 2010). In the cell body of adhesive cells, developing granules are closely associated with Golgi membranes and rough endoplasmic reticulum cisternae, suggesting that these organelles are involved in the synthesis of the granule contents (Harrison and Philpott 1966, Engster and Brown 1972, Flammang et al. 1994). Granules are conveyed to the apex of the cell, probably with the help of the longitudinal microtubules occurring in the adhesive cell distal process. They are then extruded through a duct delimited by a ring of microvilli and opening onto the knob or disc surface as a pore (Flammang 1996).

De-adhesive cells are narrow and have a centrally located nucleus. They are filled with small homogeneous electron-dense secretory granules whose ultrastructure is remarkably constant from one species to another. The cytoplasm of these cells also contains numerous rough endoplasmic reticulum (RER) cisternae, a small Golgi apparatus, and longitudinally arranged microtubules. The basal end of these cells is tapered and contacts the nerve plexus. Their apical process ends with a bulge just beneath the cuticle. This bulge lacks microvilli and generally has a short subcuticular cilium except in paxillosid tube feet (Flammang 1995, 1996).

Sensory cells are narrow and have a centrally located nucleus. Their cytoplasm encloses longitudinally arranged microtubules, elongated mitochondria, and small apical vesicles (Flammang 1996). These cells are characterized by a single short cilium whose apex protrudes into the fuzzy coat of the cuticle (described in detail below). These cilia have the regular $(9 \times 2) + 2$ microtubule arrangement (Flammang 1996, Ameye et al. 2000). The basal part of sensory cells terminates within the nerve plexus (Flammang 1996).

Support cells have a centrally located nucleus. Their apical surface has numerous microvilli. Their cytoplasm encloses one longitudinal bundle of intermediate filaments that traverses the cell and joins its apical and basal membranes (Flammang 1996). In tube feet that end in reinforced discs, these bundles of filaments connect the distal protrusions of connective tissue to the cuticle. At this level, the cuticle consists of three main sublayers: an internal fibrous layer, a middle granular layer, and an external layer made up of numerous fibrils (Holland and Nealson 1978, Ameye et al. 2000). This external layer, the so-called fuzzy coat, is the thickest layer of the cuticle and may have a role of protection and shock absorption (Ameye et al. 2000).

Tube Foot Adhesive Material
Structure

In all asteroid species investigated, the adhesive secretion usually remains firmly bound to the substratum as a footprint after detachment of the tube foot. The material constituting these footprints can be stained, allowing the observation of their morphology under a light microscope (Chaet 1965, L. A. Thomas and Hermans 1985, Flammang 1996, Hennebert et al. 2008). Footprints have the same shape and the same diameter as the distal surface of the tube foot discs. The adhesive material always appears as a sponge-like meshwork deposited on a thin homogeneous film (Flammang et al. 1994, 1998, Flammang 2006, Hennebert et al. 2008). This aspect has been observed by LM, conventional SEM, cryo-SEM, TEM, and atomic force microscopy (AFM). It does not differ according to whether the footprint has been fixed or not and whether it is observed partially hydrated or dried (Flammang et al. 1998, Flammang 2006, Hennebert et al. 2008). Based on TEM observations, the thickness of the adhesive layer ranges from 1.4 to 10.3 μm, with that of the homogeneous film varying between 0.3 and 3 μm. The thickness of the adhesive layer may also vary between different areas in a same footprint, giving different aspects to the adhesive material. In thin areas, meshes ranging from 1 to 5 μm in diameter are clearly distinguishable, whereas in thick areas these meshes are obscured because of the accumulation of material (Flammang et al. 1994, Hennebert et al. 2008). In TEM, the meshes do not appear empty as they do in SEM but appear to be filled with a loose electron-lucent material. Comparison of the images obtained with the different techniques therefore suggests that, during drying (as for instance for sample preparation for SEM), the loose material collapses, leaving empty mesh components on the thin homogeneous film. At the nanometer scale, both the material forming the meshwork and the homogeneous film are composed of a succession of globular nanostructures with diameters ranging from 50 to 200 nm (Hennebert et al. 2008).

Composition

The composition of the material constituting footprints was first investigated by using various dyes. The footprints of the asteriids *A. forbesi, A. rubens, L. hexactis,* and *M. glacialis* contain both proteins and acid mucopolysaccharides (Chaet 1965, L. A. Thomas and Hermans 1985, Flammang et al. 1994). Data on the biochemical composition of asteroid footprints are only available for *A. rubens.* The water content of the adhesive material has never been measured but, in dry weight, the footprints are made up mainly of proteins (20.6%), carbohydrates (8%), and a large inorganic fraction (about 40%). This composition agrees with the results obtained by histochemical tests in which the secretory granules of adhesive cells stain for acid mucopolysaccharides and proteins (Defretin 1952, Chaet and Philpott 1964, Chaet 1965, Souza Santos and Silva Sasso 1968, Engster and Brown 1972, Perpeet and Jangoux 1973, Flammang et al. 1994). Lipids were also detected in footprints, although they have not been detected in the secretory granules of adhesive cells by histochemical tests (Perpeet and Jangoux 1973, Flammang et al. 1998). This lipid fraction might come from the membranes of the adhesive granules or could be a contaminant (Flammang et al. 1998).

The protein moiety contains slightly more polar (55%) than non-polar (45%) amino acids. The polar fraction contains more charged (34%, of which 22% are acidic) than uncharged residues (21%). Moreover, it contains higher levels of glycine, proline, isoleucine, and cysteine than do the average eukaryotic proteins (Flammang et al. 1998). The importance of this protein fraction in the adhesiveness and cohesiveness of the footprint is clearly demonstrated by the removal of footprints from the substratum following treatment with trypsin (L. A. Thomas and Hermans 1985, Flammang 1996). Once solubilized, the protein fraction appears to be made up of 11 major protein bands that have no homology with any known protein in databases. These novel proteins might therefore correspond to the tube foot adhesive proteins and were named accordingly Sea star footprint proteins (Sfps; Hennebert 2010, Hennebert et al. 2011, 2012).

Colorimetric assays showed that the carbohydrate moiety is made up of neutral sugars (3% of the footprint dry weight), amino sugars (1.5%), and uronic acids (3.5%; Flammang et al. 1998). This fraction was investigated using lectins, which specifically recognize oligosaccharide motifs (Hennebert et al. 2011). A fraction of the carbohydrate content of the footprints is linked to two of the Sfps, one bearing mostly N-linked oligosaccharides and the other one bearing O-linked oligosaccharides. The sugar chains of both glycoproteins include galactose, N-acetylgalactosamine, fucose, and sialic acid residues. Another part of the carbohydrate fraction of the footprints is in the form of larger molecules, such as sialylated proteoglycans (Hennebert et al. 2011).

Sulfation also is a modification of the adhesive material. Sulfate groups have been detected in the footprints of *A. rubens* (about 2.5% dry weight; Flammang et al. 1998). It is not known whether these groups are attached directly to the proteins or to the carbohydrate residues.

Polyclonal antibodies raised against the footprint material of *A. rubens* strongly label the adhesive cells and the fuzzy coat but not the de-adhesive cells, demonstrating that the footprint material is mainly made up of the secretions of the adhesive cells and contains elements from the cuticle, whereas de-adhesive secretions are not incorporated into the footprints (Flammang et al. 1998). These antibodies were used to investigate the adhesive secretions from 14 asteroid species representing five orders and 10 families. In every species, there was a very strong and reproducible immunolabeling of the numerous granule-containing adhesive cells and cuticle of the disc adhesive area. This immunoreactivity is independent of the taxon considered, tube foot morphotype or function, species habitat, and ultrastructure of the adhesive cell secretory granules. It indicates that asteroid adhesives are closely related, probably sharing many identical molecules or, at least, many identical epitopes on their constituents (Santos et al. 2005b). Differences in the adhesive secretion composition not detected by the antibodies used could account for the differences observed in the structure and function of asteroid tube feet (Santos et al. 2005b).

A Model for Temporary Tube Foot Adhesion

A model for the temporary adhesion of asteriid tube feet can be proposed from all the information available (Fig. 3.6). Based on TEM observations, Flammang et al. (1994, 2005) demonstrated that different epidermal secretory cells of tube feet that end in a reinforced disc are involved in adhesion and de-adhesion and function as a duo-gland system as proposed by Hermans (1983). Adhesive cells in tube feet that were am-

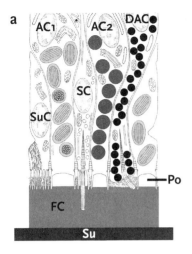
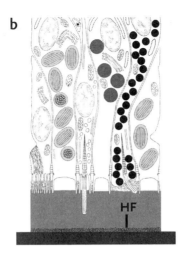
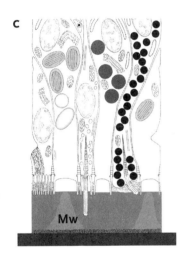
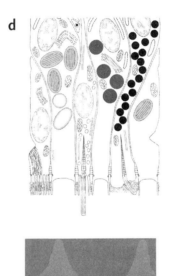
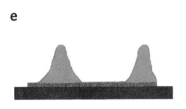

Fig. 3.6. Diagrammatic representations of longitudinal sections through the disc epidermis of a tube foot illustrating the proposed events in the attachment and detachment of asteroid reinforced disc-ending tube feet. When a tube foot attaches to the substratum, it first contacts the surface with its thick fuzzy coat (*a*). Upon contact, type 2 adhesive cells release their contents, presumably one or several surface active molecules, which form a homogeneous film covering the substratum (*b*). Almost simultaneously, type 1 adhesive cells start to release their content, molecules with a bulk function, which expand and form a thick meshwork structure (*c*). Release of the secretion from de-adhesive cells allows the tube foot to detach itself, the adhesive material and the fuzzy coat being left on the substratum as a footprint (*d*). Upon drying, the fuzzy coat collapses giving the footprint its characteristic aspect (*e*). AC1, type 1 adhesive cell; AC2 type 2 adhesive cell; DAC, de-adhesive cell; FC, fuzzy coat; HF, homogeneous film; Mw, meshwork; Po, pore; SC, sensory cell; Su, substratum; SuC, support cell. Not to scale. *Adapted from Flammang et al. 1998 and Hennebert et al. 2008.*

putated and fixed while they were firmly attached to a substratum have always released some secretory granules, but de-adhesive cells do not. However, de-adhesive cells in tube feet that were cut off and fixed just after they voluntarily detached from a substratum have released their most apical secretory granules (Flammang 1996, Flammang et al. 2005).

When a tube foot attaches to a substratum, it first contacts the surface with its thick fuzzy coat, and the compliant disc adhesive pad adapts its distal surface to the substratum profile (Santos et al. 2005c; Fig. 3.6a). Contact with the substratum is presumably detected by the cilium of the sensory cells. Their stimulation could trigger the release of the adhesive material from the two types of adhesive cells via the nerve plexus (Flammang et al. 1994). These secretions penetrate the fuzzy coat and bind the tube foot to the substratum. TEM observations of attached tube feet suggest that type 2 adhesive cells are the first to release their contents, being responsible for the formation of the homogeneous film covering the substratum (Fig. 3.6b). The material released by these cells has the same appearance as the one constituting this film (Hennebert et al. 2008). Meanwhile, type 1 adhesive cells start to release their contents, a heterogeneous electron-dense material that derives directly from the rods constituting the granules (Fig. 3.6c). This material expands gradually and fuses with the material released from other type 1 adhesive cells, initiating the formation of a meshwork pattern. The arrangement of the secretory pores of type 1 cells on the disc surface could act as a template for the formation of this pattern (Hennebert et al. 2008).

It is not known which Sfps constitute the homogeneous layer and which make up the meshwork. However, Sfps may allow adhesion through polar and electrostatic interactions between their amino acid side chains and the substratum surface, with the cohesiveness of the adhesive being reinforced through intermolecular disulphide bonds (Flammang et al. 1998, Flammang 2006). In addition, the high proportion of small side-chain amino acids, characteristic of elastomeric proteins that can withstand significant deformation without rupture, may provide compliance to the adhesive (Flammang 2006, Santos et al. 2009). The carbohydrate fraction of footprints presumably functions in the same way as the protein fraction in adhesion, the functional groups of glycan chains, such as hydroxyls, carboxylates, and amines, forming polar and electrostatic interactions with the substratum surface (Hennebert 2010, Hennebert et al. 2011).

The detachment of the tube foot is initiated by the release of secretory granules by the de-adhesive cells (Fig. 3.6d), possibly through a stimulation of their subcuticular cilium (Flammang 1996). The nature of this de-adhesive material is not known, but ultrastructural and immunocytochemical data suggest that it might function enzymatically to detach the fuzzy coat, thereby releasing the tube foot (Flammang et al. 1994, 1998). Once the tube foot is detached, the adhesive material and the fuzzy coat are left on the substratum as a footprint. Upon drying, the fuzzy coat collapses giving the footprint its characteristic aspect (Fig. 3.6e).

Acknowledgments

The authors would like to thank Prof. J. M. Lawrence for providing the opportunity to write this review. E. H. and P. F. are, respectively, a postdoctoral researcher and research director of the Fund for Scientific Research of Belgium (F.R.S.-FNRS). Work supported in part by a FRFC Grant no. 2.4532.07. This study is a contribution from the Centre Interuniversitaire de Biologie Marine (CIBIM).

4

Reproduction in Asteroidea

Annie Mercier and
Jean-François Hamel

Reproduction is the process by which individuals produce off-spring. It refers not only to gametogenesis but also to pro-cesses that occur at all levels of organization, from intra- and inter-population aspects to behavioral and molecular mechanisms at the organism level. Reproduction is also affected by the environment, both proximally and ultimately in terms of life history strategies. Pro-cesses involved in reproduction and the role of environmental factors at various stages of reproduction are considered here.

Sexuality and Modes of Reproduction

Both gonochoristic and hermaphroditic species of asteroids can be found (Giese et al. 1991). Asexual reproduction by fission occurs in *Coscinasterias* (Crump and Barker 1985, Sköld et al. 2002; see Chapter 19) and other genera (Ottesen and Lucas 1982, Achituv and Sher 1991, Karako et al. 2002, Rubilar et al. 2005), but most species reproduce sexually.

The majority of asteroids have broadcast spawning and external fertilization without any postrelease parental care, although external brood protection is relatively widespread (Fig. 4.1). Some species brood embryos laid on the substrate (Hamel and Mercier 1995). Others brood their young in a chamber formed by their arched arms, just beneath the mouth (Mercier and Hamel 2008), and some keep them on the ab-oral surface of the body wall (Lieberkind 1926). Internal fertilization, intragonadal incubation, and viviparity are less common (Byrne et al. 2003, Byrne 2005). It is probable that brooding and small size co-evolved as proposed by B. Menge (1975) and as suggested by the small size of most species in the genus *Leptasterias* (Lawrence and Herrera 2000). Brooding in *L. polaris*, a much larger species than predicted for brood-ers, may be adaptive or an example of phylogenic constraint and the inability to re-evolve the planktonic feeding stage (R. R. Strathmann and Strathmann 1982; see Chapter 18).

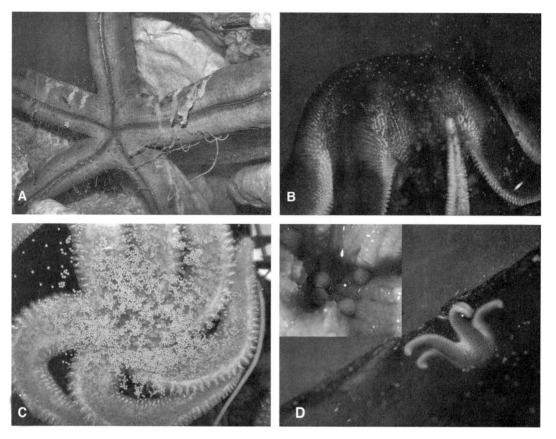

Fig. 4.1. *A*, Spawning of female *Linckia laevigata* (~20 cm diameter) through multiple gonopores along the arms; *B*, Female *Solaster endeca* (~28 cm diameter) releasing large buoyant oocytes; *C*, Female *Leptaserias polaris* (~19 cm diameter) brood-protecting its young, which it lays on the hard substrate; *D*, *Henricia* spp. showing the typical bell-shaped posture of brooding female (~6 cm diameter) with insert showing details of the growing embryos underneath the mouth. Note that dual brooding and broadcasting has been observed in *Henricia lisa*. *Photographs by J.-F. Hamel and A. Mercier.*

Sexual Reproduction

Reproductive success depends on highly synchronized processes between individuals of a population at the level of gamete synthesis and spawning. This synchronization is generally believed to be mediated by external cues (i.e., environmental factors) that may act either directly or indirectly, alone or in synergy. Increasing evidence is pointing to the transduction of environmental signals via the endocrine system, but the depth and complexity of the pathways and mechanisms involved and the possible role of chemical communication in this scheme have not been fully resolved. Different species may exhibit markedly different reproductive strategies and periodicities, based on their specific morphology and life history. Furthermore, different populations of the same species often breed at different,

locally suitable times while using the same suite of synchronizing cues.

Understanding how exogenous and endogenous factors interact to ensure that gametes are produced at the right time of year is important not only from an ecological standpoint but also for assessing how echinoderm populations may be affected by, and respond to, natural and anthropogenic disturbances (e.g., fisheries, climate change, endocrine-disrupting chemicals). The control of reproduction in echinoderms has been reviewed (Mercier and Hamel 2009).

Gametogenesis

The reproductive system of asteroids is organized either as a single proximal tuft of tubules with one gonoduct or as multiple tufts distributed along each arm (Chia and Walker 1991), in which case either each tuft has its own gonoduct and gonopore or a common gonoduct unites them all, leading to a single gonopore. Some species (e.g., *Asterias rubens* and other forciculates) have elongated gonads or branched sacs in each ray (Chia and Walker 1991). During the gametogenic cycle, gonads of male and female undergo a series of changes that lead to the development of spermatozoa and mature oocytes. Although the sequence may vary among species, they have the same main steps and general trends.

Chia and Walker (1991) reviewed spermatogenesis in asteroids. At the onset of spermatogenesis, a reorganization of the germinal epithelium occurs. Spermatogonia are released and undergo repeated cellular divisions forming the primary spermatocytes that produce spermatogenic columns. Spermatocytes move from the base of the columns toward the lumen of the gonad. Subsequently, when primary spermatocytes become dissociated from the columns, they elongate and go through the prophase / meiosis I transition. The resulting spermatozoa are stored in the lumen of the gonad. After gamete release, residual gametes and accessory cells are degraded by somatic phagocytes.

Smiley (1990) described oogenesis in asteroids. The ovaries have three layers: the germinative inner epithelium, a connective tissue with a genital hemal sinus, and an outer epithelium. Oocytes at all stages of development are generally found adjacent to one another within the inner epithelium of the same ascinus. Primordial germ cells are stem cells that produce oogonia through mitotic divisions. Oogonia divide repeatedly to produce large numbers of primary oocytes, each surrounded by follicle cells. Thereafter, the oocytes accumulate nutrient reserves during vitellogenesis and reach their maximum size. To become fully competent (fertilizable), oocytes must be ovulated at the time of spawning. As in males, nutritive phagocytes degrade residual gametes after spawning.

ENDOGENOUS MEDIATION

The timing of reproductive processes in asteroids is controlled by the nervous and endocrine or neuroendocrine systems, but external factors likely act as proximate cues (see next section) that induce biochemical changes within the gonads. Shirai and Walker (1988) discussed the perception and interpretation of environmental information. How external cues are perceived and interpreted by relevant internal effectors is not well understood. The nervous system and structures such as ocelli at the arm tips of asteroids may be involved in perception and may produce neuro-

chemicals. Alternatively, the entire surface of the body may respond to environmental stimuli (Shirai and Walker 1988).

Steroid hormones are found in asteroids (see Chapter 9) and affect gametogenesis. Fluctuations in progesterone, androstenedione, testosterone, estradiol-17β, and estrone have been studied (Barbaglio et al. 2006, Georgiades et al. 2006, Lavado et al. 2006). In *Sclerasterias mollis* and *Asterias rubens* (as *Asterias vulgaris*), levels of progesterone, estrone and testosterone vary with the reproductive cycle and exhibit a sex-specific relationship (Xu and Barker 1990a, Hines et al. 1992a; Fig. 4.2).

Final oocyte maturation occurs shortly after the gonad-stimulating substance (GSS, formerly known as gamete-shedding substance) is released from the nervous system into the coelomic cavity. When stimulated by GSS, the ovarian tissues produce the meiosis-inducing substance (MIS) 1-methyladenine, or 1-MA (Kanatani 1969). However, the action of 1-MA is probably restricted to the processes of ovulation and spawning.

Variations in gonad size in asteroids have been associated with fluctuations in steroid levels (Schoenmakers and Voogt 1981, Voogt 1982). Estrogens and progesterones have been measured in the gonads and pyloric ceca of several species such as *A. amurensis, A. rubens, Astropecten irregularis pentacanthus,* and *S. mollis* (Kanatani et al. 1971, Colombo and Belvedere 1976, Voogt et al. 1984, Xu and Barker 1990a). They presumably have several functions in reproduction and nutrition during the annual reproductive cycle (Voogt et al. 1984, Watts and Lawrence 1987). However, it is not clear whether or not the synthesis of intermediate steroid hormones is under the control of environmental factors such as temperature or photoperiod (Pearse and Eernisse 1982).

Schoenmakers and Dieleman (1981) showed seasonal patterns of progesterone and estrone in female *A. rubens* related to stages of oogenesis. They proposed an antagonistic relationship in the regulation of vitellogenesis. Voogt and Dieleman (1984) found varying levels of estrone and progesterone over the annual gametogenic cycle of males of *A. rubens*, suggesting that these steroids are involved in spermatogenesis. Progesterone metabolism (Voogt et al. 1986) and 3-β-hydroxysteroid dehydrogenase (3β-HSD) activity (Schoenmakers 1981) fluctuate in the ovaries and pyloric ceca during the reproductive cycle of *A. rubens*. Injections of estradiol-17β into the coelom increase the estrone level in the ovaries and were associated with larger oocytes in treated females (Schoenmakers

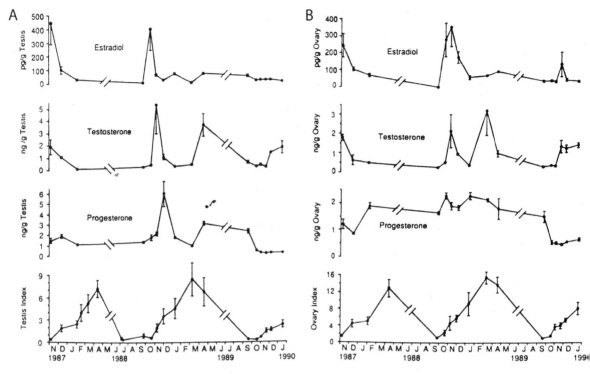

Fig. 4.2. Mean (±SEM) fluctuations of sex steroid levels and indices in the testes (A) and ovaries (B) of *Asterias rubens* (as *Asterias vulgaris*). *Hines et al. 1992a.*

et al. 1981). The testes and ovaries of *A. rubens* display annual growth cycles during which testicular or ovarian mass increases as production of gametes progressed (Hines et al. 1992a). In males, transient increases in the levels of estradiol coincided with spermatogonial mitotic proliferation, whereas increases in the levels of testosterone and progesterone coincided with spermatogenic column formation and spermiogenesis, respectively. In the ovaries, levels of estradiol and testosterone peaked at the onset of oogenesis, but progesterone levels did not change significantly throughout the annual cycle. The authors suggested that transient increases in the levels of sex steroids during gametogenesis could serve as endogenous modulators of reproduction (Hines et al. 1992a; Fig. 4.2).

Both *A. rubens* and *Asterina pectinifera* exhibited an increase in oocyte diameter and in lipid amount in vivo, as well as an increase in RNA levels in vitro following exposure of ovarian fragments to estradiol (Van Der Plas and Oudejans 1982). N. Takahashi (1982a) reported a significant increase in ovarian protein in *A. pectinifera* following the administration of estrone but not estradiol-17β or testosterone. Furthermore,

progesterone seemed to depress ovarian growth and reduce ovarian protein levels.

Seasonal changes in the ovarian levels of estradiol-17β, estrone, and progesterone of *S. mollis* are closely related to the stages of the reproductive cycle (Xu and Barker 1990a, Barker and Xu 1991a). Daily injection of estradiol-17β or estrone caused an increase in the estrone and progesterone levels in the ovaries in the early stage of the gametogenic cycle, and oocyte diameter and protein level in the ovaries increased (Barker and Xu 1993). This suggests estrogens and progesterone are involved in a regulatory system to control reproduction (Barker and Xu 1993).

Ikegami et al. (1972) showed that asterosaponins inhibit spawning in *A. amurensis* by repressing maturation of oocytes. Ovarian follicle cells were the site of action of the saponins (Ikegami 1976), indicating that regulation of oocyte maturation and spawning may be an important function of asterosaponins (Voogt and Huiskamp 1979). The seasonal variation in saponin levels are related to the reproductive cycle in *A. rubens* (Voogt and Huiskamp 1979) and asterosaponins in *A. amurensis* are related to the gametogenic cycle and spawning (Yasumoto et al. 1966, Ikegami et al.

1972). Because this pattern is not always observed, inhibition of spawning may not be the primary function of asterosaponins in all asteroid species. Burns et al. (1977) observed peak levels of asterone in *A. rubens* prior to spawning, and Mackie et al. (1977) reported high levels of asterosaponins in the gonads during the spawning season of *Marthasterias glacialis*.

Watts et al. (1990a) found that the specific activity of ornithine decarboxylase (ODC) in *A. rubens* (as *A. vulgaris*) was low when testis growth was minimal and increased with maximum testis growth. Sible et al. (1991) suggested that seasonal regulation of ODC transcription was an essential component of the reproductive cycle, because cells in the inactive spermatogenic epithelium of *A. rubens* were competent to undergo mitotic proliferation when supplied with polyamines.

Depletion of cytoplasmic reserves and catabolism of whole cells to mobilize nutrients for reproduction is suggested from the work of Wasson and Klinger (1994). They reported a decrease in total DNA content and an increase in DNA concentration in the pyloric ceca prior to spawning in *Asterias forbesi*. DNA concentration decreased post spawning but DNA content increased, indicating that nutrients were being stored by increasing the number and size of the cells of the pyloric ceca.

CORRELATION WITH EXOGENOUS FACTORS

Reproduction should be timed to environmental factors that maximize fertilization success or survival of the offspring. For this to happen, gametogenesis must be initiated before conditions become optimal, and the organisms must detect environmental changes that act as cues or zeitgebers that synchronize reproduction with the subsequent favorable conditions. These environmental changes could exert proximate exogenous control on reproduction. They need not be restricted to only one factor but may consist of several factors that interact in synchronizing reproductive activities. One factor may be dominant, at least under some circumstances (Giese and Pearse 1974, Mercier and Hamel 2009).

Photoperiod and temperature are the most commonly proposed proximate mediators of reproduction in echinoderms. They are usually examined concurrently, since day length directly influences sea surface temperatures. Although most studies are correlative (establishing parallels between reproductive status and seasonal fluctuations in photoperiod or temperature), a few laboratory investigations confirm the influence of these factors on the gametogenesis of asteroids.

Dehn (1980a) combined laboratory and field studies to show that temperature and food supply are correlated with the reproductive activity of *Luidia clathrata*. She noted that the gametogenic activity began in fall when temperatures were rapidly decreasing and suggested that this triggers metabolic processes associated with activation of gametogenesis.

Pearse and Eernisse (1982) showed that longer day lengths entrained the initiation of gametogenesis and gonadal growth in *Pisaster ochraceus*. Oogenesis in *P. ochraceus* was triggered 6 months in advance (midwinter) following laboratory exposure to long day lengths (>12 h). Because this out-of-phase oogenic cycle was not sustained for more than a few months they suggested that short day lengths may be required as well to maintain later phases of oogenesis. In contrast, a fixed photoperiod (short, neutral, or long) did not affect gametogenesis, implying that the gametogenic cycle is under the control of an endogenous annual rhythm (Pearse et al. 1986; Fig. 4.3). *Leptasterias* sp. maintained under two different photoperiod regimes shifted out-of-phase with respect to in-phase animals (Pearse and Beauchamp 1986). Pearse and Walker (1986) documented the role of photoperiod on the reproduction of *A. rubens* (as *A. vulgaris*) and proposed that control of reproduction by photoperiod might be a general phenomenon, at least for shallow-water species.

Xu and Barker (1990b) examined the regulation of oogenesis in *S. mollis*. Well-fed individuals kept 6 months out-of-phase of the ambient photoperiod gradually shifted their breeding periodicity after a few months, based on the gonad index, the size of oocytes, and the steroid levels. Individuals maintained in-phase did not entirely follow the field conspecifics, presumably because of the different light, temperature, and feeding regimes. Nevertheless, the two experimental groups displayed clear out-of-phase reproductive rhythms in response to the photoperiod regime, while their feeding, body size and pyloric ceca were unaffected.

Pearse and Bosch (2002) also demonstrated the photoperiodic control of gametogenesis in *Odontaster validus*. Gametogenesis in individuals maintained on a photoperiod 6 months out-of-phase with ambient (constant light in winter, constant dark in summer) adopted the out-of-phase regime within a year, unlike asteroids kept on an ambient photoperiod or collected from the

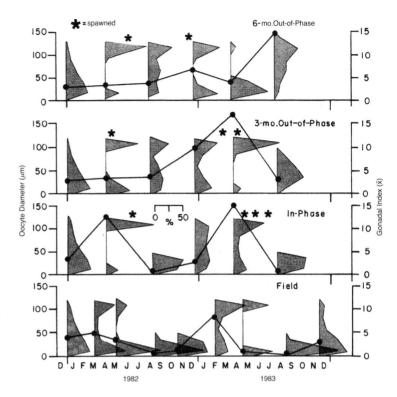

Fig. 4.3. Changes in gonad index (*solid lines*) and size distributions of oocytes (*shaded polygons*) of *Pisaster ochraceus* maintained under seasonally changing photoperiods with different phases. Individuals in 3-month out-of-phase treatment were initially shifted ahead from mid-December photoperiod to that equivalent of mid-March; those in 6-month out-of-phase treatment were shifted to equivalent of mid-June. *Pearse et al. 1986.*

field. Gametogenesis of individuals in constant light or on a 12 h L:12 h D photoperiod was maintained more or less continuously, whereas day lengths >12 h seemed to stimulate gametogenesis. Inversely, gametogenesis of individuals in constant darkness generally followed that of asteroids kept under ambient photoperiod or collected from the field, suggesting an underlying circannual rhythm (Pearse and Bosch 2002).

Although nutrition can influence reproduction indirectly in various ways, laboratory and field studies have yet to clearly demonstrate that changes of food levels can directly affect the timing of reproduction. One possible exception is the periodic pulse of detritic food to the deep seabed, which has been suggested to help time reproduction of echinoderms at depths.

The pyloric cecum is a processing and storage organ unique to Asteroidea within the Echinodermata phylum (Lawrence and Lane 1982, Lawrence 1985, 1987) that stores glycogen, lipid, and protein (Greenfield et al. 1958, Jangoux and van Impe 1977). An inverse relationship between pyloric cecum and gonad indices is often (but not always) observed in asteroids (e.g., Chia and Walker 1991, Byrne 1992, Ventura et al. 1997, Rubilar et al. 2005, Georgiades et al. 2006), sug-

gesting nutrient transfer from pyloric ceca to gonads during gametogenesis (Fig. 4.4).

A few studies have predicted that gametogenic activity in asteroids, particularly vitellogenesis, depends directly on feeding. Comparative studies of *Odontaster validus* (Pearse 1965) and *Patiriella regularis* (Crump 1971) suggest gametogenic output may be related to the abundance of food. Worley et al. (1977) found that peak activity in the reproductive cycle was positively correlated with the period of active feeding in *L. tenera*. Bouland and Jangoux (1988a) found that the course of gonad development in *A. rubens* was similar in both field animals and experimentally starved individuals, suggesting that reproduction has a greater priority than somatic maintenance during gametogenesis.

Tyler et al. (1990) observed that the availability of labile organic material favors vitellogenesis in *Dytaster grandis* from the Porcupine Abyssal Plain. Tyler et al. (1993) found parallels between the diets of two sympatric deep-sea asteroids. The diet of the aperiodic breeder, *Bathybiaster vexillifer*, showed a high prey diversity index, and there was no seasonal cycle in the organic carbon content of the sediment residue in its stomach. In contrast, the diet of the seasonal

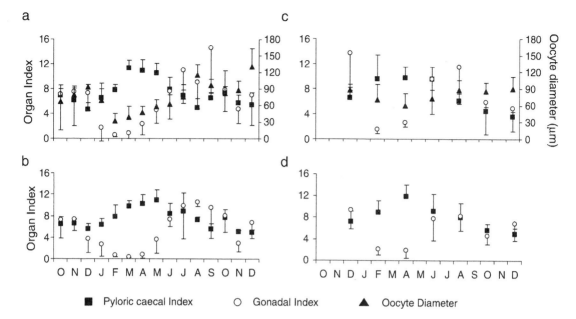

Fig. 4.4. Changes in reproductive and physiological parameters including pyloric cecal index, gonad index, and oocyte diameter (Mean ±SD) in female and male *Coscinasterias muricata* from Australia showing the inverse relationship between the gonad/gamete synthesis and the pyloric ceca. *a*, Female Governor's Reef; *b*, Male Governor's Reef; *c*, Female South Channel Fort; *d*, Male South Channel Fort. *Georgiades et al. 2006.*

breeder, *Plutonaster bifrons,* showed significantly lower prey diversity and seasonal patterns in organic carbon content of the sediment residue in its stomach (Tyler et al. 1993).

Surface productivity above the Porcupine Seabight and the Porcupine Abyssal Plain undergoes significant seasonal changes, leading to a marked pulse of phytodetritus to the seabed (Rice et al. 1994). Benitez-Villalobos et al. (2007) suggested *Henricia abyssicola* utilizes phytodetritus as an energy source to develop oocytes, the May peak in gonad index being followed by a marked drop, probably indicative of spawning between June and July.

BEHAVIOR AND INTERINDIVIDUAL COMMUNICATION

Both direct and indirect evidence supports that chemical exchanges occurs between individuals during gamete synthesis. Members of the genus *Archaster* commonly aggregate during the spawning season (Run et al. 1988, Keesing et al. 2011, Lawrence et al. 2011), and spawning pheromones are known or suspected to be produced by other species (Beach et al. 1975, R. L. Miller 1989, Hamel and Mercier 1995). However, studies of chemical signalling in asteroids

have focused more on spawning than on gametogenesis. Hamel and Mercier (1995) found that contact between *L. polaris* before spawning produced gametogenic synchronization, mostly during the late stages of gamete maturation. Run et al. (1988) documented increased male movement and greater numbers of male-on-female pair assemblages of *A. typicus* approximately 2 months before the peak spawning period, suggesting that males recognize the sex of other individuals and that some form of communication occurs during gametogenesis. Similar male-on-female pairing and behavior was observed in *Archaster angulatus* (Keesing et al. 2011, Lawrence et al. 2011).

Several abyssal and bathyal echinoderms form aggregations or small discrete groups that may facilitate breeding (Mercier and Hamel 2009). Because it is generally assumed that deep-sea invertebrates do not cue reproduction to celestial factors, interindividual communication is believed to play a role. For instance, aggregations of the bathyal species *H. lisa* occur in the laboratory only during the prespawning phase of the biannual spawning period (Mercier and Hamel 2008). But support for interindividual communication prior to breeding comes mostly from the study of deep-sea echinoids (Young et al. 1992).

Spawning
ENDOGENOUS MEDIATION

Mechanisms of final oocyte maturation and spawning have been widely investigated in asteroids (Kanatani and Nagahama 1983). Chaet and McConnaughy (1959) reported that the injection of an extract of the radial nerve of *A. forbesi* induced gamete release in ripe individuals. A radial nerve substance responsible for inducing gamete release has been confirmed for several species (Kanatani 1979). The radial nerve factor (RNF) typically acts in a non-species-specific manner (Kanatani 1973).

Responses to RNF are positive in whole animals injected with it (Noumura and Kanatani 1962, Chaet 1964) and in isolated ovaries and testes (Kanatani 1964, Chaet 1966). The substance was named the gamete-shedding substance (GSS) and is apparently present in the radial nerve throughout the year in nearly constant concentration (Chaet 1967). The GSS level in the radial nerve was equal in both sexes assayed with isolated ovarian fragments in vitro and also constant along the radial nerve and circumoral nerve ring (Kanatani and Ohguri 1966). Body components with well-developed nervous tissue (epidermis, tube feet, cardiac stomach) also contain GSS, although in a less active form (Kanatani and Ohguri 1966). GSS is a hormone, since it is present in the coelomic fluid only in asteroids undergoing natural spawning (Kanatani and Shirai 1970). Granules in *A. pectinifera* containing GSS occur in the supporting cells located just beneath the outer sheath of the radial nerve (Atwood and Simon 1971). Similar structures are found in the subepithelial plexus of the tube feet, body wall, and cardiac stomach, but not in the pyloric cecum (Kanatani and Ohguri 1966, Atwood and Simon 1971).

GSS was first believed to be a direct inducer of spawning (Chaet 1967), but experiments later revealed the action is indirect. It acts on the follicle cells of the ovary to produce a second active substance (Kanatani et al. 1969, Cloud and Schuetz 1973, Hirai et al. 1973). The latter has been isolated from GSS and ovarian fragments of *Asterias amurensis* and identified as 1-MA (Kanatani et al. 1969). The gamete-shedding substance was renamed the gonad-stimulating substance, retaining the same GSS acronym (Kanatani 1967) and later determined to be a relaxin-like peptide (Mita et al. 2009).

Ovarian fragments release eggs in seawater containing 1-MA, and ripe asteroids spawn when injected with 1-MA (Kanatani 1969). Kanatani found that oocyte release in *M. glacialis* and *A. forbesi* occurred within ~30 min of injection of 1-MA, whereas sperm release began after a slightly shorter latent period. This rapid reaction indicates that 1-MA production, presumably in the follicle cells, begins immediately after GSS is detected (Kanatani and Shirai 1970). Mita (1993) found that the amount of 1-MA produced by the follicle cells of *A. pectinifera* was sufficient for meiosis initiation and release of oocytes. However, Schuetz (2000) provided evidence of extra-follicular mediation of oocyte maturation by RNF in *P. ochraceus*, suggesting that other ovarian components may be a source of 1-MA.

The induction of ovulation and spawning by 1-MA, either through intra-coelomic injection or soaking of the extracted ovaries, has been demonstrated in many species (Stevens 1970, Bosch and Pearse 1990, Byrne and Barker 1991, Hamel and Mercier 1995, Stanwell-Smith and Peck 1998, Babcock et al. 2000, Komatsu et al. 2006). The action of 1-MA is reportedly through dissolution of the cementing substance between the follicle cells and between the follicles and the oocytes. Whether the action of 1-MA on the cementing substance is direct or indirect (i.e., activating an enzyme responsible for dissolving the cement) is not known (Kanatani and Nagahama 1983). Once the follicular envelopes around the oocytes are removed by the action of 1-MA, the denuded oocytes become free within the lumen of ovary and are expelled by contraction of the ovarian wall (Kanatani and Shirai 1969).

Ovarian fragments of *A. amurensis* placed in seawater devoid of Mg^{2+}, begin to release oocytes after ~30 min, whereas ovarian fragments kept in Ca^{2+}-free seawater for ≥ 30–45 min shed oocytes when Ca^{2+} is added (Kanatani and Shirai 1969). This suggests that lack of Ca^{2+} causes dissolution of the cement and breakdown of the follicular envelope and that contraction of the ovarian wall, caused by the addition of Ca^{2+}, then forces out the loosened oocytes. The fact that isolated ovaries do not release oocytes when placed in seawater devoid of Ca^{2+} even in the presence of GSS (Schuetz and Biggers 1968) indicates that, while contraction of the ovarian wall is essential for spawning, dissolution of the cement leading to follicular disintegration is a prerequisite (Kanatani and Shirai 1969).

While the natural mitogen that converts the vitellogenic oocytes into fertilizable oocytes was identified as 1-MA, extensive investigations have failed to isolate its receptor and early transduction pathway. Nevertheless, several elements of the signalling pathway that link the 1-MA receptor to the maturation-promoting factor, or MPF, activation and maturation have been reported (e.g., Tadenuma et al. 1992, Sadler and Ruderman 1998).

CORRELATION WITH EXOGENOUS FACTORS

Day length is often considered as a synchronizer for biological events because it is so invariant from year to year for the same latitude and season (Giese and Pearse 1974). It is more probable that photoperiod could serve as a cue at higher latitudes, where greater seasonal differences in day length occur (Giese and Pearse 1974). However, for species that spawn annually in a particular season but at different times from year to year, photoperiod is an unlikely spawning cue (Himmelman 1999). Furthermore, factors that vary along with photoperiod often obscure proposed correlations. For instance, summer solstice was proposed to cue spawning in *Coscinasterias muricata* based on gonad indices and oocyte size frequencies that showed readiness to spawn around this period (Georgiades et al. 2006). However, the authors also noticed that the main spawning event coincided with the lowest seawater temperatures.

Daily variations in the light regime are among the most obvious factors to affect spawning directly. Although this factor can only be investigated by monitoring gamete release (as opposed to indirect sampling methods) and data can be anecdotal if observations are fortuitous rather than planned, convincing evidence has been gathered.

M. glacialis spawns between 1330 and 2130 h on sunny days during summer in Ireland (Minchin 1987). On the Great Barrier Reef, spawning of *Acanthaster planci* generally occurs between early afternoon and late evening (Babcock and Mundy 1992). Finally, several species from eastern Canada (e.g., *Solaster endeca*, *Crossaster papposus*, *A. rubens*, *Pteraster militaris* and *H. anguinolenta*) consistently spawn within species-specific diel windows (Mercier and Hamel 2010).

There is no agreement on the direct influence of temperature on spawning (Giese 1959, Giese and Pearse 1974, H. Barnes 1975), but it is among the factors most often discussed as a potential spawning cue, especially since marked temperature changes induce ripe animals to shed gametes in the laboratory (see reviews by Himmelman 1999, Mercier and Hamel 2009). Costello and Henley (1971) induced spawning in gravid *A. for-*

besi and *A. rubens* (as *A. vulgaris*) by keeping them in tanks of refrigerated seawater and increasing the temperature to 20°C. However such "temperature shocks" are unlikely to occur in situ and convincing evidence of the role of temperature in gamete release is lacking. Although studies have demonstrated that gametogenesis can be stimulated by an increase in temperature in numerous species, few have shown that gamete release occurs when a particular temperature is attained (e.g., Hamel and Mercier 1995, Mercier and Hamel 2008). More evidence relates to spawning being coincident with significant temperature variations (reviewed by Himmelman 1999, Himmelman et al. 2008).

Spawning of *A. forbesi* in eastern USA began when seawater temperature was approximately 15°C in one study (Loosanoff 1964) and 16–18°C in another (Franz 1986). Hancock (1958) noted that *A. rubens* around Essex spawned synchronously when the seawater temperature reached ~15°C. During a long-term study of the deep-sea species *H. lisa* from a depth of ~600 m that spanned three breeding seasons, the onset of spawning in the laboratory consistently took place at 3–4°C (when ambient seawater temperature was either increasing or decreasing), leading to biannual reproduction (Mercier and Hamel 2008; Fig. 4.5a).

Spawning of male *L. littoralis* in Maine is probably induced by the decrease of seawater temperature (O'Brien 1976). Spawning of *L. polaris* from eastern Canada also appears to be correlated with decreasing temperature (Hamel and Mercier 1995). Conversely, Minchin (1987) concluded that the rapid increase in seawater temperature helped synchronize spawning in *M. glacialis* in Ireland. Seasonal spawning of *P. gunnii* and *P. calcar* in Australia may also be cued by increasing temperature (Byrne 1992). Temperature may serve as a proximate cue for the onset of spawning as gamete release started when sea temperatures increased and terminated well before the maximum temperature was reached, *P. gunnii* spawning out at 19–20°C (Byrne 1992). In situ observations of *A. rubens* (as *A. vulgaris*) in eastern Canada have shown that spawning events coincide with seawater temperature increases resulting from the downwelling of warm surface water (Himmelman et al. 2008).

Conditions that favor development and survival of offspring (e.g., presence of planktonic food) are often proposed to be the ultimate mediator of spawning. Again, this factor is more commonly proposed for species living at temperate high latitudes where

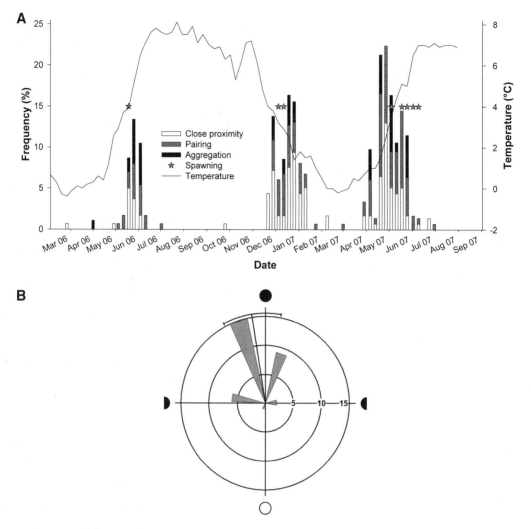

Fig. 4.5. *A,* Aggregative behavior and spawning of *Henricia lisa* in correlation with prevailing seawater temperature between March 2006 and August 2007. *B,* Circular plot depicting reproductive occurrences of spawning events in male and female *H. lisa* relative to the lunar period. Filled circle represents new moon and open circle represents the full moon. Mean angle (*line*) and 95% confidence interval (*curved bracket*) are shown. *From data in Mercier and Hamel 2008.*

seasonal peaks in phytoplankton abundance are especially pronounced. It should be noted that synchronization of the planktonic larval phase with peak phytoplankton abundance can be achieved by cuing gamete release on the algal bloom itself or on other correlated environmental factors (i.e., temperature, photoperiod; Himmelman 1999). Furthermore, seasonal spawning by Antarctic species without pelagic larvae suggests that spawning coinciding with algal blooms can be promoted by factors other than the presence of suitable food for planktotrophic larvae (Pearse et al. 1991a). Several species with lecithotrophic development (e.g., *S. endeca, C. papposus*)

spawn simultaneously with planktotrophic species in the spring in eastern Canada (Mercier and Hamel 2010). Further research is required into the possible function of phytoplankton as a spawning cue. Increasing food supply for larvae (and decreasing photoperiod) have been proposed to trigger spawning in *Cosmasterias lurida* in Patagonia (Pastor de Ward et al. 2007), overall, but there is little evidence that seasonal food supplies can directly trigger spawning in asteroids.

Several aspects of the cyclical behavior of marine organisms, including gametogenesis, spawning, mating, and the release of larvae or juveniles, are correlated with phases of the moon (Omori 1995, Naylor 1999). Yet the underlying principles of lunar reproductivé cycles are still unclear beyond the fact that it provides a means of synchronization (Naylor 1999). There are few examples of lunar periodicity in asteroid reproduction. Mating in *Archaster typicus* in the Philippines is suggested to coincide with the new moon (Janssen 1991). Yamaguchi and Lucas (1984) observed that spawning in *Ophidiaster granifer* from Guam coincided with the full moon. In situ observation of *A. planci* on Davies Reef revealed the occurrence of a mass spawning during the three-quarter moon simultaneously with numerous other species of corals and echinoderms (Babcock and Mundy 1992, Babcock et al. 1992). Clear lunar patterns of spawning were described in boreal and deep-sea asteroids. Mercier and Hamel (2010) determined that spawning of *S. endeca, C. papposus, A. rubens,* and *H. sanguinolenta* occurred around the full moons of March and April for 2 consecutive years. Mercier et al. (2011) also found statistical correlation of spawning events with the new moon in *H. lisa* collected at depths between 500–1000 m (Fig. 4.5b).

BEHAVIOR AND INTERINDIVIDUAL COMMUNICATION

Although synchronous gamete release can increase fertilization success if spawners are near one another, synchronization alone may not be sufficient if gametes are emitted by distant individuals (Levitan 2004). Mechanisms to detect conspecifics, especially those of the opposite sex, may evolve under such selective pressures (e.g., Painter 1992, Hardege et al. 1996, Soong et al. 2005). Four main types of evidence of chemical signalling related to reproduction can be distinguished: intraspecific breeding aggregation, interindividual communication prior to or during spawning, asynchronous gamete release by males and females, and simultaneous heterospecific mass spawning events.

It is generally believed that aggregative behavior, even if it is independent of spawning, will increase spawning success in broadcast spawners (Levitan 1988), in part because gametes have limited longevity and disperse quickly in seawater (Pennington 1985). However, breeding aggregations are common not only in broadcasting species but also in brooding and non-brooding species that exhibit demersal lecithotrophic development.

Long ago, Ludwig (1882) described arm interlocking in spawning *Asterina gibbosa*. The brooding species *L. ochotensis similispinus* has a similar behavior (Kubo

1951). Males and females of other brooding species aggregate before or during spawning. Examples include *L. hexactis* (Chia 1968a), *L. littoralis* (O'Brien 1976), *L. polaris* (Hamel and Mercier 1995), and *H. lisa* Mercier and Hamel 2008; Fig. 5a). Free-spawning species may also aggregate at the time of reproduction. Komatsu et al. (1979) and Tominaga et al. (1994) described annual spawning aggregations of *A. minor*. They believe that aggregation serves to increase the number of individuals before spawning (maybe to produce a threshold concentration of pheromones). Minchin (1987) correlated aggregations of *M. glacialis* with time of spawning and postulated that aggregations might be elicited by a pheromone released by females. Other free-spawning dioecious species such as *A. typicus* (Run et al. 1988), *Archaster angulatus* (Keesing et al. 2011, Lawrence et al. 2011) and *A. planci* (Beach et al. 1975, Komatsu 1983) aggregate for breeding. In situ observations showed that many individuals of *A. rubens* (as *A. vulgaris*) were tightly grouped during gamete release in eastern Canada, especially during mass spawning (Raymond et al. 2007, Himmelman et al. 2008), although it remains unclear whether this gregarious species truly aggregates to spawn. The bathyal species *H. lisa* consistently aggregated in the laboratory during the summer and winter breeding periods (Mercier and Hamel 2008). Presumably this brings males and females together to coordinate gamete maturation, synchronize spawning, and ultimately promote fertilization success.

R. L. Miller (1989) suggested chemically mediated sexual communication in *Orthasterias koehleri* and *A. forbesi* via a long-lived sperm chemo-attractant. Although females of both species emit sperm attractants into the surrounding seawater, a sex-specific release by ripe females has only been directly demonstrated for *O. koehleri* (R. L. Miller 1989). The author stated that detection of the sperm attractant in laboratory seawater was correlated with spawning events in both species. A pheromone from testes and ovaries was demonstrated to elicit aggregation and synchronize spawning in both sexes in *A. planci* (Beach et al. 1975).

Pseudocopulation (paired superposition) is a special case of aggregation. It has been observed in the broadcast spawning *A. typicus* and *A. angulatus* (Komatsu 1983, Run et al. 1988, Keesing et al. 2011, Lawrence et al. 2011) and in the brooding *Neosmilaster georgianus* (Slattery and Bosch 1993). Heterosexual recognition by contact chemoreception via superposition is suggested to occur in *A. typicus* (Komatsu 1983). Males are stimulated to spawn directly onto the oocytes released by the females (Komatsu 1983, Run et al. 1988), indicating that A. typicus is capable of initiating both mating behavior in response to a contact stimulus and male spawning in response to female spawning. Spawning behavior and superposition of the brooding species N. georgianus were observed in the field and the laboratory in Antarctica during the austral spring of 1991 (Slattery and Bosch 1993). Pre-mating activity of males in the field was triggered by the spawning of a nearby conspecific female. In the laboratory, similar behavior was elicited by the presence of spawning conspecific males. Thus, the pseudocopulatory behavior exhibited by N. georgianus is apparently mediated by chemical signals released from spawners of both sexes (Slattery and Bosch 1993). Chemoreception, particularly via terminal sensory tube feet, occurs in asteroids and may account for intra- and interspecific responses (Mayo and Mackie 1976, Sloan 1984). In their study of A. typicus, Run et al. (1988) observed that pairings favored contact of the marginal and furrow spines. In contrast, contact occurs between the arm tips of males and the gonopore region (i.e., arm base) of female N. georgianus. Contrary to the non-specific initiation of searching behavior, sperm release is only elicited in the presence of females, presumably by a second sex-specific hormone (Slattery and Bosch 1993).

In A. angulatus, pseudocopulation can occur even in spent individuals, when few gametes are present in the gonads (Keesing et al. 2011, Lawrence et al. 2011).

R. H. Smith (1971) speculated release of a substance (neural extract) might cause the pre-spawning aggregations of L. pusilla and L. aequalis. He further argued that (1) the neuro-secretory-like vesicles are located on the external surface of the radial nerve (Unger 1962, Uter 1967), (2) asteroids can detect chemical compounds in concentrations of a few parts per billion (Blumer 1969), and (3) they are capable of absorbing molecules from the external environment at low concentrations by means of active transport (e.g., Ferguson 1967a,b, Fontaine and Chia 1968).

Chia (1968a) indicated that gametes from congeners could induce spawning in L. hexactis. O'Brien (1976) noted that males of L. littoralis spawned before females while individuals of both sexes were in close proximity. McClary and Mladenov (1988) suggested that female P. militaris spawn in response to sperm or substances released with it. In contrast, spawning by female A. typicus was closely followed by spawning of its paired male, whereas male spawning did not induce spawning in the female (Run et al. 1988). Male L. polaris release sperm before females spawn (Hamel and Mercier 1995). In this species the longevity and

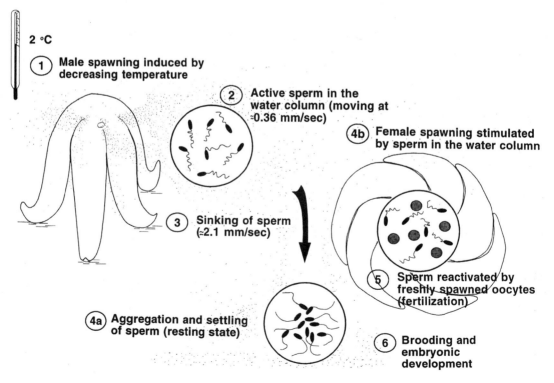

2 °C

(1) Male spawning induced by decreasing temperature

(2) Active sperm in the water column (moving at ≈0.36 mm/sec)

(4b) Female spawning stimulated by sperm in the water column

(3) Sinking of sperm (≈2.1 mm/sec)

(5) Sperm reactivated by freshly spawned oocytes (fertilization)

(4a) Aggregation and settling of sperm (resting state)

(6) Brooding and embryonic development

Fig. 4.6. Schematic illustration of spawning sequence in male and female *Leptasterias polaris. Hamel and Mercier 1995.*

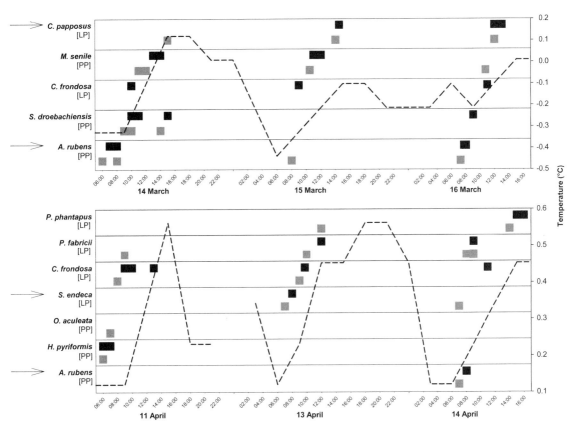

Fig. 4.7. Species-specific sequences recorded during same-day multispecies spawnings in March (*top*) and April (*bottom*) in eastern Canada. Arrows show the asteroids involved. Gray indicates males; black indicates females. Development mode is noted as PP for planktotrophic pelagic and LP for lecithotrophic pelagic. Seawater temperatures are also shown (*dashed line*). *Adapted from Mercier and Hamel 2010.*

egg-activation of sperm (Fig. 4.6) is presumed adaptive. Experimentally, the best fertilization success (>75%) occurred when the delay between the male and female spawnings was no more than 11 h, although success was still ~50% after 30 h (Hamel and Mercier 1995). Males consistently spawning before females also occurs in *C. papposus, S. endeca,* and *A. rubens* (Mercier and Hamel 2010).

S. L. Pain et al. (1982) suggested that spawning in male deep-sea asteroids is stimulated by oocyte release during chance encounters with mature females. In the deep-sea *H. lisa,* males in a reproductive aggregation always spawned 30–60 min before females, suggesting that females respond to a cue emitted by males (Mercier and Hamel 2008).

Apart from intraspecific interactions, there is growing evidence of heterospecific spawning events involving asteroids and other marine invertebrates, which raise the possibility of heterospecific spawning cues. Simultaneous spawning of holothuroids and *Henricia* sp. was observed in situ by Sewell and Levitan (1992). Spawning by *A. planci* in the field with two holothuroids during the mass spawning of acroporid corals was reported by Scheibling and Metaxas (2008). In situ observations reported by Himmelman et al. (2008) include cases of *A. rubens* (as *A. vulgaris*) spawning on the same day as ophiuroids (*Ophiopholis aculeata* and *Ophiura robusta*) and an echinoid (*Strongylocentrotus droebachiensis*). The spring spawning of more than 25 species of marine invertebrates monitored in large communal flow-through tanks with natural fluctuations of photoperiod and food supply showed that up to 11 species from various phyla can spawn in a single day, including the asteroids *C. papposus, S. endeca, A. rubens,* and *P. militaris* (Mercier and Hamel 2010; Fig. 4.7).

Acknowledgments

We would like to acknowledge the huge contribution of early and contemporary researchers who have poked and probed the intricate reproductive processes and strategies of asteroids, including those we could not cite directly due to space constraints. We also extend our warmest thanks to John Lawrence for inviting our contribution and providing helpful feedback.

5

Asteroid Evolutionary Developmental Biology and Ecology

Maria Byrne

The developmental biology of the Asteroidea is well studied, especially for the shallow water species used as model animals for developmental biology (Reviews: R. R. Strathmann 1985, Chia et al. 1993, Byrne 1999, McEdward and Miner 2001, McEdward et al. 2002). In recent research, the remarkable life history diversity in the Asteroidea has been the focus of evolutionary developmental (evo-devo) biology, developmental ecology, and phylogeny, phylogeography, and population genetics studies (Hart et al. 1997, Byrne 2006, Raff and Byrne 2006, Prowse et al. 2008, 2009, C. D. H. Sherman et al. 2008, Barbosa 2012). This chapter focuses on insights generated in these fields through studies of the diversity of larval strategies in the Asteroidea.

Egg Size, Location of Fertilization, and Development

Most asteroids, including all the species covered in this volume, are free-spawners that release their gametes into the water column for fertilization followed by development through a dispersive larva (Fig. 5.1). Other asteroids have benthic development. Some species deposit eggs that adhere to rocks on the shore. The embryos complete their development on the substratum unattended (e.g., *Asterina gibbosa, Aquilonastra minor, Parvulastra exigua*) or brooded by the adult (e.g., *A. gibbosa*; Komatsu et al. 1979, Emson and Crump 1979, Crump and Emson 1983, Byrne 1995). Benthic egg masses are fertilized in situ by direct deposition of sperm onto eggs (Komatsu et al. 1979, Byrne 1992). Other asteroids that care for their young retain their eggs in brood chambers on or in their body. *Pteraster* species have a specialized nidamental chamber on the aboral surface where the embryos develop (McEdward and Janies 1993). Self-fertilization of benthic embryos has also been observed (Barbosa et al. 2012).

Leptasterias species from temperate Northern Hemisphere waters and *Anasterias rupicola* and *Neosmilaster geogianus* from the Southern Ocean are oral brooders that incubate their young near the mouth (Chia 1966, 1968a, Hendler and Franz 1982, Blankley and Branch 1984, Bosch and Slattery 1999). Embryos on the outside of the clutch are often developmentally more advanced than the centrally located ones (Bosch

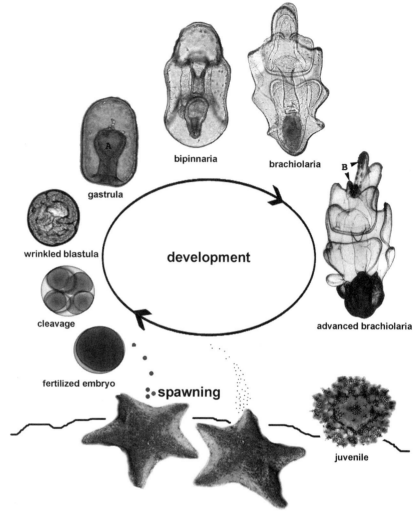

Fig. 5.1. Life cycle and developmental stages of a free spawning asteroid as illustrated by *Patiriella regularis*. *A*, archenteron, *B*, brachia. *Byrne and Barker 1991.*

and Slattery 1999), which is likely due to reduced oxygen levels in the middle of the embryo mass (R. R. Strathmann and Strathmann 1995). The Australian temperate asteroid *Smilasterias multipara* incubates its embryos in folds in its stomach (O'Loughlin and O'Hara 1990).

In asteroids that brood the young on the oral side, eggs spawned from the gonopore are picked up individually by the tube feet and transferred to the mouth or the digestive tract (O'Loughlin and O'Hara 1990). Fertilization is achieved by exogenous (out crossed) or endogenous (self-fertilization) sperm. Most asteroids that brood suborally do not resume feeding until their offspring have dispersed (Chia 1966, 1968, Himmelman et al. 1982, Bosch and Slattery 1999, Raymond et al. 2004).

Asteroid eggs range in diameter from 100 to 1100 μm (Chia et al. 1993). Species that have small eggs (ca. 100–150 μm diameter) develop through feeding (planktotrophic) bipinnaria larvae with a well-developed digestive tract. In some species with planktotrophic development, the bipinnaria develops into a feeding brachiolaria larva (Figs. 5.1, 5.2). Planktotrophic bipinnaria and brachiolaria larvae feed on phytoplankton and have two ciliary bands that loop around the body. The ciliary bands are used for capturing food and for locomotion (R. R. Strathmann 1985). Asteroids with planktotrophic larvae are often large-bodied species with high fecundity, producing many thousands to millions of eggs. These species are also typically dioecious.

Asteroids with large eggs (ca. >300 μm diameter) have non-feeding (lecithotrophic) larva fully provi-

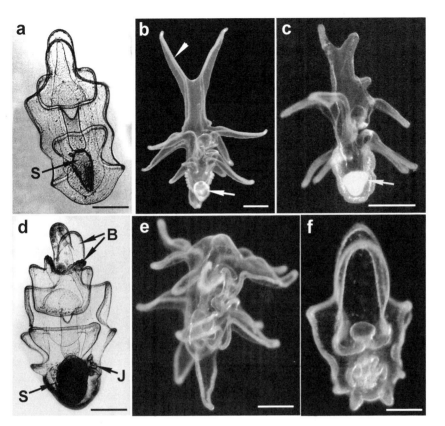

Fig. 5.2. Plankotrophic bipinnaria and brachiolaria larvae. *a*, Bipinnaria, *Patiriella regularis*. *b*, Bipinnaria, *Luidia* sp. with a clone developing at the posterior end and antenor bracchia (arrow). *c*, Unidentified brachiolaria (likely from the Ophidiasteridae) with a clone (*arrow*). *d*, Brachiolaria, *Patiriella regularis*. *e, f*, Unidentified larvae. Parent larva (*e*) with anterior end removed and regenerating clone (*f*). B, brachia, J, juvenile, S, Stomach. Scales: a = 100 μm, b,c,e = 500 μm, d = 200 μm, f = 250 μm. a,d, Byrne and Barker 1991; c, Knott et al. 2003; b,e,f, E. Balser, Illinois Wesleyan University, with permission.

sioned by the egg. Lecithotrophic developers tend to be smaller species, although this varies considerably among asteroid clades. Small size is suggested to be associated with a history of stress, because stress decreases ability to obtain food and thus decreases production (Lawrence and Herrera 2000). Asteroids with large eggs have comparatively lower fecundity than do closely related species with small eggs. Species with lecithotrophic development also include the smallest known asteroids, the viviparous hermaphroditic brooders that give birth to juveniles (Byrne 1996a, Byrne and Cerra 1996, Byrne et al. 2003). They retain their eggs in the gonad, where fertilization occurs followed by intragonadal development. Hermaphroditism is common in asteroids with lecithotrophic development, particularly in species that brood their young and in those with development in benthic egg masses (Komatsu et al. 1979, Byrne 1992, 2005; Fig. 5.3).

In asteroids with small eggs and planktotrophic development, the major energetic lipid in the eggs used to fuel development to the feeding larval stage is triglyceride (TG; Prowse et al. 2009). This fast-burning lipid dominates the small eggs of echinoderms with planktotrophic larvae (Reviews: Emlet et al. 1987, Jaeckle

1995, Prowse et al. 2009). By the time the gut develops most of egg reserves are exhausted, but this varies among species depending on egg size and nutrient quality (Miner et al. 2005, Prowse et al. 2009).

Increased size and organic content of the large eggs of asteroids with lecithotrophic development is largely accomplished by enhanced deposition of lipid (Turner and Lawrence 1979, Emlet et al. 1987, Jaeckle 1995, Byrne et al. 1999, Villinski et al. 2002, Prowse et al. 2009). A considerable portion of the egg reserves is set aside to support development of the early juvenile (Byrne et al. 1999, Byrne and Cerra 2000, Villinski et al. 2002, Prowse et al. 2008). Evolution of a large egg freed larvae from the necessity to feed, resulting in the reduction and loss of superfluous feeding structures (Raff and Byrne 2006). Lecithotrophic larvae lack a functional gut and have a simplified pattern of ciliation and may be planktonic or benthic (Chia et al. 1993, McEdward and Miner 2001, McEdward et al. 2002, Byrne 2006). Some viviparous asteroids are the exception to the egg size–larval nutrition dichotomy. Eggs in the intragonadal brooder *P. vivipara* are secondarily reduced in size (150 μm diameter); the larvae are highly reduced and growth of the intragonadal

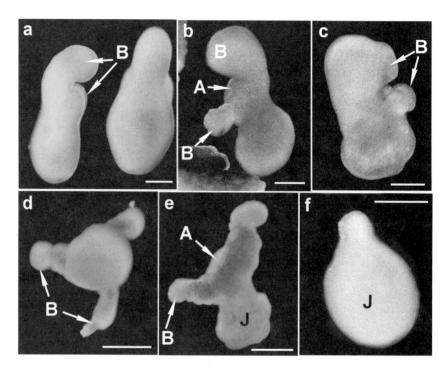

Fig. 5.3. Lecithotrophic brachiolaria from the Asterinidae. *a,b,* Planktonic brachiolaria of *Meridiastra calcar* and *M. gunni,* respectively. *c,* intragonadal brachiolaria of *Cryptasterina hystera. d,e,* benthic brachiolaria of *Parvulastra exigua* and *Asterina gibbosa,* respectively. *f,* intragonadal vestigial brachiolaria of *Parvulastra vivipara.* A, adhesive disc, B, brachia, J, juvenile. Scales: *a–c, f* = 100 μm, *d,e* = 200 μm. Byrne 2006.

juveniles is supported by sibling cannibalism (Byrne 1996a, Byrne and Cerra 1996).

In asteroids with planktonic lecithotrophic larvae, increased egg size is largely accomplished by enhanced deposition of a long-term storage lipid, diacylglycerol ether (DAGE; Prowse et al. 2009). This lipid is a slow-burning fuel. There are also low levels of TG. The DAGE conveys buoyancy to the eggs and so is likely to influence the dynamics of fertilization and larval dispersal. Asteroids that lay benthic egg masses have eggs with high yolk protein content and low levels of DAGE and TG (Prowse et al. 2009). Hypertrophic deposition of yolk protein in the eggs conveys negative buoyancy to the egg and larvae. The dichotomy in nutritive profile of the planktonic and benthic eggs of asteroids—lipid rich and protein rich, respectively—is intuitive with respect to the environmental setting of fertilization and development. This indicates a fine-tuning of egg biochemistry with respect to larval ecology.

Development

Like all echinoderms, asteroids have the deuterostome pattern of development (see Fig. 5.1 and Table 19.1), beginning with radial cleavage and development through the blastula, a hollow ball of cells with a gel-filled blastocoel (R. R. Strathmann 1989). In some as-

teroids, the blastula stage develops deep folds in the ectoderm, giving the embryo an unusual contorted appearance (Fig. 5.1). The wrinkled blastula stage occurs both in species with small and with large eggs. It has been suggested that the folds are required to accommodate the increase in surface area of the developing embryo within the constraint of a tight-fitting fertilization envelope (Cerra and Byrne 1995). Hatching occurs at the blastula or gastrula stage. Gastrulation involves invagination of one pole of the embryo to form the archenteron (Fig. 5.1). The external opening of the archenteron, the blastopore, forms the anus. The archenteron grows anteriorly to form the mouth and then differentiates into the larval digestive tract (Fig. 5.1). Coelom development begins with formation of a pair of pouches that bud from the anterior end of the archenteron. Although coelom development varies to some extent among asteroids, formation of the left coelom gives rise to the hydrocoel, and the foundation of the pentameral adult asteroid body plan is conserved (V. B. Morris et al. 2009, 2011). The gastrula develops into a bilaterally symmetrical larva, the structure of which differs among asteroid orders and is influenced by mode of development (see below). The morphology of the settlement stage larva also differs among asteroid orders (see below). The juvenile asteroid develops at the posterior end of the larva (Figs. 5.1–3).

Larval Type

In addition to the dichotomy of feeding versus non-feeding larvae, asteroid developmental categories are also based on the number of larval stages (McEdward and Miner 2001). The order Paxillosida has only one larval stage (bipinnaria), whereas the other major orders have two (bipinnaria and brachiolaria). Various hypotheses have been proposed on the likely ancestral pattern (Chia et al. 1993, McEdward and Janies 1993, McEdward and Miner 2001). The bipinnaria larva (Fig. 5.1) is shared by most asteroid orders, supporting the notion that this is the basal-type larval stage for the Asteroidea. Moreover, the bipinnaria is considered to represent the ancestral deuterostome dipleurula-type larva that is also present in the Holothuroidea and Hemichordata (Raff and Byrne 2006). With the order Paxillosida in a basal position (Lafay et al. 1995), the brachiolaria (Fig. 5.1, 5.2c,d) is suggested to have evolved secondarily as settlement stage larvae (McEdward and Miner 2001).

In the order Paxillosida (e.g., *Luidia*), the bipinnaria is the settlement stage larva (Fig. 5.2b). In these asteroids, the juvenile tube feet are used for attachment. In the Forcipulatida, Spinulosida, Velatida, and Valvatida, the bipinnaria develops into the brachiolaria larva with formation of an attachment complex at the anterior end (Fig. 5.1, 5.2c,d, 5.3a–e). This complex consists of three brachia, usually a large central brachium flanked by two smaller ones, and an adhesive disc located at the base of the brachia (Byrne and Barker 1991, Gondolf 2000, Haesaerts et al. 2005). In competent larvae, the brachia test the substratum to find a suitable place to attach. When this is identified, the brachia attach. The adhesive disc secretes a material that attaches the larva (Barker 1978a, Haesaerts et al. 2005). This is followed by development of the juvenile and resorption of the larval body (Chia and Burke 1978). As the tube feet develop, they take over the function of benthic attachment (Chia and Burke 1978, Byrne and Barker 1991, Gondolf 2000, Haesaerts et al. 2005).

Paxillosid asteroids with lecithotrophic development lack a bipinnaria and develop through a uniformly ciliated yolky barrel-shaped larva. Forcipulatid, spinulosid, velatid, and valvatid species that have lecithotrophic development also lack a bipinnaria and develop through a uniformly ciliated yolky brachiolaria (Fig. 5.3a–c). The morphology of lecithotrophic brachiolaria differs depending on where development occurs (planktonic or benthic; Fig. 5.3a–f). Planktonic lecithotrophic brachiolaria have a large central brachium flanked by two small brachia, similar to that seen in planktotrophic brachiolaria (Figs. 5.1, 5.2b,d, 5.3a–c). In contrast, the attachment complex of benthic brachiolaria is highly modified to form a tenacious attachment device (Fig. 5.3d,e). *Leptasterias* and *Parvulastra* species have a tripod-shaped brachiolaria due to hypertrophic growth of the lateral brachia resulting in three brachia of similar length (Chia 1968, Komatsu et al. 1979, Byrne 1995, Haesaerts et al. 2005). These are highly muscular and look like tube feet (Fig. 5.3d). The benthic brachiolaria of *A. gibbosa* is a sole-shaped larva (Fig. 5.3e), a morphology that appears to have evolved through fusion of two brachia (Haesaerts et al. 2006).

A rare larval form, the facultative feeding larva, occurs in *Porania antarctica* (Bosch 1989, Bosch et al. 1991). This species has a yolky bipinnaria with a functional digestive tract. It may be able to develop to the juvenile stage in the absence of feeding. These yolky feeding larvae appear to be an intermediate form in evolution of obligate lecithotrophy. The rarity of these larvae in the Asteroidea and in echinoderms in general suggests that they are evolutionary unstable (Wray 1996, McEdward 1997, J. D. Allen and Pernet 2007).

A striking feature of echinoderm development is metamorphosis of the bilaterally symmetrical larval into a radially symmetrical juvenile (Fig. 5.1). Echinoderms arose from a bilateral ancestor, and so the radial adult body plan is a highly derived feature. Explaining the evolutionary change to radial symmetry remains a major challenge in echinoderm biology (Raff and Byrne 2006). Metamorphosis involves complex morphogenetic changes. The larval body degenerates and is resorbed into the oral region of the juvenile as the juvenile forms (Chia and Burke 1978, Burke 1989).

Life History Evolution and Speciation

Although the feeding planktotrophic larva is well supported as a plesiomorphic character for modern Asteroidea, a great deal of larval evolution has taken place among closely related clades (R. R. Strathmann 1985, A. S. Smith 1997, McEdward and Miner 2001, Raff and Byrne 2006). The use of molecular phylogeny to explore the patterns and timing of larval evolution has revealed the important finding that trees based on sequence data differ from trees based on larval morphology (Hart et al. 1997, 2004, Byrne 2006; Fig. 5.4). Lecithotrophy in its various forms (e.g., planktonic, benthic) has arisen independently with considerable frequency in asteroid lineages (Hart et al. 2004). Moreover, once lecithotrophic development has evolved in a clade, subsequent radiation generates new species

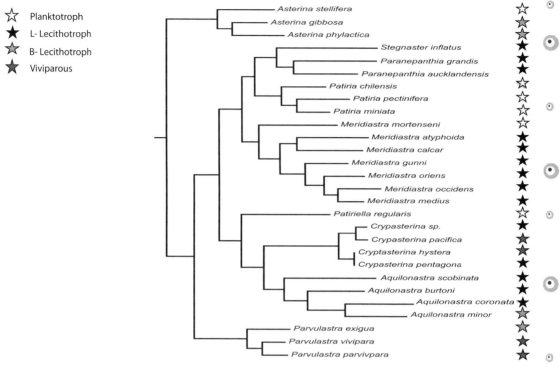

Fig. 5.4. Phylogenetic relationships among asterinid sea stars, mode of development and comparative egg size.

with this life history mode (Jeffrey et al. 2003, Hart et al. 2004). The great diversity of life history modes within closely related species indicates that evolution of development has played a major role in speciation in the Asteroidea. Careful attention to reproductive and developmental traits often reveals previously undetected biological diversity in cryptic species that are morphologically similar as adults.

Extreme Developmental Diversity in the Asteroidea: Evo-Devo in the Asterinidae

The best illustration of the influence of evolutionary development, or evo-devo, on species divergence in the Asteroidea is seen in the asterinid sea stars (Fig. 5.4). Rapid evolution of development in the Asterinidae has resulted in diverse larval phenotypes within cryptic morphospecies complexes (Hart et al. 1997, Dartnall et al. 2003, O'Loughlin and Waters 2004). With the planktotrophic larva as the plesiomorphic character, molecular phylogeny indicates at least six independent origins of lecithotrophy (Hart et al. 1997, Byrne 2006). Most asterinid clades contain species with at least two developmental modes (Fig. 5.4). Even the

most extreme life history mode, intragonadal viviparity and juvenile birth through the gonopore, has evolved at least twice. Although adult asterinids can be so morphologically similar to each other that it is difficult to discern species, attention to ontogeny has uncovered new species and rapid evolution of larval forms (Hart et al. 1997, Dartnall et al. 2003, Byrne 2006 Puritz et al. 2012).

The larval lives in the Asterinidae include the full range seen in the Echinodermata (Byrne 2006). There are species with feeding (e.g., *Patiriella*) and non-feeding (e.g., *Meridiastra*) planktonic larvae, species with strange-looking non-feeding benthic larvae (*Parvulastra, Asterina*) that maintain a tenacious hold on the seafloor, and species with larvae that swim in the gonad followed by metamorphosis and birth as nearly sexually mature asteroids. Planktonic brachiolaria, both feeding and non-feeding forms, have an attachment complex with one large and two small brachia (Fig. 5.1, 5.2d, 5.3a,b). There are two varieties of benthic brachiolaria, the tripod larva with three brachia equal in length (*Parvulastra*) and the sole-shaped larva with a two lobed attachment device (*Asterina*; Fig. 5.3d,e). There are also two varieties of intragonadal brachiolariae, minute vestigial larvae (*Parvulastra*; Fig. 5.3f), and functional brachiolaria that look no different from their planktonic cousins (*Cryptasterina*; Fig. 5.3c). The vestigial larvae of *Parvulastra* sp. (120–150 μm diameter egg) metamorphose as tiny juveniles (Byrne 1996). The first larvae to metamorphose in the gonad prey on less-developed siblings, a bizarre form of maternal provisioning. They emerge through the gonopore up to one-third of the parent's size a year or more after fertilization. In contrast, the brachiolariae of *Cryptasterina* (ca. 400 μm diameter egg) give rise to juveniles similar in size to the newly metamorphosed juveniles of congeners with planktonic larvae (Byrne et al. 2003, Hart et al. 2003).

Viviparity is the most derived life history pattern seen in the Echinodermata. In the Asteroidea, it is only known for the Asterinidae, with six species having this life history (Byrne 2006). A striking example of independent origins and different pathways in evolution of derived life histories is seen in the viviparous asterinids (Fig. 5.4). In *Parvulastra* viviparity evolved through a *P. exigua*–like ancestor that had a benthic tripod larva, but in *Cryptasterina* viviparity evolved through a *C. pentagona*–like ancestor that had a planktonic larva (Hart et al. 1997, 2003, Byrne et al. 2003 Puritz et al. 2012).

Larval Cloning

The discovery of cloning, the ability of asteroid larvae to bud off fragments that regenerate into complete larvae (Fig. 5.2b,c,e,f), adds another level to complexity to our understanding of the dynamics of the planktonic life stage (Bosch et al. 1989, Bosch 1992, Jaeckle 1994, Vickery and McClintock 1998, 2000b, Knott et al. 2003). This phenomenon was first observed in *Luidia* larvae (Bosch et al. 1989) and has since been found in field samples of other paxillosid species and *Oreaster* (Knott et al. 2003). Cloning and larval regeneration in *Pisaster ochraceus* has been described from laboratory cultures (Vickery and McClintock 1998, 2000).

Asteroid larvae that exhibit the cloning response (Fig. 5.2b,c) detach portions of their body, arms, preoral lobe, or arm tip buds (Bosch 1992, Jaeckle 1994, Knott et al. 2003). This is followed by dedifferentiation of the tissue and redifferentiation and growth into a fully formed larva. It is not known how cloning influences the dispersal dynamics of asteroids. It is possible that some larvae, potentially the large larvae of *Luidia* species, are a holoplanktonic source of asexually produced larvae, thereby extending the dispersal potential of the species.

Influence of Developmental Mode on Population Dynamics and Evolutionary Trends

In a meta-analysis of the relationship between life history mode and extreme population fluctuations in the Asteroidea, the most striking similarity among outbreaking species with large fluctuations in population density (e.g., *Acanthaster planci*) is possession of the ancestral-type planktotrophic larva and broadcast spawning (Uthicke et al. 2009). This larval type is significantly overrepresented in boom-and-bust species that exhibit population fluctuations considered anomalous with respect to the natural state (Uthicke et al. 2009). Three factors appear to render the planktotrophic larval life history a high-risk / high-gain strategy: (1) a strong nonlinear dependency of larval production on adult densities (Allee effects), (2) a low potential for compensatory feedback mechanisms, and (3) autonomy of larval and adult ecology due to the presence of a dispersive stage that requires exogenous nutrients.

For species with planktotrophic larvae, the vagaries of food supply and other ecological variables influencing feeding larvae are largely independent from those influencing the adults. There is a strong relationship between larval food supply, survival, development time, and recruitment in asteroids with planktotrophic larvae (Okaji et al. 1997, Brodie et al. 2005). For instance, survival of *Acanthaster planci* larvae increases ninefold with every doubling of phytoplankton food biomass (Okaji et al. 1997, Brodie et al. 2005). The combination of planktotrophy and broadcast spawning appears to result in positive feedback loops that lead to either rapid population increase once an outbreak cycle has been initiated or very slow recovery once populations have decreased. Disturbance and alteration of resources affecting either the adult or larval life stage can lead to large changes in population densities.

In contrast, lecithotrophic development appears to represent a more buffered life history, because compensatory feedback between adult densities and larval output is more likely than for planktotrophic developers. The direct nutritive coupling from adult to larva to the early benthic juvenile provides protection against stress and starvation. Lecithotrophic larvae are independent of planktonic food supply, and their short planktonic duration may promote local recruitment. Independence of larvae from the requirement to feed in the plankton, reduced planktonic larval duration, lower chance of mortality in the plankton, enhanced local recruitment, and maternal provisioning for the early juvenile would promote more stable population dynamics. Thus, population regulation in species with lecithotrophic development appears to be more efficient than in planktotrophs.

After 500 million years of larval evolution, approximately 68% of echinoderms with known development have the derived, lecithotrophic larval type (Uthicke et al. 2009). The evolutionary trend thus appears to be toward a lower-risk / lower-gain strategy of lecithotrophic larvae. The switch to a non-feeding larval life history, free of the vagaries of planktonic food supply, appears to have occurred in response to stressful conditions in the plankton, including during past climate change (Valentine and Jablonski 1986, Pechenik 2006, Uthicke et al. 2009). Strong regional trends are also evident related to latitudinal variations in temperature and productivity (Marshall et al. 2012). For example, lecithotrophy is the most common developmental mode in shallow-water Antarctic asteroids (Bosch and Pearse 1990).

Influence of Developmental Mode on Genetic Structure of Populations and Biogeography

The life history diversity of the Asteroidea, especially within closely related species as seen in the Asterinidae, has provided a model system to investigate the relationship between mating system (e.g., self, cross) larval type (e.g., planktonic, benthic) and planktonic larval duration (e.g., none, short, long) on biogeography and population genetic structure (Hunt 1993, Waters et al. 2004, D. J. Colgan et al. 2005, Hart et al. 2006, Sherman et al. 2008, Ayre et al. 2009, Keever et al. 2009, Hart and Marko 2010, Barbosa 2012). These studies show that—while genetic connectivity of populations can be linked, at least at the regional scale, to having planktonic larvae—population connectivity across species geographic distribution is influenced by historic biogeography (e.g., dispersal past, vicariance), demographics, fragmented habitat, and contemporary oceanographic factors (e.g., topography, currents; Grosberg and Cunningham 2000, Keever et al.

2009). This was seen in the counterintuitive result for *Patiria miniata*, in which Alaskan populations are not well connected with geographically close populations in British Columbia (Keever et al. 2009). This is believed to be due to past glacial history of the region (Keever et al. 2009). The distinct genetic structure of Indian Ocean and Pacific Ocean populations of *A. planci*, initially attributed to historic patterns of ocean circulation (Benzie 1999), may in fact be the result of incipient speciation of different *Acanthaster* clades promoted by historic sea level change (Vogler et al. 2008). For species with non-dispersive benthic larvae population genetic structure on a very fine scale (tide pools) has been observed (Barbosa 2012).

A paradox is presented by *P. exigua*, an egg-laying species with benthic development, no dispersive stage, and potential for self-fertilization (Barbosa et al. 2012). It is one of the world's most widespread asteroids, occurring in the temperate waters of the Southern Hemisphere from Australia to South Africa (Dartnall 1971, Waters and Roy 2004, Hart et al. 2006). How this benthic developer with low mobility in all life stages has been dispersed around the globe is unknown, with rafting among the most likely possibilities (Mortensen 1933, Fell 1962, Waters and Roy 2004). Biogeography indicates that both long-distance dispersal (rafting) and localized vicariance influence genetic structure of this asteroid (Waters 2008). As for *A. planci*, however, close attention to morphological characters that indicate spawning strategies (e.g., gonopore location) revealed that some southern ocean island populations of *P. exigua* might be reproductively isolated species (Hart et al. 2003, 2006). With their most extreme non-dispersive mode of development, it is not surprising that the viviparous asterinids, *Cryptasterina hystera* and *P. parvivipara*, have the most restricted distribution of any echinoderm (<100 km).

Acknowledgments
This research was supported by grants from the Australian Research Council. Many thanks to Dr. Elisabeth Balser for images of larvae and to Dr. Paula Cisternas for assistance.

6

Larval Ecology, Settlement, and Recruitment of Asteroids

Anna Metaxas

The larval and early post-settlement stages are the least understood components of the life cycle, mainly because of their relatively low abundance, small size, and cryptic behavior. However, the linkages among larval supply, recruitment, and the size of adult populations have been repeatedly acknowledged for different species and habitats. The strength of these linkages can be particularly significant in predicting population dynamics for species that are threatened or are of great ecological or economic importance, such as keystone, harvested, or invasive species. Ebert (1983) and Balch and Scheibling (2001) provided overviews of settlement and recruitment in asteroids, as one of the classes of echinoderms. Here, I provide a wider scope by considering the larval component of the life cycle.

Larval Distribution and Abundance

Larval abundance in the water column can range over several orders of magnitude (e.g., 10^{-3}–10^4 individuals m^{-3}). Spatial patterns in larval distribution have been associated with biological (e.g., layer of chlorophyll maximum) or physical (thermoclines, fronts) features in the water column (Metaxas 2001).

Larval abundance and distribution of bipinnariae or brachiolariae have been poorly quantified. In the Antarctic region, such as the Bransfield Strait, Ross Sea, McMurdo Sound, and Weddell Sea, asteroid larvae are most abundant (~2–4 individuals m^{-3}) through the winter, between July and November, and are dominated by *Odontaster validus* (Stanwell-Smith and Clarke 1998, Stanwell-Smith and Peck 1998, Bowden et al. 2009). Surprisingly, the timing of spawning and the larval planktonic period falls during the period of low abundance of phytoplankton, likely to maximize availability of resources for the new settlers (Bowden et al. 2009). While very few bipinnariae were collected in McMurdo Sound (~0–0.2 individuals m^{-3}; Sewell 2005), brachiolariae formed 23% of the meroplankton (up to 6 individuals m^{-3}) in Bransfield Strait (Vázquez et al. 2007). These spatial differences may reflect temporal variability; however, larval distribution may be associated both with particular water masses (Vázquez et al. 2007) and with surface waters (Pearse and Bosch 1986, Vázquez et al. 2007).

In temperate and subtropical regions, seasonality in abundance has been recorded. In Plymouth Sound, UK, larval abundance of *Asterias rubens* was greatest in April, following the spawning period, and decreased by July, during the period of settlement, incorporating a transition in greatest abundance from bipinnariae to brachioloariae (Barker and Nichols 1983). Similarly, in British Columbia, Canada, bipinnaria larvae were most abundant in May and September, following the main spawning periods in summer and fall (Sewell and Watson 1993). Unlike other locations, however, larval concentrations were much higher (100s to 1000s individuals m⁻³) and were attributed to larval retention in the areas of large populations of spawning adults (Sewell and Watson 1993). Lower larval concentrations (10s to 100s individuals m⁻³) of *Asterias amurensis, Asterina pectinifera, Luidia quinaria*, as well as unidentified late bipinnariae and brachiolariae, were recorded in July in the Tumer River estuary and Peter the Great Bay, Sea of Japan (Dautov 2000, 2006, Dautov and Selina 2009). Larval *A. amurensis* were most abundant in the surface layer. Their horizontal distribution coincided with that of the adults on the seafloor (Dautov 2006). For the other species, maximum larval abundance coincided with maximum phytoplankton concentration spatially (Dautov and Selina 2009) and temporally (Dautov 2000). In Puerto Rico, asteroid larvae were only present in 4 of 17 monthly sampling events, mainly in July and August, likely reflecting the spawning period, but only reaching maximum abundance of ~0.7 individuals m⁻³ and increasing in abundance with distance from shore (Williams and García-Sais 2010).

The limited information on larval abundance makes it difficult to draw any meaningful conclusions on the factors that may regulate spatial or temporal patterns.

Larval Ecology

Larval survival depends on a number of factors, such as the physical environment (temperature, salinity, and pressure), food availability (for feeding larvae), agents of mortality, as well as delivery to a suitable habitat for settlement within a period of competency. This process has been relatively poorly studied in asteroids.

The role of conditions in the physical environment on larval survival and development (which in turn enhances survival probability if accelerated) is un-

equivocal for asteroids. This role has been examined mainly for those species with great ecological significance. Examples include crown-of-thorns sea star *Acanthaster planci* (see Chapter 13) and the invasive species *Asterias amurensis* (see Chapter 17). It has also been examined for those inhabiting extreme environments, such as the Antarctic and the deep sea (see Chapter 12). Temperature has a consistently positive effect, such as on developmental rate of *A. amurensis* (Kashenko 2005a,b), where the developmental period can be shortened by 60 d for a temperature increase from 10–20°C (Paik et al. 2005), and of *Odontaster meridonalis* and *O. validus* with a $Q_{10} = 3.8$–4.5 (Stanwell-Smith and Peck 1998, Peck and Prothero-Thomas 2002). Larvae of *A. planci* are also sensitive to temperature and survival is significantly decreased above 32°C (Lucas 1973). For this species, optimum temperatures for development are 28–30°C (Lucas 1973) and larvae do not complete development in temperatures <25°C (Yamaguchi 1974).

Salinity can also have significant effects on development and survival. Survival of larval *A. planci* decreases with salinity increases from 30 to 35, but development can be completed in salinity as low as 26 (Lucas 1973). It has been suggested that bipinnariae of this species can tolerate abrupt changes in salinity from 36 to 21, but that late brachiolariae are much less tolerant to salinity changes (Henderson and Lucas 1971). An interaction between salinity and temperature effects on the development of *A. amurensis* has been observed: tolerance to lower salinities increases with increasing temperature (Kashenko 2005b). Salinity can have a greater impact on larval development than can temperature in eurythermal species, such as *Echinaster* spp. (Watts et al. 1982). In contrast, larvae of *A. rubens* tolerate a wide range of salinities that increases with larval age (Saranchova and Flyachinskaya 2001). Although short-term survival rate decreases with decreasing salinity in this species, it can still be up to 20% in salinity as low as 18 (Sameoto and Metaxas 2008a).

The effect of pressure on larval development and survival has only been addressed in asteroids in the context of adaptation to conditions in the deep sea. The percentage of developing embryos of *Plutonaster bifrons* was greatest at 200 atm, a pressure that corresponds to a depth of 2000 m; it was suggested that larval tolerance to pressure may set the bathymetric limits of this species (Young et al. 1996). For two shallow-water asteroids, *A. rubens* and *Marthasterias glacialis*, >80% larval survival over 24 h was measured in pressures as high as 200 atm, although survival decreased overall with increasing pressure (Benitez-Villalobos et al. 2006). The authors suggested that these larvae have a wider bathymetric range than do the adults. This would mean pressure tolerance of larvae does not limit the distribution of the species. The effects of ocean acidification on larval asteroids have been examined. Lecithotrophic larvae and juveniles of *Crossaster papposus* grew faster and were larger at lowered pH, and survival was not affected (Dupont et al. 2010a). A meta-analysis of studies on several species of echinoderms (including two unidentified asteroids) showed that effect size of pH on calcification, growth, and survival was greater in early life history stages, such as larvae, than it was for adults (Dupont et al. 2010b).

The extent of food availability to planktonic larvae is still unclear, as the list of potential resources that larvae may be utilizing continues to grow. Larval starvation has been invoked in some instances, but its overall magnitude has been challenged (Olson and Olson 1989). For asteroids in particular, much research on the role of food availability has focused on *A. planci*, as the outbreaks of this species have been attributed to increased larval food and consequent enhanced larval survival and recruitment (Moran 1986; see Chapter 13). In the laboratory, ingestion rate increased and filtration rate decreased with increasing phytoplankton concentration. Food quality (i.e., species composition) also had a significant effect on developmental rate (Lucas 1982, Fabricius et al. 2010). Based on those results and on measurements of phytoplankton concentration in the field, it was concluded that long-term food availability to larval *A. planci* is low and non-limiting only in periods and regions of increased river discharge (Brodie et al. 2005, Fabricius et al. 2010). However, in situ experiments found no evidence of either larval starvation or even food limitation (Olson 1985, 1987).

Additional sources of nutrition have been proposed, such as ultraplankton (Ayukai 1994) and dissolved organic matter (Hoegh-Guldberg 1994), and may be present in sufficient concentrations, when in combination with phytoplankton, to alleviate food limitation. The relative importance of the different food sources is not known, but size-selective feeding does occur in larval *A. planci*. (Okaji et al. 1997). The Antarctic asteroid *O. validus* has a long larval period in water of limited phytoplankton abundance (Rivkin et al. 1986). In situ experiments showed similar survival but slower development in the field than under controlled conditions in the laboratory and also showed no evidence of starvation (Olson et al. 1987). Additional

nutritional resources, such as the assimilation of amino acids and ingestion of bacteria, have been identified for the larvae of *Porania antarctica*, (Rivkin et al. 1986) and may also be relevant to the nutritional status of *O. validus*. The timing and extent of food deprivation during the developmental period may also play a role in survival and development. Larvae of *Asterina miniata* that were starved for 10 rather than 30 days, and late rather than early in their development, completed metamorphosis even if it was slightly delayed (Allison 1994). Parental dietary effects on egg quality can affect survivorship and developmental rate of planktotrophic larvae, as demonstrated for *Sclerasterias mollis* (Poorbagher et al. 2010) and *Pisaster ochraceus* (George 1999). For brooding asteroids, effects of maternal condition on juvenile survival and size have been recorded for two species of *Leptasterias* in Washington state, USA (George 1994a, Gehman and Bingham 2010).

During the larval stage, mortality is expected to be high (Young and Chia 1987, Rumrill 1990) but has not been measured for asteroids. One of the main sources of mortality is considered to be predation, and larvae possess different mechanisms of defense against predators (Young and Chia 1987). Chemical defense, and a consequent reduction in palatability, is one of these (Lucas et al. 1979, Cowden et al. 1984, McClintock and Vernon 1990, McClintock and Baker 1997a, Iyengar and Harvell 2001, McClintock et al. 2003). It has been observed in both lecithotrophic and planktotrophic early life stages of polar, temperate, and tropical asteroid species (see Chapter 8).

Mortality may be averted altogether through regeneration of body parts lost during predatory attacks. Larvae of *P. ochraceus* and *Luidia foliolata* that were surgically bisected across the horizontal axis, reproducing the possible effect of sublethal predation, regenerated into complete larvae within 2 weeks (Vickery and McClintock 1998). Through larval cloning, larvae may reduce mortality risk, by increasing population size, but only under favorable conditions, such as optimum temperature and high food abundance (Vickery and McClintock 2000). A number of asteroids are capable of larval cloning (Vickery and McClintock 2000b, Knott et al. 2003; see Chapter 5).

Larvae of several species of invertebrates respond behaviorally to different cues in the water column, such as thermal or salinity gradients, light, and food patches (Young 1995, Metaxas 2001, Metaxas and Saunders 2009), which may be adaptive to maximize their probability of survival. Few studies have focused on asteroids. In *Pteraster tesselatus*, buoyancy decreases between the egg and bilobed larvae, potentially enhancing dispersal of the egg and retention of the larvae (Kelman and Emlet 1999). Bipinnariae of *A. rubens* tend to swim upward and aggregate near the water surface (Sameoto and Metaxas 2008a,b), whereas those *A. planci* also show negative geotaxis but are also sensitive to mechanical stimuli, which can cause body distortions and presumably swimming modifications (Yamaguchi 1974). Exposure of larval *P. ochraceus* to microgravity makes swimming more random (Crawford and Jackson 2002).

Daigle and Metaxas (2011) examined the effect of thermoclines on the behavior of larvae of *Asterias rubens*. In the presence of thermoclines, larvae formed similar surface aggregations except when the top layer was 24°C (a temperature that rarely occurs in its habitat), and they tended to aggregate immediately above the thermocline. The magnitude of the temperature difference between water layers and the temperature in the bottom layer also had pronounced effects on the vertical distribution of sea star larvae.

Larvae of *A. rubens* also showed a strong response to the presence of haloclines in the water column: aggregation around the halocline became more pronounced with decreasing salinity in the top layer; and larvae avoided crossing into the top layer altogether, when the salinity in that layer was ≤24 (Sameoto and Metaxas 2008b). The proportion of larvae that crossed into the top layer increased linearly with survival probability in the salinity of that layer (Sameoto and Metaxas 2008b). The presence of a food patch can also affect the vertical distribution of bipinnarie of *A. rubens*, which tended to aggregate at the deeper boundary of the patch; however, no response was shown in the presence of an inhibitory gradient in salinity (Sameoto and Metaxas 2008a). Such behavioral responses can influence larval vertical position in the water column and ultimately larval dispersal.

Larval Dispersal

Larval dispersal can have a significant role in regulating population connectivity, particularly for sessile species or those with limited motility (such as asteroids). Larval transport has been conventionally attributed to advection along the dominant direction of flow, but the importance of larval behavior in this process has been recognized (Metaxas 2001, Kingsford et al. 2002, Metaxas and Saunders 2009). Modifications in larval vertical distribution are possible because larval vertical swimming speeds are on the same order of magnitude as vertical flow speeds (mm s^{-1}). Movement across water layers with differ-

ent flow velocities can modify larval direction and speed of transport. Because of inherent difficulties in measuring larval dispersal on ecological time scales, biophysical models are being used increasingly to predict dispersal direction and distance (Metaxas and Saunders 2009).

Most studies of asteroids on the role of dispersal in population dynamics have focused on *Acanthaster planci* (see Chapter 13). For this species, larval dispersal is the main factor regulating the extent and timing of secondary outbreaks (De Vantier and Deacon 1990, Johnson 1992a,b), and circulation models that incorporate larval transport have been considered successful in capturing the broad patterns of these outbreaks (Johnson 1992b). The first of these models focused on broad patterns of dispersal across the entire Great Barrier Reef (GBR), with tides, wind, and the Eastern Australian Current forcing circulation (Dight et al. 1990a). Particle tracking routines were incorporated into the physical model, where larvae were modeled as passive particles (i.e., no swimming behavior) with no mortality, and larval clouds tracked for 28 days after spawning, assuming a pre-competence period (and thus inability to settle even if a reef was encountered in that period) of 14 days (Dight et al. 1990b). The authors examined reef connectivity among 15 reefs distributed across three different geographic regions: the outer edge of the reef matrix along the edge of the continental shelf, an inner matrix bordering the lagoon, and a middle matrix. Connectivity among reefs was greatest with adjacent regions; that is, for inner reefs, it was greatest with other inner and central reefs; for central reefs, it was equal with inner and outer reefs; and for outer reefs it was greatest with middle reefs (Dight et al. 1990b). For all source reefs, connectivity was greatest with reefs to their south, rather than to more northern locations, except for a single inner reef which showed the opposite pattern (Dight et al. 1990b). A broader study using the same model incorporated all reefs in the Cairns and Central section of the marine park, as well as climatology for each spawning season (mid-November to the end of December) from 1976–77 to 1989–90 (James and Scandol 1992). The study generated a predominantly southward movement of larvae for most years, with some inconsistencies arising due to wind conditions, and a uniformly low larval incidence in the outer reefs (James and Scandol 1992). This dispersal model was subsequently incorporated in population models that predicted outbreaks of *A. planci* (Scandol and James 1992). Within individual reefs in the central GBR, the location of primary outbreaks, larval retention was greatest at the locations on the reef where adults were most abundant at the beginning of outbreaks (Black and Moran 1991). These results are in support of the suggestion that local outbreaks can occur because of increased larval retention under specific environmental conditions (Fisk 1992, Fabricius et al. 2010).

The role of larval dispersal in the population dynamics of asteroids other than *A. planci* has been examined with biophysical models in two interesting cases. Witman et al. (2003) used a biophysical model with a particle tracking subroutine to quantify the probability of local larval retention near sites in the Gulf of Maine with large aggregations of *Asterias* spp. This model treated larvae as passive particles released at different depths (1–15 m) and sites during the spawning period with a planktonic duration of 2–5 weeks (Witman et al. 2003). Larval retention varied with depth of larval release, being negligible (~0%) if released at 1 m and up to 75% if released at 15 m (Witman et al. 2003). Dunstan and Bax (2007) combined a biophysical model that predicted larval dispersal of *Asterias amurensis* with an age-structured population model that predicted larval production. Larval duration was temperature-dependent varying between 23 and 122 days, and larval mortality was 12–17% per week. Larval dispersal range varied with location (and consequently circulation patterns) and population size, which reflected the size of the larval pool. Estuarine residence time (and thus magnitude of larval retention) was positively related to population growth rate, and larval dispersal was mostly through advection.

Asteroid larvae may be capable of long-range dispersal (Scheltema 1986). Although percentage occurrence was low in a series of stations across the Central Pacific Ocean, bipinnariae were collected from the surface waters in the North and South Equatorial Currents (Scheltema 1986). Because *A. planci* shows little genetic differentiation across the Pacific Ocean, consistent with high gene flow, it was suggested that this species has a facultative teleplanic larva (Benzie 1992).

Settlement

The highly cryptic behavior of asteroid settlers has limited measures of settlement in the field to those obtained mostly using artificial collectors (e.g., Keesing et al. 1993, Balch and Scheibling 2000; but see Jennings and Hunt 2010). For the few species that have been studied, settlement is highly temporally and spatially variable. Temporal variability is best evidenced in the uniquely long-term study by Loosanoff (1964). He showed that the settlement rate of *Asterias forbesi* between 1937 and 1961 in Long Island Sound, USA, varied interannually by orders of magnitude, with pronounced peaks in only 4 of the 25 years (Loosanoff 1964). Marked interannual differences have been recorded over much shorter periods (3 years) in the abundance of settlers of *Stichaster australis* at Maori Bay, New Zealand (Barker 1979). Within a season, settlement occurs over a period of 1–2 months, usually in late spring to early summer (Loosanoff 1964, Nichols and Barker 1984a, Balch and Scheibling 2000, Jennings and Hunt 2010). In the Antarctic, a largely food-limited environment, settlement of *Odontaster validus* coincides with the onset of the summer phytoplankton bloom, likely to enhance the probability of settler survival (Bowden et al. 2009).

Spatially, settlement can vary with depth and habitat type. Settlers of *Asterias* sp. in the northwest Atlantic have been collected from depths up to 35 m, but are generally most abundant at depths up to 10 m (Loosanoff 1964, Balch and Scheibling 2000, Jennings and Hunt 2010). For this species, settlement on the seafloor did not vary between the two dominant habitats, kelp beds and barren grounds; but tended to be greater in kelp beds at ~2 m above the bottom (floating collectors), suggesting that larval availability is equal across habitats, but settlement is preferential on kelps (Balch and Scheibling 2000). Similarly, although availability of competent larvae of tropical asteroids on coral reefs was great in the water column, settlement was greater at the front than the back reef, presumably because of differences in habitat quality (Keesing et al. 1993). In most studies where multiple sites were sampled, settlement varied pronouncedly among them (Pryor et al. 1999, Balch and Scheibling 2000, Jennings and Hunt 2010). Uniform patterns in settlement across sites can occur if settlement is related to a single mesoscale oceanographic event, such as regional upwelling (Wing et al. 1995).

Many invertebrate larvae require biological, chemical, or physical cues that induce metamorphosis and eventual settlement. Some asteroid larvae prolong the period of competency, which enhances the probability of locating suitable habitat but also increases the risk of depleting resources before doing so. Decreased respiration rates during metamorphosis can allow the lecithotrophic larvae of *Mediaster aequalis* to prolong the utilization of the lipid stores and sustain them for an additional 40 days after they reach competency (Bryan 2004). Birkeland et al. (1971) suggested that these larvae are capable of delaying metamorphosis for up to 14 months. For planktotrophic larvae of

Coscinasterias calamaria and *S. australis*, metamorphosis could only be delayed by 2 weeks for successful settlement (Barker 1977). Settlement cues have been linked with tubes of polychaetes (*M. aequalis*: Birkeland et al. 1971, Bryan 2004), dead coral encrusted with coralline algae (*A. planci*: Henderson and Lucas 1971, Yamaguchi 1973), coralline algae (*S. australis*: Barker 1977; *A. rubens, M. glacialis*: Barker and Nichols 1983; *A. planci*: Henderson and Lucas 1971, Johnson et al. 1991), microbial films (*C. calamaria*: Barker 1977; *A. planci*: Henderson and Lucas 1971, Johnson et al. 1991), and kelp (*P. ochraceus*: Rumrill 1988, Sewell and Watson 1993). The cryptic nature of settlement has been demonstrated in studies where settlement tends to be higher in locations away from direct light, such as the underside of substrata (Yamaguchi 1974, Barker 1977, Metaxas et al. 2008).

During settlement, asteroid planktonic larvae exhibit searching and attachment behaviors that are consistent across several species. For *Oreaster reticulatus*, brachiolaria larvae that were competent to settle spent >50% of their time searching the substratum (Metaxas et al. 2008). While searching, larvae of *O. reticulatus* and *A. planci* swam in straight or circular paths and remained in near contact with the substratum (Henderson and Lucas 1971, Johnson et al. 1991, Metaxas et al. 2008). On stopping, *Oreaster* larvae probed the substratum with an arm; when they initially attached to the substratum, they rotated around the adhesive disc by sequentially attaching and detaching one or two brachiolar arms. A similar pattern during attachment has been observed for *Patiriella pseudoexigua* (Chen and Chen 1992), *Mediaster aequalis* (Birkeland et al. 1971), and *C. calamaria* (Barker 1977). Larvae with benthic development may show different behaviors at settlement. Brachiolariae of *Asterina gibbosa* attach themselves to the substratum as planktonic larvae fix to the substratum (Haesaerts et al. 2006). A few days after metamorphosis, they may exhibit rafting by detaching and ascending toward the water surface, perhaps as a means of dispersal (Haesaerts et al. 2006).

Larval mortality that occurs immediately before settlement, because of unavailability of suitable habitat during competency, has not been quantified. Early post-settlement mortality can be great, but it is extremely difficult to measure and is also not known for any asteroid. Using population modeling, it has been suggested that pre-settlement mortality rate of *A. planci* has to exceed 10% per day to have order of magnitude effects on the population size of settlers, whereas post-settlement mortality of as low as 1% per day can have a more pronounced effect on the adult population (Keesing and Halford 1992a). It is assumed that asteroid settlers remain cryptic until reaching a size refuge.

Recruitment

The term *recruit* is used here for individuals that are large enough to have undergone an initial phase of exponential post-settlement mortality but small enough to be reproductively immature. For the highly cryptic asteroid settlers, the lower size limit would be typically limited by detection in the field.

Recruitment is temporally and spatially variable. However, because of the subjectivity in the definition of recruits and the wide size and developmental range associated with this life history stage, the magnitude of temporal variation is difficult to compare across studies and species. Notable exceptions involve variation that entails differences in abundance of orders of magnitude, which result in population outbreaks, the most striking example being that of *A. planci*. Pulses of exceptionally high recruitment, as a result of high larval survival and settlement, have been considered responsible for periodic population outbreaks of the crown-of-thorns sea star, which in turn led to massive destruction of the coral reefs in the Great Barrier Reef (Moran 1986; see Chapter 13). A correlation between enhanced recruitment and increased rainfall or terrestrial runoff has been inferred; however, the causative mechanism of the relationship remains unclear despite several hypotheses that have been put forward (Birkeland 1982, Zann et al. 1990, Brodie 1992). Pulses of high recruitment have also been recorded in temperate species of asteroids, such as *Asterias* spp. in the northwest Atlantic (Witman et al. 2003) and *P. ochraceus* in the northeast Pacific (Sewell and Watson 1993), which were attributed to increased prey availability and larval retention, respectively. In northern Chile, recruitment of asteroids (pooled *Heliaster helianthus, Stichaster striatus, Meyenaster gelatinosus,* and *Luidia magellanica*) was lower during La Niña than during neutral conditions, likely because of a concomitant increase in the abundance of their predators (Gaymer et al. 2010).

Spatial variation in recruitment has been recorded on scales of meters to kilometers, and the mechanisms responsible for this variation are scale-dependent. Within a barrier reef or island, abundance of recruits of *A. planci* can vary from zero to tens of individuals (total number of recruits over 100s m²) among sites separated by hundreds of meters (Zann et al. 1987, 1990,

Pratchett 2005, Pratchett et al. 2009). Variation in recruitment has been recorded on similar scales for temperate species in some locations—*P. ochraceus*, San Juan Islands, USA (J. L. Menge and Menge 1974, B. A. Menge 1975), and British Columbia, Canada (Sewell and Watson 1993); *Asterina miniata*, British Columbia, Canada (Rumrill 1989); *Asterias* spp. in Nova Scotia, Canada (Balch and Scheibling 2000), and the Bay of Fundy, Canada (Jennings and Hunt 2010); *A. rubens*, Plymouth, UK (Barker and Nichols 1983); *M. glacialis*, County Cork, Ireland (Verling et al. 2003)—but not others—*Asterias* spp., Gulf of Maine, USA (Witman et al. 2003). It must be noted that in most studies, recruit abundance was very low (<0.1–1 individual m^{-2}) and spatial patterns must be considered with caution.

Spatial patterns in recruitment on these small scales may arise from different habitat characteristics, including depth, substrate type, temperature, and hydrodynamics. For several species, recruitment appears to be depth specific. In the Gulf of St. Lawrence, Canada, recruits of *A. vulgaris* and *Leptasterias polaris* were most abundant at depths of 0–12 m, whereas those of *Crossaster papposus* were at depths >13 m (Himmelman and Dutil 1991). In central Chile, recruits of *H. helianthus* were more abundant in the high and mid than in the low intertidal zone (Manzur et al. 2010). In deep-sea habitats, the distribution of juveniles does not always overlap completely with that of the adults. Sometimes it tends to be wider. For *Luidia sarsi, Plutonaster bifrons,* and *Psilaster andromeda*, juveniles were found much deeper than were adults, whereas those of *Hymenaster pellucidus* and *Brisingella coronata* were found within the bathymetric range of adults (Sumida et al. 2001).

Distinct associations of recruits with particular substrata likely arise as a result of settlement preferences or post-settlement processes. Recruits of *A. vulgaris* and *L. polaris* were found on boulders and cobbles, but not on pebbles or on sand or mud, whereas those of *C. papposus* were found only on sandy or muddy substrates in the same location (Himmelman and Dutil 1991). Recruits of *C. calamaria* were found on the undersides of rocks and boulders, whereas *S. australis* were found exclusively on *Mesophyllum insigne* (Barker 1977). *Asterias* spp. in the northwest Atlantic exhibited much higher recruitment in kelp beds than in barren grounds (Balch and Scheibling 2000). Recruits of a number of asteroid species were found in association with the tubes of the polychaete *Phyllochaetopterus* in Puget Sound, USA (Birkeland et al. 1971). Recruits of *Oreaster reticulatus* and *Protoreaster nodosus* have only been found in seagrass beds and mangroves

but not on open sandy habitats (Bos et al. 2008, Scheibling and Metaxas 2001, 2008, 2010; see Chapter 14). Recruits (and juveniles) of *A. planci* are found among basal parts of corals, typically dead and encrusted with coralline algae, in grooves and depressions on the reef slope (Yamaguchi 1973, Yokochi and Ogura 1987, Zann et al. 1987).

Larval retention has been proposed as the mechanism regulating spatial patterns in recruits for *P. ochraceus, Dermasterias imbricata* and *Pycnopodia helianthoides* (Sewell and Watson 1993) and as one of the possible factors regulating spatial patterns in *Asterina miniata* (Rumrill 1989) at Barkley Sound, British Columbia. Intense wave action may limit recruitment of *Heliaster helianthus* (Manzur et al. 2010).

Ontogenetic shifts in distribution have been observed for several species of asteroids (e.g., Manzur et al. 2010), which may provide information on the preferred habitats of new recruits. These shifts have been associated with increased survival during the early life history stages, in turn affected by availability of preferred food and escape from predation. It has been suggested that *A. planci* recruits to the deeper sections of coral reefs, at depths >50 m, because the preferred settlement substrate (coral rubble presumably covered by coralline algae) is most abundant, and then move to shallower water as adults (Johnson et al. 1991). New settlers of the crown-of-thorns initially feed preferentially on coralline algae and eventually switch to corals as they age. This opportunistic feeding promotes faster growth (Henderson and Lucas 1971, Yamaguchi 1974, Zann et al. 1987, Brodie 1992, Keesing and Halford 1992a). Similarly, newly metamorphosed *S. australis* prefer to settle on beds of coralline algae *Mesophyllum insigne* and feed exclusively on it. The switch of recruits from a herbivorous to a carnivorous diet is associated with a migration to mussel beds dominated by the preferred prey, *Perna canaliculus* (Barker 1979). Presumably to minimize risk of predation, juveniles of *Oreaster reticulatus* and *Protoreaster nodosus* found in seagrass beds and mangroves exhibit cryptic coloration, which is markedly different than that of adults and reduces the probability of detection (Bos et al. 2008, Scheibling and Metaxas 2008, 2010; see Chapter 14). Recruit (and juvenile) asteroids are rarely located in the field (Chia et al. 1984), indicating little movement and highly cryptic behavior (Zann et al. 1987, Keesing and Halford 1992a).

Mortality during this life history stage is expected to be high, and predation is considered the main cause, although abiotic disturbance may be locally important (Gosselin and Qian 1997, H. L. Hunt and Scheibling

1997). Annual juvenile mortality of *P. ochraceus* and *L. hexactis* was estimated at >99.9% (Menge 1975), whereas mortality of *A. planci* 8 and 23 months after settlement was 99.3% (Zann et al 1987). These rates may be overestimates as they were quantified using differences in the abundance of different size classes measured in the field. Given the cryptic nature of both settlers and young recruits, it is unlikely that their abundance is accurately represented in sampling efforts. *Asterias* sp. showed a positive relationship between settlers in the previous year and recruits in the year being studied, indicating that recruitment can be predicted from settlement and suggesting that mortality at these early stages may not be as significant as in other species (Balch and Scheibling 2000).

Predation has been inferred as the main agent of mortality of asteroid recruits, mainly through evidence of the presence of potential predators in the same habitat. Crustaceans, such as spiny lobsters, crabs, and pistol shrimp, have been identified as predators of *A. planci* (Zann et al. 1987, Keesing and Halford 1992b). Using experimental manipulations in the field, Keesing and Halford (1992b) showed that mortality rate due to predation decreases with age and can be 5.05 % day^{-1} for 1-month-old crown-of-thorns asteroids, 0.85 % day^{-1} for 4-month-old ones, and nondetectable for 7-month-old ones. Using a modeling approach, McCallum (1992) suggested that outbreaks of *A. planci* can be prevented, even if there is a large pulse in larval supply, if predators can remove enough juveniles. In contrast to these studies, Rumrill (1989) showed that *A. miniata* in Barkley Sound was not vulnerable to predation. In 14–39 day laboratory experiments, he showed that of two species of crabs, five species of asteroids, and one species of fish, only *Pycnopodia helianthoides* consumed juvenile *A. miniata*; however, these two species rarely co-occurred at the study sites (Rumrill 1989). Evidence of sublethal predation has been reported for *A. planci* in Fiji, where

13% of 800 sampled juveniles exhibited missing arms, damaged oral discs, and irregular growth (Zann et al. 1987). However, such damages can be repaired and may not necessarily lead to mortality. Survival through this last reproductively immature stage will ultimately determine the size and resilience of the adult populations.

Conclusions

Of the processes that are operating during the early history stages, those that regulate recruitment are the best studied in asteroids. A reasonably complete understanding of the relative importance of different life history stages in population dynamics is available only for *A. planci*.

The larval stages of asteroids in the water column are the least studied, except for more concerted efforts in Antarctic waters. Information is required on larval abundance and distribution relative to features in the water column, such as density discontinuities, fronts, or layers of chlorophyll maxima or factors such as tidal height or time of day. The presence of vertical migration has never been addressed for asteroids, despite its potentially important role in larval dispersal. No information is available on the rates or agents of mortality while larvae are in the plankton. This information would greatly enhance our ability to predict population outbreaks or potential range of dispersal.

Although considerable information on cues for larval settlement is available from laboratory experiments, the small size, low abundance, and cryptic behavior of asteroid settlers have limited our ability to quantify patterns of settlement in situ and the factors that regulate these patterns. The relationship between larval supply and settlement needs better quantification because it provides the link between the pelagic and benthic phases of life history.

7

Ecological Role of Sea Stars from Populations to Meta-ecosystems

Bruce A. Menge and
Eric Sanford

In 1982, the senior author published a review examining the community role of sea stars in marine ecosystems (B. A. Menge 1982). Major synthetic conclusions of the review addressed the role of sea stars in organizing benthic marine communities, the traits of sea stars that underpinned their role, and whether sea star importance varied geographically in a systematic manner. Sea stars are common predators in benthic communities and vary in trophic position and community importance by habitat (intertidal or subtidal), characteristic traits, and features of their prey. Sea stars commonly occupy the top trophic level, and two species (*Pisaster ochraceus* and *Stichaster australis*) were identified as keystone predators (e.g., R. T. Paine 1966, 1971, 1974), defined as consumers that have a large effect on communities and ecosystems and one that is disproportionately large relative to their abundance (B. A. Menge and Freidenburg 2001). Asteroids have a number of characteristics that were proposed to underlie their ecological importance in many communities: (1) they are relatively large compared with most prey, (2) they have indeterminate growth, (3) they have generalized digestive systems allowing them to handle diverse shapes and taxa of prey, either internally or, uniquely in the invertebrate world, externally, and (4) their mode of locomotion and morphology allows them to occupy calm to wave-beaten environments and soft to hard sediments (Menge 1982). Although other taxa of consumers can also occur in a wide range of habitats, few can do so as successfully as sea stars. Of these traits, recent theory suggests that relatively large body size is the key trait underlying species such as sea stars that have strong interaction strengths in food webs (Berlow et al. 2009). Mussels tend to be the preferred prey of sea stars and also the dominant competitors for space, both roles likely due to common traits of relatively fast growth, large size, and relative mobility (Menge 1982).

At the time of the initial review, the literature available on the dynamics of communities with sea stars playing a major role was limited, and conclusions were somewhat speculative. In this chapter, we update and extend our understanding of the ecological role of sea stars based on the substantial body of research that has accumulated in the 30 years since the earlier review. To keep the chapter at a manageable length, we restrict our focus to hard-bottom habitats in intertidal and subtidal regions. Below, we first frame our review by examining the issue of spatial scale and how ecologists have addressed the criticism that experiments are too small and local in spatial scale to be of much use when trying to understand ecosystem dynamics. We then examine the dynamics of several of the most intensively studied systems around the world and use these case study systems to examine how they have advanced ecological conceptual understanding. A primary conclusion that emerges from our review is that the effects of sea star predation in benthic communities are often driven by factors that underlie variation in sea star density, including bottom-up effects and ecological subsidies.

Spatial Scales and How to Incorporate Them: The Comparative-Experimental Approach

Over the past 50 to 60 years, community ecology has passed through several phases. Using Hutchinson's (1959) classic paper as the harbinger of the "modern" era in community ecology, the early literature (1960s and 1970s) was dominated by observational approaches with a focus on competition as the primary driver of community structure (reviewed by Robles and Desharnais 2002). At the same time, field experimentation was used increasingly, and the preeminence of the observational, competition-focused approach gradually declined (1970s and 1980s) as the power and insights provided by field experiments seized the attention of community ecology (see R. T. Paine 1994). In the 1980s and early 1990s, as the issue of scale became a focus (Dayton and Tegner 1984, Wiens 1989, Levin 1992), the field experimental approach came under criticism as being limited to small, local scales (e.g., Diamond 1986) and thus having limited capacity to yield results that offered insights into dynamics at ecosystem, landscape, or larger scales.

Thus, as ecology matured, a major issue was determining how to understand community and ecosystem dynamics at system-appropriate spatial scales. Field experimentation provides clear and relatively unequivocal answers to questions about species interaction strength, for example, but rarely can we do experiments whose replicates have plots much larger than m². Yet, determining the dynamics of communities by observation, no matter how carefully done, can lead to erroneous conclusions. Observations and measurements are crucial components of any ecological research program, so what is needed is some middle-ground approach that combines the strengths of both approaches.

One such method is the "comparative-experimental approach" (B. A. Menge 1991, B. A. Menge et al. 1994, 2002; see also Sanford and Bertness 2009, R. T. Paine 2010). This approach, first employed by Dayton (1971), uses replicated, identically designed experiments conducted simultaneously at local sites, along environmental gradients. Observational data are taken to quantify the environmental gradients (e.g., of temperature, wave action, light, depth, upwelling intensity, productivity, habitat heterogeneity). Comparisons of results along the gradient can reveal how, for example, competition, predation, facilitation, or recruitment vary and relate to the pattern of community structure. Comparisons with the measured variation in the environmental gradient, sometimes coupled with additional experimentation (e.g., in some cases the physical variables can be manipulated as well) and observation, allow strong inferences to be made about the biotic-abiotic interaction and how this influences dynamics at larger, system-relevant scales.

Case Studies: Dynamics of Large Marine Ecosystems (LMEs)

We review case studies of ecosystems where sea stars play important roles and where the larger-scale dynamics of the system have been investigated using the comparative-experimental approach. Our goal is to determine the role of sea stars in structuring these systems from local to regional to geographic scales. We focus on the mechanisms that underlie these roles, with the aim of attempting to identify generalized explanations for commonalities among some of the different systems or to determine why differences might exist among others. Throughout our discussion, we attempt to identify important gaps in our understanding.

In coastal systems, with our new appreciation of the role of spatial scale, an appropriate scale of study is the large marine ecosystem (LME). LMEs are defined as "relatively large areas of ocean space of approximately 200,000 km² or greater, adjacent to the continents in coastal waters where primary productivity is generally higher than in open ocean areas" (e.g., K. Sherman 1991; see www.lme.noaa.gov). LMEs are characterized by similar or related oceanographic conditions and biota and are distinct from other LMEs through differences in these characteristics. Although there are 64 LMEs around the world, we have insights into the community and ecosystem role of sea stars in only a few of them. Below, we highlight those few that have the greatest extent of knowledge regarding sea star roles: the California Current, Humboldt Current, Northeast U.S. Continental Shelf, and New Zealand Shelf. We also briefly consider two LMEs in which intertidal sea stars seem to have little influence (Benguela Current and East Central Australia), and the role of the crown-of-thorns sea star *Acanthaster planci* in coral-dominated reef systems of the Northeast Australia LME.

California Current LME (CC-LME)

Early studies in this system on the community role of sea stars suggested that *Pisaster ochraceus* predation was a primary determinant of community structure in the low intertidal zone (e.g., R. T. Paine 1966, 1969a, 1974, 1984, 1994, Dayton 1971). These studies have had a huge influence on conceptual advancement in ecology as well, with the ideas of keystone species, indirect effects, strong interactors, and foundation species all emerging from or being strongly emphasized by these papers. R. T. Paine (1966) and Dayton (1971) clearly felt that their results had applicability that was broader than the rocky intertidal zone of Washington state, likely applying to most of the west coast of North America. This hypothesis was not evaluated in this system until much later, however, when the issue of variation in ecological processes with increasing spatial scale became a central focus in ecology.

Beginning in the 1990s, several studies were published that indicated that (1) *P. ochraceus* does indeed play a keystone role along the shores of the California Current, ranging from southern Vancouver Island, British Columbia, to Point Conception, California, and (2) this effect varies in a context-dependent way with the intensity of wave action, seawater temperature, air temperature, and meso-scale variation in phytoplankton productivity (e.g., B. A. Menge et al. 1994, 1996, 1997, 2004, Robles et al. 1995, Navarrete and Menge 1996, Sanford 1999, Robles and Desharnais 2002, Pincebourde et al. 2008, Robles et al. 2009). For example, in Oregon, experiments at wave-exposed and wave-protected areas of two sites along the coast separated by ~65 km, and with contrasting oceanographic conditions, showed that predation rates of *P. ochraceus* on the mussel *Mytilus californianus* were higher at wave-exposed than at wave-protected sites, but when rates were estimated on a per capita (per individual) basis, sea stars at wave-exposed sites actually fed slower (B. A. Menge et al. 1994, 1996). Preda-

tion rates also differed between sites and were higher at the site that had higher prey production. At both sites, after a 3-year absence of sea stars, the lower limit of the mussel bed was lower on the shore by about 0.5 m vertically, whereas no changes in the lower limit occurred in the presence of sea stars (Menge et al. 1994), a result that is consistent with R. T. Paine's (1966, 1974) results on Tatoosh Island, Washington. Similar results have been observed on the outer coast of Vancouver Island (Robles et al. 1995, 2009) and in central California (D. Jech and P. Raimondi, pers. comm.). These results are consistent with R. T. Paine's (1966) suggestion that *P. ochraceus* plays a similar role in setting the lower limit of mussels along the shores of the North American coast.

Other results, however, have revealed that the role of *P. ochraceus* in controlling low intertidal community structure varies with wave exposure and latitude. In Oregon, per population predation rates (i.e., effects of the predator population on prey) on *M. californianus* were consistently low in wave-sheltered areas. Although sea star predation still likely ultimately controlled mussel abundance in the low intertidal zone, other factors affecting mussel abundance in sheltered areas were identified as low supply of mussel recruits and periodic sand burial (Menge et al. 1994). In southern California, sea stars can be sparse in or even absent from sheltered locations, where the distribution of intertidal mussels is instead controlled by intense predation by whelks, fish, and lobsters (Robles 1997). These and other results cited indicate that the role of *P. ochraceus* in the CC-LME can vary with context.

Although *P. ochraceus* likely sets the lower limit of mussels on wave-exposed shores across a wide geographic range, investigation of how rates of predation varied latitudinally at 14 wave-exposed sites ranging from Oregon to the Santa Barbara Channel revealed a more nuanced dynamic (B. A. Menge et al. 2004). Although per capita rates were similar across all sites north of Point Conception, California (see Discussion and Synthesis), per population rates of predation varied primarily among sites with little relationship to latitude (Fig. 7.1). Across the full range of the study, upwelling varies from moderate and less persistent from Cape Blanco, Oregon, northward to strong and more persistent between Cape Blanco and Point Conception, to weak and infrequent southeastward of Point Conception. The experiments showed that per population predation rates were low in the region with weak upwelling and stronger but more variable

Fig. 7.1. Latitudinal variation in per population interaction strength between sea stars (*Pisaster ochraceus*) and mussels (*Mytilus californianus*) in the California Current Large Marine Ecosystem. Map shows the 13 study sites located in regions where upwelling was weak (sites 1–3), persistent (sites 4–8), or intermittent (sites 9–13). At each site, mussels were translocated to the low intertidal zone and predation rates were assessed by comparing rates of mussel mortality in plots where sea stars were present at natural densities versus plots where sea stars were excluded. Data are means (+SE) from experiments conducted in summer 1999. *Modified from B. Menge et al. 2004.*

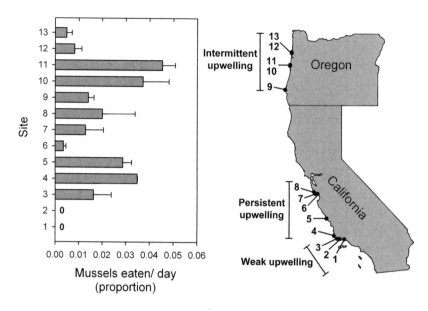

Fig. 7.2. Time series of mussel recruitment (monthly mean number of recruits per day; error bars not shown for clarity) at two sites on the Oregon coast, USA: Strawberry Hill, and Seal Rock (located 12.5 km south of the Newport South Jetty). Note that *y* axis scales differ between sites. Similar increases in mussel recruitment occurred in the early 2000s at multiple other long-term monitoring sites along the Oregon coast. *B. Menge et al. 2009.*

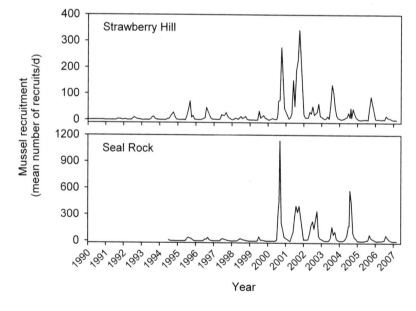

in the regions with strong and moderate-upwelling. Thus, predation on wave-exposed shores varies with upwelling regime and is generally less intense when upwelling is weak and infrequent. Further analysis showed that per population predation rate varied within regions and was most strongly influenced by sea star density (B. A. Menge et al. 2004). This indicates that the influence of *P. ochraceus* predation is likely driven by the factors that underlie variation in sea star density, including bottom-up effects and ecological subsidies.

Evidence suggests that a bottom-up enhancement of *P. ochraceus* recruitment occurred along the Oregon coast during the early 2000s. In 2000, mussel recruitment at long-term monitoring sites along the Oregon coast increased by orders of magnitude compared with recruitment recorded during the previous decade (Fig. 7.2). This increase in prey recruitment

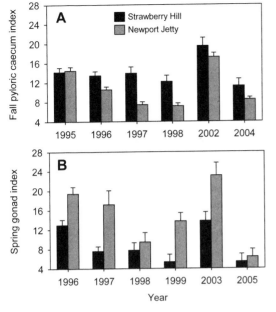

Fig. 7.3. Interannual variation in *Pisaster ochraceus* condition at Strawberry Hill and the Newport South Jetty, Oregon, USA. Bars are means (+SE) for 12 sea stars per site. *A*, Peak pyloric cecum index recorded during fall of each year. *B*, Peak gonad index recorded in May of each subsequent year. Pyloric ceca were unusually large at both sites in fall 2002, leading to enhanced gonad production during spring 2003. *Sanford and Menge 2007.*

coincided with phase shifts in the Pacific Decadal Oscillation (PDO) and the North Pacific Gyre Oscillation (NPGO) and with an increase in the abundance of phytoplankton (B. A. Menge et al. 2009). In fall 2002, the digestive organs (pyloric cecum indices) of *P. ochraceus* were 1.5 to 2.4 times larger than they had been during the late 1990s (Sanford and Menge 2007; Fig. 7.3a). This increase in stored energy was followed by increased reproductive output in spring 2003, with gonad indices that were roughly twice as large as in the late 1990s (Sanford and Menge 2007; Fig. 7.3b). Relative to the 1990s, size frequency distributions of *P. ochraceus* in 2003 showed a striking increase in the proportion of smaller size classes (<50 g wet weight, likely 1 and 2 year olds; Sewell and Watson 1993), suggesting increased sea star recruitment during the early 2000s, followed by growth of this cohort between 2003 and 2005 (Fig. 7.4). The mechanism underlying this enhanced recruitment is unknown but could be related to increased reproductive output, enhanced larval survival, reduced post-settlement mortality of newly settled sea stars, or some combination of these (see below, Discussion and Synthesis). The timing of the increased recruitment of *P. ochra-*

ceus on the central Oregon coast appeared to coincide approximately with a similar increase in *P. ochraceus* recruitment observed >1200 km to the south in the Channel Islands of California (Blanchette et al. 2005). This suggests that increased sea star recruitment during the early 2000s may have been a broad phenomenon within the CC-LME, perhaps related to phase shifts in the PDO or NPGO or both.

In summary, the sea star *P. ochraceus* clearly has a profound influence on rocky intertidal community structure in the CC-LME. This role is context-dependent, however, varying with wave exposure and oceanographic regime, which can have direct and indirect effects on sea star abundance and effect by causing variation in bottom-up processes and ecological subsidies.

Humboldt Current LME (HC-LME)

Extensive investigation of the role of sea star predation has also been carried out in this LME, as summarized in Castilla et al. (2007; see Chapter 15). Salient elements of this research program are as follows: as in the CC-LME, mussels (*Perumytilus purpuratus*, *Semimytilus algosus*) are common space occupiers, with the former inhabiting a narrow zone in the upper mid-intertidal region. Primary predators of these prey included the sea star *Heliaster helianthus* and the large whelk *Concholepas concholepas* (Castilla and Duran 1985, R. T. Paine et al. 1985). A sea star exclusion experiment demonstrated that mussel abundance increased and extended lower on the shore in the absence of *H. helianthus* (Paine et al. 1985). In partial contrast to the *P. ochraceus* exclusion experiments in North America, however, changes in mussel abundance occurred relatively slowly (increasing from ~60% to ~80% cover over 2 years).

As in the CC-LME, comparative experiments and observations across 900 km of coastline in Chile revealed a strong signal of upwelling variability on the dynamics of the ecosystem (Navarrete et al. 2005). In this system, upwelling was stronger and more persistent north of ~32° S latitude and weaker and less persistent south of this latitude. Associated with this discontinuity, prey recruitment and mussel cover were low to the north and high to the south. Experiments indicated that per population predation rates (mostly by sea stars and whelks) on transplanted clumps of mussels were similar north and south of the discontinuity (Fig. 7.5b). However, predator removals led to a strong increase in mussel cover in the south, but not in the north, where recruitment rates were consistently extremely low (Fig. 7.5c). Thus, here too, the

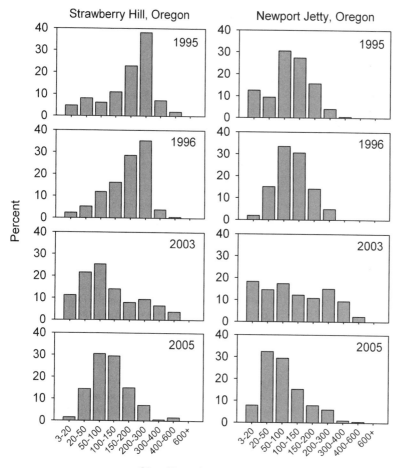

Fig. 7.4. Interannual variation in *Pisaster ochraceus* size structure at Strawberry Hill and the Newport South Jetty, Oregon, USA. The same rocky intertidal areas were sampled at each site in 1995, 1996, 2003, and 2005. At each site, the first 100 sea stars encountered along each of two belt transects (1 m wide × 30 m long) were weighed. Data are percent of *P. ochraceus* in each weight class (n = 200–225 sea stars per site per year). The proportion of sea stars in smaller size classes was greater in 2003, consistent with increased sea star recruitment in the early 2000s.

community impact of sea stars was linked to bottom-up and ecological subsidy effects (see below, Discussion and Synthesis).

Northeast U.S. Continental Shelf LME (NUSCS-LME)

In this LME, as summarized by B. A. Menge (1982), predation by the sea stars *Asterias forbesi* and *Asterias vulgaris* have an important role in controlling the abundance of mussels in the low intertidal region but little effect higher on the shore. However, in the low intertidal zone, the impact of sea stars was context dependent in space. Whereas *Asterias* spp. appeared to be dominant predators at some sites, at others, their role was subordinate to other predators including the whelk *Nucella lapillus* and the crab *Carcinus maenas* (B. A. Menge 1983).

In the subtidal zone of this LME, sea stars (*A. forbesi* and *A. vulgaris*) appear to be a major factor in control-

ling mussel abundances. In 1995, a massive recruitment of *Mytilus edulis* blanketed much of the subtidal zone (8–12 m depth) in the southern Gulf of Maine (Witman et al. 2003). The following 2 years sea stars and crabs aggregated and responded to the mussel seascape produced by this event and eliminated the mussels. After a delay of 12–14 months, the huge increase in prey led to an increase in sea star recruitment, and eventually, as prey became scarce again, to cannibalism in one of the sea stars, *A. vulgaris*. This example provides a clear illustration of the important role that ecological subsidies, in this case, prey recruitment, can have in driving the dynamics of higher trophic levels.

New Zealand Shelf LME (NZS-LME)

The shores of New Zealand harbor a community that is very similar to those in the CC-LME and the HC-LME (reviewed by B. A. Menge 1982), including the presence of another forciculate sea star, *Stichaster*

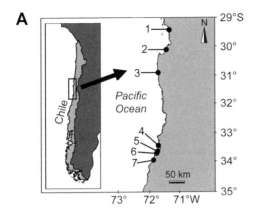

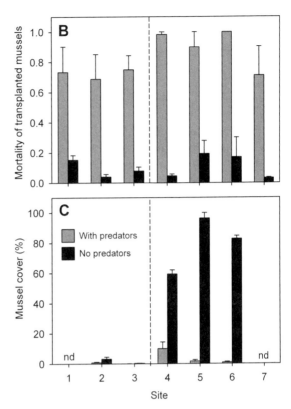

Fig. 7.5. Results of field experiments to quantify predation by the sea star *Heliaster helianthus* and gastropod *Concholepas concholepas* within the Humboldt Current Large Marine Ecosystem. *A,* Map of the seven study sites along the central coast of Chile. *B,* Mortality of transplanted mussels at sites north (1–3) and south (4–7) of the observed discontinuity in mussel recruitment at 32°S–33°S (*dashed line*). Bars are mean mortality (+SE) of mussels in replicated clumps in the presence (*gray bars*) and absence of predators (*black bars*) after 50 days in the field. *C,* Effects of predator exclusions on prey cover at a subset of the sites north and south of the discontinuity. Bars are the mean percent cover (+SE) of mussels in replicated plots in the presence and absence of predators. nd, no data. *Modified from Navarrete et al. 2005.*

australis (e.g., Morton and Miller 1968, R. T. Paine 1971). The experiment conducted by Paine (1971) demonstrated that *S. australis* had a strong controlling effect on mussels (*Perna canaliculus*) at a site on the west coast of the North Island of New Zealand.

Recent research has broadened and deepened our understanding of the role of sea stars in this system and of the factors that shape community structure in the NZS-LME (e.g., B. A. Menge et al. 1999, 2002, 2003, Rilov and Schiel 2006, Schiel 2011), especially of the influence of ecological subsidies including the supply of prey. Investigation of wave-exposed rocky benches on the east and west coasts of the South Island revealed substantial variation in community structure (Menge et al. 1999, 2003; see also Blanchette et al. 2009, Schiel 2011). While many shores showed the widely observed pattern of intertidal vertical zonation (with high, mid, and low zones dominated by barnacles, mussels, and macrophytes, respectively), this was not a universal pattern in New Zealand. Most strikingly, on some

shores, mussel beds were almost absent and barnacles dominated both the high and mid zones. On other shores, even barnacles were sparse and rock surfaces were almost barren except for low intertidal kelp beds (B. A. Menge et al. 2003).

Comparative-experimental investigations of the factors underlying these differences indicated that predation by *S. australis* was the primary source of mortality for sessile invertebrates at sites on the west coast and that predation on transplanted mussels in general was weak to completely ineffectual at sites on the east coast (B. A. Menge et al. 2003). It is interesting to note that *S. australis* was largely absent from east coast rocky shores. The cause of this near-absence of *S. australis* on east coast shores is unknown, but lack of dispersal is one potential cause. Another possible cause is low prey production; mussel and barnacle recruitment on the east coast was orders of magnitude lower than on the west coast. These differences seemed to be driven by differences in oceanography: the west coast experiences upwelling alternating with downwelling, which may result in periodic delivery of recruits to the shore, while the east coast is downwelling dominated, which inhibits blooms of phytoplankton (due to lack of delivery of nutrients to surface waters) and may thus create conditions inimical to the survival of invertebrate larvae (Menge et al. 2003).

Benguela Current LME (BC-LME) and East-Central Australia LME (ECA-LME)

Two LMEs with intensively studied rocky intertidal zones are those of the Benguela Current system of western temperate to subtropical Africa and the east-central coast of Australia. In contrast to the examples summarized above, however, these systems are notable for the lack of a dominant, strongly interacting intertidal sea star (e.g., Underwood et al. 1983, Bustamante et al. 1995, Bustamante and Branch 1996, Blanchette et al. 2009). Although sea stars do occur in the BC-LME (Branch et al. 2007), those that are predators of mussels (e.g., *Marthasterias glacialis*; see Penney and Griffiths 1984) are mostly limited to subtidal regions and appear to have weak effects on mussel abundance (G. M. Branch, pers. comm.). Since the BC-LME is highly productive, with strong bottom-up influences on herbivore abundance (Bustamante et al. 1995) and high abundances of mussels and barnacles (Blanchette et al. 2009), this absence of ecologically important sea stars is curious.

Strongly interacting sea stars also seem absent from the intertidal region of the East-Central Australia LME. Extensive research in this region has examined predation effects, but the predators involved are whelks (e.g., *Morula marginalba, Thais orbita*; e.g., Fairweather et al. 1984, Fairweather 1985, 1988, Fairweather and Underwood 1991). The sea star *Coscinasterias muricata* (formerly *C. calamaria*) preys on mussels in South Australia (e.g., Day et al. 1995) and occurs in the ECA-LME, but to our knowledge its role as a predator in this LME has not been investigated. The lack of research on sea star impacts in this region suggests to us that sea stars are not strong interactors in this coastal marine ecosystem.

Northeast Australia LME (NA-LME)

Predation can have strong effects on the structure of subtidal coral reef communities in the Northeast Australia LME. Corallivorous predators on the Great Barrier Reef (GBR) include a variety of fish and invertebrates (reviewed by Rotjan and Lewis 2008). The crown-of-thorns sea star *Acanthaster planci* has attracted considerable attention because its population dynamics are characterized by mass outbreaks that can devastate large areas of coral reefs (reviewed by Moran 1986, Birkeland 1989). In some regions, outbreaks have consumed 80 to >90% of all live coral along many kilometers of coastline (Moran 1986, Birkeland and Lucas 1990). The causes of *A. planci* outbreaks continue to be debated (see Chapter 13) but appear to be related to the bottom-up influence of terrestrial runoff on the survival of sea star larvae (see Discussion and Synthesis). The geographic distribution of *A. planci* spans multiple LMEs throughout the Indo-Pacific including Panama, Hawaii, the Mariana Islands, the GBR, the Red Sea, and other locations. Outbreaks on the GBR appear to be the most extensive in the world (Moran 1986).

Although there has been little experimental work on the effects of *A. planci*, a growing body of data from surveys and long-term monitoring provides insight into the role of this predator in the Northeast Australia LME. Surveys indicate that the extent of outbreaks, the density of sea stars, and their effects on coral reefs are highly variable (Lourey et al. 2000, Pratchett 2010, K. Osborne et al. 2011). For example, across 1300 km of the GBR, coral cover did not decline systematically during the period 1995–2009 but rather increased in some subregions (10–100 km) and decreased in others (K. Osborne et al. 2011). In areas where coral cover declined, outbreaks of *A. planci* and storm damage were the primary causes. In finer scale surveys conducted within a subregion of the northern GBR, densities of *A. planci* were also highly variable among sites around the perimeter of Lizard Island. The extent of coral declines differed among sites and was strongly related to sea star density. Even moderate densities of *A. planci* (~1–2 sea stars per 200 m²) were associated with a 46% decline in coral cover over 2.5 years (Pratchett 2010), suggesting that the per capita effects of *A. planci* on corals can be large. Estimates indicate that one *A. planci* can consume 5 to 6 m² of live coral per year (Birkeland 1989).

As documented in other asteroids, the effects of *A. planci* can be highly context-dependent. Even during mass outbreaks, *A. planci* predation on corals is generally weak in the shallow subtidal zone (<2 m depth). Sea stars appear to avoid these turbulent habitats due to risk of dislodgment, creating a refuge from predation for shallow-water corals (Moran et al. 1985, M. W. Colgan 1987). The community importance of *A. planci* also appears dependent on the local structure of coral communities (Moran 1986). *A. planci* strongly prefer *Acropora* corals, and reefs that are dominated by *Acropora* experience greater loss of coral cover during sea star outbreaks (Pratchett 2010). Initial studies suggested that preferential consumption of fast-growing

Acropora corals by *A. planci* might reduce competition for space and increase coral diversity (reviewed by B. A. Menge 1982). Surveys on the GBR suggest that sea star predation may increase the relative abundance of less-preferred corals (and thus increase evenness; Pratchett 2010). However, because *A. planci* generally includes some less-preferred corals in its diet, outbreaks tend to have a net negative effect on coral diversity (Pratchett 2010). Recovery rates following an outbreak of *A. planci* are also quite variable and context-dependent. In areas of high coral recruitment and favorable abiotic conditions, *Acropora* corals can recolonize and grow quickly, leading to recovery in 10–15 years (M. W. Colgan 1987, Lourey et al. 2000, Bruno and Selig 2007, Pratchett 2010). In other areas, recovery may proceed at a much slower rate, or not at all, leading to persistent changes in coral composition (Lourey et al. 2000, Pratchett et al. 2011). In addition to effects on the coral assemblage, *A. planci* outbreaks can lead to a variety of cascading effects on populations of macroalgae and fish (reviewed by Birkeland 1989).

Discussion and Synthesis
Causes of Variation in Per Population Effects and Sea Star Abundance

Many of the summarized studies document strong effects of bottom-up and top-down processes on the structure of benthic marine communities. Variation in coastal upwelling regimes can lead to regional differences in nutrients, phytoplankton, and prey recruitment that have major bottom-up influences on community dynamics (B. A. Menge et al. 1997, 1999, 2003, Navarrete et al. 2005). Overlain on these patterns, variation in the abundance of sea stars and other predators often exerts a strong top-down influence on community structure. In some cases, variation in sea star abundance is correlated with regional differences in oceanographic regimes (e.g., B. A. Menge et al. 2003, 2004; but see Navarrete et al. 2005). However, in many cases, sea star abundance varies greatly within regions. Indeed, studies of the community impact of sea stars generally find that population interaction strength varies strongly over scales of tens of kilometers, driven largely by variation in sea star density (B. A. Menge et al. 1994, 2003, 2004, Navarrete et al. 2005, Pratchett 2010). Thus, an unresolved issue concerns the processes that generate spatial variation in sea star abundance. In particular, do differences among sites in sea star abundance reflect a bottom-

up influence of nearshore productivity on the survival of sea star larvae? Is sea star abundance influenced by the availability of benthic prey, and, if so, over what spatial scale and via what mechanisms does this coupling occur?

Recruitment of sea stars might be linked to nearshore productivity if the abundance of phytoplankton influences the survival of planktotrophic sea star larvae. For example, phytoplankton blooms have been implicated as a cause of periodic outbreaks of the crown-of-thorns sea star *A. planci* in the coral reef ecosystems of the Indo-West Pacific (Birkeland 1982, Houk et al. 2007). Birkeland (1982) suggested that during years of heavy rainfall, increased nutrients associated with terrestrial runoff fueled phytoplankton blooms that in turn increased survival of *A. planci* larvae in the plankton. Although alternative hypotheses have been proposed (reviewed by Moran 1986), evidence has generally supported a connection between factors leading to phytoplankton blooms (terrestrial runoff, oceanographic fronts) and sea star outbreaks (Brodie et al. 2005, Houk et al. 2007, Fabricius et al. 2010, Houk and Raubani 2010; see Chapter 13).

Although data are limited, there is evidence that larval survival might influence recruitment success in other sea stars as well (Uthicke et al. 2009). A 25-year study with twice-weekly sampling in Long Island Sound (NUSCS-LME) found that settlement of *A. forbesi* larvae varied among years by over four orders of magnitude (Loosanoff 1964). Given this frequent sampling schedule, variation in settlement likely reflected strong temporal differences in the supply or delivery of competent larvae, rather than in post-settlement mortality (Ebert 1983). Although the cause of variation in larval supply was unknown, year-to-year variation in larval settlement was unrelated to the adult population size, suggesting the importance of larval survival rather than larval production (Loosanoff 1964). In a more recent study, the abundance of *P. ochraceus* recruiting to artificial collectors along the Oregon coast was greater at sites with higher phytoplankton productivity, also consistent with the larval survival hypothesis (B. A. Menge et al. 2004). Despite considerable variability among years, it appears that settlement of sea star larvae is often consistently higher at some sites than others (Loosanoff 1964, J. L. Menge and Menge 1974, B. A. Menge et al. 2004). Although more data are needed, this pattern is consistent with the hypothesis that persistent differences in phytoplankton productivity or ocean currents lead to differences among sites in the survival or delivery of sea star larvae or both.

Oceanographic variation might also exert a bottom-up influence on sea star abundance by influencing juvenile and adult life stages. A comparison of two sites on the central Oregon coast suggested that higher nutrient levels fueled phytoplankton blooms, which increased the growth and abundance of filter-feeding prey, in turn supporting higher densities of sea stars (B. A. Menge 1992, B. A. Menge et al. 1996). Similarly, a strong association between the abundance of mussel prey and sea stars was observed in the NUSCS-LME (Witman et al. 2003). After a massive recruitment of mussels in the Gulf of Maine, the recruitment of sea stars (Asterias spp.) subsequently increased dramatically to the highest level ever recorded in 16–18 years of monitoring. The authors hypothesized that the enhanced recruitment of sea stars was a bottom-up effect of increased prey availability on the reproduction of sea stars. In particular, they suggested that greater consumption of prey may have supported greater reproductive output of sea stars, leading to greater larval production, and a resulting increase in the recruitment of juvenile sea stars throughout the region (Witman et al. 2003).

However, a strong association between prey abundance and sea star abundance is not supported by all analyses. Wieters et al. (2008) analyzed patterns of sea star and prey abundance generated by studies in the Humboldt Current LME (22 sites across 900 km of coastline) and the California Current LME (20 sites across 2100 km of coastline) and found no relationship to either prey recruitment or density of adult prey. The authors suggest that such a pattern is to be expected for predators like sea stars with broadly dispersed larvae. Although abundant prey might lead to increased production of larvae, these larvae would likely be dispersed many kilometers, decoupling the abundances of predator and prey at any given site (Wieters et al. 2008). These studies thus suggest contradictory trends in the strength of sea star–prey coupling. In the HC-LME, sea star and prey abundance and recruitment were uncorrelated, in the NUSCS-LME mussel recruitment and increases in sea star density were strongly correlated, and in the CC-LME evidence is conflicting. Wieters et al. (2008) emphasize that a positive association between the abundance of sea stars and their prey is expected to occur as one approaches the scale of dispersal of the sea star.

Additional analyses along the coast of Chile found no evidence that sea star (H. helianthus) abundance was correlated with prey abundance over scales ranging from a few kilometers to 200 km (Navarrete and Manzur 2008). In contrast, in the Gulf of Maine, mussel recruitment and sea star recruitment both increased at multiple sites across a region of 120 km. Navarrete and Manzur (2008) suggested that these differing responses might reflect variation in the dispersal potential of the sea star species. Both H. helianthus and P. ochraceus have long planktonic durations of 2 to 3 months or longer (Navarrete and Manzur 2008, M. Strathmann 1987). In contrast, Asterias spp. have a shorter planktonic duration of 2 to 5 weeks (Witman et al. 2003), and thus sea star larvae produced in response to a massive recruitment of mussel prey might be retained within a smaller region.

These considerations highlight large gaps in our knowledge of the early life history of sea stars (reviewed by Ebert 1983). In particular, if increased prey recruitment triggers a reproductive response (sensu Witman et al. 2003) in sea star populations, that response could be driven by at least two mechanisms: (1) an increase in larval production by adults or (2) an increase in successful recruitment of larval sea stars or both. In the first mechanism, sea stars feed voraciously in response to abundant prey and build up nutritional reserves in the pyloric ceca. At least in some sea stars, this stored energy is converted into gonad production (Lawrence and Lane 1982). Recent work with Pisaster ochraceus confirms that the size of the pyloric ceca varies with the quality of prey available and that interannual differences in the size of the pyloric ceca accurately predict variation in reproductive output (Sanford and Menge 2007). Thus, in P. ochraceus, and presumably other sea stars (e.g., Crump 1971, Barker and Xu 1991a), there is a plausible direct link between increased prey and increased reproductive output.

In the second mechanism, increased recruitment could result from higher survival of larvae (see above) or from the influence of prey on post-settlement mortality. Mortality of newly settled sea stars during the first year of life is estimated to be very high (e.g., in P. ochraceus, greater than 97–99%; B. A. Menge 1975, Sewell and Watson 1993; see also Hancock 1958, Ebert 1983). Recently metamorphosed P. ochraceus prey on very small mussels, barnacles, and gastropods (J. L. Menge and Menge 1974, Sewell and Watson 1993, Sanford unpublished observations). Although largely untested, it seems plausible that high recruitment of tiny prey could reduce post-settlement mortality in some sea stars (Birkeland 1974, B. A. Menge et al. 2004, Wieters et al. 2008). Such a mechanism might also help explain why the abundance of Asterias spp.

responded to a recruitment pulse of mussels (Witman et al. 2003), whereas the abundance of *H. helianthus* in Chile was unrelated to mussel or barnacle recruitment (Navarrete and Manzur 2008). In *Asterias*, post-settlement mortality is high, and very small prey such as barnacles and mussels are required by newly settled sea stars (Hancock 1958). In this case, an abundance of small mussel recruits in the subtidal zone might increase the survival of co-occurring sea star recruits, leading to a coupling of predator and prey abundance. In contrast, *H. helianthus* recruits are found in high or mid-high intertidal boulder fields and crevices on wave-protected shores and appear to shift to lower tidal levels and wave-exposed habitats as they grow (Manzur et al. 2010). Moreover, these juvenile sea stars feed mostly on small periwinkles, with small mussels and barnacles consumed in lower frequencies (Manzur et al. 2010). Thus, the ontogenetic shift in habitat and diet that occurs in *H. helianthus* (but not in *Asterias*) might contribute to variation in the strength of coupling between mussel recruitment and sea star recruitment observed in different sea star species (Witman et al. 2003, Wieters et al. 2008). Birkeland (1974) also proposed that sea stars with generalized juvenile diets might exhibit more regular recruitment than sea stars with more specialized juvenile diets, an intriguing hypothesis that merits further attention (see also Ebert 1983).

The early life history of sea stars remains a critical gap in our understanding of the role of asteroids in marine communities. Larval sea stars and newly metamorphosed juveniles are notoriously difficult to locate and study. However, detailed studies of these early life stages will be essential to further our understanding of the mechanisms that might link primary productivity, prey recruitment, and other bottom-up mechanisms to the size of sea star populations.

Context-Dependent Predation: The Role of Environmental Factors

Sea star density often drives variation in effects on prey by the sea star population (termed "per population" effects), but recent research suggests that per capita effects of sea stars are highly context-dependent and can also contribute to variation in the effects of sea stars on benthic communities (B. A. Menge et al. 1994, 1996, Power et al. 1996). Although some earlier researchers clearly suspected that the community-level effects of sea stars might vary spatially (R. T. Paine 1980), it was not until the 1990s that the causes and

consequences of such variation in interaction strength became a major focus of study.

Consistent with environmental stress models (B. A. Menge and Sutherland 1987), this research suggested that the per capita effects of predatory sea stars can vary along environmental gradients. B. A. Menge et al. (1994, 1996) quantified the effects of the *P. ochraceus* on transplanted clumps of mussels along a wave-exposure gradient. Per capita interaction strength was lower in wave-exposed areas, suggesting that wave forces reduced the feeding efficiency of sea stars (see also Sanford 2002). The strength of the interaction between *P. ochraceus* and mussels is also highly sensitive to water temperature. During periods of cold-water upwelling on the Oregon coast, *P. ochraceus* became less active, and per capita predation rates on transplanted mussels declined sharply (Sanford 1999, 2002; Fig.7.6).

These results prompted the hypothesis that per capita predation rates of sea stars might also vary latitudinally, with stronger predation intensity occurring in warmer, low-latitude waters (Sanford 1999). This hypothesis was subsequently tested by quantifying *P. ochraceus* predation rates on transplanted clumps of mussels at 14 sites within the CC-LME (B. A. Menge et al. 2004). Per capita interaction strength did not differ consistently between Oregon and central California (Fig. 7.7). However, the temperature

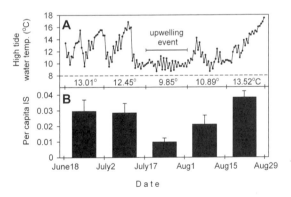

Fig. 7.6. Seawater temperature and intensity of predation by *Pisaster ochraceus* on mussels (*Mytilus californianus*) during consecutive 14-day periods (x axis). All data are means (+SE) of values recorded at three rocky benches spanning 0.75 km along the central Oregon coast, USA. A, High tide water temperatures (the mean from 2 h before to 2 h after each high tide). Temperatures above the x axis are the overall means for the 27 high tides of each period. B, Mean per capita interaction strength (IS) of sea stars on transplanted mussels during each 14-day period. Per capita IS during upwelling was significantly lower than during the other four periods. *Sanford 1999.*

difference between these regions is slight; water temperatures in central California are generally only 1–2°C warmer than in Oregon during the summer months (Sanford and Menge, unpublished data). In southern California (south of Point Conception), *P. ochraceus* occurs at low densities and thus per capita predation was estimated at only a single location in this region. However, these data did not reveal greater per capita effects in southern California, where water temperatures are often 5–8°C warmer than in Oregon. In sea stars and other ectotherms, exposure to warmer temperatures is expected to increase rates of metabolism and feeding (Sanford 2002). However, prolonged exposure to warmer temperatures may result in acclimation that would minimize physiological variation

among sea stars in different regions. In *P. ochraceus*, the general consistency of per capita effects across a broad latitudinal gradient (B. A. Menge et al. 2004) suggests that sea stars may in fact be acclimatized to temperatures within each region.

In contrast, regional variation in water temperature might have an important effect on predation by other sea stars, such as the crown-of-thorns sea star *A. planci*. An outbreak of *A. planci* in Guam killed 90% of corals inhabiting 38 km of coastline in less than 2 years (Birkeland and Lucas 1990). In contrast, an outbreak of *A. planci* on Hawaiian reefs moved little and had minimal impacts. Further studies showed rates of movement of sea stars in Guam that were five times greater, and coral consumption rates that were twice as great, as individuals in Hawaii. Birkeland and Lucas (1990) hypothesized that these differences in per capita predation rates might be driven by regional differences in water temperature. Water temperatures in Guam are typically 28–29°C, whereas temperatures in Hawaii are often below 25°C. Although this hypothesis has not been tested directly, water temperature may drive regional differences in the predation intensity of *A. planci*, with little modification by acclimatization. Collectively, these studies of *Pisaster ochraceus* and *A. planci* suggest that there is still much to learn regarding the effect of water temperature in driving regional variation in the community importance of sea stars.

Recent work also suggests that exposure to warm aerial temperatures can influence the per capita effects of intertidal sea stars (Pincebourde et al. 2008, 2012). In laboratory experiments, *P. ochraceus* were submerged for 18 h per day and then exposed daily to a 6-h low tide period with body temperatures maintained at one of four treatment levels. Daily exposure to low tide body temperatures of ≥23°C led to a decrease in the per capita effects of sea stars on mussels (Pincebourde et al. 2008). In New Zealand, exposure to temperature and desiccation also appeared to modify the per capita interaction strength between the intertidal sea star *S. australis* and its mussel prey (B. A. Menge et al. 1999, 2002). In the field, mussel clumps were transplanted to narrow, horizontal rocky benches in two positions: either at the edge or near the center (70–110 cm from the edge). Per capita interaction strength was reduced in the center of the bench, possibly because sea stars feeding on these mussels risked exposure to increased temperature and desiccation (Menge et al. 1999, 2002). In New Zealand, *S. australis* were observed to migrate upshore during high tide, remove mussels from the mid-zone, and retreat to

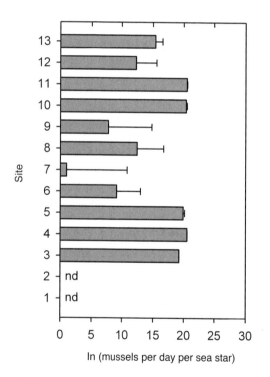

Fig. 7.7. Latitudinal variation in per capita predation rates of sea stars (*Pisaster ochraceus*) feeding on mussels at 13 sites in the California Current Large Marine Environment (see Fig. 7.1 map for site locations). Predation rates were assessed by comparing rates of mussel mortality in plots where sea stars were present at natural densities with plots where sea stars were excluded. Per capita predation rates were calculated using average sea star density in each plot. To make values positive, values of ln(mussels per day per sea star) were coded by adding 21 to each value. Data are means (+1 SE) from experiments conducted in summer 1999. nd, no data.
Modified from B. Menge et al. 2004.

lower zones with the falling tide, a strategy that presumably reduces exposure to environmental stress.

In the studies described above, per capita interaction strengths between sea stars and mussels were quantified using identical methods, providing a novel opportunity to compare directly the relative effects of intertidal keystone predators in two LMEs. This comparison suggests that, in general, the per capita effects of *P. ochraceus* in Oregon were roughly three times greater than those of *S. australis* in New Zealand (B. A. Menge et al. 2002). This difference might reflect, in part, differences in tidal regimes between these two LMEs. NZS-LME tides are semidiurnal and often expose *S. australis* to two periods of aerial exposure each day. As mentioned above, *S. australis* generally remove prey from the mid-intertidal zone during high tides before moving with the prey to lower zones, presumably to minimize periods of aerial exposure. In contrast, mixed semidiurnal tides occur within the CC-LME and generally expose *P. ochraceus* to at most a single period of aerial exposure per day. Although *P. ochraceus* sometimes migrate vertically with tidal cycles, they also frequently suspend these migrations and remain hunched over patches of prey for periods of a week or longer (Robles et al. 1995, Sanford 2002; see Chapter 16). These differing behaviors, perhaps linked to differences in tidal regimes, presumably increase the per capita effects of *P. ochraceus* relative to *S. australis* (B. A. Menge et al. 2002). Variation in tidal regimes might also help explain why intertidal sea stars play an important role in some LMEs but not in others. For example, in the NUSCS-LME, where semidiurnal tides also occur, the effects of *Asterias* spp. are generally restricted to the low intertidal zone (B. A. Menge 1976, Lubchenco and Menge 1978).

The studies summarized above emphasize that species interactions can be highly context-dependent and that per capita effects of sea stars may be tightly linked to environmental variation. Given the central role that sea stars play in many benthic communities, these considerations suggest a pathway through which climate change may affect these communities. In particular, interactions between sea stars and competitively dominant prey that are sensitive to environmental variation may act as leverage points through which small climatic changes might lead to large changes in benthic marine communities (Sanford 1999).

Context-Dependent Predation: The Role of Biotic Factors

Although environmental factors can clearly modify predation intensity by sea stars, it is also increasingly clear that the community importance of a given sea star species is strongly shaped by characteristics of its prey and the surrounding community. For example, along the coast of Chile, predation intensity by *H. helianthus* on transplanted mussels was similar at both southern and northern sites that spanned a discontinuity in coastal upwelling regimes (Navarrete et al. 2005). However, there were striking differences in the community response to *H. helianthus* removal. At southern sites, *H. helianthus* removal led to 60–95% mussel cover in experimental plots in just 3 months (Fig. 7.5). In contrast, at the northern sites, neither barnacle nor mussel cover increased in predator removal plots. This difference was driven by strong variation in prey recruitment; barnacles and mussels recruit heavily to southern sites but rarely settle north of the oceanographic discontinuity. Thus, variation in the community importance of *H. helianthus* was driven not by variation among sea star populations but by characteristics of the prey community itself (see Estes and Duggins 1995 for a related discussion).

Many sea stars have broad geographic ranges and across these spatial scales, a given sea star species will frequently span communities characterized by differences in the recruitment, abundance, or composition of prey species. For example, on the GBR, the impacts of *A. planci* were reduced at sites where preferred prey (*Acropora* corals) were uncommon (Moran 1986, Pratchett 2010). Similarly, R. T. Paine (1980) suggested that in Torch Bay, Alaska, the keystone predator *P. ochraceus* assumes the role of "just another starfish" because the mussel *M. californianus* was uncommon in the northern region of this sea star's geographic range. In the absence of *M. californianus*, *P. ochraceus* at Torch Bay preyed on the mussel *M. edulis* and on barnacles. However, because these species were not dominant competitors for space, the community importance of *P. ochraceus* was likely diminished. These observations emphasize that the feeding preferences of the predator and the competitive status of prey species will strongly influence the role of a given sea star in its community. In New Zealand, *S. australis* spans >1000 km of the western coasts of the North and South Islands where different mussels dominate (*P. canaliculus* and *Mytilus galloprovincialis*, respectively). However, despite this difference in species composition, experiments demonstrated that *S. australis* played a keystone role on both the North and South Islands (R. T. Paine 1971, B. A. Menge et al. 1999). In this case, the community importance of *S. australis* appears to be general throughout its geographic range because it consistently preys on mussels that are dominant competitors for space.

The importance of prey characteristics is further demonstrated by a comparative-experimental study that maintained long-term removals of intertidal sea stars in the CC-LME, the HC-LME, and the NZS-LME (R. T. Paine et al. 1985). In all cases, the community response was similar: sea star removal led to an increase in cover of the dominant mussel, and, in two LMEs (California Current, Humboldt Current), to a decrease in the diversity of species occupying primary space in the low intertidal zone. The sea star exclusions were eventually terminated and subsequent changes in the community were monitored. In the CC-LME and NZS-LME, the mussels that colonized and overran the low zone had attained a size refuge and thus coexisted with their predators for >14 years. In contrast, when sea stars were allowed to return to the removal plots in Chile, they rapidly eliminated the low zone mussels and the community returned to its control state. The difference in these recovery patterns was attributed to differences in the mussels that occupy these LMEs. In North America and New Zealand, *M. californianus* and *P. canaliculus* can grow to large sizes that are essentially immune to sea star predation. In contrast, on the Chilean shore, *P. purpuratus* is a small-bodied mussel that rarely exceeds 4 cm in length and is thus readily consumed (R. T. Paine et al. 1985). These studies emphasize that the role of sea stars in benthic communities depends not just on the abiotic environment and on characteristics of the predators themselves but also on prey characteristics. Despite these insights, further research is needed to address the roles that prey characteristics (including recruitment, competitive ability, and body size) and food web structure play in modifying the effects of sea stars and other predators in marine ecosystems (Berlow et al. 2009).

Conclusions

Sea stars are important predators in many intertidal and subtidal ecosystems, and have been demonstrated to be keystone predators in several large marine ecosystems. In other LMEs, they seem much less important. These differences in role are likely due to variation in one or more of several factors: abundance, body size, diet and prey preferences, system productivity, rates of subsidies of prey recruits, and characteristic prey size. In the intertidal zone, however, additional factors may come into play that are related to the tidal regime, and how environmental stress varies with tide cycles and climate. In subtidal habitats, sea star impacts seem influenced by a similar suite of factors, except for exposure to air. With the advent of human-caused climate change, an important research focus will be on responses to thermal variation and stress and on how this might influence the structure and integrity of benthic ecosystems in which sea stars play a major role. Other important research foci should include increasing our understanding of the influence of larval production, transport, and recruitment on the ecology of juvenile stages.

More generally, major advances in understanding the community and ecosystem roles of sea stars have burgeoned in recent years. These insights grew out of an expansion of research focus from straightforward emphasis on top-down effects to a more integrative approach that merged investigation of the roles of top-down effects, bottom-up factors, ecological subsidies, and spatial and temporal scale into a more synthetic view of the dynamics of LMEs.

8 Chemistry and Ecological Role of Starfish Secondary Metabolites

James B. McClintock,
Charles D. Amsler, and
Bill J. Baker

Chemistry
What Are Secondary Metabolites?

Metabolism can be thought of as having primary and secondary roles. Primary metabolism is what we generally regard as biochemistry. It is responsible for fundamental, life-sustaining processes such as protein synthesis, transcription, and cell signaling cascades. Primary metabolites include amino acids, nucleotides, carbohydrates and lipids, and polymers thereof, but also include a large number of small molecule cofactors. They are ubiquitous in living organisms and function in similar fashion across phyla. Secondary metabolism does not necessarily sustain physiological function but nonetheless can have a role in supporting life. Building blocks of secondary metabolites are isoprene, acetate, and amino acids, resulting in, respectively, terpenes, polyketides, and small peptides and alkaloids. However, some secondary metabolites are modified primary metabolite monomers, which include oxidation or reduction products or esterified or glycosylated products. Regardless of biosynthetic origin, what distinguishes secondary metabolites are their functions, which include defense and communication, and their phylogenetic distribution, which is often species specific. The term *natural products* is often used by chemists to describe secondary metabolites, reflecting the origin in nature of this group of small molecules.

Starfish are prolific producers of secondary metabolitic lipids. These include cholesterol derivatives (steroids) and fatty acids, the former are often modified by attached sugar groups (glycosylation), while the latter are primarily amides of sphingosine. Glycosylated steroids from starfish bear one to six sugar groups and are known collectively as saponins, reflecting their soap-like tendency to form emulsions. The most common group of related metabolites in starfish is the sulfated saponins known as asterosaponins. The major group of sphingosine amides (sphingolipids) is glycosylated ceramides known as cerebrosides. More than 800 compounds have been reported as secondary metabolites of starfish, the vast majority (>80%) of which are steroids, with sphingolipids comprising a minor (14%) component. About two dozen miscellaneous compounds complete the current extent of what is known about starfish secondary metabolites. A recent comprehensive review documents the considerable breadth of starfish lipophilic chemistry (Dong et al. 2011).

Steroids

Echinoderms were among the first marine organisms studied chemically, chosen as they were for their known toxicity, easy access, and conspicuous presence in the intertidal. Following closely on a discovery of saponins from holothurians (Chanley et al. 1959), the first report of starfish chemistry detailed the asterosaponins (Yasumoto et al. 1964), distinguishing echinoderms and plants as the only known sources of saponins. Subsequent identification of saponins in sponges (Schmitz et al. 1988, Takada et al. 2002, Regalado et al. 2010) has mitigated that distinction, but saponins are still considered rare in animals and in the marine realm in general. Saponins represent 27% of reported starfish secondary metabolites and their sulfated analogs represent 29%. Polyhydroxylated sterols (17%) sulfated polyhydroxylated sterols (9%) make up the remainder of the steroid complement. Polyhydroxylated sterols and their sulfate derivatives can be obtained by hydrolysis of saponins and are termed aglycones.

Asterosaponins A (Fig. 8.1a) and B (Fig. 8.1b) from *Asterias amurensis* were disclosed in a series of papers by Hasihmoto's group nearly 50 years ago (Yasumoto et al. 1964, Yasumoto and Hashimoto 1965, 1967). The structural analysis was complicated by the challenge of elucidating the constitution of previously unde-scribed sugars. The aglycone, asterone (Fig. 8.1c), was reported separately, setting the stage for the class of asterosaponins as being characterized by the 6α-hydroxy and 9(11)-olefin structural features. There are now many starfish saponins described as asterosaponins, some of which lack those characteristics, though the term asterosaponin has come to represent any saponin isolated from Asteroidea. The aglycone of asterosaponin B is known as thornasterol (Fig. 8.1d), having been isolated from the hydrosylate of saponins from the crown-of-thorns starfish, *Acanthaster planci* (Kitagawa et al. 1975).

The asterosaponins are representative sulfated saponins, but starfish glycosides are nearly equally distributed between sulfated and non-sulfated. Structural variability of these compounds includes sulfated sugars, as featured in picasteroside A (Fig. 8.2a), a saponin isolated from *Pisaster* sp. (Zollo et al. 1989). Picasteroside A also highlights the polyhydroxy nature of starfish aglycones. The locus of glycosylation in starfish saponins can be highly variable, including the side chain glycosylation observed in picasteroside A and C-3 glycosylation observed in echinasteroside A (Fig. 8.2b). Echinasteroside A, from *Echinaster sepositus* (Zollo et al. 1985), also provides an example of O-methylation, which is not uncommonly found on starfish saponin glycosidic hydroxyl groups. Starfish saponins typically display one to six sugars, predominantly glucose and galactose but also xylose and quinovose, in one to three glycosidic chains.

Steroids, aglycones and polyhydroxylated steroids, and their sulfated derivatives, represent more than a quarter of reported starfish secondary metabolites.

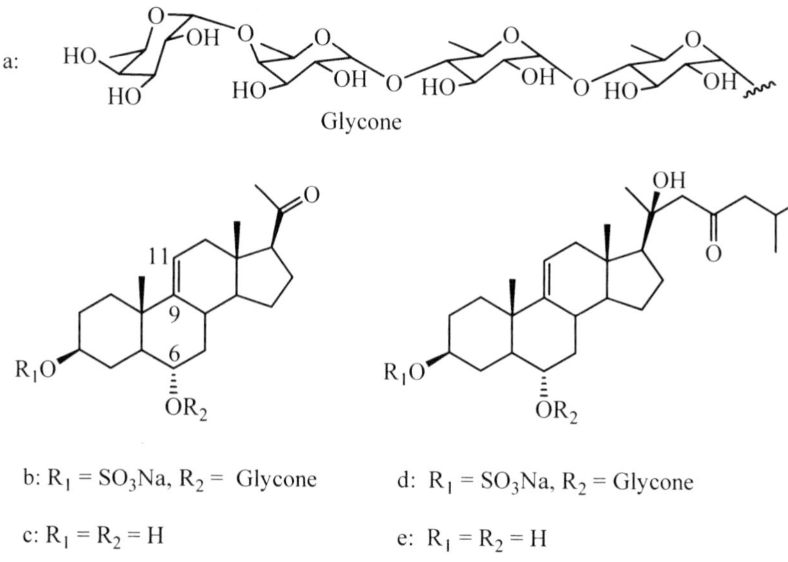

Fig. 8.1. Chemical structure of asterosaponins and their aglycones. *a*, Glycone; *b*, Asterosaponin A; *c*, Desulfoaglycone of asterosaponin A, asterone; *d*, Asterosaponin B; *e*, desulfoaglycone of asterosaponin B, thornasterol.

a: Glycone

b: $R_1 = SO_3Na$, $R_2 = $ Glycone

c: $R_1 = R_2 = H$

d: $R_1 = SO_3Na$, $R_2 = $ Glycone

e: $R_1 = R_2 = H$

Derivatives of thornasterol (Fig. 8.1d) are highly representative, though steroids bearing carbons with as many as nine or ten hydroxyl or sulfate groups have been reported, exemplified by polyhydroxy steroids from *Archaster typicus* (Riccio et al. 1986 [Fig. 8.2c] and Yang et al. 2011 [Fig. 8.2d].

Forbeside (Fig. 8.3a), from *Asterias forbesi* (Findlay et al. 1989), is representative of sulfated saponins bearing multiple sulfate groups. One or two sulfate groups is the norm for starfish saponins, though one example of three sulfate groups has been reported (Tang et al. 2009). Tremasterol A (Fig 8.3b), from *Tremaster novaecaledoniae* (De Riccardis et al. 1992), illustrates a rare example of a saponin phosphate.

Occasional unique structural motifs are found in starfish steroids, though fewer than a dozen are reported from >800 known starfish-derived compounds. Consider, for example, the chlorohydrin (Fig 8.3c) found in *Echinaster sepositus* (Minale et al. 1979). Taken with the two chlorohydrin derivatives from the same starfish, these are the only halogenated compounds reported from the Asteroidea. Equally unique is asterasterol A (Fig. 8.3d), from an Antarctic species of the Asteriidae (De Marino et al. 1997), as the only representative secosteroid among starfish; asterasterol A also has the rare 3α-hydroxy configuration. Taurine amides, exemplified by triseramide (Fig. 8.3e), from *Astropecten triseriatus* (Stonik et al. 2008), occur infrequently in starfish steroids and other lipids.

Sphingolipids

Long-chain fatty acid amides of sphingosine, known collectively as sphingolipids, are widespread in Asteroidea as ceramides (sphingolipids aglycones) and glycosphingolipids (cerebrosides and gangliosides, depending on the nature of the sugar), representing 14% of compounds reported from starfish. Cerebrosides (monoglycosylated ceramides) are the most common sphingolipid, especially glucosides, although a number of galactosides are also reported. Gangliosides are not uncommon, but other multiglycosylated ceramides are rare. Sphingolipids are well known in mammalian biochemistry as cell wall components, offering rigidity and occasional cell-recognition motifs. Physiological disruption of biosynthesis, accumulation, or distribution of sphingolipids causes a number of human diseases. Similar physiological roles in starfish are not well understood, though some starfish sphingolipids have pharmacological activity.

The ceramides related to AC-1-6 (Fig. 8.4a), from *A. planci* (Inagaki et al. 1998) were the first sphingolipid aglycones reported from starfish, though one report of a peracetylated ceramide series predates this report (Sugiyama et al. 1990). More than a dozen ceramides are known from starfish, including examples from *Luidia maculata* (Inagaki et al. 2006), *Asterias amurensis* (Ishii et al. 2006), and *Distolasterias nipon* (Rho and Kim 2005).

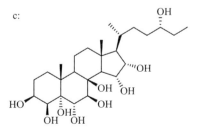

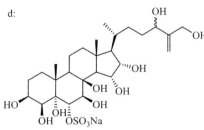

Fig. 8.2. Chemical structures of various compounds derived from asteroids. *a*, Picasteroside A; *b*, Echinasteroside A; *c*, A polyhydroxysteroid; *d*, A sulfated polydroxysteroid.

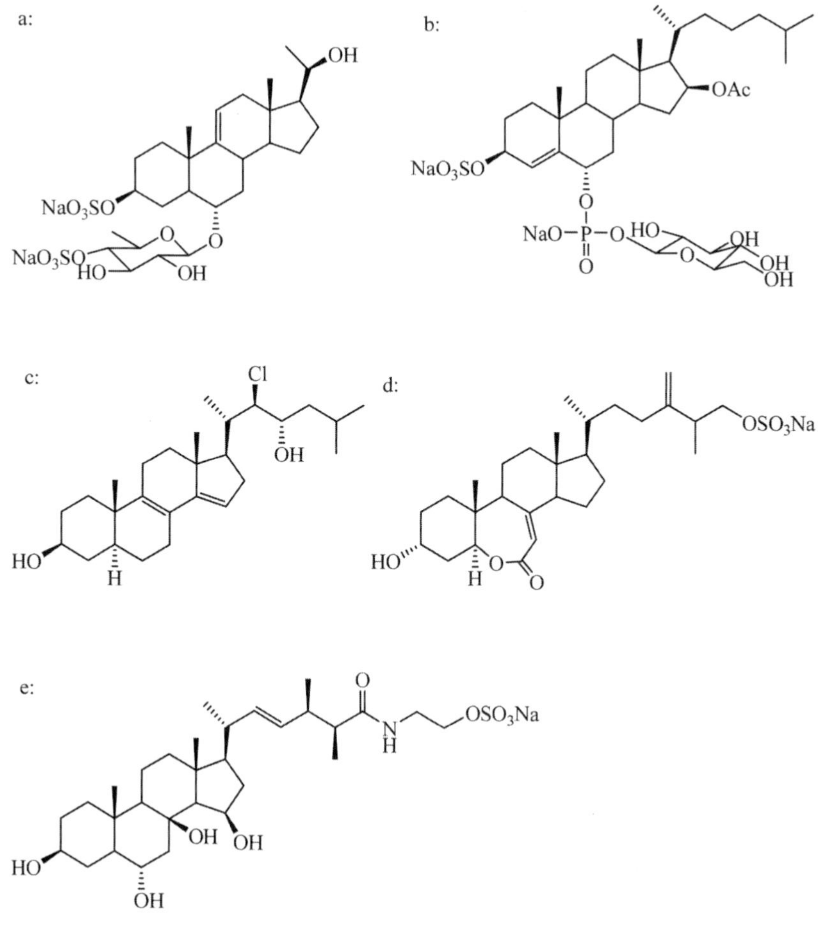

Fig. 8.3. Chemical structures of various compounds derived from asteroids. *a*, Forbeside; *b*, Tremasterol A; *c*, 22-(*R*)-Chloro-5α-cholestan-8(9),14-diene-3,23-diol; *d*, Asterasterol A; *e*, Triseramide.

Glycosidic ceramides feature glysosylation on C-1 of the sphingosine moiety. Cerebrosides from starfish are exemplified by the glucosphingolipid linckiacerebroside (Fig. 8.4b), from *Linckia laevigata* (Maruta et al. 2005). The same species is the source of a representative ganglioside, LLG-5 (Inagaki et al. 2005; Fig. 8.4c). Note the characteristic sialic acid units of LLG-5 distributed among the glycosides.

Other Classes of Starfish Secondary Metabolites

Metabolites deriving from other than cholesterol and sphingosine are rare in starfish (~6% of known starfish compounds). However, three intriguing examples of alkaloids have been reported. Perhaps the most notable, for its biological properties, is imbricatine (Fig. 8.5a), an isoquinoline alkaloid from *Dermasterias imbricata* (Pathirana and Andersen 1986). A similar isoquinoline, fuscusine (Fig. 8.5b) was reported from *Perknaster fuscus antarcticus* (Kong et al. 1992), and the mercapto-*N*-methylhistidine portion of imbricatine,

vis ovothiol A (E. Turner et al. 1987; Fig. 8.5c) was found in the eggs of *Evasterias troschelli*.

A fourth alkaloid from a starfish is 1-methyl-1,2,3,4-tetrahydro-α-carboline-3-carboxylic acid (salsolinic acid; Fig. 8.5d), which accompanies three alkaloidosteroids in *Lethasterias nanimensis chelifera* (Kicha et al. 2003a. The alkaloidosteroids are salts of salsolinol and sulfated asterosaponin aglycones. Salsolinol is an endogenous human neurotransmitter that mimics the effects of dopamine in catecholaminergic regulatory processes and may have a role in the etiology of Parkinson's disease (Mravec 2006).

Ecological Role

In contrast to the relatively large number of secondary metabolites that have been isolated, structurally elucidated, and identified in the body tissues of representatives of the Asteroidea (Blunt and Munro 2011), our knowledge of the ecological significance of these compounds remains in its infancy (Slattery

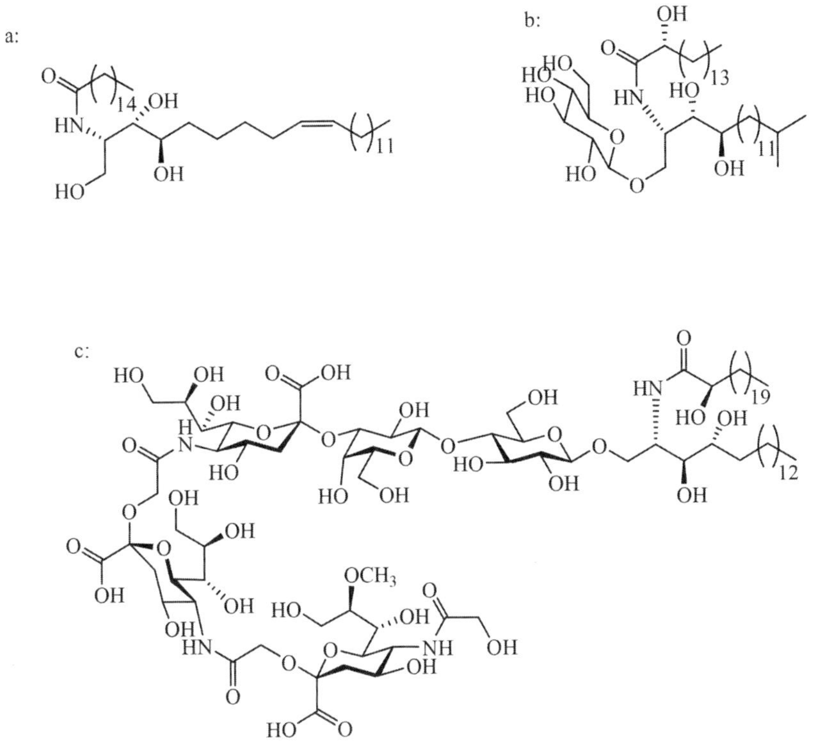

Fig. 8.4. Chemical structure of various compounds derived from asteroids. *a*, AC-1-6; *b*, Linckiacerebroside; *c*, LLG-5.

2010). To date, no studies have examined the ecological significance of known secondary metabolites from starfish. Rather, the relatively small number of studies outlined below have employed an a particular class of compounds (e.g., crude saponins), a crude lipophilic and hydrophilic tissue extract, or fresh ripe ovaries, eggs, embryos, juveniles, or intact adults in their evaluation whether chemistry is responsible for an ecological attribute (e.g., feeding deterrence, antifoulant properties, etc.). It should be noted that such studies, if properly conducted with protocols that employ natural tissue-level concentrations of extracts and, preferably, ecologically relevant predators or fouling organisms, can provide ecologically meaningful information despite the lack of the identity of the responsible bioactive secondary metabolite or metabolites.

Feeding Deterrents

The first study to directly examine the feeding deterrent properties of adult starfish was that of Bryan et al. (1997). Here, fresh body wall tissue and aqueous ethanol body wall extracts (targeted to extract asterosaponins) of 12 species of starfish from the Gulf of Mexico were tested in feeding deterrent bioassays using two species of generalist fish (the pinfish *Lagodon rhomboides* and the killifish *Cyprinidon variegatus*) and

one species of spider crab (*Libinia emarginata*). All 12 species had bite-size pieces of fresh body wall tissue rejected significantly more often than dried krill controls by at least one of the two species of fish. In contrast, only two of the species, *Henricia downeyae* and *Tamaria halperni*, had fresh body wall tissues that deterred against spider crabs. Bioassays with agar food pellets containing krill as a feeding stimulant and natural tissue-level concentrations of body wall ethanol extracts presented to pinfish indicated that deterrent chemistry explained fish-feeding deterrence in 75% (8 of 12) of the species. In contrast, krill pellets with starfish body wall extracts were readily consumed when presented to spider crabs. The most parsimonious explanation is that crabs are primarily deterred by the physical properties of starfish body wall (e.g., ossicles), whereas fish are more commonly deterred by chemistry (Bryan et al. 1997).

McClintock et al. (2006) examined the chemical antifeedant defenses in two species of brooding starfish, *Ganaster nutrix* and *Neosmilaster georgianus*, common to the Antarctic Peninsula. Both the diminutive *G. nutrix* and larger, hand-sized *N. georgianus* were deterrent when presented as live individuals to the Antarctic starfish *Odontaster validus*. It is important to note that the ubiquitous Antarctic keystone starfish,

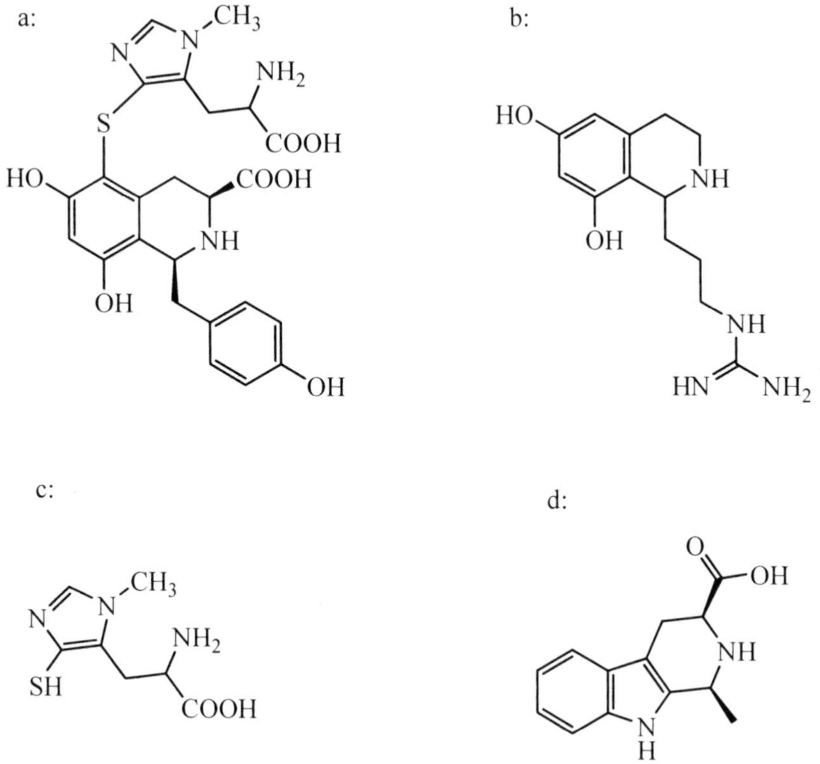

Fig. 8.5. Chemical structures of various compounds derived from asteroids. *a*, Imbricatine; *b*, Fuscusine; *c*, Ovothiol A; *d*, 1-methyl-1,2,3,4-tetrahydro-β-carboline-3-carboxylic acid (salsolinic acid).

O. validus, has been used extensively as a model predator to evaluate chemical defenses in sympatric Antarctic marine invertebrates (studies cited in McClintock and Baker 1997a, Amsler et al. 2001, McClintock et al. 2010). Its use as a model predator is based on its broad omnivorous feeding habit (Dearborn 1977, McClintock 1994), high abundance and keystone status (Dayton et al. 1974), and its behavioral propensity to climb the sides of seawater tanks and once reaching the seawater-air interface, rolling its arms backward so as to expose the ambulacral grooves. This allows the placement of food models containing pure secondary metabolites or crude organic extracts, pieces of tissue, or small live individuals halfway between the distal arm tip and the oral opening and observing whether the item is moved by the tube feet to the arm tip or pushed off the side of the arm and dropped (rejection) or, in contrast, carried to the mouth and held against the cardiac stomach (acceptance; McClintock and Baker 1997a, Amsler et al. 2005). Some investigators have preferred to conduct feeding deterrent bioassays with *O. validus* by placing them in buckets with food items and followed feeding behavior over a period of up to 24 h (Avila et al. 2000). Other investigators have employed a sensory tube-foot retraction response as an indirect measure of feeding deterrence. For example, the Ant-

arctic spongivorous *Perknaster fuscus* was used in tube-foot retraction assays as a model predator to evaluate the deterrent properties of sponges (e.g., McClintock et al. 1994, 2000a).

Granaster nutrix placed directly on the exposed ambulacral groove of *O. validus* were rejected significantly more often than controls consisting of dried krill (McClintock et al. 2006). Similarly, live *N. georgianus* placed into buckets with adult *O. validus* positioned on their aboral surfaces were neither attacked nor consumed. Alginate food pellets containing dried krill powder and tissue level concentrations of methanolic body wall extracts of the two starfish were rejected significantly more often than control pellets without extracts. Moreover, the copious mucus that is released from the body wall of *N. georgianus* coated onto alginate pellets was deterrent to *O. validus*. Chemically mediated starfish-starfish interactions, such as those elucidated in this study, are important in understanding determinants of benthic community structure, particularly as some starfish (e.g., *O. validus*, *Pisaster ochraceus*) play keystone roles in community dynamics (R. T. Paine 1966, Dayton et al. 1974).

Studies of the potential role of starfish secondary metabolites of early life history phases (eggs, embryos, larvae, juveniles) as antifeedants are somewhat more extensive and predate by a decade those studies that have examined adults. This is likely attributable to the general consensus that early life stages of marine invertebrates are more vulnerable to predation (McClintock et al. 2000b). Lucas et al. (1979) initiated these studies with their examination of the role of crude saponins as fish antifeedants in eggs and larvae of *A. planci*. Aqueous gelatin particles prepared with yeast as a feeding stimulant and with various concentrations of crude saponin extracted from eggs of *A. planci* were tested in fish antifeedant bioassays using four species of common planktivorous fish. One fish species, *Acanthochromis polyacanthus*, was deterred at crude saponin concentrations two and four orders of magnitude below natural levels detected in the eggs and larvae of *A. planci*. The authors suggest that as saponins are relatively soluble in seawater, that it is possible that fish may detect waterborne saponins prior to ingestion or find the eggs and larvae unpalatable and learn to avoid them. It is important to note that the authors also demonstrate that deterrence may be predator-specific (some species of fish were less deterred by crude saponins than others), that the nutritional condition of the fish impacted deterrence (hungry fish were less discriminatory), and that the level of food stimulant added to the gelatin pellets influenced the ability of crude saponins to act as an effective deterrent (higher levels of yeast made a given level of crude saponin less of a deterrent). These are important observations that have been borne out over the subsequent 3 decades across a large number of studies in marine chemical ecology. To date, only one other study has reported a presumptive chemical defense in larvae of a starfish with a planktotrophic mode of reproduction. The bipinnaria of an unidentified asteroid were deterrent to the marine fish *Leiostomus xanthrus,* while being consumed readily by polyps of the hard coral *Oculina arbuscula* (Bullard et al. 1999).

McClintock and Vernon (1990) examined chemical defenses in the eggs and embryos of eleven species of Antarctic starfish from McMurdo Sound, Ross Sea. Eggs and embryos were freeze-dried, ground into a powder, and a standardized 5% mass was added to agar pellets containing 5% fish meal and presented to an allopatric fish model (the marine killifish *Fundulus grandis*). They found the pelagic lecithotrophic eggs of *P. fuscus* and *Porania antarctica* were deterrent, as were the brooded embryos of *Diplasterias brucei* and *Notasterias armata*. The presence of defensive chemistry in brooded embryos of these starfish indicates

that provisioning early crawl-away juveniles with chemical defenses may be a selective advantage in their survival. Their experimental protocol had the disadvantages of loading pellets with levels of tissue both below and above that corresponding to single eggs or embryos and the use of an allopatric model fish. Therefore further experiments are warranted using sympatric predators.

McClintock and Baker (1997b) greatly expanded on this earlier study by examining chemical defenses in the ovaries, eggs, embryos, or larvae of five species of Antarctic starfish. They found the brooded embryos of *D. brucei* and the pelagic lecithotrophic embryos and larvae of *P. fuscus,* and *P. charcoti* were unpalatable to *O. validus* and to the large omnivorous sea anemone *Isotealis antarctica*. Feeding assays using agar pellets containing natural concentrations of a hydrophilic extract confirmed the chemical basis for rejection in the embryos of *D. brucei*, and presumptive secondary metabolites were deemed likely responsible for deterrence as early life stages lack structural or behavioral defenses. The sympatric omnivorous amphipod *Paramoera walkeri* was also deterred from grazing on the embryos and on the larvae of *P. fuscus* and *P. charcoti*. In contrast, the ripe ovaries of the planktotrophic *O. validus* and the planktotrophic eggs of the sea urchin *Sterechinus neumayeri* were readily consumed by all three predators. McClintock and Baker (1997b) hypothesized that that selection for chemical defenses is high in the offspring of those Antarctic starfish that have a lecithotrophic mode of development, especially as embryonic and larval development is greatly protracted in starfish in polar waters (Pearse et al. 1991a).

This hypothesis relating chemical defense in larvae to reproductive mode is further supported by a study of the presumptive chemical defenses in the early life history stages of a suite of lecithotrophic starfish from the Pacific Northwest (Iyengar and Harvell 2001). Ripe ovaries or eggs of five of seven species (*Crossaster papposus, Leptasterias* sp., *Solaster dawsoni, S. endeca,* and *S. simsoni*) were found to be unpalatable to at least one of three predators, the sea anemone *Anthopleura elegantissima*, the shrimp *Pandalus danae*, or the sculpin *Oligocottus maculosus*. The ripe ovaries or eggs of *Henricia leviscula* were unpalatable to all three predators, whereas those of *Mediaster aequalis* were readily consumed. Using sculpin as a model fish predator, Iyengar and Harvell (2001) detected no intraspecific differences in the deterrent properties of various stages of development (egg, larva, juvenile) in *S. simsoni, H. leviscula,* or *Leptasterias* sp. McClintock et al. (2003) extended the examination of the chemical defenses of lectithotrophic species to the brooded embryos and juveniles of *N. georgianus* and *Lyasterias perrieri* from the Antarctic Peninsula. Both embryos and juveniles of both species were unpalatable to *O. validus*. A chemical defense was confirmed in *L perrieri* as *O. validus* rejected alginate food pellets containing a natural concentration of a methanol extract of its embryos significantly more often than control food pellets. A physical defense cannot be ruled out as the basis for the lack of palatability of juveniles to *O. validus*. Alginate food pellets containing a natural concentration of methanol extract of juveniles of *L. perrieri* was not deterrent.

Antifoulants

Similar to most sessile or sluggish marine invertebrates and despite often possessing pedicellariae that may help clean the outer epithelium, starfish are vulnerable to fouling organisms. Only three studies have been conducted on the potential antifoulant properties of secondary metabolites in starfish. This small number of studies does not reflect a lack of chemical antifoulant activity. Bryan et al. (1996) examined the body wall aqueous ethanol extracts (to best isolate crude saponins) of a large number of starfish species from the northern Gulf of Mexico. They found that 13 species had body wall extracts that significantly inhibited settlement of barnacle and bryozoan larvae, some at concentrations as low as 0.12 mg/ml seawater. They also found that body wall extracts from *Goniaster tesselatus* and *L. clathrata* inhibited the attachment of cells of the marine bacteria *Deleya marina* and *Alteromonas luteoviolacea*. Ethanol body wall extracts from *A. articulatus, Linckia nodosus, Tamaria halperni, Narcissia trigonaria,* and *Tethyaster grandis* actually enhanced marine bacterial settlement. Bryan et al. (1996) suggest that enhanced bacterial attachment could reflect a tolerance to bacteria or serve to enhance nutrient uptake by facilitating bacterial colonization. Nonetheless, the widespread presence of antifoulants in these starfish active against marine settlement in invertebrate larvae reflects strong selection to avoid problems when outer tissues become fouled and restrict the movements of ions, nutrients, waste, and seawater across epithelial surfaces.

Greer et al. (2003, 2006) examined the ability of desalted aqueous extracts of the starfish *A. articulatus* and *L. clathrata* with known bioactivity to influence spore settlement and germination in the brown alga *Hincksia irregularis*. They found that both starfish extracts had a significant negative effect on both germi-

nation and settlement at levels below those of estimated tissue-level concentration. Higher doses of extracts caused rapid toxic effects on spores. Iken et al. (2003) expanded this analysis to include an evaluation of the swimming behavior of zoospores of *H. irregularis* exposed to echinoderm extracts. Spores exposed to both aqueous and organic body wall extracts of *A. articulatus* and *L. clathrata* at levels well below natural tissue levels showed significant effects on spore swimming behaviors (as determined by the ratio of rate in direction change over swimming speed) when compared with controls. Iken et al. (2003) conclude that these quantifiable changes in spore behavior in response to bioactive extracts such as those present in starfish provide a novel laboratory antifouling bioassay technique capable of screening for antifoulant compounds at very low concentration. Greer et al. (2006) utilized this bioassay-guided methodology to further purify antifouling compounds from the starfish extracts.

Alarm Pheromones and Escape-Eliciting Compounds

Pheromones are defined as secreted or excreted chemical factors that trigger a specific reaction response in conspecifics. Within this broad definition, there are two examples of starfish detecting chemical cues from conspecifics that alter their social behavior; in both cases triggering a flight response (Lawrence 1991, McClintock et al. 2008a). The specific chemical or chemicals serving as pheromones were not determined. Lawrence (1991) observed in the large multiarmed *Pycnopodia helianthoides* that an injured individual (simulated by fresh pieces of body tissue, pyloric ceca, and tube feet) released a waterborne chemical that elicited an alarm response (flight) in conspecifics. Time lapse videography revealed that *O. validus* elicited an escape response when contacted by *P. aurorae* (McClintock et al. 2008a). Fleeing *O. validus* in turn triggered a flight response in conspecifics after physical contact.

A large number of studies have examined the escape responses elicited in prey when encountering starfish. Most have found the response to be tactile, but several have demonstrated a chemical response. For example Mahon et al. (2002) found that crude hydrophilic extracts of *N. georgianus* induced escape behaviors in the sympatric limpet *Nacella concinna*. Mackie (1970) found that a steroid glycoside with surface-active properties from *Marthasterias glacialis* triggers avoidance and escape reactions in a suite of prospective molluscan and ophiuroid prey Mackie and Turner (1970) partially characterized this biologically active steroid glycoside. *D. imbricata* causes a flight response in the sympatric sea anemone *Stomphia* sp. (Elliott et al. 1989), a potential prey of the starfish.

Conclusions

It is evident that starfish are capable of synthesizing a rich array of secondary metabolites, the majority of which are saponins and their sulfated analogs. The saponin class of natural products is most commonly associated with terrestrial plants that are bitter to the taste and toxic at sufficient concentration. Similar to plant saponins, asterosaponins serve as antifeedants and are likely to also have antifoulant properties against marine invertebrate larvae and bacteria and other microbes. Despite our knowledge of secondary metabolites in starfish, there is considerable work ahead to establish their definitive ecological functions. The information garnered to date from ecological studies employing crude tissue extracts indicates that this will likely be a rich area for future research.

Acknowledgments
Manuscript preparation was supported in part by National Science Foundation awards ANT-0838773 (JBM, CDA), ANT-0838844 (JBM), and ANT-0838776 (BJB) from the Antarctic Organisms and Ecosystems program. JBM also wishes to acknowledge support from an Endowed University Professorship provided by the University of Alabama at Birmingham.

9

Steroids in Asteroidea

Stephen A. Watts and
Kristina M. Wasson

C hemical communication among cells, tissues, organs, and even among organisms has been an active area of research for decades. In addition, a recent resurgence in the study of metabolomics has occurred because of a renewed interest in bioactive compounds and their effects on metabolism. Most of the knowledge of these bioactive compounds is derived from studies of vertebrates. However, functions for many of these bioactive compounds would have arisen much earlier in metazoan evolution, most likely among the invertebrate phyla. Despite their origin in these phyla, we have only a cursory understanding of the role of bioactive compounds in regulating physiological, cellular, and biochemical processes in most invertebrates.

Steroids represent one important class of bioactive compounds. Since the award of the Nobel Prize in Chemistry to Adolf Friedrich Johann Butenandt for his work on sex steroids in 1939, steroids have been evaluated for their roles in sex-specific processes including reproduction and sex differentiation and as regulators of physiological processes including metabolism, ion balance, and immune function. By the 1940s and 1950s, the sex steroids, including progestogens, androgens, and estrogens were known to regulate mammalian gonadal function and sex differentiation.

By the early 1960s, significant progress had also been made toward understanding the biosynthesis and metabolism of cholesterol, the precursor to the sex steroids (Bloch 1965, Vance and van den Bosch 2000). Bloch (1965) found that acetate was the precursor to cholesterol and that mevolanic acid and squalene were intermediates synthesized during cholesterol formation. Further work revealed that cholesterol, whether obtained from the diet or produced by de novo synthesis, can be metabolized to form the classical sex steroids progestogens, androgens, and estrogens, as well as glucocorticoids, mineralocorticoids, and vitamin D (Fig. 9.1). The synthesis and function of these various steroids are well understood in many vertebrates. The roles of steroids in invertebrates, however, are not well understood.

Examination of the presence of sex steroids in echinoderms began not long after their discovery in vertebrates. Using a simple vertebrate

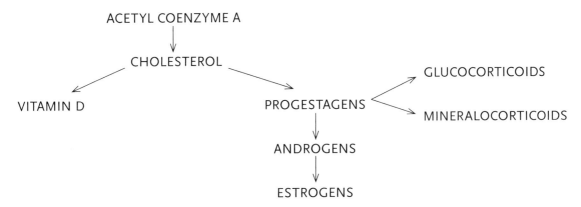

ACETYL COENZYME A

CHOLESTEROL

VITAMIN D   PROGESTAGENS   GLUCOCORTICOIDS

MINERALOCORTICOIDS

ANDROGENS

ESTROGENS

Fig. 9.1. Synthetic relations of steroids in vertebrate tissues. Vitamin D, glucocorticoids, and mineralocorticoids have not been identified in asteroids.

bioassay, Hagerman et al. (1957) detected the presence of minor amounts of an estrogenic-like compound in the ovaries of *Asterias forbesii* as well as species from other invertebrate phyla, suggesting that sex steroids might be present in invertebrates. However, the authors did not speculate on the role of such a steroid in the ovaries, most likely because an earlier study by Hagerman (1956) reported no evidence of an estradiol effect on the rates of oxygen consumption, glucose, pyruvate or glycogen utilization, or lactate production in the ovaries of *A. forbesii*. Subsequent work by Botticelli et al. (1960) tentatively identified mammalian-type estrogens and progesterone in ovaries of *Pisaster ochraceus*. They found that the ovaries had 17β-estradiol present at the time of spawning, but they could only tentatively conclude that progesterone was also present.

These early studies were able to identify an estrogenic-like compound despite the lack of knowledge concerning reproductive cycles. Farmanfarmian et al. (1958) first described an annual reproductive cycle in several species of asteroids. Giese (1959) reviewed the first reports of the cyclic nature of the gonads and the inverse relationship between the sizes of the gonads and the pyloric ceca, suggesting the need for chemical regulators of both reproduction and nutrient allocation. Subsequent work sought to evaluate steroids in relation to reproduction, and it continues today.

The specific roles of steroids in echinoderms are unclear, but their synthesis in echinoderm tissues suggests they are active regulatory molecules. In this chapter we provide a review of the biosynthesis, metabolism, and possible functions of steroids and their derivatives in the Asteroidea. We begin with a summary of the studies on sterols and then present a synopsis of the studies on sex steroids in asteroids.

Presence of Sterols
Δ7 and Δ5 Sterols

More than 100 years ago, Dorée (1909) suggested that asteroids do not have the same sterols commonly found in vertebrates. Kossel and Edlbacher (1915) identified a sterol unique to the *Astropecten aurantiacus* and named it stellasterol. Bergmann (1962) determined that asteroids as well as holothoruids contained a mixture of Δ7 sterols, whereas echinoids, crinoids, and ophiuroids contained mixtures of Δ5 sterols. He speculated that these sterols were derived from dietary sources but also attributed phylogenetic differences to their occurrence within their respective taxonomic group. However, this hypothesis was disputed later.

Confirming Bergmann's work, Gupta and Scheuer (1968) demonstrated that whole-body extracts of *Acanthaster planci* had a mixture of six Δ7 sterols: two primary components including cholest-7-en-3β-ol and 24-methyl-cholest-7-en-3β-ol (α7-ergostenol), and four minor sterols comprising cholesta-7,22-dien-3β-ol (C27 sterol), acanthasterol (a C30 sterol), and a mixture of 24-ethyl and 24-ethylidinecholest-7-en-3βol (C29 sterols). Analyzing animals collected at different times of the year revealed some small variations in the proportions of these C27, C28, and C29 sterols, but the number of animals that they used was too small to be able to identify any significant changes. They also identified sterols in other echinoderm species and came to a similar conclusion that Bergmann (1962) did: asteroids contained primarily Δ7 sterols.

Subsequently, Goad et al. (1972) reported the presence of 5α-cholest-7-en-3β-ol as the major sterol in *A. rubens* and *Henricia sanguinolenta* from British coastal waters. A survey of several asteroid species resulted in the identification of C27, C28, and C29 sterols that varied throughout the year. They also reported the presence of a Δ7, 22 C26 sterol and a Δ22 C26 sterol, and suggested that observed sterol patterns may reflect the dietary preferences of individual species. However, a parallel survey of the sterol composition of the diets of these species was not reported. A. G. Smith et al. (1973) studied *A. rubens* and identified 22 Δ7 sterols, comprising 4-demethyl sterols, which were a majority of the mixture, and C-4 methyl sterols, which only existed in minor amounts. In describing the sterols from asteroids found in the Indo-West Pacific Ocean, Sheikh et al. (1973) reported that *Linckia multiflora, Protoreaster nodosus, Protoreaster linki, Culcita schmideliana, Nardoa variolata,* and *A. plankii* contained cholestenol in addition to C27 to C30 mono- and disaturated sterols.

Kicha et al. (2001) identified 5α-cholest-7-en-3β-ol as the primary sterol found in *Asterina pectinifera*. Seldes and Gros (1985) reported the presence of mono- and diunsaturated C27 and C28 derivatives as major components as well as minor quantities of C29 and C26 compounds in *Comasterias lurida*. Asterosterol and amurasterol were also present.

In a review of sterols in asteroids, Dong et al. (2011) indicated that asteroids were one of the most abundant sources of sterols, particularly polyhydroxysterols. Most of these were derived from 3β-hydroxysterols and were generally assumed to have a role as secondary metabolites instead of a primary physiological role (Kicha et al. 2004).

It was not until the early 1980s that the seasonality of the sterol composition was examined. Burnell et al. (1982) examined the sterol content of *A. rubens* (as *A. vulgaris*) throughout its reproductive cycle and found that the concentration of sterols was at its lowest in June after spawning. By July the concentration of sterols, mostly the C28 and C29 sterols, increased 200-fold, suggesting these sterols were obtained from a dietary source. This was one of the first reports of the presence of cholesterol in an asteroid, yet cholesterol, in contrast to other sterols, did not vary in concentration throughout the year. However, these researchers did not distinguish between male and female asteroids. They simply combined three to seven individuals each month to extract sterols.

Little work on the presence of Δ5 sterols in asteroids has been reported. As described previously, Burnell et al. (1982) detected cholesterol in *A. rubens* (as *A. vulgaris*) but reported it did not vary during the annual reproductive cycle. Another study (De Simone et al. 1980) examined the sterol composition of *Echinaster sepositus* and found that cholesterol was a minor

component of the sterols; most were Δ7 sterols. The animals were collected at various times during the year, but no mention of whether any of the sterols varied during the annual reproductive cycle was made (De Simone et al. 1980).

Dietary Origin of Sterols

The early work by Bergmann (1962) found a complex mixture of sterols that was predominantly Δ7 sterols, suggesting that they were converted from dietary sources of Δ5 sterols. However, Gupta and Scheuer (1968) found the sterol composition of *A. planci* from two different locations were similar. They also analyzed the sterol composition of echinoderms from the four other classes and concluded that sterol composition was genus dependent and independent of their environment. In contrast, Fagerlund (1969) suggested that sterols in asteroids are of dietary origin. In addition, Kanazawa et al. (1976) reported that the sterol composition of *A. planci* was similar to that of other invertebrates in a coral reef environment and concluded that these sterols were derived and modified via the food chain.

De Novo Synthesis of Sterols
Δ7 STEROLS

A. G. Smith and Goad (1971a) found both *A. rubens* and *H. sanguinolenta* converted [2-¹⁴C] mevalonic acid into squalene and that *A. rubens* also produced 5α-cholest-7en-3β-ol, suggesting that *A. rubens* could synthesize this compound de novo (Smith and Goad 1971a). Similarly, Kanazawa et al. (1974) reported that *Leiaster leachii* and *Henricia ohshimai* could convert mevalonate into sterols, indicating that all echinoderms can synthesize sterols de novo. Voogt and Schoenmakers (1973) demonstrated that Δ7 sterols can be synthesized from radioactively labeled Δ5 sterols when injected into the coelom of *A. rubens* but at low efficiency. They suggested that Δ5 sterols are primarily converted to other lipid classes or metabolized to carbon dioxide. Fagerlund and Idler (1960) and later Fagerlund (1969) showed that ingested Δ5 sterols are efficiently converted into Δ7 sterols in *P. ochraceus*. Additional work by Voogt and van Rheenen (1976a) demonstrated that *A. rubens* was not able to synthesize C28 and C29 sterols from cholesterol, but that it could synthesize cholesterol, albeit in minor amounts, from mevalonate. In another study, they further suggested that C28 and C29 sterols were of dietary origin (Voogt and van Rheenen 1976b). In addition, they confirmed the ability of *A. rubens* to convert Δ5 sterols to Δ7 sterols. Other investigators have also shown that many asteroid species can synthesize sterols from acetate or meval-onate, including *A. rubens* (Goad et al. 1972, Walton and Pennock 1972, Voogt 1973), *H. saguinolenta* (A. G. Smith and Goad 1971a, Goad et al. 1972), *Marthasterias glacialis*, and *A. aurantiacus* (Voogt 1973). Compara-tively, Teshima and Kanazawa (1975) reported that the asteroid *Laiaster leachii* was capable of synthesizing squalene, lanosterol, and desmethylsterols from mev-alonate and that Δ7 cholestenol is the primary sterol synthesized. They further indicated that this asteroid had no ability to alkylate at C24 or introduce a dou-ble bond at C22 of Δ7 cholesterol.

Further work by A. G. Smith and Goad (1975) con-firmed the ability of de novo sterol synthesis in *A. ru-bens*. Injection of [2-^{14}C]-mevalonic acid into *A. rubens* resulted in the synthesis of squalene, lanosterol, 4,4-dimethyl-5α-cholesta-7,24-dien-3β-ol; 4,4-dimethyl-5α-cholesta-8,24-dien-3β-ol; 4α-methyl-5α-cholesta-7,24-dien-3β-ol; 4α-methyl-5α-cholest-7en-3β-ol; and 5α-cholest-7en-3β-ol, showing that *A. rubens* is capable of de novo synthesis. Both the gonads and the pyloric ceca were efficient at incorporation of radioactivity into 5α-cholest-7-en-3β-ol.

Δ5 STEROLS

Little work has been reported for the de novo synthe-sis of Δ5 sterols. Identification of the metabolites pro-duced after injection of *A. rubens* with [4-^{14}C]-cholest-5-en-3β-ol resulted in 5α-cholestanol, an intermediary compound in the synthesis of 5α-cholest-7-en-3β-ol (A. G. Smith and Goad 1971a). The use of the two-radiotracer precursor, [4-^{14}C, 3α-^{3}H]-cholest-5-en-3β-ol, allowed the identification of 5α-cholestan-3β-ol, con-firming the formation of a 3-oxosteroid intermediate (A. G. Smith et al. 1972).

Function of Sterols

Despite the extensive research on sterol synthesis in asteroids, the function of these sterols in asteroids has not been generally investigated. Their function is pre-sumed to be similar to those of other sterols in cells and tissues of other organisms, either as components of cell membranes or as precursors to other sterol de-rivatives including saponins (see Chapter 8), or C21, C19, and C18 steroids (Kicha et al. 2003b). In some spe-cies, sterols are purported to be involved in lipid emul-sification (similar to bile acids), but this has not been evaluated in asteroids. Kicha et al. (2004) suggested that most sterols are likely secondary metabolites and do not have a physiological function. In vertebrates, vitamin D is derived from 7-dehydrocholesterol and has pleiotrophic effects. However, vitamin D synthe-sis, presence, or function has not been described in any asteroid species.

Presence of Progestogen, Androgen, and Estrogen Steroids

A tentative identification of 17β-estradiol and proges-terone in *P. ochraceus* was first reported by Botticelli et al. (1960) and progesterone in *A. amurensis* by Ike-gami et al. (1971). Advances in the detection of ste-roids in asteroids, however, required the utilization of the more sensitive radioimmunoassays (RIA) for specific steroids. The first report of a validated RIA for use with asteroid tissues was by Dieleman and Schoenmakers (1979). Using a single individual of *A. rubens*, they developed RIAs for both progesterone (P4) and estrone. They found that the ratio of estrone to P4 was 30-fold higher in the ovaries than in the pyloric ceca and that P4 levels were tenfold lower in the ovaries than in the pyloric ceca. Subsequent anal-yses involved measuring P4 and estrone levels in the testes (Voogt and Dieleman 1984), ovaries, and py-loric ceca of *A. rubens* throughout the starfish's annual reproductive cycle (Schoenmakers and Dieleman 1981, Voogt and Dieleman 1984). For males, P4 levels were tenfold higher in the pyloric ceca than in the testes and the onset of a new reproductive cycle (as determined by a decrease in the pyloric ceca index) coincided with a significant decrease in P4 levels and a significant increase in estrone levels in the pyloric ceca. These data suggested that P4 and estrone in the pyloric ceca may be involved with mobilizing nutri-ents for gametogenesis (Voogt and Dieleman 1984).

For females, P4 and estrone levels varied with the reproductive cycle. P4 levels in the ovaries were highest during stage 1, decreased during stage 2, and remained low during stages 3 and 4 of the re-productive cycle. In contrast, P4 levels in the pyloric ceca were highest during stage 4 just before spawn-ing and lowest during stage 3 (Schoenmakers and Dieleman 1981). Estrone levels in the ovaries were highest at the start of vitellogenesis (stage 1), while the estrone levels in the pyloric ceca remained low throughout the cycle. It is interesting to note that the peak of estrone in the ovaries coincided with the maximal protein content in the pyloric ceca and was followed by a peak of protein levels in the ovaries. These changes in the protein content and estrone levels led Schoenmakers and Dieleman (1981) to hy-pothesize that estrone or other estrogens affect the biosynthesis of proteins and possibly their incorpo-ration into oocytes.

Although the size of the pyloric ceca of *A. rubens* (as *A. vulgaris*) did not exhibit any seasonal trends, it did exhibit transient increases in estradiol (E2), P4, and testosterone similar to the transient increases observed in the gonads (Hines et al. 1992a). In males, the highest levels of E2, P4, and testosterone in the testes occurred at the onset of spermatogenesis, while transient increases in the levels of E2, T, and P4 coincided with spermatogonial mitotic proliferation, spermatogenic column formation, and spermiogenesis, respectively. In the ovaries, the highest levels of E2 and testosterone occurred at the onset of oogenesis, but P4 levels did not change significantly throughout the cycle (Hines et al. 1992a). Similar to Schoenmakers and Dieleman (1981) and Voogt and Dieleman (1984), Hines et al. (1992a) hypothesized that transient increases in the levels of sex steroids during gametogenesis serve as endogenous modulators of reproduction in *A. rubens* (as *A. vulgaris*).

In *Sclerasterias mollis*, the indices of the pyloric ceca and gonads both have annual cycles and have the typical inverse relationship (Xu and Barker 1990a, 1993). In general, estrone levels as measured by RIA were higher in pyloric ceca than in ovaries. Estrone levels in the pyloric ceca peaked 2 months earlier than in the ovaries. This led Xu and Barker (1990a) to conclude that estrogens promote oogenesis and protein synthesis in ovaries, but P4 may be inhibitory. Nutritional status (fed vs. starved) did not affect the relative levels of estrone in the ovaries compared with the testes, although ovarian levels were significantly higher (Xu and Barker 1990b). Thus, Xu and Barker (1990b) concluded that P4 was involved in nutrient translocation from the pyloric ceca to the gonads.

Microscopic examination of the ovaries of *A. rubens* allowed Schoenmakers et al. (1977) to view ultrastructural features that were characteristic of steroid-producing cells during the annual cycles when 3β-hydroxysteroid dehydrogenase (3β-HSD) activity was present. These data provided the first histological evidence for the presence of steroid-producing cells in an asteroid.

De Novo Synthesis of C21, C19, and C18 Steroids

The origin and presumed functional significance of progestogens, androgens, and estrogens in echinoderms has often been questioned. Historically, there was substantial doubt that vertebrate steroids were synthesized in invertebrates (hence the classical term *vertebrate steroids*), and anecdotal discussions arising at scientific conferences suggested that their presence

in asteroid (and other invertebrate) tissues originated from the diet. Alternatively, others have suggested that steroid-like compounds are produced in asteroids, having similar function, but differing in structure. Definitive identification required further evidence. Initially, Gaffney and Goad (1974) indicated that progesterone could not be metabolized to 17α-hydroxyprogesterone or testosterone in *A. rubens* or *M. glacialis*, suggesting vertebrate metabolic pathways were not present. Later, Schoenmakers et al. (1976) and Schoenmakers (1977) showed that progesterone could be synthesized from pregnenolone, demonstrating the presence of 3β-HSD, which is essential for steroid biosynthesis. Activity levels of 3β-HSD were higher in the testis than in the ovary (Schoenmakers et al. 1976). Further evaluation of steroid precursors including cholesterol, progesterone, and androstenedione indicated that *A. rubens* could produce all classes of steroid metabolites, with the exception of estrogens (Schoenmakers et al. 1976, Schoenmakers 1979a,b, Schoenmakers and Voogt 1980, 1981), using the same pathways found in vertebrates. Voogt et al. (1986) evaluated differences among studies of steroid metabolism and found that progesterone metabolism was highly dependent on experimental conditions and season. Furthermore, Voogt et al. (1990) showed that Δ5 steroids could be converted to Δ4 steroids in pathways in ways similar to in higher vertebrates. Collectively, these studies showed that 3β-HSD, 17β-HSD, 17α-hydroxylase, 20α-HSD, 20β-HSD, 21-hydroxylase, C17-C20 lyase, 5α-reductase, and Δ5 ketoisomerase were active in asteroid tissues. Similar conclusions were reached by Colombo and Belvedere (1976) following incubations of tissues with labeled steroid precursors in *A. irregularis* (as *A. irregularis pentacanthus*). A comparison of these synthetic pathways was reviewed in Voogt and Schoenmakers (1980). Additional evidence of steroid synthesizing capacity was reported in body tissue extracts of *A. rubens* (as *A. vulgaris*), in which androstenedione was converted to testosterone, 5α-androstane-3β,17β-diol and 5β-androstane-3β, 17β-diol, as well as 5α-androstanedione and epiandrosterone (Hines et al. 1992b). It was not until the use of gas chromatography and mass spectroscopy analysis that the definitive identification of 17β-estradiol was confirmed in the pyloric ceca of *A. rubens* (Voogt et al. 1992).

Voogt and van Rheenen (1986) discovered the production of a novel fatty acyl testosterone in tissues incubations of radiolabeled androstenedione. They postulated that conjugation of testosterone to a fatty acid might represent a regulatory activity that controls the activity of biologically active steroids.

Schoenmakers (1979b) further evaluated changes in the biosynthetic capacity of steroid synthesizing enzymes in relation to the state of gonadal maturation. He indicated that low levels of 3β-HSD were found after spawning but increased dramatically during the early growth stages of the ovary during the annual cycle. Similar temporal changes in steroid synthesizing capacity were observed in the pyloric ceca during the annual cycle. In addition, Schoenmakers et al. (1977) indicated that cells with ultrastrucutural characteristics of steroid synthesis were found in the germinal epithelium near invaginations of the hemal system into the ovarian lumen, but these cells were only observed at the period of high synthetic activity. Later, Voogt et al. (1991) found that conversion rates of the steroid precursors dehydroepiandrosterone, progesterone, and androstenedione varied during the annual reproductive cycle in both the gonad and pyloric ceca of *A. rubens*. The combined patterns of the conversion rates for the three steroids during the reproductive cycle suggested a strict sequence in each organ, which was both sex- and organ-specific. Steroid metabolism was maximal at the onset and end of gonadal growth. Collectively, these authors supported the supposition that steroids have a function in the regulation of asteroid reproduction.

Regulation of steroidogenesis is presumed to be cued either by endogenous (intrinsic) or exogenous (environmental) factors (see Chapter 4). Citing the effects of changing photoperiods on the gametic stage of asteroids (Pearse et al. 1986), Voogt et al. (1991) evaluated the effects of photoperiod change on progestogen and androgen synthesis. The production of steroid metabolites appeared to be controlled somewhat by changes in the photoperiod (decreasing photoperiod); however, confounding factors and variability among test specimens did not allow definitive conclusions.

Exogenous Application of Steroids

Both the presence and synthesis of steroids, particularly as they relate to the annual reproductive cycle, provide strong corroborative evidence of the role of steroids in regulating reproductive processes in asteroids. Induction of specific physiological effects by the application (injection or ingestion) of steroids provides further evidence of their role.

One of the first reports of the direct effects of steroids on asteroid physiology was published by H. Takahashi and Kanatani (1981). Oocyte diameters increased significantly when ovarian fragments obtained from *A. pectinifera* were incubated in the presence of 17β-estradiol for 3 days in the culture medium. Simi-

larly, Schoenmakers et al. (1981) found that the diameters of oocytes from *A. rubens* increased with injections of 17β-estradiol for 16 days; no changes in oocyte diameters were detected in the sham-injected animals. However, they concluded that oocytes had to be of some minimal size before estradiol had an effect on oocyte growth. In addition, ovarian weight increased rapidly and there was accelerated oogenesis and active vitellogenesis (based on increased PAS-positivity and increased basophility in treated oocytes). Finally, estradiol injections resulted in a tenfold increase in estrone levels in the ovary; however, levels in the pyloric ceca remained unchanged. N. Takahashi (1982b) injected *A. pectinifera* for 16 days with estrone and androstenedione and reported significant increases in ovarian protein content. In contrast, injection of progesterone appeared to decrease growth of the ovary.

Van der Plas et al. (1982) reported RNA levels increased in explants of pyloric cecal tissues of *A. rubens* exposed to estradiol. However, similar increases in RNA levels were not observed in the pyloric ceca when individuals were injected with estradiol in vivo. Lipid levels increased in the pyloric ceca with injections of estradiol in vivo, but only in November and December. Some of the results of this study were in contrast to those published earlier (Schoenmakers et al. 1981) and the authors suggested that the timing (seasonality) of administration of the steroid affects the response because the tissues may be insensitive to steroid administration during some phases of the annual cycle.

Watts and Lawrence (1987) injected *Luidia clathrata* with estrogens for 16 days. Specific activities of the metabolic enzymes glycerol phosphate dehydrogenase (GPDH) and 6-phosphogluconate dehydrogenase increased in the pyloric ceca of individuals fed a high ration and injected with estradiol, but only GPDH increased in those fed a low ration. These results indicated that estrogens have a role in regulating cellular metabolism in the pyloric ceca, and that nutritional condition of the individual influences the response of the pyloric ceca to steroids.

Xu and Barker (1993) injected estradiol or estrone into the coelomic cavity of *S. mollis* for 16 days. Estradiol injections increased estrone levels in the ovaries and pyloric ceca. Estrogen injections also resulted in increased incorporation of protein into the oocytes. Xu and Barker (1993) concluded that the gonads are the primary target organs of estrogens.

At the cellular level, steroids will most likely exert their effects on the genes regulating processes related to protein synthesis or cell proliferation. Since previous studies have suggested a direct effect of estrogens on protein synthesis and gamete production, A. G. Marsh and Walker (1995) evaluated mitotic activity in the testis of *A. rubens* (as *A. vulgaris*) in vitro in response to estradiol or progesterone exposure. Mitotic activity was estimated by radioactive thymidine incorporation and c-myc RNA expression in testicular explants. Estradiol exposure did not stimulate mitotic activation, nor did progesterone exposure. In fact, progesterone was found to be inhibitory to mitotic activity. However, significant increases in mitotic activity were found when testis explants were exposed to progesterone prior to estradiol exposure. The overall mitotic response increased with the length of time of the progesterone pre-treatment. These data suggest that a metabolic cascade, possibly involving multiple steroids or other activators, is required for direct (or indirect) activation of reproductive activities, including nutrient translocation and gamete development.

Possible Functions of Steroids

The possible functions of steroids in asteroids are presumed to be similar to those found in vertebrates. Despite corroborative evidence suggesting sex steroids have a role in regulating reproductive processes in asteroids (see Chapter 4), evidence of their role or roles within a regulatory cascade is implied but has not been confirmed. The reported activities of steroid synthesizing enzymes are important in suggesting the role of steroids, but we lack knowledge of definitive intrinsic or extrinsic (environmental) cues. We do not understand the inductive genomic pathways associated with steroid biosynthesis or any downstream control mechanisms. Ultimately, the function of steroids must be mediated through the activity of steroid-binding proteins or receptors. DeWaal et al. (1982) characterized a 17β-estradiol receptor binding protein in the pyloric ceca of *A. rubens* and suggested that it is comparable with that found in higher vertebrates. Proposed interactions among the various organs (pyloric ceca and gonads) suggest intimate communication, most likely chemically mediated. However, other data support the premise of local (organ-specific) steroid production and activity. These data support the hypothesis that steroids are involved in regulating reproduction and potentially nutrient translocation, but much more information is needed before regulation of these and other physiological processes can be confirmed in asteroids.

Other Steroids

The typical vertebrate pathways include the conversion of cholesterol to other sterols, saponins, sex steroids, gluco- and mineralocorticoids, and vitamin D. Conversion of precursors to a variety of sterols, saponins, and sex steroids has been confirmed in many asteroids. We were unable to find any studies on the presence, synthesis, or function of gluco- and mineralocorticoids or vitamin D. It is possible that these compounds are not synthesized or active in asteroids and that they may not be required for normal physiological function. However, the absence of information cannot confirm or refute their potential role in asteroids, and additional research should be done. We suggest that a molecular approach, evaluating those genes potentially involved in their synthesis and activity, is needed.

Conclusions

Lipid metabolism is not well characterized in most invertebrate phyla. Consequently, we believe it is fortunate that a large number of reports concerning sterol and steroid structure, metabolism, and function are available for a number of asteroids, and that a wide variety of sterols has been identified in a number of asteroids. Whether these sterols are intermediates in the synthesis of other compounds, secondary metabolites, have as yet undetermined pharmacological effects on asteroids, or are antipredatory compounds is not firmly established (see Chapter 8). Sterols such as cholesterol or 7-dehydrocholesterol are well known to be involved in membrane function in many species, and similar functions would be expected in asteroids.

The role of sex steroids in regulating physiological processes is of considerable interest. If we accept the basic premise that sex steroids have normal physiological roles in the regulation of gamete synthesis and maturation and in nutrient translocation in asteroids, then we can suggest that steroid regulation of reproduction evolved millions of years before the Chordata. There is supporting evidence that similar steroid pathways function in echinoids (Wasson and Watts 2007), suggesting other echinoderm classes use sex steroids in the regulation of reproduction.

Definitive evidence of the role of steroids should be forthcoming. A host of new technologies have emerged in the previous decade, most of which involve genomic and transcriptomic approaches. Sequence information is becoming available and could provide a platform for renewed efforts in understanding not only the expression of genes responsible for steroid production but also those genes involved in signal transduction, including receptors and steroid response elements. Of particular interest will be the role of estrogens. Although 17β-estradiol is the primary estrogen associated with estrogenic effects in higher vertebrates, estrone appears to be an effector of similar processes in asteroids. There will be many opportunities for advancing knowledge of asteroid steroid functions and makeup, and we are fortunate that an appropriate framework has been developed by many interested and capable investigators over the past century.

Dedication

This chapter is dedicated to the memory of Professor Peter Voogt, University of Utrecht, for his pioneering contributions to the study of the role of steroids in asteroids.

PART II • INTEGRATIVE BIOLOGY

10

Astropecten

Carlos Renato R. Ventura

W hat features must the *Astropecten* species have to be considered ecologically and evolutionarily important? At first, some of these features could make somewhat contradictory statements. In general, *Astropecten* species show high larval dispersal and the adults are morphologically very much alike, in spite of the fact that the genus is extremely diverse. *Astropecten* are widespread and play an analogous ecological role in different latitudes. Like other paxillosids, they have a particular development and morphology among asteroids, and there are several controversial interpretations of their evolutionary and phylogenetic status.

Astropecten GRAY, 1840, is a cosmopolitan genus of the order Paxillosida and of the family Astropectinidae, comprising some 150 extant species, all with highly similar morphology. Such morphological similarity has resulted in a complex and ill-defined taxonomy (Zulliger and Lessios 2010). Although many invalid species have been described (A. M. Clark and Downey 1992; World Asteroidea Database 2012) and misidentifications have been made (Chao 1999), the genus is considered as one of the most diverse among sea stars (Clark and Downey 1992, World Asteroidea Database 2012, Zulliger and Lessios 2010). This great diversity seems unexpected for a genus with mostly broadcast-spawning species with potentially large-scale dispersal of planktotrophic larvae (Zulliger and Lessios 2010).

Astropecten species live on soft bottoms (especially sandy or muddy substrata) from shallow to deeper (1500 m) water (A. M. Clark and Downey 1992, Rowe and Gates 1995, World Asteroidea Database 2012). Their morphology is very functional and adaptive to these environments: flattened body, small to moderate discs, and five arms with conspicuous block-like marginal plates bearing fringes of prominent erect (or appressed) spines around the ambitus (Fig. 10.1a,b,c,d,e,g). Their abactinal spines are small, crowded, and shaped like a column supporting an apical cluster of spinelets or granules (i.e., paxilliform spines; Clark and Downey 1992; Fig. 10.1g). *Astropecten* species have knob-ending tube feet arranged in two lines along ambulacral furrows (Fig. 10.1b,e,f). Unlike in the majority of sea stars, each tube foot has two bulbs (ampullae; Heddle 1967, Nichols 1969, Lawrence 1987; Fig. 10.1d). The ampullae provide the necessary power to displace sediment grains, to give support to movement and burrowing, and to bring infaunal food to the mouth. *Astropecten* species are able to regenerate their arms (Fig. 10.1h; see also Chapter 2) but cannot reproduce asexually.

In general, in contrast to other sea stars, *Astropecten* species are not able to evert their stomachs. Consequently, only intraoral digestion occurs, and the size

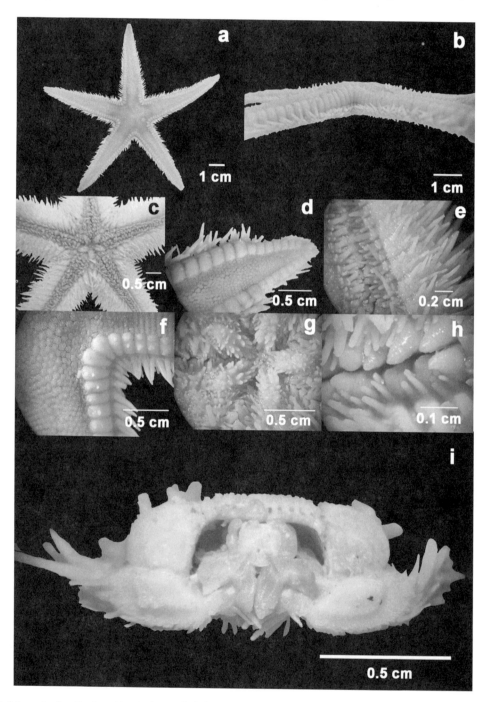

Fig. 10.1. *Astropecten brasiliensis. a,* Abactinal view of whole specimen; *b,* Lateral view; *c,* Actinal view; *d,* Regenerating arm tip; *e,* Knob-ending tube feet, ambulacral furrow, ambulacral plates; *f,* Detailed view of abactinal superomarginal plates and spines, paxillosid spines; *g,* Mouth plates; *h,* Knob-ending tube feet; *i,* transversal section of arm.

of all ingested food is limited by the coelomic space. Their digestive system is incomplete (without anus), therefore, indigestible food remains have to be eliminated through the mouth (Jangoux 1982a, Lawrence 1987).

Functional Morphology

In contrast to other paxillosid genera, Astropecten species have smaller discs, fewer actinal plates in a single short series (Fig. 10.1a,b,e), and are able to live in more turbulent habitats (Blake 1990a, A. M. Clark and Downey 1992). In addition, they show just subtle morphological differences regarding the width of arms, the position and covering of superomarginal plates, the shape of adambulacral furrow spines, the shape and size of subambulacral spines, and the shape of inferomarginal ossicles (Blake 1973, Schmid 1981, Schmid and Schaerer 1981, A. M. Clark and Downey 1992). These subtle differences in morphology (called homeomorphy by Blake 1989) are not enough to produce functional differences. Therefore, the morphological similarities among Astropecten species suggest that they are strongly adapted to the environment in mostly the same way (Blake 1989).

Body shape can reflect internal and external needs but it can also indicate constraints (Blake 1989). In the case of Astropecten species, the flattened body (Fig. 10.1c) and the ambitus rounded by block-like marginal plates (Fig. 10.1a,c,d,h) reduce, respectively, the internal space and, to some degree, the flexibility of their arms. The reduced coelomic space limits gonad enhancement and food intake, both considered as important features for fitness. In fact, the gonad index in Astropecten species reaches about 2.5% during the mature stage (Ventura et al. 1997, Ventura et al. 1998), which corresponds to about half (or even a quarter) of the gonad index for other asteroids (Mauzey 1966, McClintock et al. 1995, Carvalho and Ventura 2002, Georgiades et al. 2006, Barker and Scheibling 2008). However, a flattened body is more successful than a voluminous one for an infaunal predator living in soft-bottom environments, since the flattened body functions as a snowshoe on more unstable soft substrata (Thayer 1975, Blake 1989). Moreover, the relatively low flexibility of their arms does not limit predation; they swallow their prey completely (intraoral digestion) compared with other sea stars that have extraoral digestion (Jangoux, 1982a). In addition, Astropecten species have relatively small discs and unfused abactinal ossicles that give their bodies some flexibility (Blake 1989). It is not rare to find specimens of Astro-

pecten with discs deformed by the ingestion of large prey or of many small prey, usually mollusks (A. M. Clark 1977, Lawrence 1987, Blake 1990). Changes in body flexibility may be used both during predation (Christensen 1957) and defense against predation. For example, body stiffness has been observed in Astropecten bispinosus when attacked by Astropecten aranciacus (Schmid and Schaerer 1981, Blake 1989). Marginal spines are common and can protect against predators as well (generally larger paxillosids, such as Tethyaster species). Usually, the presence of pedicellariae on asteroids is also effective against predation (Blake 1989). Although pedicellariae are unusual in Astropecten species (occurring in 27% of Atlantic species according to A. M. Clark and Downey 1992), it is remarkable that they appear on the less armored species, such as Astropecten alligator, Astropecten americanus, Astropecten cingulatus, and Astropecten nitidus (A. M. Clark and Downey 1992). However, the effective role of pedicellariae against predation has not been demonstrated for Astropecten species.

Like other paxillosids, Astropecten species are able to move and bury themselves in the soft bottom. Their knob-ending tube feet (Fig. 10.1e,f) are highly adaptive to the soft-bottom environment. These feet have a convex apex suitable for burying in the sediments and digging into sandy or muddy substrata, as well as moving on them. However, Astropecten species cannot move laterally below the sediment surface and can burrow only a few centimeters into the bottom (Christensen 1970). Although some differences in inferomarginal ossicles have been highlighted for some Astropecten species (Blake 1973, Schmid and Schaerer 1981), locomotion and burrowing seem to follow the same pattern in all species of the genus studied (Heddle 1967).

The burrowing movement starts with marginal plates bending upward, which opens the ambulacral furrows. After that, the tube feet are thrust through the substratum and dig vertically into the sediment. Next, they bent laterally to remove grains from under the body. These movements are synchronized along the furrows, and the sea star eventually sinks into the sediment (Heddle 1967).

In Astropecten species, the convex shape of tips of the tube feet seems to be the most adaptive feature to soft-bottom environments, since the adhesive properties and ultrastructure characteristics of the three types of tube feet found in asteroids (i.e., knob-ending, simple disc-ending, and reinforced disc-ending tube feet) show no significant differences among them (Santos et al. 2005b). Because of the adhesive properties of their tube feet, the *Astropecten* species are able to climb glass tank walls for a short time (Santos et al. 2005b). Despite the adhesive properties of their knob-ending tube feet, this shape does not allow *Astropecten* species to perform well on hard-bottom environments. Consequently, they are not found in such environments.

Locomotion of *Astropecten* species on soft-bottom environments is quite efficient. They are among the asteroids with the highest rate of movement (Feder and Christensen 1966). The locomotion rate seems to be related to sediment type, seawater temperature, and body size (Pabst and Vicentini, 1978). It may vary among species depending on these parameters: *A. aranciacus* (ranging from 8 to 13 cm radius) move on average 9.3 m day^{-1} on coarse sand at 24°C (Burla et al. 1972), whereas *Astropecten jonstoni* (ranging from 4 to 6 cm radius) move on average 4.5 m day^{-1} on fine sand at 12°C to 16°C (Pabst and Vicentini 1978). The movement of *A. jonstoni* and *A. aranciacus* on soft-bottom substrata is highly directional. *Astropecten jonstoni* (5 cm radius) can move a maximum of 28 m per day (Pabst and Vicentini 1978), and larger specimens of *A. aranciacus* (14 cm radius) can move 22.9 m in one night (their most active period; Burla et al. 1972).

Paxilliform ossicles are also well-adapted to soft-bottom environments. They are columnar, have a broad round apex covered by spinelets, and are arranged in close rows on the abactinal region, forming a somewhat continuous and uniform surface. Consequently, when specimens burrow, grains of sediment are retained on their ossicle apex and seawater can circulate among the columns, providing cleansing and respiration (Blake 1989).

Considering the high homeomorphy reported among *Astropecten* species, taking into account body shape, spines, and tube feet, it is likely that locomotion and burrowing behaviors are similar for all species in the genus.

Evolution and Phylogeny

The evolution and phylogeny of the Asteroidea are still controversial and unclear, even at the level of order, family, or genus (Blake 1987, Gale 1987, Knott and Wray 2000, Blake 2010, Janies et al. 2011; see Chapter 1). Problems result from different sources of evolutionary information. Fossil data are scanty, due to the disarticulation of the skeleton after death (Fell 1963, Blake 1973, Blake 2010). Morphological and molecular analyses frequently provide conflicting hypotheses concerning either separate or combined data

(Lafay et al. 1995, Knott and Wray 2000, Vickery and McClintock 2000a). The development of embryos and larvae sheds some light on evolutionary trends but also shows convergence. For these reasons, more data are necessary (McEdward and Janies 1993, Hart et al. 2004, Janies 2004).

For many years, paxillosids have been a focus of evolutionary debate, which yielded different interpretations of the same features (MacBride 1921, MacBride 1923a,b, Mortensen 1922, 1923, Blake 1987, Gale 1987, Knott and Wray 2000, Matsubara et al. 2005, Blake 2010). As in other asteroid groups, there are some puzzling questions concerning the evolutionary history of the genus *Astropecten* (Zulliger 2009, Zulliger and Lessios 2010).

Döderlein (1917) was the first to propose a hypothesis on the evolutionary history of *Astropecten* species based on morphological features. After more than 90 years, Zulliger (2009) and Zulliger and Lessios (2010) revisited this subject, making a phylogeny of the genus inferred from molecular data (mitochondrial DNA sequences of three regions, 12S rRNA, 16S rRNA, and cytochrome oxidase c subregion I, or COI).

Two main evolutionary hypotheses might be raised regarding the early radiation and the current high diversity within the genus: (1) *Astropecten* is an ancient group that had undergone speciation gradually for a long time and, in contrast, (2) *Astropecten* has evolved quickly, generating new species only recently (Zulliger 2009). In order to test these hypotheses, it is necessary first to recognize *Astropecten* lineages and to estimate their divergence times based on fossil records, geological events, and the molecular clock.

According to Zulliger (2009), the oldest fossil record of *Astropecten* recognized at the species level (*Astropecten granulatus*) dates back to the Late Eocene (ca. 40 million years ago [mya], Bartonian, UK). There are some doubts as to the identification at the species level (Zulliger 2009) of older fossil records dated from ca. 65 mya (the Late Maastrichtian and the Early Paleocene; Jagt 2000) and from ca. 62 mya (the Upper Danian; Rasmussen 1972).

The phylogeny proposed by Döderlein (1917), based on morphology, distinguishes three major groups of species related to different geographic regions: Indo-Pacific, Pacific, and Atlantic-Mediterranean. No common species inhabits at least two of these large regions, suggesting phylogenetic gaps between them. Moreover, Döderlein (1917) argued that some extant species are closely related and represent a connection between these regions, for instance: *Astropecten marginatus* (from the Atlantic), *Astropecten regalis* (from the

Pacific coast of America), and *Astropecten latespinosus* (from Japan).

The molecular phylogeny proposed by Zulliger and Lessios (2010) also recognized three distinct groups of species (clades) distributed in a somewhat corresponding way to the three large geographic regions highlighted by Döderlein (1917), Indo-Pacific, Neotropical (east and west coast of America), and East Atlantic–Mediterranean. However, molecular phylogenies do not corroborate the close relation of *A. marginatus*, *A. regalis*, and *A. latespinosus*. Therefore, molecular phylogenetic analyses show that such similarity in morphology should be interpreted as convergences, because it has evolved independently. In addition, the relationships within each major clade are quite different from those proposed by Döderlein (1917). The molecular phylogenies carried out by Zulliger and Lessios (2010) demonstrated that several species from the East Atlantic and Mediterranean are not monophyletic (e.g., *Astropecten irregularis*) and that *A. aranciacus* is distinct from all other species, including those having a similar geographic distribution (East Atlantic and Mediterranean).

Zulliger (2009) used two main geological events (the closure of the Tethys Sea and the rise of the Isthmus of Panama) to calibrate the divergence times of extant *Astropecten* lineages found in the phylogenetic analysis. Molecular phylogeny data show that the evolutionary history of *Astropecten* starts from a single common Indo-Pacific ancestor lineage and that the first divergence occurred in the Mid Miocene (ca.13.5 mya). Zulliger (2009) suggested that divergence between Indo-Pacific and Atlantic-Mediterranean lineages is probably related to the closure of the Tethys Sea. However, it means that at least an older lineage of *Astropecten* became extinct since the first fossil recognized as an *Astropecten* species dates from the Late Eocene (ca. 40 mya). Zulliger (2009) argues that it is unlikely that an old lineage species is missing in the sampling, because specimens of all world seas and some of the deep sea were included in the phylogenetic analysis. This seems to be an open question, similar to the other concerning the evolutionary history of Asteroidea.

Another aspect that should not be disregarded is that, because of *Astropecten*'s unfused and meshy skeleton, well-preserved fossils are rare. Identification at the species level is usually difficult when an entire specimen is not available (Blake 1973). That is the reason much knowledge of the genus comes from ossicles analysis (Blake 1973).

Divergence time estimates between the Atlantic and Pacific lineages of American species show it occurred before the rise of the Isthmus of Panama (ca. 5.3 mya), as was the case for other marine species (Collins et al. 1996, Marko 2002, Zulliger 2009). Ancient divergent currents and salinity differences between the Atlantic and the Pacific are likely to have been effective in disrupting gene flows through America seaways (Zulliger 2009).

The present idea about the evolutionary history of the genus *Astropecten* corroborates the hypothesis that this genus has undergone rapid speciation processes only recently. Both the molecular phylogeny analysis (Zulliger and Lessios 2010) and the earlier morphological approach (Döderlein 1917) clearly demonstrate that related species are distributed within the same large geographic regions. This supports the relation between divergence times and major geological events through the evolutionary pathways of the *Astropecten* lineages. However, some puzzling questions remain as a result of limited sources of evolutionary information as mentioned above.

Reproduction and Development

Species of *Astropecten* reproduce sexually only, are gonochoric with equal proportion of males and female in natural populations. Their reproductive system has the same general features as most asteroids. It is composed of gonads, gonoducts, and gonopores. Gonads are sac-like organs, that is, gonadal walls are typically formed by outer and inner sacs as in other echinoderms (Walker 1982). A genital coelomic sinus is present between the outer and inner sacs. The outer sac is involved in maintaining general activities, such as the circulation of perivisceral fluid on external gonadal walls (provided by their flagellated cells), the shape of gonads, and the contraction of circular muscles during gamete release. The inner sac is more directly involved in gamete production and contraction of gonad longitudinal muscles during spawning. It is formed by an expanding genital sinus of the hemal system that has a germinative epithelium covering its inner wall (Walker 1982).

Moreover, the reproductive system of *Astropecten* species may show two different arrangements: (1) as serial sacs along each arm of the coelomic space, connected to a common gonoduct that opens abactinally in a single gonopore (e.g., *Astropecten irregularis*) or (2) as a single tuft of tubules near each interradial region and linked to one gonoduct that also opens in an abactinal gonopore (e.g., *Astropecten articulatus*, *Astropecten brasiliensis*, and *A. cingulatus*; Walker 1982, Chia and Walker 1991, Marion et al. 1998, Mercier and Hamel 2009).

During the gametogenic cycle of *Astropecten* species, gonads increase in size with the production of gametes, as typically occurs in sea stars. This variation in size is related to fluctuations in steroids levels (estrogens and progesterones) in the asteroid gonads, as in *A. irregularis pentacanthus* (Colombo and Belvedere 1976). These two hormones are likely to carry out several functions in reproduction, as well as in other vital activities (nutrition, for instance) during the annual reproductive cycle of sea stars and other echinoderms (see Chapter 9). However, it is still unclear whether their synthesis is driven by environmental cues such as daylight periodicity and seawater temperature (Mercier and Hamel 2009; see also Chapter 4).

Seasonal reproductive cycles have been described to temperate species of *Astropecten* in Japan (*A. latespinosus* and *Astropecten scoparius*, Nojima 1983), the Mediterranean (*A. irregularis*, Grant and Tyler 1986, Freeman et al. 2001), and the Northern Gulf of Mexico (*A. articulatus*, Marion et al. 1998). For all these temperate species, spawning time takes place during the summer, when seawater temperature is higher. In a tropical upwelling region of the Brazilian coast (Cabo Frio), *A. brasiliensis* shows a marked seasonal reproduction when compared with *A. cingulatus* in the same area (Ventura et al. 1997, 1998). In this case, these two sympatric species seem to react differently to environmental factors: *A. brasiliensis* spawns synchronously during the upwelling period (springtime, November, when seawater temperature can drop from 22°C to 14°C), whereas *A. cingulatus* spawns from May to August (fall to winter), when coastal tropical water is present on the continental shelf and, consequently, the temperature is often higher than 18°C. An explanation for these unexpected differences in the reproductive cycle of these congeneric species may be related to the range of their geographic distribution: the Cabo Frio region is the southern boundary of *A. cingulatus* and the northern boundary of *A. brasiliensis* (A. M. Clark and Downey 1992). In fact, *A. cingulatus* shows some tropical features in its reproductive rhythm, with longer spawning periods. However, *A. brasiliensis* has a quite synchronous reproductive cycle that is typical of temperate species (Ventura et al. 1998). Spawning season is important, because it implies an effective chance of larval survival and successful recruitment.

Astropecten species are broadcast spawners and most have planktotrophic bipinnaria larvae. However, at least two species develop non-feeding barrel-shaped larvae (*A. latespinosus* and *Astropecten gisselbrechti*; Komatsu 1975, Oguro et al. 1976, Komatsu and Nojima 1985). Relatively few studies describe the embryonic and larval development of *Astropecten* species considering the high species richness in this genus. Therefore, generalizations about the most common mode of development of *Astropecten* species must wait further investigation. Most information on developmental modes now available is inferred from egg diameter (approximately 100 to 150 μm). It is well known that the amount of maternal energy available in such small eggs does not allow larvae to achieve the metamorphosis phase without external sources of food (McEdward and Janies 1997).

Considering the available information on *Astropecten* species developmental modes, Wakabayashi et al. (2008) carried out a set of hybridization assays between species with feeding (bipinnaria) and non-feeding (barrel-shaped) larvae. Their results have some interesting morphological and evolutionary implications. The authors performed cross-fertilization between gametes of *A. scoparius* (which develops feeding larvae) and *A. latespinosus* (which develops barrel-shaped larvae). Moreover, they analyzed offspring genetically (sequence data of partial 18S rDNA) to determine whether genuine hybrids were formed. Their results demonstrate that heterospecific crosses yield hybridization and parthenogenesis. Only genuine hybrids were produced by crosses between eggs of *A. scoparius* and sperm of *A. latespinosus*. Reciprocal crosses did produce both hybrids and parthenogenetic embryos. All hybrids show bipinnaria morphological features, but the authors detected differences in gut formation. They argue that sperm of *A. scoparius* (the bipinnaria developer) may induce parthenogenesis in eggs of *A. latespinosus*. In addition, Wakabayashi et al. (2008) corroborated the hypothesis that barrel-shaped larvae derive from the bipinnaria (Oguro 1989, McEdward and Miner 2001) and also that the presence of bipinnaria features in reciprocal hybrids may be interpreted as a restoration of ancestral condition.

However, despite the fact that hybridization assays shed some light on the evolutionary pathways of *Astropecten* species, interpretation should be done carefully and generalization avoided. Experiments of reciprocal fertilization with at least four species (two bipinnaria developers and two barrel-shaped developers) might contribute to the understanding of ancestral condition and mechanism that induces parthenogenesis in *Astropecten* species.

Feeding and Ecological Interactions

All species of *Astropecten* are active predators on infaunal invertebrates. The genus is considered to be one of the most trophically specialized among Asteroidea (Jangoux 1982a), although variation in the food niche breadth occurs when congeneric species are sympatric (Ventura et al. 2001). They prey mainly on mollusks (bivalves and gastropods), small crustaceans (cumaceans), ophiuroids, and polychaetes (see Jangoux 1982a for a review). Although the number of prey consumed by *A. latespinosus* increased with body size, the number of prey consumed per body weight decreased (Doi 1976). In addition, some larger species (such as *A. aranciacus*) prey on congeneric sea stars (*A. bispinosus*) (Schmid 1981, Schmid and Schaerer 1981). Cannibalism has been reported for *A. articulatus* (Wells et al. 1961).

The digestive system of *Astropecten* species is well-adapted to prey on soft-bottom environments. It is composed of the mouth (with strong armored plates, Fig. 10.1g), the cardiac stomach, and five pairs of pyloric ceca that extend along the coelomic cavity of each arm. There is no rectum or anus (Jangoux 1982b).

Like other paxillosids, *Astropecten* species swallow their prey whole and digest it inside the stomach cavity (intraoral digestion). Therefore, all food remains are regurgitated after digestion. *Astropecten* species are not able to evert their stomachs to eat larger prey. Although *A. irregularis* can evert its stomach slightly, it shows no sign of effective extraoral digestion (Christensen 1970, Jangoux 1982a).

This feeding behavior of swallowing the whole prey has facilitated many reliable studies on the diet of some *Astropecten* species based on their stomach contents (see Jangoux 1982a for a review; see also Franz and Worley 1982, Nojima 1988, Nojima 1989, Ventura et al. 1994, Lemmens et al. 1995, Freeman et al. 1999, Ventura et al. 2001, Brogger and Penchaszadeh 2008, Caregnato et al. 2009). Since these studies were conducted with different species of *Astropecten* from several localities, dietary comparison may provide valuable information about their ecological function in benthic communities. The trophic classification of prey inferred by stable isotopes (carbon and nitrogen) is a useful approach for this purpose (Carlier et al. 2007, Jeffreys et al. 2009). Some genera of prey are remarkably frequent in the gut contents of different *Astropecten* species, for instance, the bivalves *Corbula*, *Donax*, *Mactra*, *Spisula*, *Pitar*, *Venus* (all active suspension feeders), *Nucula* (surface deposit-feeder), and *Tellina*

(interface feeder); the gastropods *Natica* and *Nassarius* (predators), *Turritella* (active suspension feeder), and *Cerithium* (surface deposit-feeder); and the scaphopod *Dentalium* (predator). Therefore, *Astropecten* species can play a wide range of trophic roles in soft-bottom benthic communities. They may either be unselective deposit feeders that exploit organic matter from sediment and have a mixed diet (Nojima 1988, Jeffreys et al. 2009) or selective predators on consumers of several trophic levels, such as bivalves and gastropods (Christensen 1970, Beddingfield and McClintock 1993, Freeman et al. 1999, Ventura et al. 2001, Ganmanee et al. 2003).

The selective feeding behavior of these sea stars seems to be related to prey densities and competition with other infaunal predators. As *Astropecten* species are able to migrate easily (Pabst and Vicentini 1978, Nojima 1983, Ventura and Fernandes 1995, Freeman et al. 2001), they can avoid strong competition with congeneric species (Lemmens et al. 1995, Ventura et al. 2001). Sea stars can shift their food niche breadth in sympatry, and individuals similar in size do not share the same region. These interactions have been reported for *A. aranciacus* and *A. bispinosus* from the Mediterranean (Schmid and Schaerer 1981), for *Astropecten zebra* and *Astropecten velitaris* from Australia (Lemmens et al. 1995), and for *A. brasiliensis* and *A. cingulatus* from Brazil (Ventura et al. 2001).

Growth and Longevity

In contrast to other asteroids, investigations on the growth rates of *Astropecten* species in natural environments are few. In some studies, growth rates have been estimated based on size-frequency distribution (arm length). However, a comparison of the few data available should be done carefully because of differences in the estimation of growth parameters. For instance, the growth constant (k) has been estimated either as a monthly variable (Nojima 1982, Freeman et al. 2001) or an annual variable (Christensen 1970, Ventura 1999).

In terms of bioenergetics, somatic growth and reproduction depend on the acquisition of nutrients (Lawrence and Lane 1982, Lawrence 1987). Morphological and behavioral features of organisms limit the food intake rate (intrinsic limitation). Moreover, nutrient acquisition may be limited below this level by environmental conditions (Lawrence 1987). In the case of *Astropecten* species, the intraoral digestive mechanism, the flattened body, and the ability to hunt are intrinsic limitations for nutrient acquisition, whereas seawater temperature and density of optimal prey, for instance, are extrinsic limitations.

Estimates for *Astropecten* species have shown that growth rate is highly seasonal. It decreases during the period of gonad maturation of *A. latespinosus* in Japan (Nojima 1982), *A. irregularis* in Europe (Freeman et al. 2001), and *A. brasiliensis* in Brazil (Ventura 1999). However, growth rate estimates for *A. cingulatus* show some seasonal reduction during the recovering stage of gametogenesis. For these two sympatric species from Brazil (*A. brasiliensis* and *A. cingulatus*), seawater temperature seems to be the environmental factor (extrinsic limitation) that causes a decrease in somatic growth, since it coincides with the upwelling period in both species.

Longevity can also be estimated from size-frequency distribution data (arm length). Therefore, comparison of the life span of different species is limited when different methods are applied. However, available data for longevity are, at least, informative. *A. latespinosus* and *A. brasiliensis* survive for approximately 4 years (Nojima 1982, Ventura 1999), *A. cingulatus* for 2.5 years, and *A. irregularis* for 3.5 years. Christensen (1970) stated *A. irregularis* has lived for up to 10 years in the laboratory.

Acknowledgments
I am thankful to John M. Lawrence for suggestions and criticism that improved the chapter. D. Zullinger, H. A. Lessios, and R. Santos kindly shared scientific information. A. D. Martins and M. R. Tavares edited the figures and gathered some published data. B. F. Fonseca, E. M. Lopes, and M. Contins provided general assistance. A. B. Carvalho provided some references. M. S. Martins helped with the English.

Plate 1. Odontaster validus. Two individuals at Palmer Station, Antarctica, at a depth of ca. 40 m feeding on the sponge *Mycale acerata*, one deep within the sponge. See Chapter 12. *Photo courtesy of B. Baker.*

Plate 2. Archaster angulatus. Two individuals off Perth, Australia, at a depth of 6 m in the pseudocopulatory position with the male above and female below. See Chapter 4. *Photo courtesy of J. K. Keesing.*

Plate 3. *Asterias amurensis.* Aggregation feeding on mussels (*Mytilus sp.*) removed from ships' hulls in the upper intertidal zone, Derwent River Estuary, Tasmania. See Chapter 17. *Photo courtesy of M. Byrne.*

Plate 4. *Coscinasterias murica.* An individual, which has undergone fission and is regenerating a portion of the disc and its arms, at 1 m depth in Otago Harbour, New Zealand. See Chapter 19. *Photo courtesy of M. Barker.*

Plate 5. *Dermasterias imbricata*. An individual exposed at low tide at Bodega Bay, California. *Photo courtesy of J. Sones.*

Plate 6. *Acanthaster planci*. Individuals feeding on coral at Keeper Reef, central Great Barrier Reef, Australia. White areas are feeding scars. See Chapter 13. *Photo courtesy of K. Fabricius.*

Plate 7. Henricia leviuscula. Individual at Sares Head, north of Rosario Beach, Washington. The exposed mouth with red tentacles are of buried holothuroids (*Psolus chitonoides*). *Photo courtesy of J. Nestler.*

Plate 8. Heliaster helianthus. Individual on a boulder pursuing sea urchins (*Tetrapygus niger*) at Cisnes Bay, northern Chile. Note the raised arms. See Chapter 15. *Photo courtesy of C. F. Gaymer.*

Plate 9. Leptasterias
polaris and Asterias forbesi.
Individuals with six arms
are *L. polaris* and individuals
with five arms are *A. forbesi*.
Aggregation feeding on a bed
of mussels (*Mytilus edulis*) at
a depth of 6 m, the Mingan
Islands, northern Gulf of St.
Lawrence, Canada. See
Chapter 18. *Photo courtesy of
C. F. Gaymer.*

Plate 10. (Right) Patiriella
regularis. Individuals collected
from Dunedin, Otago, New
Zealand. See Chapter 5. *Photo
courtesy of M. Byrne.*

Plate 11. *Patiria miniata*. Individual on surf grass (*Phyllospadix scouleri*) exposed at low tide at Bodega Bay, California. *Photo courtesy of J. Sones.*

Plate 12. *Pycnopodia helianthoides*. Individual on surf grass (*Phyllospadix scouleri*) exposed at low tide at Bodega Bay, California. *Photo courtesy of J. Sones.*

Plate 13. Pisaster ochraceus. Aggregation on a rocky shore at Barkley Sound, British Columbia, removing gooseneck barnacles (*Pollicipes polymerus*) and small mussels (*Mytilus californianus*) from the surface of a patch of larger mussels. See Chapter 16. *Photo courtesy of C. Robles.*

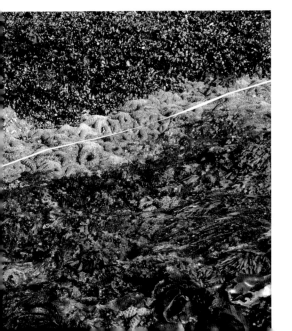

Plate 14. Pisaster ochraceus. Aggregation at Barkley Sound, British Columbia, at a site of a natural episode of massive recruitment of the mussel *Mytilus trossulus.* The mean length of the major radius (from the center of the disc to the tip of the center arm of the trivium) is ca. 16 cm. See Chapter 16. *Photo courtesy of C. Robles.*

Plate 15. *Stichaster australis.* Individuals exposed at low tide near Greymouth, west coast of South Island, New Zealand. *Photo courtesy of E. Sanford.*

Plate 16. *Oreaster reticulatus.* Individual below an undercut mangrove bank at a sand cay off southern Belize. See Chapter 14. *Photo courtesy of R. E. Scheibling.*

11

Luidia

John M. Lawrence

Forbes (1839) named the genus *Luidia* after Edward Lhuyd (1660–1709), an accomplished Welsh naturalist and keeper of the Ashmolean Museum. Sladen (1889) established the luidiids as a taxon in his report from the HMS *Challenger*. *Luidia* is the only genus in the Luidiidae with 49 currently accepted species (Mah and Hansson 2012a).

Like all paxillosids, the tips of the tube feet of luidiids narrow to a blunt point or small knob (Vickery and McClintock 2000a; see Chapter 3, Fig. 3.4a). A. M. Clark and Downey (1992) listed the characteristics of the family, including the flat, strap-like, gently tapering arms (Fig. 11.1a,c). The extraordinary *Luidia (Platasterias) latiradiata* instead has a strongly petaloid arm (Blake 1982; Fig. 11.1b,d). The superomarginal plates of *L. (P.) latiradiata* are slightly elongated transversely (Fig. 11.1f) and inferomarginal plates greatly so (Fig.11.1h). No observations have been made on living *L (P.). latirardiata*. Like *Astropecten,* (Fig. 10.1b,d,f), luidiids have conspicuous inferomarginal plates but in contrast the superomarginal plates are reduced so they are indistinguishable from the adjacent lateral paxillae (A. M. Clark and Rowe 1971; Fig. 11.1e,f). The functional consequences of this are not known.

The number of arms in *Luidia* ranges from 5 to 11 (A. M. Clark and Downey 1992). In the seven-rayed *Luidia ciliaris,* the two supernumerary rays develop in the two interradii on either side of the interradius with the madreporite (Hotchkiss 2000; see Chapter 2). In the nine-rayed *Luidia senegalensis,* four supernumerary rays appear at the same time in four interradii (Lawrence and Avery 2010).

Luidiids in the Atlantic range in size from a major radius of 72 (*Luidia patriae*) to 300 mm (*L. ciliaris*; A. M. Clark and Downey 1992). The largest in the Pacific are *Luidia savignyi* (major radius = 370 mm), *Luidia magnifica* (major radius = 380 mm), and *Luidia superba* (major radius = 415 mm; Downey and Wellington 1978).

Distribution

Luidiids are found in oceans from the tropics to high temperate latitudes (Sladen 1889). Some species span the entire range, while others

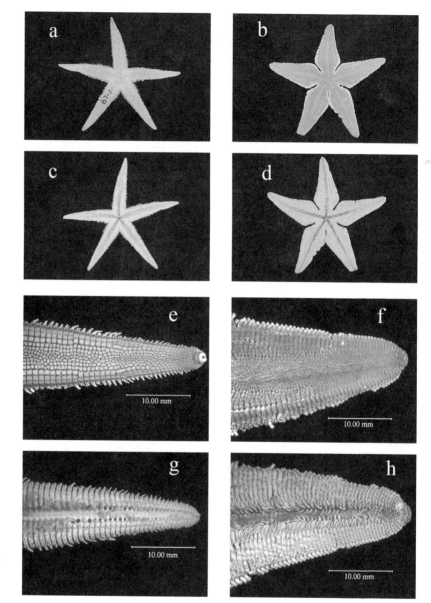

Fig. 11.1. *Left column, Luidia clathrata.* Specimen from the Fish and Wildlife Research Institute, Florida Fish and Wildlife Commission. Radius = 65 mm. *a,e,* aboral view; *c,g,* oral view. *Right column, Luidia latiradiata.* Specimen from the Colección Nacional de Equinodermos, Universidad Nacional Autónoma de México. Radius = 71 mm. *b,f,* aboral view; *d,h,* oral view.

are more restricted in distribution (A. M. Clark and Downey 1992). *Luidia alternata alternata* occurs from the east coast of North America (North Carolina) through the tropics to northern Argentina. Others are more restricted. *Luidia atlantidea* occurs only from Atlantic Morocco to the coastal part of the Republic of the Congo, including the Cape Verde Islands, and *L. patriae* only off northern Argentina.

Luidiids occur primarily in shallow water and on the continental shelf (A. M. Clark and Downey 1992). Of the nine species collected by the *Challenger,* all were found in the "littoral" zone, only three (*Luidia sarsi,*

Luidia elegans, and *Luidia barbadensis*) on the continental slope and none in the abyssal zone (Sladen 1889).

Habitat

Except for *Luidia clathrata,* which is found in estuaries at low salinities, luidiids occur in oceanic salinities. Sladen's (1889) comment that the definite character of the bottom inhabited by luidiids is recorded in very few instances still holds. If recorded, it is usually only in general terms as particulate substrates, generically described as sand, sandy mud, mud, clay, sand and shell

(Table 11.1). Klinger (1979) reported *L. clathrata* occurred in Charlotte Harbor, Florida, on sediment composed of very fine sand mixed with varying amounts of shell fragments and mud. The range of mean grain sizes was small, ranging from 2.86 to 3.57 phi. Their rate of burrowing in the laboratory in homogeneous sediment ranging from 0 to 2 phi (= coarse sand to very coarse sand) did not differ. Ventura and Fernandes

Table 11.1. Substrate of luidiids

Species	Substrate type	Reference
Luidia alternata alternata	Sandy mud, mud, shell and calcareous algae	A. M. Clark and Downey 1992
Luidia alternata numidica	Sandy mud, shell and calcareous algae, mud and rocks	Clark and Downey 1992
Luidia alternata	Mosaic of sand, rocks, boulders	Netto et al. 2005
Luidia aspera	Green mud and coral mud	Sladen 1889
Luidia atlantidea	Mud, sand and calcareous algae, muddy sand, broken shell, mud or sand with stones, fine sand	Clark and Downey 1992
Luidia barbadensis	Calcareous pebbles	Sladen 1889
Luidia barbadensis	Sand	Clark and Downey 1992
Luidia ciliaris	Gravel and sand, gravel	Holme 1984
Luidia ciliaris	Shell gravel, sand	Heddle 1967
Luidia ciliaris	Shell gravel, shell gravel and mud sand	Brun 1972
Luidia ciliaris	Mud, gravel	Thrush and Townsend 1986
Luidia ciliaris	Sand or gravel, often without mud	Clark and Downey 1992
Luidia clathrata	Shelly mud, medium-grain sand, fine sand	Hulings and Hemlay 1963
Luidia clathrata	sand	Schwartz and Porter 1977
Luidia clathrata	Fine sand mixed with shell fragments and mud	Klinger 1979
Luidia clathrata	Fine sand mixed with mud	Hopkins and Knott 2010
Luidia elegans	Calcareous pebbles	Sladen 1889
Luidia foliolata	Fine sand to mud	Mauzey et al. 1968
Luidia foliolata	Sand	Birkeland et al. 1982
Luidia foliolata	Sand	Sloan and Robinson 1983
Luidia forficifer (= *hardwicki*)	Green mud and coral sand	Sladen 1889
Luidia hardwicki	Sandy mud	A. M. Clark 1982
Luidia hardwicki	Soft bottoms	Chiu et al. 1990
Luidia heterozona heterozona	Mud, sandy mud	Clark and Downey 1992
Luidia lawrencei	Sand with shell hash	Hopkins and Knott 2010
Luidia longispina	Mud	Sladen 1889
Luidia longispina	Sandy mud	Clark 1982
Luidia longispina	Soft bottoms	Chiu et al. 1990
Luida ludwigi scotti	Moderately well-sorted sand Mean phi = 0.59 to 0.68	Ventura and Fernandes 1995
Luidia maculata	Rock	A. M. Clark 1982
Luidia maculata	Sand	Thompson 1982
Luidia magellanica	Rocky bottoms with *Lithophyllum,* shell or coarse sand	Viviani 1978
Luidia quinaria	Silt clay, sand, muddy sand, sandy gravel	Ganmanee et al. 2003
Luidia sagamina aciculata	Sand, mud and shells, rocks, hard bottom with calcareous algae, broken shell	Clark and Downey 1992
Luidia sarsi	Coarse sand, mud and clay	Ursin 1960
Luidia sarsi	Sand, coarse shell sand	Sladen 1889
Luidia sarsi	Fine sand, fine gravel	Allen 1899
Luidia sarsi	Sand and mud	Fenchel 1965
Luidia sarsi	Soft bottom	Holme 1984
Luidia sarsi	Fine sand	Guillou 1990
Luidia sarsi sarsi	Mud or clay	Clark and Downey 1992
Luidia sarsi africana	Mud, coarse sand and shell, rock and shell, sand	Clark and Downey 1992
Luidia sarsi elegans	Fine sand and mud	Clark and Downey 1992
Luidia savignyi	Sand	Santos et al. 2005b
Luidia senegalensis	Hard shell and quartz sand	Halpern 1970
Luidia senegalensis	Sandy mud, muddy sand with shells	Clark and Downey 1992
Luidia senegalensis	Sand, muddy sand	Monteiro and Pardo 1994
Luidia superba	Sand	Downey and Wellington 1978

(1995) found the mean grain sizes of sediment at Cabo Frio (Brazil) at two depths where *Luidia ludwigi scotti* were most abundant were smaller, 0.59 phi (45 m depth) and 0.68 phi (60 m depth). Even with a particulate substrate, luidiids are not found in shallow water with high hydrodynamics.

There are occasional reports of presence of luidiids on hard surfaces. *Luidia magellanica* is usually found on sand in protected waters but sometimes on rock covered with *Lithophyllum* (Viviani 1978). Thompson (1982) and A. M. Clark (1982) recorded *Luidia maculata* in Tolo Harbour, Hong Kong, on sand and on rock. Netto et al. (2005) reported *Luidia alternata* occurs on sand and mud but also on hard substrate in an area composed of a mosaic of sand, rocks, and boulders.

The morphological features of luidiids associated with the particulate nature of the substrate are their tube feet and paxillae (Blake 1989; see Chapter 3). Heddle (1967) suggested the rounded or pointed tip is adapted for pushing between the particles of the substrate and contribute to efficient movement on the surface and that the bilobed ampullae produce greater turgidity of the tube feet. He described the morphology of the ambulacral system of *Astropecten irregularis* and *L. ciliaris* in relation to their burrowing and locomotion on a particulate substrate. In locomotion, the animal walks with the tube feet used as stilts. Brun (1972) reported a rate of movement of large *L. ciliaris* of 3 m min^{-1}. Heddle (1967) described burrowing in *L. ciliaris*. In burrowing, the inferomarginal spines are adducted against the sides of the arms and the aboral surface of the arms are narrowed and thrown into longitudinal folds. The tube feet are protracted and make lateral movements outward, moving the sand away from beneath the animal. The animal sinks into the substrate. The irregular arrangement of the medial abactinal paxillae (Fig. 11.1e,f) may facilitate the longitudinal folding of the arm.

There are no systematic studies on the depth of burrowing but generally luidiids are described as shallow burrowers, just beneath the substrate surface. Fenchel (1965) reported that *L. sarsi* always keep the arm tips above the substratum, similar to what Christensen (1970) noted for *Astropecten irregularis*. During the day, *L. clathrata* in Old Tampa Bay are buried at a mean depth of 0.4 ± 0.3 cm SD (N = 14; Lawrence, unpublished observations).

The usual absence of luidiids from hard substrates is related to the knob-ending tube feet. Although the tube feet are adhesive (see Chapter 3), they do not provide sufficient area for strong attachment, as tube feet with discs do. Asteroids with tube feet with discs occur on particulate substrate (e.g., *Asterias rubens*; Guillou 1990) but typically do not bury themselves. Species of the Archasteridae have tube feet that end in discs (Sukarno and Jangoux 1977). They live on particulate substrate and bury just below the surface (Lawrence et al. 2010).

The plates of the aboral surface between the inferomarginal plates have paxillae, columns with small spinelets on the top that can be raised or lowered to form a covering. Hyman (1955) said they resembled a field of flowers. The number of spinelets varies among species (A. M. Clark and Downey 1992), but whether this has a functional consequence is not known. Water is presumed to move through the distinct passages (fascioles) between the marginal plates protected from sediment (Blake 1989).

Food

The food of the luidiids is variable, consisting of both epifauna (primarily other asteroids and gastropods) and infauna (holothuroids, ophiuroids, small bivalves, polychaetes; Table 11.2). O. D. Hunt (1925) reported the food of *L. ciliaris* was almost entirely echinoderms, primarily *A. rubens* and including *Marthasterias glacialis*. He noted that the mouth frame can be stretched to ingest large prey such as the spatangoid *Spatangus purpureus*.

Schwartz and Porter (1977) reported an interesting difference in food of coexisting species: *L. clathrata* fed on *Argopecten gibbus* but *L. alternata* did not. Because *L.sarsi* and *L. ciliaris* prey on *A. rubens* and *M. glacialis*, Guillou (1990) called them "super-predators." Mauzey et al. (1968) concluded that *Luidia foliolata* is generalized in diet and that prey is determined primarily by relative abundance of prey species in the particular habitat. Chiu et al. (1990) similarly described *Luidia longispina* and *Luidia hardwicki* as opportunistic feeders on a wide range of prey. They found mollusks were the primary food in one location, echinoderms in second, and substrate in a third. In the laboratory *L. clathrata* showed switching behavior, feeding on the most abundant food model (McClintock and Lawrence 1985a).

Lawrence et al. (2012) reported *L. clathrata* is distributed randomly on a sand bottom in Old Tampa Bay. They suggested this indicates homogeneity of the substrate and random distribution of the prey (small infauna). Individuals in aquaria in the laboratory showed neither attraction nor repulsion to each other but aggregated to food.

Table 11.2. Food of luidiids

Species	Primary prey	Reference
Luidia alternata	Asteroids (*Luidia clathrata, Astropecten*)	Schwartz and Porter 1977
Luidia barimae (= *heterozona barimae*)	Ophiuroids, gastropods (*Fasciolaria, Murex*), bivalves (*Amusium*)	Penchaszadeh and Lera 1983
Luidia ciliaris	Holothuroids (*Cucumaria, Thyone*), asteroids (*Asterias, Marthasterias*), spatangoids (*Echino-cardium, Spatangus*), echinoid (*Echinus*), ophiuroids, mollusks	Hunt 1925
Luidia ciliaris	Echinoid (*Psammechinus*), ophiuroids (*Ophiothrix, Ophiura, Ophiocomina*) and bivalves	Brun 1972
Luidia clathrata	Large quantities of sediment, foraminiferans, gastropods (*Odostomia*), bivalves, crustaceans	Hulings and Hemlay 1963
Luidia clathrata	Mainly cumaceans and small bivalves (*Tellina*), amphipods, ostracods, gastropods	Lawrence et al. 1974
Luidia clathrata	Scallops (*Argopecten*)	Schwartz and Porter 1977
Luidia clathrata	Ophiuroids, gastropods (*Olivella, Turbonilla, Strombiformis, Ancilla*), bivalves, polychates, asteroids (*Astropecten* spp.), echinoids, isopods, decapods, foraminferans, copepods	Pechaszadeh and Lera 1983
Luidia clathrata	Bivalves (*Mulinia*)	McClintock and Lawrence 1985b
Luidia clathrata	Polychaetes, crustaceans, ophiuroids	Tararam et al. 1993
Luidia foliolata	Mostly holothuroids (*Cucumaria, Molpadia, Leptosynapta*), bivalves (*Protothaca, Clinocardium, Neocardium, Macoma*), crabs (*Cancer*), sponges	Mauzey et al. 1968
Luidia foliolata	Bivalves, echinoids, ophiuroids, polychaetes, crustaceans, some sediment	Carey 1972
Luidia foliolata	Ophiuroids, holothuroids, bivalves	Birkeland et al. 1982
Luidia foliolata	Small bivalves, especially *Clinocardium*	Sloan and Robinson 1983
Luidia hardwicki	Bivalves (*Theora, Tellinides, Fulvia, Paphia*); echinoids (*Temonopleurus, Schizaster*); sediment	Chiu et al. 1990
Luidia lawrencei	Clypeasteroid (*Mellita*)	McClintock and Lawrence 1985a
Luidia ludwigi	Bivalves (*Pitar, Adrana*), ophiuroids (*Amphiura*), foraminiferans, amphipods	Brögger and Penchaszadeh 2008
Luidia ludwigi scotti	Ophiuroids, cumaceans	Tararam et al. 1993
Luidia magellanica	Asteroids (*L. magellanica, Stichaster, Patiria, Meyenaster, Heliaster*), ophiuroids (*Ophiactis*), echinoids (*Tetrapygus, Loxechinus*)	Viviani 1978
Luidia quinaria	Ophiuroids (*Ophiura*), asteroids (*L. quinaria, Asterias, Astropecten*), echinoid (*Echinocardium*), gastropods (*Ringicuina, Philine*), crustaceans (*Crangon, Pinnixa*), bivalves (*Raetellops*)	Ganmanee et al. 2003
Luidia sarsi	Ophiuroids (*Ophiothrix, Ophioglypha*), asteroids (*Asterias*), echinoids, bivalves (*Tellina*), crustaceans, sediment	Eichelbaum 1910
Luidia sarsi	Ophiuroids (*Ophiura, Amphiura*), spatangoid (*Echinocardium*)	Fenchel 1965
Luidia senegalensis	Bivalves (*Mulinia, Tellina, Strigilla*), gastropods (*Olivella*)	Lima-Verde and Matthews 1969
Luidia senegalensis	Small bivalves (*Abra*), gastropods, ophiuroids, polychaetes, bryozoans	Halpern 1970
Luidia senegalensis	Bivalves (*Tellina, Nuculana, Chione, Pitar*), gastropods, asteroids (*Astropecten* spp.), irregular echinoids, copepods, decapods, polychaetes	Penchaszadeh and Lera 1983
Luidia senegalensis	Gastropods, bivalves, crustaceans, polychaetes, foraminiferans, tunicates, ophiuroids, sponges	Monteiro and Pardo 1994
Luidia senegalensis	Bivalve (*Pitar*), ophiuroids	Tararam et al. 1993

Feeding

Luidiids are intraoral feeders (Jangoux 1982a). This may be related to their occurrence primarily on particulate bottoms where available prey is primarily infauna. Extraoral feeding is associated with the extrusion of the stomach onto surfaces or over large epifauna (see Chapter 2). However, McClintock et al. (1983) made an exceptional observation of extraoral feeding by *L. clathrata* in the field. They noted buried individuals had an extruded stomach. Extrusion of the stomach was induced in the laboratory by placing individuals on organic-rich sediment. Organic detritus was suggested to be the food in the field.

The depth in the substrate to which luidiids feed has not been studied thoroughly. Brun (1972) reported that most feeding *L. ciliaris* were half buried in the substrate with the oral disc and marginal plates completely covered. Mauzey et al. (1968) similarly found *L. foliolata* with the disc 5–10 cm below the substrate surface and concluded it captures prey by burying. Sloan and Robinson (1983) described *L. foliolata* burying its disc to a shallow depth to capture prey. In the laboratory, *L. clathrata* depresses the disc into the substratum over buried prey and use the tube feet to move prey to the mouth (McClintock and Lawrence 1981). This behavior contrasts with other asteroids that feed on infauna. *Pycnopodia helianthoides* excavates the substrate with the tube feet, *Pisaster brevispinus* extends the tube feet into the sediment, and *Mediaster aequalis* extrudes the cardiac stomach into the sediment (Sloan and Robinson 1983).

Luidiids cannot separate small infauna from the sediment while feeding. Sometimes their stomachs are filled with sediment. Hulings and Hemlay (1963) noted sediment often comprised more that 50% of the total stomach contents and the frequent occurrence of foraminiferans in the stomachs of *L. clathrata*.

Several asteroid species respond to various amino acids and proteins (McClintock et al. 1984). *L. clathrata* emerge from the substrate and move about when exposed in the laboratory to a large number of amino acids at 10^{-5} M concentrations (McClintock et al. 1984). They found the response was greatest with L-cysteine, L-cystine, L-isoleucine and L-glutamic acid, compounds associated with animal flesh.

Intact prey in the field must release chemostimulants. Cannibalism on intact conspecifics occurs in the field in *Luidia magellanica* (Viviani 1978) and *Luidia quinaria* (Ganmanee et al. 2003), which means the intact individuals release chemostimulants. Although cannibalism by *L. clathrata* has not been observed in

the field or in individuals maintained under good conditions in the laboratory, it occurs both in the field and the laboratory on wounded individuals (Lawrence et al., in press a).

Chemosensory responses can change with feeding history. Ingestive conditioning of feeding occurs in asteroids (Valentinčič 1983). *L. clathrata* that initially showed no difference in contact-reception or teloreception in the laboratory between bivalve and shrimp flesh showed significant preference for shrimp after being fed shrimp for 21 days (McClintock and Lawrence 1984). Individuals starved for 21 days showed no preference for either food. Switching behavior occurred even if the abundance of the bivalve and shrimp models differed (McClintock and Lawrence 1985a). Nutritional history affects chemosensitivity. Starved *L. clathrata* had a significantly lower teloreceptive response time than fed individuals and consumed significantly more food (McClintock and Lawrence 1984).

McClintock and Lawrence (1981, 1985b) described the effect of prey size on feeding by *L. clathrata*. Although *L. clathrata* consumes macroprey in the field (*Mellita tenuis* 4–5 cm in diameter), individuals presented with different sizes of artificial prey in the laboratory preferred smaller sizes (0.5 cm diameter) to large sizes (1.0 and 2.0 cm diameter) They suggested this is related to ease to manipulation that increases maximal intake per unit time. When presented equal numbers of different sizes of artificial prey, more midsize (1.0 cm diameter) prey were consumed. Prey 2.0 cm in diameter were consumed in either case. This is similar to the consumption of small prey by *L. sarsi* (Fenchel 1965) and *L. foliolata* (Sloan and Robinson 1983).

L. clathrata is nocturnal (McClintock and Lawrence 1982). It emerges at dusk and begins to decline in activity after midnight, suggesting that it becomes satiated or that potential for predation increases before dawn. *L. clathrata* shows associative learning to relate onset of darkness with food (McClintock and Lawrence 1982).

Feeding can be affected by temperature and salinity changes. *L. clathrata* in Old Tampa Bay can be exposed to seasonal temperatures ranging from 10–30°C and salinities ranging from 15–30%, with highest temperatures and lowest salinities occurring in the summer. A decrease from summer levels of temperature or salinity in the laboratory decreased food consumption (Forcucci and Lawrence 1986, Watts and Lawrence 1986, 1990a). Simultaneous decreases in both temperature and salinity had a greater effect (Watts and Lawrence 1990b).

Digestion

Jangoux (1982b) concluded the digestive system of paxillosids is one of the simplest of the Asteroidea, always lacking an intestine, rectal ceca, and anus. Unusually, each pyloric cecum attaches separately to the stomach. He noted the stomach is very large and suggested this is probably associated with their capacity to ingest more substantial prey.

Digestive enzymes in the pyloric ceca have been reported for only two luidiid species. Araki and Giese (1970) found carbohydrases for α and β-glycosidic linkages but unexpectedly no glycogenase in *L. foliolata*, although they reported it for other asteroid species. Das et al. (1971) reported proteases, trypsin, chymotrypsin, and esterase in the pyloric ceca of *L. ciliaris*. A wide variety of carbohydrases, proteinases, esterases, and lipases have been reported for the pyloric ceca of asteroid species (Lawrence 1982).

Intermediary Metabolism

Ellington's (1982) review concluded echinoderms have the basic intermediary metabolic pathways present in metazoans but that details were lacking. Shick (1983) noted that echinoderms in general are basically aerobic but that their internal organs may be largely anaerobic, even under normoxic conditions because they lack an efficient circulatory system. Ellington (1982) referred to this as "the problem of internal tissues" that would apply to the gonads and pyloric ceca of asteroids.

In luidiids, intermediary metabolism has been studied only for the pyloric ceca of *L. clathrata*. Durako et al. (1979) found the activity of malate dehydrogenase was much higher than that of lactate dehydrogenase. This led them to suggest the product of anaerobic glycolysis is not lactate and that malate dehydrogenase functioned not only in anaerobic glycolysis but also in organic synthesis. L. A. Smith and Lawrence (1987) concluded the activity of glycolyic enzymes indicate the glycolytic capacity of the pyloric ceca is lower than reported for other invertebrate tissues. A low amount of food provided to *L. clathrata* resulted in a decrease in size of the pyloric ceca and the activity of glucose-6-phosphate dehydrogenase, 6-phosphogluconate dehydrogenase, and pyruvate kinase but not glycerophosphate dehydrogenase (Watts and Lawrence 1987), indicating a general decrease in metabolic activity. The response of the enzymes to food amount is affected by steroid hormones (see Chapter 9).

The pyloric ceca of *L. clathrata* and *L. senegalensis* undergo an annual cycle, with growth occurring during the summer followed by apparent transfer of nutrients to the gonads during fall and winter (Lawrence 1973, S. R. Miller and Lawrence 1999, respectively). This growth is primarily an increase in lipid and protein. Steroid hormones are involved in the intermediary metabolism of the pyloric ceca of *L. clathrata* (see Chapter 9). Feeding also results in high RNA/DNA ratios, indicating anabolic activity (Watts and Lawrence 1990b).

Watts and Lawrence (1990b) studied seasonal variability in intermediary metabolic activity in the pyloric ceca of *L. clathrata* associated with the annual cycle in production and with feeding. They found activity of enzymes in the pyloric ceca involved in glycolysis, the hexose monophosphate shunt and the glycerol phosphate shunt, was lowest in late summer when the pyloric ceca were large and subsequently when the gonads began to increase in size in early winter. Highest activities occurred in spring during gonad maturation when nutrients are transferred from the pyloric ceca to the gonads. The activities of the enzymes of *L. clathrata* fed ad libitum in the laboratory were not affected by temperature but decreased with lowered salinity. They suggested this resulted from decreased feeding and an apparent increase in the metabolic cost of cell-volume regulation and emphasized the important effects of combinations of environmental factors on cellular responses during the annual reproductive cycle.

Different nitrogenous waste products have been reported for luidiids. Ellington and Lawrence (1974) found a decrease in amino acids in the pyloric ceca of *L. clathrata* exposed to hyposmotic conditions associated with an increase in excretion of ammonium. Diehl and Lawrence (1979) reported starvation of *L. clathrata* resulted in a substantial decrease in the pyloric ceca index and protein content associated with an increase in the rate of ammonium excretion that was greater than the rate of urea excretion. This contrasts with Stickle's (1988) finding that the rate of urea excretion exceeds ammonium excretion for a number of species of holothuroids, echinoids, and asteroids, including *L. foliolata*. The total amount of nitrogen excreted by the species Stickle (1988) studied far exceeded the sum of the amounts of urea and ammonium, indicating an unknown nitrogenous excretory product.

Predators

O. D. Hunt (1925) commented *L. ciliaris* had no known predators. The high frequency of arm loss in

L. clathrata (Lawrence and Dehn 1979, Pomory and Lares 2000) suggests predation but the predators are not known. Because *L. clathrata* is nocturnal, the predators are probably diurnal fish or crabs that use vision to locate prey. *Meyenaster gelatinosus* is a sublethal predator on *L. magellanica* (Viviani 1978). Cannibalism occurs in *L. magellanica* (Viviani 1978) and *Luidia quinaria* (Ganmanee 2003).

Secondary metabolites have been reported for *Luidia quinaria, Luidia maculata, Luidia clathrata,* and *Luidia ludwigi* (Dong et al. 2011; see Chapter 8). Most reports only describe the structure of the secondary metabolites and do not address function. Asterosaponins from *L. quinaria* have low in vitro cytotoxicity against leukemia cell lines (De Marino et al. 2003). Polyhydroxysteroids from *L. clathrata* have antifoulant and antipathogenic active and also deter predation by fish (Iorizzi et al. 1995). I have never seen fouling organisms or epidermal lesions on *L. clathrata*, which suggests the presence of antifouling and antipathogenic agents.

Physiological Ecology

Ferguson (1990) reported the perivisceral coelomic and water-vascular fluids of asteroids, including *L. foliolata,* are slightly hyperosmotic to seawater. Because the difference is small, asteroids are considered extracellular osmoconformers to the external medium (Stickle and Diehl 1987). There is a direct relation between concentrations of sodium, calcium, magnesium, and chloride in the coelomic fluid of *L. clathrata* and salinity (Diehl and Lawrence 1984).

L. clathrata shows intracellular osmoregulation, modifying the concentration of the intracellular solution to that of the coelomic fluid (Diehl and Lawrence 1982, 1985). Isosmotic intracellular regulation involves coordinated changes in intracellular concentrations of both organic and inorganic osmolytes (primarily glycine and sodium and potassium ions). Regulation of intracellular concentration of osmolytes is important. Cation concentration affects the activity of glycolytic and citric acid cycle enzymes in the pyloric ceca of *L. clathrata* (A. Marsh and Lawrence 1985, Leverone et al. 1991). *Luidia duplicatus* and *L. senegalensis* are found in the lower part of Tampa Bay, where salinities of the Gulf of Mexico occur, but not in the upper part of the bay where salinities are low. It is reasonable to conclude they lack the intracellular osmoregulatory capacity of *L. clathrata*.

Hintz and Lawrence (1994) reported intracellular osmoregulation in gametogenic cells of *L. clathrata*.

They found the optimal salinity for fertilization and development through gastrulation was that at which the parents had been held, 25 or 35%. This is one of the reasons *L. clathrata* persists in an estuarine environment. Fenchel (1965) suggested one reason for irregular recruitment of *L. sarsi* in the Baltic is that it never occurs at salinities less than 30%. He did not consider the possible effect of salinity on development.

Forcucci and Lawrence (1986) distinguished between direct and indirect effects of reduced salinity of production in the body wall and pyloric ceca of *L. clathrata* by feeding individuals maintained at field salinity the lower amount of food consumed by individuals at a lowered salinity. Production was reduced at low salinity due to both a decrease in feeding rate and to a decrease in efficiency of utilization of material and energy absorbed. The low efficiency indicates that either additional energy is required for maintenance or energy is used inefficiently in low salinity. The lower rate of arm regeneration by *L. clathrata* at 20% salinity than at 30% salinity (Kaack and Pomory 2011) is probably related to this.

Low oxygen concentrations can occur during the summer in bays and estuaries. Diehl et al. (1979) reported *L. clathrata* can tolerate hypoxia (<53 mm Hg = 2.4 mg O_2/L) for only a short time and has little ability to survive extreme low oxygen concentrations. Oxygen concentrations do not reach this low concentration in Old Tampa Bay, Florida (Grabe et al. 2003), and mortality in the summer has never been observed there (Lawrence, unpublished observations). In contrast, oxygen concentrations can be <2 ppm in the summer in Charlotte Harbor, Florida, resulting in mass mortality of *L. clathrata* (S. W. Osborne 1979). *Luidia quinaria* in Ise Bay, Japan, migrates to avoid low oxygen concentrations (<3 ppm) in the summer that results from organic pollution (Ganmanee et al. 2003).

Schram et al. (2011) tested the effect of predicted near-future ocean acidification on *L. clathrata* by comparing individuals regenerating arms at pH 8.2, approximating a pCO_2 of 380 ppm, and at pH 7.8, approximating a pCO_2 of 780 ppm. They found no effect of reduced pH on regeneration, proximate composition of the body wall and pyloric ceca, and righting behavior. This suggests the predicted future decrease in pH of the ocean water would have no direct effect on *L. clathrata*.

Reproduction

L. clathrata has seasonal reproduction (Lawrence 1973, Dehn 1980a,b, Watts and Lawrence 1990a, Pomory and Lares 2000; Fig. 11.2). The gonads begin to increase in size in the fall, reach their maximum in late winter, and decrease in mid- to late spring when spawning occurs. The pyloric ceca have a generally reciprocal cycle to that of the gonads (Lawrence 1973, Dehn 1980a,b, Watts and Lawrence 1990a). The seasonal cycle in gonad size persists although the amount of gonads produced by *L. clathrata* can vary annually, reflecting annual variation in the estuarine environment (Dehn 1980a,b).

Seasonal variation in gonad size has not been followed for other luidiids except *L. senegalensis*, which spawns in the fall (S. R. Miller and Lawrence 1999). Seasonal variation in occurrence of larvae indicates seasonal spawning of other luidiids in temperate latitudes. Larvae have been observed in the field in fall and winter for *L. sarsi* (Wilson 1978) and fall and spring (Domanski 1984) and in summer for *L. quinaria* (Dautov and Selina 2009). Mortensen (1913) simply stated the breeding season of *L. ciliaris* is June to July. *L. foliolata* is ripe in late spring (M. Strathmann 1987), *L. quinaria* from July to September (Komatsu et al. 1982), and *L. maculata* in July (Komatsu et al. 1994).

Eggs of *L. clathrata* have a volume of 2.3–2.6 10^6 μm^3, larger than the eggs of *Asterias forbesi* (0.9–1.8 10^6 μm^3) and smaller than those of *Echinaster modestus* (172–299 10^6 μm^3) (Turner and Lawrence 1979). The size of eggs of luidiids ranges from 124 to 197 μm except *L. sarsi*, which has an egg size of 30 μm (Komatsu et al. 2000). A low food ration does not affect egg size of *L. clathrata* but results in lower fertilizability and larval survival (George et al. 1991). The eggs of *L. clathrata* have proportionally more lipid and protein than the eggs of *A. forbesi* and less than the eggs of *E. modestus*. This reflects the difference in provisioning of nutrients to the eggs by the three species.

Larvae

Larval development in the luidiids is non-brachiolarian type, with the bipinnaria metamorphosing without passing through a bracholarian stage (see Chapter 5). In contrast to the bipinnarian arms of luidiids, the brachiolarian arms are adhesive and attach to the substrate at settlement (Flammang 1996; see Chapter 3). Settlement on particulate substrate may not be facilitated by adhesive brachiolarian arms.

Komatsu et al. (1991) pointed out the fully grown bipinnariae of *L. clathrata* and *L. senegalensis* are 1.8 and 1.5 mm long, respectively, much smaller than those of other luidiids such as *L. sarsi*. *L. foliolata* fed low concentrations of food have bipinnariae with arms that are larger and longer than those fed high concentrations, suggesting phenotypic plasticity in response to food availability (George 1994b).

The bipinnaria larva of *L. sarsi* can reach 35 mm in length. The rudiment normally detaches from the larval body before reaching 5 mm in length but can be 15 mm in diameter if the larva remains in midwater (Domanski 1984). With extended existence in the water column, the larva and rudiment continue to grow until the larval body can no longer support the rudiment. The rudiment absorbs the larval body and is the final stage before enforced settlement. The estimated planktonic stage is up to more than 1 year, giving the species potentially great dispersal capacity.

Time from fertilization through settling has been measured in the laboratory. It ranged from 60–140 days, with an average of 79 days for *L. foliolata* (R. Strathmann 1978), 35 days for *L. quinaria* (Komatsu et al. 1982), 28 and 18 days, respectively, for *L. clathrata* and *L. senegalensis* (Komatsu et al. 1991), and 50 days for *L. maculata* (Komatsu et al. 1994). The length of time for complete development undoubtedly depends on the quantity and quality of food available for the larvae. Delay in first feeding decreases rate of development, growth, and survival of larvae of *L. clathrata* (George et al. 1990). Larvae of *L. foliolata* fed a high food ration completed development in 85 days, while those fed a low food ration required 167 days (George 1994b). Consequently the times to settlement given here from laboratory studies may not indicate actual differences among the species and are only estimates for the time of larval development in the field.

Asteroid larvae can clone and reproduce asexually (see Chapter 5). Bosch et al. (1989) found bipinnariae of *Luidia* sp. with highly modified bilateral arms in the Sargasso Sea and Gulf Stream and concluded these were the result of fission. Jaeckle (1994) suggested asexual reproduction by larvae could increase length of larval life, number of individuals, and dispersal. Both sections of bisected *L. foliolata* larvae regenerate completely (Vickery and McClintock 1998). Bosch (1992) found symbiotic bacteria in the majority of bipinnariae of *Luidia* sp. The incidence of symbiosis was greater in actively cloning larvae. He reported the epidermal cells of the larvae of *Luidia* sp. engulf and digest the bacteria and suggested this may contribute to the nutrition of the larvae.

Growth

Field studies indicate luidiids can grow rapidly. *L. ciliaris* grew from 5 cm to more than 30 cm in diameter in

2.5 years (Renouf 1937); *L. sarsi*, from 15 to an average of 80 mm in 1.5 years (Fenchel 1965); *L. senegalensis*, from 10 to 53 mm radius from November to January (Halpern 1970); and *L. clathrata*, an average of 13 mm month^{-1} (maximum 21 mm month^{-1}) from January through July (S. W. Osborne 1979). Rates of growth in the field would be dependent upon biotic and abiotic conditions. The population of *L. clathrata* studied by Osborne (1979) was subjected to hypoxic conditions during the summer.

The laboratory studies of Dehn (1980b) showed the effect of quality and quantity of food on growth of *L. clathrata*. Growth of juveniles fed high amounts of food was allometric. The exponential growth period was 55 and 27 days for increase in weight and radius length, respectively, indicating growth was first to increase arm length followed by an increase in mass. Growth of juveniles fed a low amount of food was less but still allometric, consistent with the indeterminate growth characteristics of echinoderms. This indicates the maximum size of individuals in a population depends on the quantity and quality of food available.

Arm Autotomy and Regeneration

Forbes (1841) famously described the response of a *L. ciliaris* (as *fragillisima*) to collection: "In a moment he proceeded to dissolve his corporation and at every mesh of the dredge his fragments were seen escaping. In despair I grasped at the largest, and brought up the extremity of an arm with its terminating eye, the spinous eyelid of which opened and closed with something exceedingly like a wink of derision." Luidiids can autotomize at any position along the arm (Emson and Wilkie 1980). Lawrence and Pomory (2008) described the gross regeneration of arms by *L. clathrata*. The new arm bud appears ca. 8 days after arm loss regardless of the where the arm loss occurs. The rate of regeneration is inversely related to the position of arm loss. This is like the regeneration growth curve of an arm regenerated at its base (Lawrence and Ellwood 1991). Regeneration is rapid at first and slows as the regenerating arm nears the length of the intact arms. This trajectory is like that of the growth curve of arms of intact individuals described above. The control of this pattern is not known.

The frequency of arm loss can be high and varies with population and time (Lawrence 1992). From 1971 to 1975, Lawrence and Dehn (1979) found no *L. clathrata* regenerating arms in Old Tampa Bay. In contrast,

Pomory and Lares (2000) reported 60% of the same population had regenerating arms in 1996. If arm loss results from predation, predators were much different at the two dates. Generalizations about frequency of arm loss based on few observations cannot be made.

The functional consequence of arm loss has not been studied but the energetic cost of loss and regeneration has been considered. Laboratory studies with *L. clathrata* (Lawrence et al. 1986, Lawrence and Ellwood 1991) suggest that there is no cost of arm regeneration when food availability is low because regeneration does not occur (Lawrence et al. 1986) or when food availability is high because of unlimited resources (Lawrence and Ellwood 1991). Pomory and Lares (2000) found no difference in the dry weight of the body wall or pyloric ceca between intact arms of intact or regenerating *L. clathrata* in the field. They also attributed this to low food availability.

Population Biology and Community Ecology

The population biology of luidiids has been little studied. Ursin (1960) noted the irregular occurrence of *L. sarsi* in the North Sea and suggested it was due to irregular incurrence of water occurrence with larvae. Fenchel (1965) also suggested the sporadic occurrence of *L. sarsi* in Danish waters depended on irregular recruitment. Halpern (1970) described a similar great change in density of *L. senegalensis* due to recruitment. The literature survey by Holme (1984) indicated fluctuation of the abundance of *L. ciliaris* in the English Channel and North Sea. He suggested changed water conditions affected larval appearance and recruitment. Guillou (1996) reported low biomass of *L. ciliaris* and *L. sarsi* in the Bay of Douarnenez, Brittany, from 1981 through 1985, followed by a precipitous increase in 1986 and then a slow decline through 1988. She also attributed the increased abundance to larval supply. Ventura and Fernandez (1995) discussed variable recruitment regarding a similar abrupt change in density of *Ludia ludwigi scotti* at Cabo Frio, Brazil, occurred between 1986 and 1987. All these studies emphasized the importance of recruitment for establishment of populations of these species. In contrast to this variation, the population of *L. clathrata* in Old Tampa Bay has persisted since 1970 (Cobb et al. in press).

Studies of the role of luidiids in communities are few. In a short-term experiment, Thrush (1986) found

L. ciliaris had no major role in organizing community structure at a depth of 23 m in a sea-lough. Guillou (1990, 1996) noted the temporal sequence of abundance of ophiuroids, *Asterias rubens, Marthasterias glacialis, L. ciliaris* and *L. sarsi* in the Bay of Douarnenez. She attributed the change in community structure to the change in abundance of *L. ciliaris* and *L. sarsi* that prey on *A. rubens* and *M. glacialis*.

Acknowledgments

I thank James B. McClintock and Stephen A. Watts for comments on a draft of this chapter.

12

John S. Pearse

Odontaster validus

P enguins are icons of the Antarctic. For benthic ecologists, however, a more appropriate icon is a little red sea star, *Odontaster
validus* KOEHLER, 1906. It occurs in the Southern Ocean all
around the Antarctic continent and can be extraordinarily abundant
in shallow water (Fig. 12.1). It was present in nearly all collections
made by the pioneering expeditions of the early and mid-twentieth
century, has been a player in many benthic studies over the past 50
years, and has been a model animal for our understanding of Antarctic
marine biology. Here I cover the major works on this little sea star and
what they have revealed about the ancient and now rapidly changing
ecosystem that it inhabits.

Systematics and Distribution

Odontaster validus was described by René Koehler (1906) from specimens collected off the west coast of the Antarctic Peninsula by the
Expédition antarctique française of 1903–5. Since then it has been
found not only around the Antarctic continent but also off subantarctic islands (South Georgia, Bouvet, Marion, Prince Edward; Branch
et al. 1993) and possibly on the Burdwood Bank off Tierra del Fuego
and Tristan da Cunha in the south Atlantic (Smithsonian Institution
2011). Although some specimens in museum collections were taken
from well over 2000 m, it occurs mainly in shallower waters, 200 m or
less, right up to the shoreline. Moreover, it is abundant mainly in shallow areas of high productivity, where it feeds on a diverse variety of

Fig. 12.1. (Opposite) *Above*, Large population of *Odontaster validus* on basaltic
gravel in shallow water under annual sea ice in McMurdo Sound, Antarctica. Diver
entering through hole in ice in background. *Below*, the starfish *O. validus* with the
sea urchin *Sterechinus neumayeri* in shallow water next to the shore in southeastern
McMurdo Sound. The fuzzy objects are clusters of anchor-ice platelets; when
dislodged, they rise with specimens of *O. validus*, other animals, and sediments to
be frozen into the underside of sea ice above. *Photographs © Norbert Wu. Used
with permission.*

items, including benthic microalgae that probably provide it with its color.

Five species of *Odontaster* are described for the Southern Ocean and southern South America (Janosik and Halanych 2010, Janosik et al. 2011). Besides *Odontaster validus,* there are *Odontaster penicillatus* (PHILIPPI, 1879), southern South America including the Patagonian Plateau out to the Falkland/Malvinas Islands, subantarctic islands (South Georgia and perhaps Marion, Macquarie, and Auckland; H. E. S. Clark and McKnight 2001), and off the Antarctic Peninsula (Smithsonian Institution 2011); *Odontaster meridionalis* (E. A. SMITH, 1876), Antarctic circumpolar, subantarctic islands (South Georgia, Marion, Kerguelen, Heard), and the Patagonian Plateau; *Odontaster pearsei* JANOSIK AND HALANYCH, 2010, western side of the Antarctic Peninsula; and *Odontaster roseus* JANOSIK AND HALANYCH, 2010, South Shetland Islands off the northern tip of Antarctic Peninsula. These five species are morphologically very similar and specimen identification can be tentative. The last two species were first detected by sequence analyses of their mitochondrial genes; their full distribution remains to be determined, but they have probably been confused with *O. meridionalis* and *O. validus* in earlier studies. Like *O. validus,* all are found mainly in less than 200 m depth, although there are specimens of both *O. penicillatus* and *O. meridionalis* that were taken from more than 2000 m (Smithsonian Institution 2011).

Janosik et al. (2011) showed that *O. meridionalis* is the sister species to the other four species. It is sympatric with *O. validus* over most of its range. Both are almost entirely restricted to the Southern Ocean south of Drake Passage, but a few specimens of both species have been collected north of Drake Passage off eastern Tierra del Fuego (Smithsonian Institution 2011). In contrast, *O. penicillatus* is almost entirely subantarctic in distribution off Argentina, although it is apparently not common and its biology has not been studied (Brogger et al. 2012). It co-occurs with the other two species mainly off South Georgia Island in the Southern Ocean. A few museum specimens that are probably correctly identified (C. Mah, personal communication) were collected from off the west coast of the Antarctic Peninsula (Smithsonian Institution 2011).

O. validus, O. meridionalis, and *O. roseus* have long-lived planktotrophic larvae (Bosch and Pearse 1990; Stanwell-Smith and Peck 1998; Janosik at al. 2011), as do almost certainly the other two species. Larvae of all these species must cross Drake Passage from time to time, particularly in eddies that carry parcels of water across it in both directions (A. Clarke et al. 2005).

Yet Drake Passage appears to delineate a formidable barrier to these species. The Antarctic Polar Front, which dissects Drake Passage today, separates the Antarctic biota from those to the north (A. Clarke and Johnston 2003). In benthic trawls taken from islands of the Scotia Arc across the Antarctic Polar Front, S. Kim and Thurber (2007) found *O. validus* to be abundant in the southern samples and absent in the northern samples. The Antarctic Polar Front has a steep temperature gradient going below 4°C to the south and above 5°C to the north (Griffiths 2010), which may pose a barrier for many Antarctic fishes and invertebrates (Peck et al. 2004). Whether temperature limits the distribution of *O. validus,* however, is uncertain. Kidawa et al. (2010) reported loss of motor coordination and other activities in *O. validus* at 4–5°C, but Peck et al. (2008) found that with acclimatization, adults are much more eurythermal, acclimatizing to 6°C and continuing to be active up to 9°C. Temperature sensitivity is inversely related to size, and smaller individuals survived exposure up to 12–14°C (Peck et al. 2009). Nevertheless, the animals showed no Hsp70 heat shock response at even higher temperatures (M. S. Clark et al. 2008). In addition, embryonic developmental rate of both *O. validus* and *O. meridionalis* is highly sensitive to temperature (Hoegh-Guldberg and Pearse 1995, Stanwell-Smith and Peck 1998, Peck and Prothero-Thomas 2002). The reason *O. validus* and *O. meridionalis* are rare north of Drake Passage, and *O. penicillatus* to the south, remains unresolved.

Janosik et al. (2011) suggest that *O. penicillatus* diverged from *O. pearsei* when glacial maxima in the Pleistocene shifted the Antarctic Polar Front north of Drake Passage, allowing Antarctic species to become established on the Patagonian shelf. Presumably, the ancestor of *O. penicillatus* was left behind when the front moved south, and it and *O. pearsei* diverged on the two sides of Drake Passage. This scenario may be true, but the mechanism of speciation among the other Antarctic species remains even more obscure. *O. validus* and *O. meridionalis* often are found together, spawn at the same time (late winter, early spring; Bosch and Pearse 1990), and readily hybridize in the laboratory producing viable larvae (I. Bosch, pers. comm.). They differ, however, in trophic ecology; *O. meridionalis* is a specialist feeding on sponges, whereas *O. validus* has a wide and varied diet (see below).

In addition to the five species in the Southern Ocean and off South America, nine other species are recognized in the genus *Odontaster,* most occurring in the Southern Hemisphere and none in the Arctic. The genus is within the family Odontasteridae VERRILL, which contains six genera and a total of 28 species, occurring mostly on continental shelves and slopes in the Southern Hemisphere (Mah and Hansson 2012). Mah and Foltz (2011) placed Odontasteridae + Chaetasteridae as the sister clade to the other families in the order Valvatida, while Janies et al. (2011) placed Odontasteridae + *Pseudarchaster parelii* as the sister clade to other species in the order. Most of the species in the family occur in the Antarctic or adjacent regions with those in the Northern Hemisphere limited to species of *Odontaster*, suggesting Gondwana origins. The single fossil reported for the family, the putative species *Odontaster priscus* FELL, 1954, was found in the Jurassic of New Zealand (Fell 1954), further supporting the suggestion that both Odontasteridae and the genus *Odontaster* are very old clades with southern origins. Indeed, Aronson et al. (2011) consider *O. validus* one of the species that gives the nearshore communities in the Antarctic a "distinctly anachronistic, Paleozoic character."

Anatomy

O. validus is a "cushion" star with arms wide at the base, a thin body wall impregnated with small ossicles, and ample coelomic fluid so that the animal is soft and, indeed, cushion-like in appearance and texture. There are typically five rays, rarely four or six. Four-rayed specimens show no sign of a missing ray, but the few six-rayed animals on record have a forked ray, suggesting that the growing ray tip had been split by injury (H. E. S. Clark 1963, Pearse 1965).

The marginal ossicles are small and inconspicuous. Clusters of spinelets on small stalks (paxillae) are carried by the small abactinal ossicles, giving the sea star a velvety appearance (Fig. 12.2). The number and shape of the paxilla spinelets, which are nearly microscopic, are the main characters separating the Antarctic and southern South American species (Janosik et al. 2011), so it is no wonder that two of the species were distinguished only recently. Numerous papulae occur between the paxillae, and there is a single conspicuous madreporite. The actinal ossicles each have up to seven short spines, while small two- or three-valved pedicillariae are found in the oral angles and rarely also abactinally. In each oral angle, pointing away from the mouth, is a single, recurved, hyaline spine, the feature that gives the genus its name; larger specimens often carry one or two enlarged accessory spines on each side of the primary oral spine.

Although they are usually a shade of red, a wide range of colors have been recorded for live animals

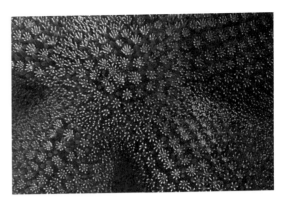

Fig. 12.2. Abactinal surface of *Odontaster validus* in Ross Sea, Antarctica, showing multispined paxillae and papulae. *Photo by John Weller, with permission.*

taken from various areas, ranging from dark brown, dark carmine, beet red, and deep red to pale pink abactinally and orange-yellow to white actinally (H. E. S. Clark 1963). Color almost certainly depends on carotenoids and other algal pigments in the diet. Animals from an area that was mostly ice free in the summer in McMurdo Sound had a much richer color than those from an area mostly covered with thick sea ice and snow, restricting algal growth (see photos in Clark 1963, Pearse 1965). A similar range of colors is present in the internal organs (gonads, intestinal ceca, pyloric ceca). Pearse (1965) found a wide variation in the amount of chlorophyll derivatives in acetone extracts of the pyloric ceca, apparently depending on diatoms and other microalgae in the diet.

The internal anatomy of *O. validus* as described by Pearse (1965) is similar to that of *Odontaster crassus* of the northeast Pacific (Fisher 1911; Fig. 12.3). Particularly notable are the small cardiac stomach, the baglike Tiedemann's pouches on the oral side of the pyloric ceca, and the large intestinal ceca, all indicative of ciliary-mucoid feeding (Gemmill 1915, Anderson 1960, Scheibling 1982a). A pair of lobular gonads is located in the arm angles, each gonad with a gonoduct opening on the edge of the abactinal surface, typical of broadcast spawners. Most animals have ten gonads, but in one study about 10% had only eight or nine gonads (Pearse 1965).

Body weight and volume vary enormously both among individuals in the same sample and within a single individual weighed successively (Pearse 1967). Some individuals in a sample are notably swollen while others are markedly shriveled (Pearse 1965). The variation depends on the volume of fluid in the perivisceral coelom. When "relaxed" in isosmotic magnesium

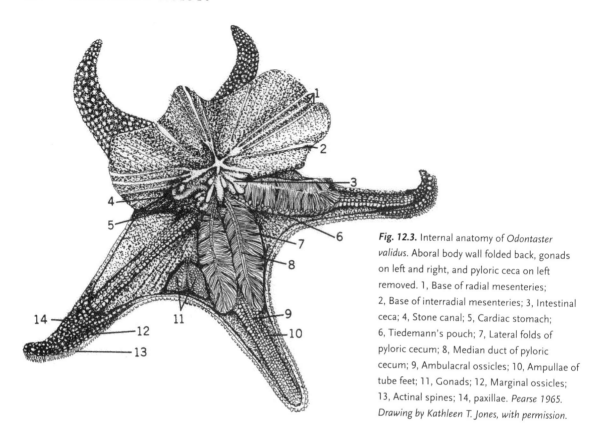

Fig. 12.3. Internal anatomy of *Odontaster validus*. Aboral body wall folded back, gonads on left and right, and pyloric ceca on left removed. 1, Base of radial mesenteries; 2, Base of interradial mesenteries; 3, Intestinal ceca; 4, Stone canal; 5, Cardiac stomach; 6, Tiedemann's pouch; 7, Lateral folds of pyloric cecum; 8, Median duct of pyloric cecum; 9, Ambulacral ossicles; 10, Ampullae of tube feet; 11, Gonads; 12, Marginal ossicles; 13, Actinal spines; 14, paxillae. *Pearse 1965.* Drawing by Kathleen T. Jones, with permission.

chloride solutions, most animals lose weight, which they regain when returned to seawater, suggesting active control of coelomic fluid volume. Animal wet weight also fluctuates with a rhythm whose period increases with size (Pearse 1967). Gemmill (1915) found that body weight increases during ciliary-mucoid feeding in *Porania pulvillus*. This might be responsible for the variation in weight seen in *O. validus*. However, coelomic fluid volume fluctuated seasonally at McMurdo Sound and was lowest in late summer during maximal microalgal production, presumably when ciliary-mucoid feeding would be most intense, and when the pyloric ceca and gonads were growing. The relationship between feeding and coelomic fluid volume remains unclear (Pearse 1965).

Food, Feeding, and Nutrient Stores

O. validus is omnivorous, feeding on a wide variety of foods, live and dead, animal and algal, microscopic and prey larger than themselves (Dearborn 1977, McClintock 1994). Individuals can be collected, often in very large numbers, in traps baited with seal and fish meat (Dearborn 1965, Pearse 1965, Arnaud 1970, 1974). Arnaud (1970) reported that animals were trapped

only in the winter at the Dumont d'Urville Station on Terre Adélie (66°S); perhaps they fed on microalgae in the summer and were not attracted to carrion bait. Large aggregations of *O. validus* are commonly seen gathered around prey by divers and photographed, e.g., on a large isopod (Peckham 1964), a heart urchin (Pearse 1969a), and a sea urchin (Brueggeman 1999; see also McClintock and Baker 1997a). Zamorano et al. (1986) found that *O. validus* would aggregate on and consume unburied individuals of the bivalve *Laternula elliptica*, which might be exposed when icebergs scour the bottom. In a quantitative survey in McMurdo Sound, Dayton et al. (1974) reported that 33% of the prey of *O. validus* were bivalves and 30% were sponges. Similarly, McClintock (1987) found that *O. validus* fed on the three most abundant sponges in McMurdo Sound, *Rosella rocovitzae, R. nuda,* and *Tetilla leptoderma,* selectively for the last two. These species were the ones in their study that were mildly or not toxic in trials on goldfish. Dayton (1989) also documented *O. validus* feeding extensively on the sponge *Homaxinella balfourensis* when it was abundant in shallow water in McMurdo Sound.

Dayton et al. (1974) classified most of their remaining feeding observations of *O. validus* as filter feeding (rays curled aborally) and detritus feeding (presumably with cardiac stomach everted over the substrate). Such activity corresponds well with the gut contents documented by Dearborn (1965) and Pearse (1965), which included diatoms and bits of algae, sponge spicules, small gastropods, remains of polychaetes and crustaceans, and unidentified particles and sediment. Based on color, nutritional condition, and reproductive output, microalgae (diatoms) are a major food source (McClintock et al. 1988). In addition, in an area with a seal rookery, the sea stars were seen aggregated on seal feces, which filled their cardiac stomachs (Pearse 1965). However, O. validus was conspicuously absent from an area in McMurdo Sound that was enriched with human sewage and covered with anaerobic bacteria (S. Kim et al. 2007).

Gut contents of O. validus were enmeshed in mucus (Pearse 1965), consistent with ciliary-mucoid feeding, as indicated by their internal anatomy (see above). In addition, many amoebocytes occurred, perhaps aiding in digestion and nutrient transport. Finally, several types of ciliates are present in the stomach, perhaps acting as commensals, as is well documented for sea urchins (Lynn and Strüder-Kypke 2005).

In addition to scavenging and feeding on microalgae, sponges, bivalves, small animals, detritus, and feces, individuals of O. validus were reported by Dayton et al. (1974) to attack and completely consume individuals of the large sea star Acodontaster conspicuus. A single individual of O. validus would climb on to and digest an opening into the aboral surface of an individual of A. conspicuus, which moved rapidly away. Other individuals of O. validus joined in, along with the nemertean Parbolarsia corrugatus, until the victim was completely subdued and consumed. Within 2 weeks the scavengers dispersed, leaving a star-shaped pile of A. conspicuus ossicles (see photos in Brueggeman 1999). During the summer of their study, Dayton et al. (1974) observed 14 piles of ossicles of A. conspicuus, which they estimated represented 3% of the prey population in the area. In addition, about 20% of the healthy individuals of A. conspicuus in the population, had scars that were probably the result of such attacks.

Whether attacks by O. validus were on healthy, undamaged individuals of A. conspicuus or on those that had been damaged previously, even in a minor way, releasing coelomic fluid that could be sensed by individuals of O. validus, is unclear. For example, in an experiment to label individuals of the echinoid Sterechinus neumayeri, Brey et al. (1995) collected animals, injected them with tetracycline, and returned them to where they had been collected. The next day

all were being attacked and consumed by individuals of O. validus, which had ignored them previously. Only when the sea urchins were injected and then held in cages to heal for several days before release, did they avoid attack by O. validus. In addition, McClintock (1994) observed O. validus attacking wounded individuals of the sea star Perknaster fuscus antarcticus in the laboratory.

Sensitivity to chemicals that promote feeding by O. validus has been examined by Kidawa (2001, 2005a,b) and Kidawa et al. (2008). In an elegant series of laboratory experiments using agar blocks of different size and nutritional value, Kidawa (2009) showed that individuals varied in their selectivity toward food; smaller starved animals were less selective, probably because they had fewer reserves than did well-fed or larger animals. Kidawa (2001) also reported that individuals of O. validus could distinguish between the sizes and nutritional states of conspecifics, presumably preventing cannibalism, even within large feeding aggregations.

McClintock et al. (2008b) also demonstrated with time-lapse videography not only detection of wounded prey (sea urchins) by O. validus, but fencing bouts between individuals over possession of the prey, with the largest individual usually prevailing. Individuals of O. validus did not respond to two other species of sea stars but showed a flight response upon contact with a third species, Perknaster aurorae (McClintock et al. 2008b). Moreover, when an individual of O. validus contacted one fleeing from P. aurorae, it too took flight, apparently an alarm response. The diet of P. aurorae is unknown, and it is also not known whether it is a competitor or predator of O. validus, so these behaviors remain unexplained.

In contrast, many potential prey species appear to have chemical defenses against predation by O. validus. Extracts of sponges, soft corals, brachiopods, ascidians, gastropods, and sea stars cause tube-foot retraction, inhibit righting, or deter feeding by O. validus (reviewed by McClintock and Baker 1997a). In additional work, O. validus was used to test palatability and ingestion of potential prey, finding them to be extensively defended (e.g., McClintock and Baker 1997b, Mahon et al. 2003, McClintock et al. 2006, Peters et al. 2009, reviewed by McClintock et al. 2010; see Chapter 8). Moreover, like many other gastropods confronted with asteroids, the limpet Nacella concinna, displays an escape response to O. validus that is mediated by both chemical and tactile stimuli (Mahon et al. 2002).

Like other sea stars, O. validus stores nutrients in the pyloric ceca, which presumably serve as a source of nutrients supporting gametogenesis. The mass of the pyloric ceca varies with the amount of food available. Biochemical contents of the pyloric ceca has been measured by Pearse (1965), McClintock et al. (1988), and Stanwell-Smith and Clarke (1998). Protein and lipid showed little variation in percent dry weight, either seasonally or among different populations (40–60% protein, 15–20% lipid). In contrast, carbohydrate was more variable, fluctuating around 2–5% throughout the year at McMurdo Station but reaching more than 10% at the end of the summer in the more productive area off Cape Evans (Pearse 1965). Carbohydrate levels were even higher at Signy Island (about 13%), perhaps indicating greater algal productivity (Stanwell-Smith and Clarke 1998).

Where there is strong seasonal production of microalgae in the summer, the pyloric ceca of O. validus show strong seasonal fluctuations in mass, while in areas under sea ice during the summer or in deeper water with limited seasonal microalgal production, the pyloric ceca remain relatively small throughout the year (Pearse 1965, McClintock et al. 1988, Stanwell-Smith and Clarke 1998, L. J. Grange et al. 2007). Because pyloric ceca shrink in the winter, Stanwell-Smith and Clarke (1998) suggested that Odontaster validus does not feed then. However, as Grange et al. (2007) pointed out, seasonal feeding is unlikely for such an opportunistic omnivore, and, in fact, animals are readily trapped throughout the year with carrion bait (Dearborn 1965, Pearse 1965). Nevertheless, the animals are almost certainly food limited, and when kept in the laboratory for 2 years with ample supplies of carrion, both animal weight and pyloric ceca increase remarkably (McClintock et al. 1988, Pearse and Bosch 2002).

Reproduction

O. validus is gonochoric, with close to a 1:1 sex ratio. Hermaphrodites are rare. Of 350 animals dissected, Pearse (1965) found only one male hermaphrodite with a single gonad containing both sperm and eggs, with the other nine having only sperm. Similarly, of 221 animals dissected and sexed, Stanwell-Smith and Clarke (1998) found six hermaphrodites. Of 407 animals dissected, L. J. Grange et al. (2007) found three hermaphrodites, suggesting that the incidence of hermaphrodites may decrease with increasing latitude, being lowest at McMurdo Sound, but they recognized the need for more data.

Stanwell-Smith and Clarke (1998) also reported that 38% of the gonads of O. validus were infested by what they identified as the ascothoracid Dendrogaster antarctica GRYGIER, 1980. Parasites in the gonads were not mentioned in other studies. Dendrogaster antarctica was described from specimens taken from the coelomic cavity (not the gonads) of the sea star A. conspicuus in McMurdo Sound, with an infection rate of <2% (Grygier 1981). In addition, Grygier (1987) found an immature ascothoracid probably of the same species inside the body cavity of O. validus, as well as another ascothoracid, Gongylophysema asetosum, embedded in the body wall.

Gametogenesis was thoroughly described for populations in McMurdo Sound by Pearse (1965) and is similar to that in populations studied off Signy Island (Stanwell-Smith and Clarke 1998) and Rothera (L. J. Grange et al. 2007). Spermatogonia are present in the testes throughout the year and produce spermatocytes nearly all year, except in fall (April–June). Spermatogenesis takes about 6 months, and sperm are abundant in fall and winter. Oogonia likewise are present in the ovaries throughout the year, producing oocytes continuously except in late summer and early fall. Oocytes accumulate for up to 1 year before undergoing active vitellogenesis. Most oocyte growth occurs between spring (August–September) and fall (March–April), after which full-grown oocytes are held until spawned during the following winter and early spring. Consequently, oogenesis normally takes about 18 months to 2 years. After the first cohort of oocytes is formed, there always are two or three cohorts of oocytes in the ovaries at the same time. Moreover, many more oocytes are formed than reach full growth. Indeed, most of the early oocytes disintegrate, probably providing nutrition to those remaining and undergoing vitellogenesis (Pearse 1965).

In Odontaster validus, unlike many other sea stars, pyloric cecum growth does not alternate seasonally with gonad growth. Pyloric ceca and gonads both may grow in summer, or only the gonads or the pyloric ceca fluctuate seasonally while the other organ remains of constant size, varying from year-to-year (McClintock et al. 1988, Stanwell-Smith and Clarke 1998, L. J. Grange et al. 2007). The relative constancy seen in gonad size in some populations may reflect the use of some oocytes to support the growth of other oocytes.

The level of the main biochemical constituents in the gonads is similar to that of the pyloric ceca (see above), as measured by Pearse (1965), McClintock et al. (1988), Stanwell-Smith and Clarke (1998), and Chiantore et al. (2002), with little variation seasonally or among sites. As expected, the proportion of lipid in the ovaries was nearly twice that in the testes, presumably because of lipid stores in the oocytes (Pearse 1965, McClintock et al. 1988).

The seasonal cycle of gametogenesis is the same at Signy Island in the South Orkney Islands (60°S) near the northern limit of the distribution of *Odontaster validus* (Stanwell-Smith and Clarke 1998) as it is at McMurdo Sound (78°S) at its southern limit (Pearse 1965), as well as at Adelaide Island (67°S) in between (L. J. Grange et al. 2007). While seasonal sea temperature ranges from −1.9°C to +1.5°C at Signy Island, it is nearly constant at around −1.9°C in McMurdo Sound; temperature fluctuations almost certainly do not synchronize the timing of reproduction in this species.

The populations in which reproduction has been studied were all in shallow water, less than 30 m depth. However, oocyte distributions and spermatogenic stages in samples taken off Victoria Land (71°S) and the Balleny Islands (66°S) at 274 m depth were remarkably similar to those from McMurdo Sound (Pearse 1966). This depth is below the euphotic zone, making photoperiodic control unlikely, and Pearse (1966) suggested that the seasonal phytoplankton bloom synchronizes the timing of reproduction in *O. validus*. Later, experiments clearly showed that gametogenesis can be shifted out of phase within a year after animals were put on a photoperiod cycle 6 months out of

phase with ambient, that is, continuous light in winter and continuous dark in summer (Pearse and Bosch 2002; Fig. 12.4). Such synchronization could function throughout the species' range in shallow water if the critical photoperiod was near 12 h L: 12 h D, because the equinox is the same everywhere. However, it remains difficult to understand what synchronizes populations below the euphotic zone.

Pearse and Bosch (2002) also found that gametogenesis appears to be continuous when the animals are kept under constant light or on a photoperiod of 12 h L: 12 h D, indicating that light stimulates gametogenesis. However, animals kept in near constant darkness retained the annual cycle, indicative of an underlying circannual rhythm, as demonstrated for *Pisaster ochraceus* (Pearse et al. 1986). Pearse and Bosch (2002) suggested that the circannual rhythm might be set in larvae when they are swimming in the euphotic zone, and its timing is maintained throughout the life of the animal.

O. validus spawns in winter and early spring, and fertilization occurs in the water column. Embryonic and larval development is slow; feeding larvae appear in laboratory cultures 28 to 35 days after fertilization (Pearse 1969b, Stanwell-Smith and Clarke 1998), developing through a typical bipinnaria stage and settling and metamorphosing after 167 days (Bosch and Pearse 1990). Although Pearse (1969b) thought that the larvae

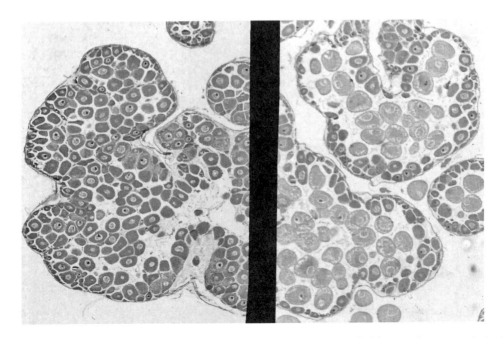

Fig. 12.4. Histological sections of ovaries of *Odontaster validus*. The animals were held in the laboratory for 1-year under in-phase (*left*) and 6-month out-of-phase (*right*) photoperiods and then dissected in December. *Pearse and Bosch 2002.*

were demersal—mainly because of their swimming on the bottom of culture dishes, failure of collecting them in plankton samples, and the influence of Thorson's rule at the time (see Pearse 1994)—they are planktotrophic larvae, typical of valvatid asteroids, that were found in the plankton in McMurdo Sound with more intensive sampling (Pearse and Bosch 1986; Fig. 12.5). Sea star embryos and larvae have been found in the plankton in subsequent studies in the Antarctic, albeit usually in low to very low numbers (Stanwell-Smith et al. 1999, Sewell 2005, Freire et al. 2006, Bowden et al. 2009), and based on laboratory cultures, these larvae were almost certainly those of *Odontaster*. Indeed, asteroid larvae sampled from the plankton have been identified as those of both *O. validus* and *O. roseus* by molecular sequences (Janosik et al. 2011, Sewell and Jury in press).

With spawning occurring in the dark winter and early spring, feeding larvae would be in the plankton before much phytoplankton is available for food. However, Rivkin et al. (1986 as modified by Pearse et al. 1991a) showed that the larvae are able to feed on bacteria, which are more uniformly present throughout the year. Because the larvae can also utilize dissolved organic material (Shilling and Bosch 1994, Shilling and Manahan 1994), their metabolic rate is extremely low (Peck and Prothero-Thomas 2002; see Pace and

Manahan 2007 and Ginsburg and Manahan 2009 for comparable work on the Antarctic sea urchin *Sterechinus neumayeri* and *O. meridionalis*, respectively), and the eggs are relatively well endowed with lipids (Moore and Manahan 2007), very little food is required to support their nutritional needs. Moreover, in situ experiments demonstrated that larvae of *O. validus* developed normally in ambient conditions in the field (Olson et al. 1987). Consequently, obtaining adequate nutrition for larval survival and development is probably not a problem.

That one of the most abundant animals in the Antarctic benthos has pelagic, planktotrophic larvae is somewhat surprising considering the abundance of species with lecithotrophic larvae or non-pelagic development (Pearse 1994, Pearse et al. 2009). However, species with pelagic, planktotrophic larvae, such as *O. validus*, occur mainly in shallow water in the Antarctic, which can be highly disturbed by anchor ice and icebergs. Pearse et al. (1991b) and Palma et al. (2007) suggested that small dispersive planktotrophic larvae would be favored to colonize such disturbed areas but not so favored in stable deeper waters with well-developed communities of sessile suspension feeders. In support of this hypothesis is the fact that the ovaries (and presumably eggs and larvae) of *O. validus* lack chemical defenses, unlike the large leci-

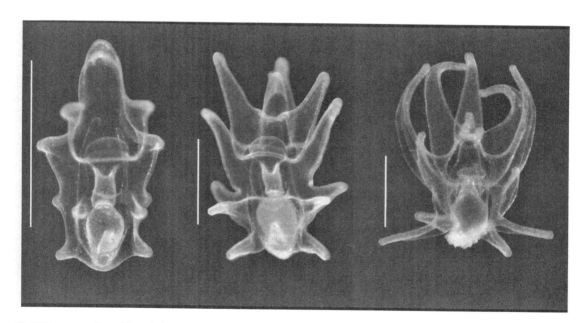

Fig. 12.5. Larvae collected from drift tows near the South Shetland Islands (62°57'43" S; 59°22'57" W) in 2004. The two bipinnaria larvae (*left* and *center*) are identical to those of larvae of *Odontaster validus* reared in the laboratory, approximately 40 and 90 days old. The brachiolaria larva (*right*) has longer arms than larvae reared in the laboratory, perhaps because of less food availability. Scale bars 1 mm. *Bosch and Pearse 1990; photo by E. Balser, with permission.*

thotrophic eggs and larvae of most Antarctic species (McClintock and Baker 1997b), suggesting that selection is weaker on recruits into disturbed areas without many predators.

Extensive year-long sampling of the plankton of Signy Island in the South Orkneys (Stanwell-Smith et al. 1999), Adelaide Island off the west coast of the Antarctic Peninsula (Bowden et al. 2009), and McMurdo Sound (Sewell and Jury in press) found sea star larvae, almost certainly mostly *O. validus,* in winter and spring only. Late bipinnaria larvae were present in spring until December, when the summer phytoplankton bloom began. This timing corresponds to winter–early spring spawning and the 5.5 month developmental time to settlement as found in the laboratory (Pearse and Bosch 1986). As Bowden et al. (2009) point out, the timing also indicates that what selects for the restricted winter–early spring spawning period is not larval food. Instead summer conditions may be critical for recently settled juveniles. Their data indicate that many other species settle in summer and fall, and by settling earlier, juveniles of *O. validus* would be in a position to prey on other new recruits. Recruits also will be able to feed on benthic microalgae growing on the bottom during the summer, at least in shallow water where *O. validus* is most abundant.

Growth and Age

As with other sea stars, it is very difficult to age individuals of *O. validus,* and with such slow-growing animals, also difficult to estimate growth rates. However, considering the weight of the smallest animals with only one cohort of small oocytes (3–5 g wet weight) and the smallest with full-grown oocytes (4–8 g), together with observations that vitellogenesis takes 6 months to a year to complete, Pearse (1969a) estimated that early growth in the field was likely only 1 to 2 g per year, and animals would be between 3 and 6 years old at first spawning. The majority of animals in McMurdo Sound weighing 25–30 g, therefore, would be approximately 20 years old. Some animals reach nearly 100 g, and because growth must slow as animals get larger, they could exceed 100 years in age.

Dayton et al. (1974) took a more direct approach to estimate growth rates of *O. validus* in McMurdo Sound. They tagged animals ranging in size from 7.8 to 33.8 g, released them, and recollected 19 individuals after 10.5 to 12.5 months. Average growth rate was 5.4%. However, similarly tagged and caged animals at an area with more microalgal food doubled their weight, indicating that growth is food limited.

In another study, caged animals held for 1 year at different depths in McMurdo Sound showed no overall change, but large fluctuations in individual weight, from −31% to +39% (McClintock et al. 1988), perhaps reflecting only body fluid content, a problem with all studies based on wet weights (see above, Pearse 1967). But animals of three distinct size groups, held in the laboratory and fed carrion ad libitum for 2 years, showed substantial growth, although at a much slower rate than temperate and tropical species reared under similar conditions (Fig. 12. 6). The pyloric ceca of these lab-reared animals also grew much more than did the field animals (Pearse and Bosch 2002), indicating food limitation in the field. Clearly, growth rate in the laboratory, even though much slower than in sea stars at lower latitudes, is much faster than is likely achieved in the field under most conditions. Moreover, elemental and stoichiometric analyses of *O. validus* by A. Clarke (2008), did not support the hypothesis that high RNA content, with high phosphorus content, should be linked to high growth rates. Levels of phosphorus and other elements in *O. validus,* as well as in

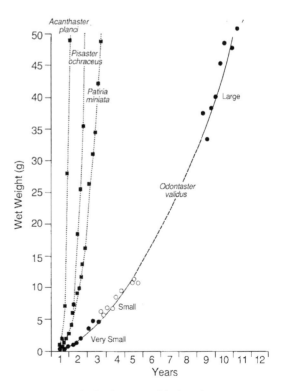

Fig. 12.6. Growth of *Odontaster validus* based on measurements of individuals held in the laboratory with unlimited food for 2 years compared with growth measurements of two Californian and one tropical species. *McClintock et al. 1988.*

other Antarctic benthic species, were similar to those in comparable animals in lower latitudes with higher growth rates.

As with most other sea stars, there is no evidence of senility in *O. validus* or that death is ever age-related, instead of being purely accidental. Individuals in shallow water are sometimes ensnared in anchor ice that is dislodged from the bottom and frozen into the undersurface of the overlying sea ice (Dayton et al. 1969, 1970; see Fig. 12.1), and dried individuals have been found on the shore of Ross Island, apparently swept onshore by waves when the sea is ice free (Pearse, pers. obs.). Predators of *O. validus* include the relatively large and rare asteroids *Macroptychaster accrescens* (Dayton et al. 1974) and *Labidiaster annulatus* (Dearborn 1977), and the large sea anemone *Urticinopsis antarcticus*, which snares prey when they contact the tentacles (Dayton et al. 1970, 1974, Brueggeman 1999).

Ecological Relationships

Populations of *O. validus* can be very large in some shallow-water locations in the Antarctic (e.g., about 2 m^{-2} at McMurdo Station and more than 5 to 25 times that at Cape Evans on the Ross Island side of McMurdo Sound, McClintock et al. 1988, Thrush and Cummings 2011; about 6 m^{-2} at King George Island of the South Shetlands, Palma et al. 2007). Because they are omnivores feeding on benthic microalgae (mainly diatoms) and a wide range of animal prey, both dead and alive, they might be expected to have a major impact on community structure when they are so abundant. Dayton and his colleagues are the only researchers that have attempted to establish such an impact.

Dayton et al. (1974) focused on the regulation of the sponge community dominating the bottom between 30 and 60 m depth of the southeastern corner of McMurdo Sound. *O. validus* preyed in nearly equal amounts on detritus or suspended particles, bivalves (two species), and sponges (three species) and may have had an impact on the last two primarily due to its abundance (about 3 m^{-2}). However, several other sea stars, especially *A. conspicuus* and *O. meridionalis*, and the nudibranch *Austrodoris mcmurdensis*, also consumed sponges, specializing on them. Most of the sponges grew extremely slowly, yet had high-standing crops that could withstand predation for many years; indeed, the standing crops alone of species preyed on by *O. validus* could supply food to all their predators for centuries. The major sponge predator in the community was *A. conspicuus,* and Dayton et al. (1974) suggested that occasional predation on this sea star by *O. validus* partially limited its population and impact on the sponge community. Dayton (1972) and Dayton et al. (1974) further suggested that by preying on recently settled recruits and juveniles of the spongivorous sea stars and nudibranchs, among other animals, as they fed on detritus on the bottom, the large population of *O. validus* could further limit the population of spongivorous predators, allowing the sponge community to flourish. Whether *O. validus* actually preyed significantly on recent recruits and juveniles, however, was not demonstrated.

Dayton (1989) further explored the impact of *O. validus* on a more transient sponge community in shallower water (<30 m) in southeastern McMurdo Sound. At such depths anchor ice can seasonally strip most sessile organisms from the bottom, which is dominated by mobile animals, including *O. validus* (Dayton et al. 1969). Anchor ice formation was limited in the 1970s, and large populations of the fast-growing sponge *Homaxinella balfourensis* became established (Dayton 1989). Despite predation by *O. validus* (which reached a density of about 5 m^{-2}), as well as by *O. meridionalis* and *P. fuscus antarcticus*, both sponge specialists, the population of *H. balfourensis* persisted until a heavy formation of anchor ice developed in 1984, effectively eliminating it.

Because of its apparent impact on sponge communities in southeastern McMurdo Sound through feeding on sponges and one of the sponge predators, and putatively on recruits and juveniles of sponge predators, *O. validus* has been viewed as an important regulator of the sponge community in McMurdo Sound (Barnes and Conlan 2007), and termed a "keystone species" (e.g., McClintock et al. 1988, 2008a,b; esp. 2008b, chap. 8; Kidawa et al. 2010). However, though often used very loosely, the keystone species concept generally refers to "species that exert influences on the associated assemblage, often including numerous indirect effects, *out of proportion* to the keystone's abundance or biomass" (Paine 1995, p. 962, italics mine). As Dayton (1972, 1989) and Dayton et al. (1974) demonstrated, predation on sponges by *O. validus* has little impact on the sponge populations, at least in part because so many of the sponges are chemically defended (Peters et al. 2009). Moreover, as pointed out by McClintock et al. (2006), the impact of *O. validus* on sponge predators, either on adults or putatively on recruits and juveniles, is due to its own high abundance. In contrast, Dayton (1972) and Dayton et al. (1974) argued that, by preying on and limiting the extent of the fast-growing, competitively dominant sponge *Mycale acerata,* the relatively scarce (~1 100 m⁻²) sea star *P. fuscus antarcticus* is closer to being a keystone species in the system they studied. *O. validus,* in contrast, does not feed on *M. acerata.*

Rather than resembling the classic keystone species, *P. ochraceus* (see Chapter 7), *O. validus* may be more comparable to *Patiria miniata,* an abundant, omnivorous asteroid in Californian kelp forests, which putatively feeds on recruits and juveniles (including its own) but has little or no demonstrable ecological role (Harrold and Pearse 1987). Clearly abundant and conspicuous in many shallow-water habitats in the Antarctic (although conspicuously missing in others, e.g., western McMurdo Sound, Dayton and Oliver 1977; Weddell and Lazarev Seas, Gutt and Starmans 1998), ecologically important because of its omnivorous diet and abundance, and an established model species for the Antarctic benthos, *O. validus* nonetheless needs much more in-depth study to establish its actual ecological role.

Acknowledgments

I thank John Lawrence for inviting me to write this chapter on the little sea star I first met and began studying 51 years ago. At the time, I did not realize that that asteroid would become such a model species for understanding life in the Antarctic. I also thank Carlos Barboza, Martin Brogger, Mark Grygier, and Chistopher Mah for help when I was writing the manuscript; Isidro Bosch, John Weller, and Norbert Wu for use of their photographs; Kathy Jones for the drawing; Vicki Pearse for editing and advice; and Isidro Bosch, Andrew Clarke, Paul Dayton, Alexis Janosik, James McClintock, and Lloyd Peck for comments on the draft of the chapter.

13

Acanthaster planci

Katharina Fabricius

Distribution and Biology of *Acanthaster planci*

Acanthaster planci (family Acanthasteridae), also known by the common names crown-of-thorns sea star and crown-of-thorns starfish, is a large coral-eating asteroid that has gained notoriety due to the enormous amount of loss it can inflict on coral reefs and to its large population fluctuations (Fig. 13.1). *A. planci* occurs throughout the tropical and subtropical Indo-Pacific, from the Red Sea, Gulf of Oman and east African coast in the west, to the tropical eastern Pacific including the Galapagos, Panama coast, and the Gulf of California in the east. Its latitudinal limit is about ~33° latitude north and south, being confined by the distribution of coral reefs. *A. planci* does not occur in the Atlantic Ocean. The asteroids typically have 14 to 18 arms and range from 20 to 40 cm in diameter, although individuals >70 cm in diameter have been recorded. They are covered with 4 to 5 cm long poisonous spines and contain several types of saponin toxins that act as anti-predatory agents (see Chapter 8). Depending on geographic region or diet, individuals are purple, purple-blue, reddish grey or brown with red spine tips, or green with yellow spine tips.

Earlier genetic analysis showed the presence of only slight genetic differences among *A. planci* across the Indo-Pacific, indicating long-distance larval dispersal, with genetic similarities matching patterns in connectivity by ocean currents (Yasuda et al. 2009). However, other recent genetic studies based on mitochondrial DNA suggested substantial genetic structure in *A. planci* populations at regional to within-reef scales, indicating that larvae rarely realize their dispersal potential and that outbreaking populations are genetically not different from non-outbreaking populations (Vogler et al. 2008, Timmers et al. 2012).

Due to its environmental relevance, the biology of *A. planci* is relatively well known. Moran (1986) and Birkeland and Lucas (1990) provide comprehensive reviews of the biology and ecology of this asteroid, summarizing the results of 3 decades of significant research investment. Since the 1990s, scientific attention has increasingly shifted to other threats to coral reefs, slowing scientific progress toward understanding this keystone species. *A. planci* are extremely fertile, with a ~40 cm large female sea star producing ~60 million very small eggs

Fig. 13.1. A feeding front of a large population of *Acanthaster planci* on a reef patch in the Great Barrier Reef in 2002. The white patches indicate skeletons of recently eaten corals. *Photo by K. Fabricius.*

during one spawning season. Spawning occurs in midsummer, when seawater temperatures and the probability of monsoonal rainfall are the highest (e.g., Great Barrier Reef: November to January, with a peak in December; Lucas 1982, Babcock et al. 1993). They, therefore, do not participate in the common spring mass spawning event of corals and other invertebrates. Nevertheless, individual *A. planci* aggregate to spawn and synchronize their spawning by releasing pheromones (Birkeland and Lucas 1990). Aggregation and synchronized gamete release contribute significantly to their high fertilization rates, which can reach 80% if males are located <10 m downstream of the females (Babcock and Mundy 1992). *A. planci* larvae are 0.6–1.2 mm in length, planktonic and phytoplanktotrophic. Larval survival is only moderately sensitive to salinity and water temperature, with up to threefold higher survival at 30 ppt salinity and 28°C, compared with higher and lower salinities and temperatures (Lucas 1973). Larval survival is highly dependent on

food availability, as also recorded for some echinoid larvae (Reitzel et al. 2004; see Chapter 6). The larvae feed on nano- and microplankton-sized phytoplankton (6–20 μm cell size), while picoplankton, bacteria, and dissolved organic matter are not used efficiently (Okaji et al. 1997). Laboratory experiments, in which freshly hatched *A. planci* larvae were reared in seawater with natural phytoplankton, showed that the growth and survival rates of larvae increased strongly with increasing concentrations of nanoplankton and microplankton. At low plankton concentrations (below ~0.25 μg chlorophyll L^{-1}, with chlorophyll being used as a proxy measure for phytoplankton), only few larvae developed from the bipinnaria to the early brachiolaria stage, and most started to regress after ~10 days (Lucas 1982, Okaji 1996). Above 0.25 μg chlorophyll L^{-1}, the odds of completing development increased by a factor of eight for each doubling of chlorophyll concentrations, up to ~3.0 μg chlorophyll L^{-1}, at which level completion of development was nearly certain (Fabricius et al. 2010). Also, the time for 50% of larvae to complete development was 14 days at 2.0 µg chlorophyll L^{-1}, compared with 41 days at 0.5 µg chlorophyll L^{-1}. Larvae reared at low chlorophyll levels metamorphosed to significantly smaller asteroids than did larvae reared at high chlorophyll (0.44 mm vs. 0.66 mm diameter; Okaji 1996). A prolonged pelagic phase and small size of post-metamorphosis juveniles may also negatively affect their survival probability in the field (although the larvae appear partially protected against predation as they, like the other life stages of this asteroid, contain predator-deterrent saponins). The following chlorophyll thresholds for larval development were identified (Okaji 1996, Fabricius et al. 2010): At <0.25 µg chlorophyll L^{-1} (<220 eukaryotic cells mL^{-1}), a negligible proportion of larvae complete development, suggesting starvation; at 0.25–0.8 µg chlorophyll L^{-1} (220–670 eukaryotic cells mL^{-1}), the proportion of larvae completing development is still low, development is slow and

body sizes of larvae and juveniles remain small, suggesting severe food limitation; finally, at >2 µg chlorophyll L^{-1} (>1700 eukaryotic cells mL^{-1}), larval developmental success is high, development is fast, and both larvae and juveniles grow to maximum possible sizes, suggesting release from trophic limitation.

During their pelagic phase, the larvae travel passively with the ocean currents. After becoming competent for settlement, they become negatively buoyant and, after coming into contact with a surface, begin to show searching behavior (Okaji 1996). Settlement appears to be chemically triggered by molecules produced by the biofilms on crustose coralline algae (Johnson and Sutton 1994). After metamorphosis into juvenile asteroids (five-armed, ~0.5 mm in diameter), they feed on crustose coralline algae for ~6 months (up to 18 arms, reaching 5–10 mm in diameter; Yamaguchi 1974). The asteroids then begin to feed on corals, a diet which allows them to grow more rapidly, reaching

Fig. 13.2. *Acanthaster planci* feed by everting their stomach through the mouth and extraorally digesting their prey. This image shows the semi-translucent stomach membranes wrapped around a branch of *Acropora* corals. *Photo by K. Fabricius.*

60–70 mm in diameter at the end of their first year. They become sexually mature at the end of their second year (~200 mm in diameter), and have a life expectancy estimated to be about 6 years (Kenchington 1977, reviewed in Moran 1986, Scandol 1999).

A. planci is a specialist coral feeder (corallivore). To digest coral tissue, the asteroid everts its membranous stomach through its mouth (Fig. 13.2), secreting digestive enzymes and absorbing digested coral tissue (extraoral feeding). After 4 to 6 hours of feeding, the stomach is retracted and A. planci moves off, leaving behind a white feeding scar of exposed coral skeleton. An adult asteroid consumes 5 to 6 m² of coral tissue per year.

A. planci has a strong feeding preference for corals of the families Acroporidae and Pocilloporidae (De'ath and Moran 1998a). Acropora is the most preferred coral genus (Fig. 13.3), favored 14 times more than the least preferred coral genus, Porites. A. planci also shows preferences for coral growth forms, with tabular growth forms being preferred 35 times more than massive growth forms, and branching, submassive and foliaceous forms all preferred seven times more than massive forms (De'ath and Moran 1998a). However, A. planci will feed on the less preferred groups of corals and may even occasionally graze on other organisms, including octocorals or algae, if the preferred types of coral food are largely depleted.

Toward the end of a population outbreak, the coral cover on a reef may be reduced from more than 60% to less than 5 or 10%. By that time, the asteroids are assumed to starve and die or move off in the quest for areas with abundant food. Adult asteroids can move up to 10 m/h when food is sparse. In captivity, individuals can live for several months without food; however, the fate of starving asteroids in the field, and their ability to move from one reef to the next over large areas of inter-reefal sandy seafloor remain poorly understood (Ormond et al 1973).

Fig. 13.3. *Acanthaster planci* much prefer corals of the genus *Acropora* to other corals, leading to dense aggregations when food becomes sparse. *Photo by K. Fabricius.*

Ecology of *Acanthaster planci* and the Formation of Population Outbreaks
Spatial and Temporal Distribution of Outbreaks

Typically, *A. planci* are rare, with only a few individuals observed along a kilometer of reef front (e.g., 0.06 asteroids ha^{-1}; Endean 1974, 5–20 per km reef face; Ormond et al. 1973). However, like some other echinoderms with planktotrophic larvae, the species can show large population fluctuations (Uthicke et al. 2009). It is through population outbreaks that *A. planci* causes widespread damage to Indo-Pacific coral reefs. Population outbreaks of *A. planci* were first recorded throughout the Indo-Pacific in the late 1950s and early 1960s; earlier outbreaks are likely but have not been reported (Moran 1986, Birkeland and Lucas 1990). Outbreaks are defined as the density of *A. planci* at which their rate of consumption exceeds the typical long-term average rate of growth of the coral prey. This threshold outbreak density was calculated as >15 individuals per hectare (Moran and De'ath 1992). Depending on reef size and coral cover, an outbreaking population may vary between hundreds and millions of *A. planci*.

A. planci population outbreaks are often classified as one of two types, depending on their origin. *Primary outbreaks* (Endean and Stablum 1973) are sudden population emergences after a period of rarity, probably due to some changing environmental conditions (Moran 1986, Birkeland and Lucas 1990, Timmers et al. 2012). The question as to what causes such primary outbreaks has remained contentious in the scientific community (see below). *Secondary outbreaks* are those that have resulted from massive larval supply from a primary outbreak. Because these secondary outbreaks are easily explained by the large number of larvae produced by a primary outbreak and by hydrodynamic modelling of *A. planci* larval transport (Black et al. 1995), the causes of these secondary outbreaks are not contentious. Secondary outbreaks are well documented from the Great Barrier Reef and the Ryukyu Islands. In other regions, the distinction between primary and secondary outbreaks is less clear, due to a lack of knowledge about their connectivity to potential source populations that may have produced the larvae and the capacity of the larvae to travel over long distances.

On the Great Barrier Reef, three series of *A. planci* outbreaks have been recorded, with their presumed primary outbreaks first observed in 1962, 1979, and 1993–94. Each of these outbreaks started at a relatively small section of the central Great Barrier Reef between latitudes 14.5°S and 17°S (Moran et al. 1992, Sweatman et al. 2001, Brodie et al. 2005, Fabricius et al. 2010), the only section where the reef matrix is close enough to the land to frequently encounter river plumes (see below). The first two outbreaks were observed and recorded on one single reef (Green Island), whereas more comprehensive surveys during the onset of the third outbreak (1993–94) showed that populations of *A. planci* had been building up simultaneously on several reefs north of Cairns: the implication is that primary outbreak centers may not be as discrete as previously assumed.

The Great Barrier Reef comprises ~2900 individual reefs within an area of 349,000 km^2, spread along >2300 km of coast. Hydrodynamic models are available to estimate the extent of larval self-seeding, source, and sink levels for reefs in the central and northern Great Barrier Reef (Dight et al. 1990a, James et al. 2002). In agreement with these hydrodynamic models, secondary outbreaks originating from these primary outbreaks gradually spread in predominantly southward progressing waves throughout the central regions until, about 12–15 years later, they reached the southern end of the Great Barrier Reef (Moran 1986, Moran et al. 1992, Sweatman et al. 2008).

There is no doubt that population sizes of *A. planci* can vary by orders of magnitude simply due to the extremely high fecundity of this asteroid, with mature females producing up to 60 million eggs per year (Moran 1986 and references therein). Natural fluctuations in food availability, water temperature, and salinity could all lead to stochastic fluctuations in the planktonic larval survival. The question, therefore, is not whether outbreaks per se are a natural phenomenon (Birkeland and Lucas 1990) but whether the presently observed frequencies are greater than could be expected in conditions without human interference. At least three lines of argument suggest the latter:

1. The presently observed outbreak frequency of about 15 years does not allow for full recovery of coral cover between outbreaks, and consequently coral cover has been gradually declining on many Indo-Pacific reefs (Lourey et al. 2000, Fabricius et al. 2010), with *A. planci* being the most important cause of coral mortality on the Great Barrier Reef (K. A. Osborne et al. 2011)

2. Outbreak waves are less well formed now compared with the 1960s, akin to model predictions of a gradual desynchronization of waves of epidemics, suggesting that the resilience of

reefs has been damaged over time (Seymour and Bradbury 1999)

3. *A. planci* feeding scars on massive *Porites,* which provide historic records of outbreaks, appear to have substantially increased in frequency since the 1960s compared with earlier decades (DeVantier and Done 2007), as have the densities of *A. planci* ossicles in the more recent layers of reef sediment cores when compared with deeper and earlier sediment layers (Fabricius and Fabricius 1992).

However, identifying the main factor or factors controlling *A. planci* populations and hence identifying the cause or causes of the present population outbreaks has remained contentious for decades. The two main theories about what has caused an increase in the frequency and severity of outbreaks are the predator removal hypothesis and the terrestrial runoff hypothesis. The arguments for and against both hypotheses are covered below in detail.

Causes of *Acanthaster planci* Outbreaks: The Predator Removal Hypothesis

The predator removal hypothesis states that *A. planci* populations are controlled by predation and that the removal of predatory fishes through increased fishing pressure has resulted in increased numbers of *A. planci* individuals surviving to maturity (Cameron and Endean 1981). Intense predation of juveniles undoubtedly occurs (Birkeland and Lucas 1990, Keesing et al. 1996). Missing and regenerating arms have been found in 17–60% of juvenile and adult individuals (Moran 1986 and references therein). Small individuals are strictly nocturnal, only emerging from crevices at night when visually oriented fish are least active, suggesting they avoid fish predators (De'ath and Moran 1998b). In contrast, large individuals often also emerge and feed openly during the day, especially during outbreaks, suggesting that their large size and long poisonous spines enable them to partially escape predation.

Known fish predators of juvenile *A. planci* include the humphead maori wrasse (*Cheilinus undulates*), a puffer (*Arothron hispidus*), and two trigger fishes (*Balistoides viridescens* and *Pseudobalistes flavimarginatus*). All of these fish species are presently uncommon, and only *C. undulatus* is targeted by fishers. Emperors (Lethrinidae) and snappers (Lutjanidae), which are abundant and targeted by fishers, may also contribute to predation on juvenile asteroids (Ormond et al. 1990). However, feeding experiments and gut content analyses have shown that these fish take juvenile *A. planci* only reluctantly and often reject them (Moran 1986, Birkeland and Lucas 1992, Sweatman 1995, Mendonca et al. 2010). There is, therefore, little experimental or observational evidence to indicate that predatory fishes targeted by humans actually feed on *A. planci.*

In contrast, invertebrates are known to eat this starfish. Very small *A. planci* (<10 mm in diameter, when still feeding on coralline algae) are even vulnerable to attack by the mesenterial filaments of corals (Yamaguchi 1974), and mortality is estimated to be 99% in the first year after settlement (Zann et al. 1987, Keesing and Halford 1992b). Known invertebrate predators of *A. planci* include the giant triton snail (*Charonia tritonis*), the harlequin shrimp (*Hymenocera picta*), and the annelid worm (*Pherecardia striata*). Recently, the sessile corallimorpharian *Paracorynactis hoplites* has been added to this list of invertebrate predators, with polyps capable of swallowing entire *A. planci* up to 35 cm in diameter (Bos et al. 2011). But all of these are too uncommon or feed too slowly to appear capable of controlling *A. planci* population outbreaks. However, coral reefs host tens of thousands of omnivore or carnivore invertebrate species. This makes it possible that some of these species may have suddenly declined in numbers and, thus, may have played a critical role in *A. planci* outbreaks. If some, as yet unidentified, invertebrate species were responsible for controlling populations, they would need to be widely distributed and abundant and would have to have undergone a worldwide decline in numbers to account for the observed global outbreak patterns in *A. planci.*

There is no doubt that most coral reefs are severely overfished and that the permanent removal of predators can have unexpected consequences for food webs and ecosystems. Three field studies have added support to the fish predator removal hypothesis. Ormond et al. (1990) showed that potential fish predators had lower densities on reefs with recent *A. planci* outbreaks compared with adjacent reefs that had not suffered outbreaks. Dulvy et al. (2004) showed a relationship between *A. planci* densities and human population density across six islands in Fiji and also some relationship between human population density and fish predator densities. Sweatman (2008) also demonstrated that fewer *A. planci* outbreaks occurred at some of the no-take zones of the Great Barrier Reef compared with adjacent zones open to fishing. Since intense research to identify potential predators has led to the conclusion that no commercially exploited fish like to eat *A. planci*, alternative explanations for causal chains are needed. Potentially, the reduced populations of predatory fishes may have triggered some complex trophic cascades, eventually resulting in fewer bottom-dwelling invertebrates that eat juvenile *A. planci* (Keesing et al. 1996, Dulvy et al. 2004, Sweatman 2008). Alternatively, large predators may have stopped eating the toxic and spiny *A. planci*, because intraspecific competition for food between heavily fished predators is now too low to trigger prey-switching behavior. The three apparent correlations between *A. planci* and predator densities therefore require a complex line of argument. It also remains unknown how the removal of predatory fishes from coral reefs contributes to fluctuations in the population densities of *A. planci* that often appear synchronised across regions.

Causes of *Acanthaster planci* Outbreaks: The Terrestrial Runoff Hypothesis

The second hypothesis relating to the causes for an increased frequency in *A. planci* population outbreaks is the terrestrial runoff hypothesis (Birkeland 1982), also known as the larval survival hypothesis (Lucas 1982). As shown above, a large proportion of *A. planci* larvae starve in the absence of phytoplankton blooms. This hypothesis argues that chronic or episodically enhanced nutrient supply (e.g., from terrestrial runoff, the upwelling of nutrient rich deep water, or nutrient resuspension during storms) leads to a high larval survival probability and is an essential factor in the sudden emergence of a large population of a rare but highly fertile animal (Fabricius et al. 2010). The larval survival hypothesis has at least four lines of evidence (Brodie et al. 2005).

The first line of evidence postulates that echinoderm species with numerous planktotrophic larvae, like *A. planci*, have the propensity to exhibit boom and bust population dynamics, whereas those with fewer lecitotrophic larvae do not, confirming larval supply as a key factor in the population dynamics of echinoderms (Uthicke et al. 2009). The second line of evidence is that more outbreaks occur on reefs near high Pacific islands or continental coasts from which terrestrial runoff occurs than on reefs near low atolls without terrestrial runoff. Tellingly, indigenous peoples from high Pacific islands, but not from atolls, have specific names for *A. planci* (Birkeland 1982). The spatial pattern may be related to the fact that nutrients and sediments only run off high continental land masses but not off low-lying coral atolls free of rivers and deep soils. Outbreaks tend to be noted about 3 years after periods of high (especially drought-breaking) rainfall, with the 3 year delay due to the time it takes the metamorphosed sea star to grow to a size where they leave conspicuous scars on the coral skeletons and are less cryptic.

On the Great Barrier Reef, the three known primary *A. planci* outbreaks were all observed 3–5 years after the three largest river flood events on record, in the only region along the Great Barrier Reef where offshore reefs are exposed to river plume derived phytoplankton blooms with chlorophyll concentrations >2 μg L^{-1} on an almost annual basis (reviewed in Brodie et al. 2005, Fabricius et al. 2010). Those outbreaks that were only recorded 5 years after floods contained hundreds of thousands of large individuals that may have been several years old. Data on the 1958–59 floods are sparse, but the 1974 and 1991 floods reduced salinity for >60 days during *A. planci* spawning times, and primary outbreaks were detected a few years later where the floods had intersected the reefs.

The third line of evidence is that the timing and location of outbreaks in oceanic waters appears to coincide with years and locations of high winter ocean productivity fronts (Houk et al. 2007).

Finally, the fourth line of evidence is that chronic *A. planci* outbreak populations are often found in areas of deepwater upwelling or at high latitudes, where naturally high chlorophyll concentrations are found even in the absence of human activities. For example, the reefs in the Gulf of Oman and in the southern offshore Great Barrier Reef both have frequent upwelling events and appear to have near-chronic *A. planci* outbreaks (Sweatman et al. 2008, Mendonca et al. 2010). Other Indo-Pacific regions, including Suva in Fiji (Zann and Weaver 1988; Dulvy et al. 2004) and the Ryukyu Islands in Japan (Nishihira 1987), now also appear to have chronic *A. planci* outbreaks. However, the latter two regions are exposed not only to severe terrestrial runoff but are also subject to overfishing, coral destruction, dredging, and other stressors.

In non-upwelling areas, the apparent increase in *A. planci* outbreak frequencies is attributed to increased runoff of fertilizers, sewage entering the ocean, and erosion of nutrient-rich soils, as recorded throughout many parts of the Indo-Pacific in modern times. Around the world, river loads of nutrients have increased markedly since preindustrial times; anthropogenic fluxes of nitrogen are now considered to be at least twice as high, and of phosphorus two to three times higher than those occurring before the industrial and agricultural revolutions (Howarth et al. 2011). For the Great Barrier Reef, it is estimated that river exports of suspended sediments, total nitrogen and total phosphorus have increased six- to ninefold since colonization (Kroon et al. 2011). River floods, the largest source of new nutrients for the inshore Great Barrier Reef, trigger phytoplankton blooms that average 2 µg chlorophyll L^{-1} and at times exceed 4 µg chlorophyll L^{-1} (Brodie et al. 2005).

An increase in nutrient levels leads to increases in phytoplankton biomass and a shift in the composition of the phytoplankton towards larger eukaryotic size classes such as diatoms, dinoflagellates, and chlorophytes. Based on the experimental results of Okaji (1996), the probability of successful development of *A. planci* larvae could be as much as 60-fold higher during floods with 2.0 µg chlorophyll L^{-1} compared with the long-term average of 0.53 µg chlorophyll L^{-1} for the central and northern Great Barrier Reef (Fabricius et al. 2010). Such a strong and non-linear dose-response relationship would confirm the important link between concentrations of natural phytoplankton in the water column (which can vary by an order of magnitude or more between flood and ambient conditions in tropical waters) and sudden increases in population densities of *A. planci*. Due to the enormous numbers of *A. planci* larvae produced by the large number and dense aggregations of individuals in primary outbreak populations, secondary outbreaks are then almost inevitable—even at only moderate phytoplankton abundances—on hydrodynamically connected reefs with sufficient coral cover (Moran 1986, Moran et al. 1992, Sweatman et al. 2008).

An *A. planci*–coral population model, based on field data of *A. planci* and chlorophyll in the Great Barrier Reef, on estimates of historic increases in river nutrient loads and on the laboratory larval survival results, also showed that river floods and regional differences in phytoplankton availability are related to spatial and temporal patterns in *A. planci* outbreaks (Fabricius et al. 2010). The model showed that, if chlorophyll levels remained <0.4 µg L^{-1}, outbreaks would be infrequent (one per ~60 years). Floods could initiate outbreaks, but levels of chlorophyll had to be moderate for the outbreak waves to be sustained. This study, together with previous evidence (Birkeland 1982, Brodie et al. 2005, Houk et al. 2007), has added new and strong support to the hypothesis that *A. planci* outbreaks are predominantly controlled by three drivers, all food-resource related: (1) high chlorophyll concentrations during spawning time that allow for a large number of *A. planci* larvae to complete development in the water column, (2) moderate long-term chlorophyll availability that maintains secondary outbreak waves, and (3) sufficiently high coral cover to

provide food for growth of a large population of *A. planci* after metamorphosis.

Whether the availability of crustose coralline algae (the trigger for settlement and metamorphosis and the initial food for freshly metamorphosed *A. planci*) is a fourth driver for *A. planci* populations remains unclear. On the Great Barrier Reef, outbreaks are rare on muddy inshore reefs (Lourey et al. 2000, Sweatman 2008), where crustose coralline algae are limited by high levels of sedimentation, but other factors—such as small reef size and poor connectivity to larval source reefs—may equally contribute to the observed difference in outbreak frequencies between inshore and midshelf reefs.

Effects of *Acanthaster planci* Outbreaks on Coral Reefs

In many Indo-Pacific coral reefs, coral cover has been declining at a rate of 0.2–1.5% per year since the 1960s (Bruno and Selig 2007). A comprehensive long-term monitoring program has shown that predation by outbreaking populations of *A. planci* was responsible for 37% of all recorded loss in coral cover on the Great Barrier Reef between 1995 and 2009. This makes predation by this asteroid a greater cause of coral mortality than any other form of disturbance, including storms (33.8%), coral diseases (6.5%), and coral bleaching (5.6%) (Osborne et al. 2011). However, the extent of coral loss from *A. planci* outbreaks varies within and between reefs. Sixty-five percent of surveyed reefs in the central Great Barrier Reef had experienced a recent *A. planci* outbreak in the 1980s, with a mean loss in coral cover on outbreaking reefs from >30% to ~9% (Lourey et al. 2000).

Coral reefs subjected to *A. planci* outbreaks follow classic Lotka-Volterra cycles in coral cover and *A. planci* abundance: When sufficient coral food is available (high coral cover), the increase in numbers of asteroids marks the beginning of an outbreak, leading to declining coral cover. At the height of the outbreak, there are large numbers of *A. planci* and large amounts of recently dead coral (low live coral cover), which is typically followed by rapidly declining numbers of *A. planci*, due to food shortage. Such outbreaks may last for 3–5 years on larger reefs, but small coral patches may be depleted within a few months. The recovery phase of a reef is characterized by very low numbers of *A. planci*, while live coral cover gradually increases. The separation of about 15 years between outbreaks has been attributed to the time it may take a reef to reestablish sufficiently high coral cover to feed another cohort of asteroids.

The rate of coral recovery strongly depends on coral growth, varying markedly between species, depth zones and latitudes, and the connectivity of a reef to sources of coral larvae (Scandol 1999, Lourey et al. 2000). Some shallow-water reefs may only require 10–15 years to recover from *Acanthaster planci* predation (e.g., Done et al. 2010), especially where the fast-growing tabulate *Acropora* are prolific and high densities of grazing fish prevent a phase shift to algal dominance. Recovery may, however, still be incomplete in deeper areas and in less current-exposed regions after 20 or even 50 years. On the Great Barrier Reef, long-term monitoring data suggest recovery trajectories varying between reefs from 5 to >1000 years, with a median of 23 years ($N = 56$ reefs; recovery being defined as an increase in coral cover by 30%; Lourey et al. 2000).

Not only the location but also the feeding behavior of the asteroids can determine the speed of recovery. Small colonies, and patches of tissue on corals with complex colony shapes or on the bases of massive corals, are often left behind by the "sloppy" feeding of the asteroids. As corals are modular organisms, they can often regenerate from small patches of surviving tissue. Tissue remnants may rapidly spread over freshly grazed coral skeleton, facilitating much faster reef recovery than would be achieved by way of larval settlement, metamorphosis, and growth. The role of tissue remnants is especially important on reefs that are hydrodynamically poorly connected to larval sources from other reefs.

The *A. planci*–coral population model (Fabricius et al. 2010) showed that, at present concentrations of chlorophyll, outbreaks frequencies on the Great Barrier Reef may have increased from one in 50–80 years to one every ~15 years. This indicates that, given the measured coral recovery rates, coral cover on reefs in the central Great Barrier Reef might have been nearer to 60% than to 20% in the past and that cover may now be only 30–40% of its potential value, all else being equal.

Mitigation of *Acanthaster planci* Outbreaks

A 15-year return interval of *A. planci* population outbreaks is clearly unsustainable for most reef systems, especially because coral mortality from a warming climate (leading to increasing damage from coral bleaching and cyclones), acidifying seawater, and land-based pollution is also increasing. Bleaching, cyclones and acidification are not amenable to local management. There is therefore an urgent need to investigate where, when, and how active intervention to manage

A. planci populations can be effective, to minimize further coral loss by these asteroids (last reviewed in Moran 1986, Zann et al. 1990).

At individual sites, the most cost-effective way to remove starfish is for divers to inject sodium bisulfate (also known as sodium hydrogen sulfate, $NaHSO_4$, a biodegradable granular acid, commonly available as swimming pool "dry acid") into each of the arms. The injection of copper sulfate has been stopped to prevent copper pollution, which can, even in small doses, affect the health of other reef organisms. Injection is a relatively effective method because it minimizes handling of the spiny sea stars, which are often in crevices or wrapped around corals. Alternatively, *A. planci* are being collected by divers, placed into large bags and lifted to a barge or similar floating platform. A major problem with collecting *A. planci* is the real potential for a stress release of gametes, which, if male and female individuals are packed together, leads to high fertilization success and the reseeding of the reef with large numbers of larvae. Collected animals and all fluids seeping from them therefore have to be retained in containers and taken back to the shore or heat destroyed, otherwise the collections could become extremely counterproductive. Cutting individuals into pieces and leaving them in situ has been dismissed as a control mechanism due to concerns that it might aid propagation, as many asteroids are capable of regeneration from body parts (Moran 1986).

Not surprisingly, mitigation programs are very expensive. It is also unclear whether any control programs, despite significant investment of resources and the removal of large number of individuals, has significantly changed the trajectory of outbreaks. The chance of success would appear to improve if removal programs were focussed on preventing primary outbreaks, that is, aggressively reducing population sizes within 1–2 years after drought-breaking floods in flood-affected reefal and inter-reefal habitats, before populations become sexually mature. Such mitigation program would entirely focus on reducing the likelihood of developing a primary outbreak rather than fighting secondary outbreaks. Successful mitigation, therefore, would need to be preplanned, funding-guaranteed, initiated as soon as possible after the early signs of an aggregation have been detected, and have a commitment to continue the program until sea star densities are low in the whole high-risk area. Even a 3-year reduction in the frequency of a primary outbreak, e.g., from a 15- to 18-year interval, might make a significant difference for the resilience of coral reef systems, because it would give reefs more time to move into a later successional stage with higher coral cover and coral larval production. However, depending on resource availability, such searches are probably feasible only at the scale of a few tens of reefs. To protect corals from predation once outbreaks have established is even more difficult and probably feasible only at the scale of hectares.

Other long-term strategies are needed to reduce the frequency of *A. planci* outbreaks. First, improved soil and nutrient retention on land, leading to a reduction in phytoplankton biomass to summer values of <0.4 µg L^{-1} appears the best option to reduce the frequency of primary *A. planci* outbreaks and the persistence of secondary outbreaks (Fabricius et al. 2010). Second, active interventions are needed to minimise starfish numbers in high-risk areas before aggregations form. Third, as an additional safeguard, permanent fishing closures might be established in high-risk areas that are most commonly inundated by floods. This three-pronged approach appears the best presently available strategy to protect coral reefs from unsustainable *A. planci* predation and to prevent further coral loss on Indo-Pacific reefs at a time of increasing climate instability.

14

Oreaster reticulatus

Robert E. Scheibling

*O*reaster reticulatus (LINNAEUS, 1758) is one of two members of the tropical family Oreasteridae known from the Atlantic Ocean. A. M. Clark and Downey (1992) give the distribution of *O. reticulatus* as western Atlantic, ranging from South Carolina to Brazil and throughout the Caribbean, and the distribution of *Oreaster clavatus* as Cape Verde Islands and west coast of Africa. *O. reticulatus* commonly occurs on open sand bottoms or seagrass beds at depths of meters to tens of meters (Scheibling 1980a,b, Scheibling and Metaxas 2001). It also is found in mangrove lagoons, rubble beds surrounding coral reefs, and habitats characterized by interspersed patches of seagrass and coral (Hendler et al. 1995, Wulff 1995, 2000, Guzmán and Guevara 2002, Scheibling and Metaxas 2010). *O. reticulatus* is the largest member of the Oreasteridae, with a radius (from the center of the disc to the tip of the arm, R) of up to 23 cm (Scheibling 1980c). It has a high, arched disc and is typically five-armed. Four- or six-armed individuals are uncommon, and seven-armed ones are rare. Adults are brightly colored, with red, orange, tan or yellow papular areas usually contrasting with the reticulate pattern of tubercles and short spines (Plate 16). Juveniles (R <12 cm) are usually green with a mottling of orange, brown, tan, or gray (Scheibling 1980b, Guzmán and Guevara 2002, Scheibling and Metaxas 2010). Unlike most tropical sea stars (with the exception of *Acanthaster planci*), the biology and ecology of *O. reticulatus* is fairly well known. Nearly all of this knowledge has been acquired within past 35 years in studies that are centered in the Grenadines, U.S. Virgin Islands, Bahamas, Panama, and Belize and that form the basis of this review.

Population Abundance, Spatial Distribution, and Size Structure

Across its geographic range, *O. reticulatus* generally occurs at densities of 1 to 10 individuals 100 m^{-2} in populations ranging from hundreds to thousands of individuals, even on open sand bottoms with no discernable boundaries (Scheibling and Metaxas 2001, 2010, Guzmán and Guevara 2002). Maximum densities of 13 to 18 individuals 100 m^{-2}

have been recorded at a few sites in the U.S. Virgin Islands (Scheibling 1980b), Panama (Guzmán and Guevara 2002), and Belize (Scheibling and Metaxas 2010). Biomass (wet weight) estimates range from 1 to 5.5 kg 100 m^{-2}, and reflect variation in both density and size structure among populations (Scheibling and Metaxas 2010).

Within populations, individuals tend to be randomly dispersed (at scales of 10–100 m^2) in relatively homogeneous habitats such as open sand flats, and aggregated in patchy habitats such seagrass and coral substrata, reflecting spatial variation in food resources and shelter (Scheibling and Metaxas 2001, 2010). A shift to a more aggregated dispersion was observed during the reproductive period in populations on sand bottoms in St. Croix, U.S. Virgin Islands (Scheibling 1980b) but not in the Bahamas (Scheibling and Metaxas 2001). In a large sand patch amid a dense seagrass bed in St. Croix, *O. reticulatus* formed migratory feeding aggregations, a phenomenon that may be associated with spatial containment.

Size-frequency distributions of *O. reticulatus* on open sand bottoms generally approximated normal distributions, whereas size distributions in patchy seagrass, coral, and mangrove habitats were skewed to the right by the presence of juveniles and small adults; this resulted in smaller mean size in patchy habitats compared with populations on sand bottoms (Scheibling and Metaxas 2001, 2010, Guzmán and Guevara 2002). The average size of *O. reticulatus* appears to be inversely related to population density both within locations and throughout the Caribbean (Scheibling and Metaxas 2001, 2010). This relationship is strongly evident on sand bottoms (Scheibling and Metaxas 2010) but also was observed across a range of seagrass and coral habitats in Panama (Guzmán and Guevara 2002). A similar relationship exists between average size and population biomass but as a negative exponential function (Scheibling and Metaxas 2010). This pattern may result from intraspecific competition for particulate food resources, resulting in reduced or even negative growth (Scheibling 1980a). The relationship is less evident for populations in habitats such as patches of sparse seagrass adjacent to dense seagrass beds, coral reefs or mangroves, where macrofaunal prey such as sponges and sea urchins augment the sea star's microphagous diet (Scheibling 1980a, Wulff 1995, Scheibling and Metaxas 2010). Alternatively, an inverse relationship between body size and density may be related to differences in recruitment rate among populations (Guzmán and Guevara 2002), consistent with the high proportion of juveniles

found at sites where density was greatest (Scheibling and Metaxas 2001, 2010).

Populations of *O. reticulatus* in the Grenadines changed little in density, spatial distribution, and size structure over a 20-year interval, suggesting a high degree of stability in the absence of significant human perturbation (Scheibling and Metaxas 2001).

Food and Feeding Behavior

O. reticulatus is an omnivore that feeds on a wide variety of benthic organisms by extra-oral eversion of the cardiac stomach (Scheibling 1980d, 1982a, Wulff 1995). The feeding mode varies among habitats in relation to the availability of macrofaunal prey, and can be categorized by the size of food items as (1) microphagous substratum grazing or deposit feeding and (2) macrophagous predation or scavenging. In the Grenadines and St Croix, Scheibling (1982a) found that *O. reticulatus* primarily fed on epipsammic and epiphytic microorganisms and particulate detritus on sand bottoms, filamentous algal mats, and seagrasses (mainly *Halodule wrightii*). Microphagy accounted for 99–100% of feeding observations in seven of eight populations. In the remaining population, predation and scavenging of sea urchins and sponges accounted for 10% of feeding observations (total observations: 10,328). In the San Blas Islands, Panama, Wulff (1995) observed a much greater frequency of macrophagy (65% of 840 feeding observations), primarily on sponges, in populations on sand, seagrass, and coral rubble substrata surrounding coral reefs. *O. reticulatus* also feeds opportunistically on carrion, including dead fish and crabs, although this was a negligible component of the diet in sampled populations (Scheibling 1982a, Wulff 1995).

Microphagous Grazing and Deposit Feeding

In the microphagous mode, *O. reticulatus* accumulates organic-rich substrata beneath the disc through the raking action of the tube feet along each ray (Scheibling 1980d). This moves sediment or grass blades toward the mouth, while rotating the animal about its oral/aboral axis, to form a mound or blade cluster on which the cardiac stomach is everted. This specialized feeding behavior enables *O. reticulatus* to draw on food-rich substrates over an area defined by its radius while retaining the vulnerable stomach beneath the protective disc. On carbonate sand, the chlorophyll content of the "feeding mound" is two to four times greater than that of the raked area, which appears as a lighter colored halo around the mound (Scheibling 1980d). A diverse array of particulate matter—including diatoms, dinoflagellates, cyanobacteria, meiofaunal crustaceans and polycheates, and detrital aggregates—have been observed in the cardiac stomachs of individuals feeding on sand bottoms (Scheibling 1979, 1982a).

The rotary-raking behavior of *O. reticulatus* may be initiated or terminated by chemical or tactile cues from the organic content of sediment in contact with the tube feet (Scheibling 1980d). The unusually large and extensively eversible cardiac stomach of *O. reticulatus* (Anderson 1978) maximizes contact between this flagellary-mucus feeding organ and the substratum, and the accumulation of sediments in a mound beneath the disc may serve to further increase the stomach-substratum interface (Scheibling 1980d). Anderson (1978, 1979) describes specialized structures in the digestive system of *O. reticulatus*, including enlarged Tiedemann's pouches, highly flagellated and folded structures in the pyloric ceca, and pronounced intestinal ceca, as adaptations to enhance water circulation through the gut. These specializations may function in microphagy to pump particle-laden interstitial water from sediments, clusters of grass blades or filamentous algal networks into the gut, as opposed to suspension feeding from the water column as Anderson (1978) proposed. *O. reticulatus* is maximally inflated while substratum feeding and periodically expels water forcefully from the anus (Scheibling 1982a). The increased grain size of sediments in the feeding mound formed by raking may facilitate a putative pumping action by increasing porosity (Scheibling 1980d).

Macrophagous Predation and Scavenging

The incidence of macrophagous predation and scavenging by *O. reticulatus* is determined by the availability and capturability of suitable prey (Scheibling 1982a, Wulff 1995). In seagrass habitats in the Grenadines, *O. reticulatus* preys on the sea urchin *Tripneustes ventricosus* in proportion to its availability (Scheibling 1982a). At a site in Union Island, predation increased with the density of sea urchins in patches of sand and sparse seagrass, and accounted for 23–25% of feeding observations where *T. ventricosus* was abundant (Scheibling 1982a). At a site in Carriacou, predation rarely occurred due to the scarcity of sea urchins in the seagrass beds (*Halodule wrightii*) inhabited by *O. reticulatus*. *T. ventricosus* was abundant in adjacent beds of *Thalassia testudinum* but inaccessible to *O. reticulatus* due to the limited mobility of the sea star in dense

stands of this seagrass (Scheibling 1980a). Sea urchins that occasionally migrated into beds of *H. wrightii*, or were experimentally transplanted there, were readily consumed by the sea star (Scheibling 1979, 1982a). Long-spined diadematid sea urchins (*Diadema antillarum*, *Astropyga magnifica*, *Centrostephanus rubricingulus*), occasionally observed among populations of *O. reticulatus*, are not preyed upon (Scheibling 1982a, Wulff 1995).

In a large sand patch in St. Croix, *O. reticulatus* co-occurred with the spatangoid sea urchin *Meoma ventricosa* (Scheibling 1982b). Although *M. ventricosa* formed dense aggregations at the sediment surface, predation of the sea urchin by *O. reticulatus* rarely was observed (<1% of 1162 feeding observations), possibly due to a chemical deterrent released by the *M. ventricosa* (Kier and Grant 1965, Scheibling 1982a). *O. reticulatus* occasionally scavenged *M. ventricosa* that were being consumed by cassid gastropods. Observations of predation by *O. reticulatus* on other mobile macrofaunal species are few (Scheibling 1982a, Wulff 1995) and include other sea urchins (*Lytechinus variegatus*, *Clypeaster* sp.), a juvenile sea cucumber (*Holothuria mexicana*), and a sea hare (*Aplysia dactylomela*). Cannibalism is rarely observed in natural populations, although a few sea stars may bear epidermal scars that reflect previous cannibalistic encounters (Scheibling 1982a). Under food-limited conditions in the laboratory, the incidence of cannibalism is greater (Scheibling 1979).

O. reticulatus feeds opportunistically on sponge fragments that are dislodged by wave action and deposited amid sea star populations in sand, rubble, and seagrass habitats (Scheibling 1979, 1982a, Wulff 1995, Scheibling and Metaxas 2010). In the San Blas Islands, Panama, sponges (excluding species that bore into coral rubble) accounted for 39% of feeding observations (Wulff 1995). The high incidence of sponge feeding in populations of *O. reticulatus* in the San Blas Islands appears to be related to an abundant source of sponge fragments in the surrounding reefs (Wulff 1995). In sand cays off Belize, the prop roots of mangroves (*Rhizophora mangle*) also are a rich source of sponge and ascidian fragments, which are consumed by *O. reticulatus* in surrounding sand and seagrass habitats (Wulff 2000, Scheibling and Metaxas 2010).

Herbivory

Benthic diatoms and other microalgae are important components of the largely microphagous diet of *O. reticulatus* (Scheibling 1980d). In seagrass habitats, the sea star frequently envelops grass blades (mainly

Halodule wrightii) and macroalgae within the everted cardiac stomach (Scheibling 1980a, 1982a). The blades are not noticeably altered, suggesting they are primarily a source of epiphytes or extraneous microbial or particulate organic matter. However, various species of filamentous and fleshy red algae are conspicuously discolored (becoming bright orange) by contact with the stomach, indicating chemical degradation (Scheibling 1979). *O. reticulatus* also feeds intensively on mats of filamentous green algae (primarily *Chaetomorpha* sp. and *Enteromorpha* sp.) during periodic blooms in seagrass beds (Scheibling 1980a). These mats are laden with particulate detritus and meiofauna, which likely enhances their nutritional value.

In the laboratory, the feeding frequency of *O. reticulatus* on 10 species of marine macrophytes, individually offered to groups of juvenile sea stars over 4-day trials, was positively correlated with a digestibility index based on the degree and rapidity of change in physical properties (color, cell structure and content, rigidity, tensile strength) during incubation in an enzyme extract of the pyloric ceca (Scheibling 1979). The species most frequently fed upon, and most susceptible to enzymatic breakdown, were fine-structured epiphytic and filamentous green and red algae that likely are utilized directly as a food source. These species also were somewhat sensitive to buffered seawater of the same pH (5.8) as the extract, suggesting the acidity of digestive fluids can enhance chemical lysis of algal tissue. Fleshy and turf-forming branching red algae also were frequently fed upon and moderately digestible. In contrast, seagrasses were infrequently fed upon or avoided, as were a fleshy brown alga (*Dictyota divaricata*) and a filamentous cyanobacterium (*Microcoleus lyngbyaceus*); these species also appeared to be resistant or refractory to enzymatic degradation. In food-choice experiments, *O. reticulatus* strongly preferred an epiphytic green alga (*Chaetomorpha crassa*) above its seagrass hosts (Scheibling 1979).

In another study, juvenile *O. reticulatus* were fed *Chaetomorpha crassa* for up to 14 weeks in the laboratory (Scheibling 1979, 1981c). Bundled filaments enveloped by the cardiac stomach were noticeably degraded and regularly replenished. The pyloric ceca of algal-fed sea stars were dark green and pulpy (compared with a fish-fed group) by the end of the experiment and contained chlorophyll degradation products, indicating digestion of algal tissue. Despite the abundance of food, algal-fed sea stars decreased measurably in size and pyloric cecal mass, indicating that an algal diet alone does not meet the nutritional requirements of *O. reticulatus*.

Nine different carbohydrases were identified in extracts of the pyloric ceca of *O. reticulatus* and enzymatic activity quantified by reducing sugar and viscometric methods (Scheibling 1980e). Oligosaccharides (sucrose, trehalose) and reserve polysaccharides (amylose, laminarin) were readily hydrolyzed, while structural (cellulose, alginate) and mucilaginous polysaccharides (agar, carragheenan) were refractory. Although the carbohydrase spectrum of *O. reticulatus* reflects its capacity to utilize algal material as a food source, it cannot be strictly interpreted as an adaptation for herbivory in this species. Similar enzymes are found in a various sea stars, including carnivorous species (Scheibling 1980e).

Feeding Rates and Activity Cycles

O. reticulatus exhibits a distinct diel cycle of microphagous grazing and deposit feeding in seagrass and sand habitats (Scheibling 1979). Foraging activity is lowest in the early morning, when ~40% of individuals are feeding and increases to a maximum level (~80%) by midday. This peak level is sustained throughout afternoon and probably most of the night before decreasing again before dawn. On sand bottoms in St. Croix, strong tidal currents (20 cm s^{-1}) appear to inhibit feeding activity in the morning but not in the afternoon. Feeding frequency in the morning also may vary in relation to the reproductive cycle. Juveniles in a seagrass habitat in Carriacou were significantly less active than adults were in the morning (25% feeding at 1000) but not during peak activity in the afternoon and dusk. In St. Croix, morning (0900) feeding rate of adult populations on sand bottoms increased during the period of gametogenesis from January to May, while afternoon (1500) feeding remained at a peak level throughout the year. The highest frequency of feeding, ranging from 60–90% during the day, was observed in a large, isolated sand patch where *O. reticulatus* likely was food-limited. While light intensity likely is the primary cue that regulates the periodicity of feeding activity in *O. reticulatus*, food availability appears to determine the amplitude of the diel cycle.

O. reticulatus maintained in a large outdoor flowing-seawater tank and observed at 4-hour intervals for 5 consecutive days fed mainly during the night (91% of all observations) on fine sand, silt, and detritus that settled on the tank bottom (Scheibling 1979). A similar pattern was observed in laboratory feeding experiments among juvenile sea stars provided only with a sand substratum. The diel cycle was obscured when *O. reticulatus* preyed on sea urchins (*Tripneustes ventricosus*) in the laboratory, likely because of the time required for extraoral digestion of large prey (8–12 hours for sea urchins 5–7 cm test diameter; Scheibling 1979).

On sand bottoms in St. Croix, *O. reticulatus* fed at a different location approximately once every hour during 3- to 5-hour periods of continuous microphagous foraging at the peak of the diel cycle (Scheibling 1981a). Given the time allocated to moving between feeding sites (5%; Scheibling 1980d) and sediment handling (mound-building; 15%; Scheibling 1981a), this leaves 80% of foraging time for feeding (extraoral digestion and ingestion). This concurs with independent estimates of feeding frequency at the peak of the activity cycle (Scheibling 1979). Adjusting this rate for diel variation in feeding activity gives estimates of average daily feeding rates of 18–20 feeding sessions individual^{-1} day^{-1}, or 15–17 h feeding individual^{-1} day^{-1}, for these populations (Scheibling 1979). The frequency of feeding and incidence of microphagy is lower at sites where macrofauna are a substantial part of the diet (Scheibling 1979, 1980a, Wulff 1995).

Consumption rates of macrofauna have been measured in the laboratory (Scheibling 1979). In a 7-month experiment in which relative prey abundance was varied at 1- to 2-week intervals, adult *O. reticulatus* preyed on sea urchins (*Tripneustes ventricosus*) in proportion to their availability, both in the presence or absence of sponge fragments (Scheibling 1979). When only sea urchins were provided (1–5 per sea star), predation rates increased linearly with prey availability from 5 to 25 g wet weight consumed individual^{-1} day^{-1}. When sponges were provided ad libitum and when sea urchins were at relatively low abundance (0.4–0.8 per sea star), sea stars fed primarily on sponges, consuming 11–23 g individual^{-1} day^{-1}. When sponges were provided ad libitum and when sea urchins were maintained at high abundance (two per sea star), the consumption of sea urchins (15–25 g individual^{-1} day^{-1}) was two to four times greater than that of sponges (6–7.5 g individual^{-1} day^{-1}). These results are consistent with field observations of density-dependent predation of sea urchins by *O. reticulatus* (Scheibling 1982b) and suggest that sea stars switch from scavenging sponge fragments to preying on sea urchins above a threshold level of sea urchin abundance. The rate of microphagous feeding on the sand bottom of the experimental tank (0.2–0.7 feeding sessions individual^{-1} day^{-1}) was an order of magnitude lower than that of a control group in a tank without macrofaunal food. In another experiment, juvenile *O. reticulatus* fed morsels of raw fish ad libitum consumed 3.9 g individual^{-1} day^{-1} over a 6-week period, while a group fed filamentous green algae (*Chaetomorpha crassa*) consumed 5.6 g individual^{-1} day^{-1} (Scheibling 1979).

Movement and Migration

When feeding in the microphagous mode, *O. reticulatus* exhibits a pattern of directed and limited movement that may optimize net energy gain from particulate food resources that are relatively evenly distributed but in low concentration on the substratum (Scheibling 1981a, 1985). On sand bottoms, *O. reticulatus* frequently moves a distance approximating its diameter between feeding sessions, the shortest distance that avoids overlap of feeding sites. The estimated percentage overlap between successive feeding sites (raked by the tube feet) was 1% and 7% in two populations of differing density and nutritional condition in St. Croix. This pattern of movement minimizes foraging time, allowing more time for extraoral digestion of organic matter in relatively undisturbed sediments. The short distances traveled between feeding sites occasionally were interspersed with longer excursions (Scheibling 1981a). These may occur when *O. reticulatus* encounters a tract of low organic content, perhaps due to recent grazing by conspecifics. On open sand bottoms, individuals exhibit directional but randomly oriented foraging paths (Scheibling 1981a). Directional movement may be adaptive in preventing sea stars from reencountering areas they have recently grazed and in extending the foraging range (area sampled) by maximizing displacement over time. The mean displacement of individuals during 4-hour intervals of diurnal foraging and feeding activity was 6% of the maximum possible based on the average locomotion rate on sand bottoms (~20 cm min^{-1}; Scheibling 1981a).

In a large sand patch amid a dense seagrass bed in St. Croix, *O. reticulatus* formed linear feeding aggregations (or fronts) that migrated throughout the patch (Scheibling 1980c). These fronts form as sea stars encounter sediments with relatively high organic content and reduce their foraging movements and spend more time feeding. Trailing individuals, foraging on recently grazed sediments, move further and eventually join those at the food-rich interface. Sea star density averaged 13 individuals m^{-2} along the leading edge of 3 to 5 m wide fronts and decreased to <0.7 individuals m^{-2} in the wake. Movements of individuals in a front were strongly directed toward the richer sediment. The rate of advance of the front averaged 7 m day^{-1} and varied directly with sea star density along its length. Fronts were deflected at the patch border. Sea stars that accumulated along this boundary gradually moved back across the patch and formed new aggregations in areas where microbial films had regenerated in the absence of intensive grazing over several weeks. Lauzon-Guay et al. (2008) developed an individual-based model that predicted the formation of feeding fronts at discontinuities in food abundance and their rate of advance as a function of sea star density, with a strong concordance between simulated and observed dynamics. However, these feeding aggregations may be atypical for *O. reticulatus*, occurring only among dense populations in a confined habitat (Scheibling 1980c).

Dense populations of *O. reticulatus* may arise in large isolated sand patches due to continual recruitment from and limited emigration to the surrounding seagrass beds (Scheibling 1980c). These beds may serve as nursery areas for juvenile sea stars but adults are rarely found there (Scheibling 1980a,b). In one such habitat in St. Croix, groups of adult sea stars, transplanted from the sand patch to release sites located 20 m into the surrounding seagrass at different positions around the patch perimeter, showed significant directed movement toward the patch, and more specifically towards the nearest aggregation of conspecifics after 24 h (Scheibling 1980f). Chemotaxis may account for this homing movement: sea stars moved to sites downstream of aggregations, relative to tidal currents, were most successful in reaching the sand patch. *O. reticulatus* may orient to waterborne chemical cues emanating from feeding sea stars in the patch, such as mucus or excretory products.

In a sparse seagrass bed (*Halodule wrightii*) in the Grenadines, *O. reticulatus* migrated offshore in response to passing storms and seasonal increases in wave action (Scheibling 1980a). With the resumption of calmer conditions after a tropical storm, individuals moved shoreward to reestablish the predisturbance distribution of the population within a few days.

Growth and Reproduction
Growth in the Field

Growth of *O. reticulatus* in the wild has only been measured for a single population, contained within a large sand patch in St. Croix (Scheibling 1980b). The change in R of marked individuals over 8 months was negatively related to their initial R, which ranged from 11 to 15.5 cm. Only animals below the average size in the population (R = 13 cm) grew by 1–4%; larger sea stars decreased by as much as 3–4%. A decrease in size is attributed to resorption of body-wall tissue

when consumption falls short of metabolic demands. This also was observed in juvenile sea stars fed nutritionally deficient diets in the laboratory (Scheibling 1979, 1981b). The low capacity for growth within this population was consistent with other measures of condition, suggesting food limitation (Scheibling 1980b, 1981b). The ability of *O. reticulatus* to resorb biomass under nutritional stress is of adaptive value in preventing mortality by starvation (Scheibling 1980b). The total energy requirement of an individual is directly related to its body size, and resorption may function as a population regulatory mechanism that ultimately maintains biomass at the carrying capacity of the habitat.

The characteristic unimodal and approximately normal size-frequency distribution of *O. reticulatus* (Scheibling and Metaxas 2001, 2010, Guzmán and Guevara 2002), particularly among adult populations on sand bottoms, is consistent with a low growth rate. As juveniles recruit to these populations, their growth decreases such that different cohorts are indistinguishable by size and meld together as a single adult mode. The low growth rate indicates longevity in this species that can attain a maximum R of 20–22 cm, although age or size-at-age has not been measured (Scheibling 1980a,b, Scheibling and Metaxas 2001, Guzmán and Guevara 2002).

Spatial and Temporal Patterns in Reproduction

O. reticulatus is gonochoric, and adult populations have an unbiased sex ratio (Scheibling 1981b, Guzmán and Guevara 2002). Individuals become reproductively mature at R~12 cm (Scheibling 1981b), although reproductive gonads have been observed in individuals small as ~7 cm (Guzmán and Guevara 2002). Reproductive periodicity of *O. reticulatus* has been documented using organ index methods (organ weight as a percentage of total body weight) and histological observation (Scheibling 1981b, Guzmán and Guevara 2002). In St. Croix and the Grenadines, populations exhibited an annual reproductive cycle. The gonad index (GI) increased in spring to a peak in summer or fall and declined as the populations spawned between July and October (Scheibling 1981b). The pyloric ceca index (PCI) followed a reciprocal annual cycle, increasing during the quiescent phase of the gametogenic cycle in winter and gradually decreasing with active gametogenesis and gonadal growth. The reciprocal relationship between the two indices, which is common among sea stars (Lawrence and Lane 1982), is attributed to the translocation of nutrient reserves from pyloric ceca to gonads for gametogenesis (Scheibling 1981b).

O. reticulatus on sand bottoms in St. Croix showed a high degree of spawning synchrony (a sharp decrease in mean GI within 1 month), although the amplitude and phase of the GI cycle differed among populations (Scheibling 1981b). Spawning was less synchronous in a population in a seagrass bed in the Grenadines, where the GI decreased gradually between July and September from a peak in June (Scheibling 1981b). Metaxas et al. (2002) found that *O. reticulatus* from sand and seagrass habitats in the Bahamas were either ripe or had recently spawned in August and September, consistent with the peak of annual cycle observed in St. Croix and the Grenadines. In contrast, Guzmán and Guevara (2002) observed asynchronous gonad development in a population in Panama and identified three peaks in GI (July, November, February) over a 13-month period. Individuals at different gametogenic stages, including fully mature, were observed throughout the year, suggesting multiple spawning events.

Environmental factors such as water temperature and photoperiod may regulate reproductive cycles in *O. reticulatus* (Scheibling 1981b). Guzmán and Guevara (2002) suggest that asynchronous, year-round spawning in Panamanian populations may be related to a protracted period of high temperature, compared with more northern areas of the Caribbean (Scheibling 1981b). They also note that peaks in GI in July and November coincide with periods of heavy rainfall, suggesting that changes in nutrients or salinity due to runoff may affect gonadal maturation and spawning. The low mean values of peak GI in Panama (<5; Guzmán and Guevara 2002), compared with populations in St. Croix and the Grenadines with a distinct annual cycle of reproduction (up to 17 for females; Scheibling 1981b), is consistent with multiple spawning periods in the Panamanian population.

Female *O. reticulatus*, like many sea star species (Lawrence and Lane 1982), tend to have a higher GI than males during the active phase of the reproductive cycle (Scheibling 1981b, Guzmán and Guevara 2002). This reflects a greater energetic investment by females in reproduction. Differences in maximum GI (a measure of reproductive output) among adult populations in the same geographic area are attributed to differences in nutritional conditions between habitats (Scheibling 1981b). In populations on sand bottoms in St. Croix that fed almost exclusively on particulate organic matter, maxima in mean GI and PCI (a measure of nutrient reserves), as well as mean size,

were inversely related to density, suggesting intraspecific competition for food limits reproduction and nutrient storage in these habitats (Scheibling 1981b). Differences in the degree of competitive interaction were indicated by the turnover rate of surface sediments by feeding sea stars, which increased twofold across the range of population density (Scheibling 1979). Differences in reproductive output, storage capacity, and size between populations in seagrass beds in the Grenadines were attributed to the availability of macrofaunal prey (mainly sponges and sea urchins) that enabled sea stars to supplement microphagy with opportunistic scavenging or predation (Scheibling 1981b, 1982a).

Dietary Effects on Growth and Reproduction in Laboratory Experiments

Laboratory feeding studies have shown that inclusion of macrofaunal prey in the omnivorous diet of *O. reticulatus* can substantially augment growth and reproduction. Adults (12.5–15.5 cm R) fed sponges or sea urchins (*Tripneustes ventricosus*) in the laboratory grew minimally (up to 0.05 cm month^{-1}), whereas those maintained only on sand collected from the natural habitat progressively decreased in mean R by 0.19 cm month^{-1} (Scheibling 1979). The 7-month experiment was conducted over the period of active gametogenesis (November–June) of the natural population. The mean PCI of fed sea stars did not decrease significantly as the mean GI increased to 7.9 during the experiment. In contrast, the mean PCI of sea stars deprived of macrofaunal food decreased from 9.1 to 2.1 and GI was minimal (0.3). These results indicate that a microphagous diet alone does not sustain *O. reticulatus* in the laboratory, resulting in marked resorption of body wall and cecal tissues and arrested gametogenesis. Although these sea stars spent at least 33% of their time deposit feeding, the frequency of microphagy in nature is usually two times higher and microbial and meiofaunal food resources are more rapidly replenished.

Juvenile *O. reticulatus* (5–11 cm R) fed raw fish over a 14-week period grew in proportion to the ration provided, up to 0.09 cm week^{-1} under ad libitum feeding in the final 6-week period, and with a constant growth efficiency (Scheibling 1979, 1981c). The PCI also increased on a fish diet, by 56% over the experimental period. In contrast, individuals maintained on a diet of filamentous green algae decreased in R and PCI, indicating tissue resorption. They also were prone to cannibalism. Subgroups that were switched to the alternate diet after 8 weeks grew when fed fish and shrank when fed algae in the final 6 weeks. The respiration rate of sea stars fed fish was twice that on the algal diet (15.8 vs. 8.1 mL O_2 h^{-1}). Differences in secondary production and metabolic rate between diet treatments are attributed to differences in energy content and biochemical composition of the two food types.

Spawning and Fertilization Success

O. reticulatus is a free-spawner, releasing eggs or sperm into the water column where fertilization occurs. Spawning has only been observed at night in the laboratory (Scheibling 1979), suggesting it may be a nocturnal activity. Males and females can be chemically induced to spawn in nature (Metaxas et al. 2002). Females remain stationary and rise up on their ray tips while spawning, elevating egg release by as much as 5 cm above the resting posture (personal observations). Naturally spawning females of *Protoreaster nodosus* exhibit a similar behavior (Scheibling and Metaxas 2008). Even small increases in release height within the benthic boundary layer will enhance downstream dispersal of the eggs, as they encounter greater flow rates (Metaxas et al. 2002). The eggs of *O. reticulatus* are large (224 µm diameter) and negatively buoyant, sinking to the bottom within 2–3 minutes under static conditions (Metaxas et al. 2008). Given an average current velocity (~10 cm s^{-1}) measured at height of egg release (~10 cm) on sand bottoms in the Bahamas, eggs could disperse a minimum of 20 m before sinking to the bottom (Metaxas et al. 2008).

In field experiments in sand and seagrass habitats in the Bahamas, fertilization rates of eggs decreased exponentially with downstream distance from a spawning male, from 74% at 1 m to 31% at 32 m (Metaxas et al. 2002). A coupled diffusion-fertilization kinetics model, parameterized using measures of gamete release rate in the laboratory and flow velocity in the field, gave predicted rates that were strongly concordant with the experimental results, along the downstream axis (Metaxas et al. 2002). The high downstream fertilization rates are consistent with those recorded for other sea stars (*Acanthaster planci*, Babcock et al. 1994; *Coscinasterias muricata*, Babcock et al. 2000) and indicate that successful fertilization can occur even at low population densities of *O. reticulatus*. On sand bottoms in St. Croix, populations of *O. reticulatus* showed increased aggregation during the period of active reproduction and spawning in summer and fall, which may enhance fertilization success (Scheibling 1980b).

Fertilization success (per capita zygote production) in free-spawning benthic invertebrates varies directly with the amount of gametes released and inversely with the distance between spawners (Levitan 1995). Therefore, a reduction in inter-individual spacing should increase the proportion of eggs fertilized but may decrease the absolute number of eggs produced if closely spaced individuals compete for food. This trade-off between fertilization rate and fecundity should result in an inverse relationship between body size (which determines gamete production) and population density (which determines nearest mate distance) among mobile animals, like *O. reticulatus*, that utilize a nutritionally sparse but uniformly distributed food resource (Scheibling 2001). This prediction was tested using the diffusion-fertilization kinetics model developed for *O. reticulatus* (Metaxas et al. 2002) to compare fertilization rates at the average nearest-neighbor distances for *O. reticulatus* (R = 17–20 cm) and two smaller but morphologically and behaviorally similar oreasterids, *Pentaceraster cumungi* (15–16 cm) and *Nidorellia armata* (7–8 cm; Scheibling 2001). These species had comparable predicted fertilization rates at their respective nearest-mate distances (estimated from average nearest-neighbor distances measured in the field), suggesting that the reproductive trade-off governs their spatial distribution.

Larval Development, Settlement, and Recruitment

O. reticulatus has a planktotrophic larval stage with a developmental period of 23 days at 23°C (Metaxas et al. 2008), similar to that of *Protoreaster nodosus* (Scheibling and Metaxas 2008). In the laboratory, competent larvae provided with pebbles encrusted with coralline algae explored the substratum by swimming in straight or circular paths, periodically stopping to probe the surface and attach with a preoral or brachiolar arm (Metaxas et al. 2008). Based on the frequency and net displacement of exploratory movements, searching larvae could move up to 7 m (under static conditions) before finally attaching. On final attachment, metamorphosis to the juvenile form occurred within 24 hours. Larvae settled mainly on the undersides of pebbles in both light and dark treatments, indicating a preference for cryptic microhabitats. Recently metamorphosed juveniles (520–655 µm, diameter) moved at an average speed of 0.074 cm min^{-1} over sand and coralline-encrusted pebbles (Metaxas et al. 2008). Although slow, such movements could enable settlers to locate protective microhabitats and reduce the probability of being dislodged or consumed.

Juveniles (R <12 cm) comprised 83–91% of populations of *O. reticulatus* in dense seagrass, coral, or mangrove habitats in Panama and Belize, compared with 0–9% of populations on open sand bottoms (remote from seagrass beds) and 3–29% of populations in sparse seagrass beds in the Grenadines and Bahamas (Scheibling and Metaxas 2001, 2010, Guzmán and Guevara 2002). The greater abundance of juveniles in the Central American populations suggests a higher recruitment rate in this region compared with the insular populations along the Antilles range. This in turn may be related to more continuous reproduction in Central American populations (Scheibling and Metaxas 2010).

The spatial separation of life history stages and cryptic coloration of juvenile *O. reticulatus* suggest larvae preferentially settle in biogenic nursery habitats and juveniles then emigrate to open sand or sparse seagrass flats as they approach sexual maturity and assume the more conspicuous coloration of adults (Scheibling 1980a, Scheibling and Metaxas 2001, 2010). An ontogenic shift may coincide with a size (R~12 cm) at which individuals are less vulnerable to predation and less able to maneuver in a structurally complex habitat (Scheibling 1980a). Habitat-dependent differences in recruitment rate could also be the result of differential post-settlement mortality. Given the intensity of microphagy in populations of *O. reticulatus* and lack of spatial refuges on sparse seagrass and sand bottoms, recent settlers may be particularly vulnerable to consumption by adult conspecifics in these habitats (Scheibling 1980a). A similar pattern of coloration and ontogenic habitat differentiation is observed in *Protoreaster nodosus* in the tropical Pacific (Bos et al. 2008, Scheibling and Metaxas 2008).

Agents of Mortality

Fish have been implicated as predators of juvenile *O. reticulatus* based on indirect evidence and inference. Keller (1976) observed greater recruitment of *O. reticulatus* in caged enclosures than in uncaged control plots in a seagrass bed and no recruitment in cages without tops, suggesting that fish rather than benthic invertebrates are important predators of juveniles. In the Grenadines, small juveniles (R = 2–8 cm) occasionally were found with freshly severed or regenerating ray tips, consistent with fish predation, in dense seagrass beds (Scheibling 1980a). The mottled coloration and cryptic habit of juveniles amid seagrass beds and mangrove banks, and around patch reefs, may be defensive adaptations to limit detection by fish and other visual predators (Scheibling 1980a, Scheibling and Metaxas 2010).

Reef fish may inhibit adult *O. reticulatus* from foraging on reefs where palatable sponges are abundant (Wulff 1995). In Panama, Wulff (2008) observed parrotfish biting the spines of *O. reticulatus* that ventured into reef habitats, although predation was not reported. After the parrotfish were eliminated, sea stars moved onto the reef and readily consumed large sponges that had persisted there for years. Guzmán and Guevara (2002) also found fewer *O. reticulatus* on coral reefs than on mixed reef and seagrass habitats in Panama. The rigid and heavily calcified body wall and network of small spines provide the sea star with an effective armor against most fish predators, especially once it reaches adult size.

The only known natural predator of adult *O. reticulatus* (aside from humans) is the triton *Charonia variegata*. This large snail occasionally was observed preying on the sea star or regurgitating its spines and ossicles on sand bottoms in St. Croix (Scheibling 1980b). In the laboratory, individuals of *C. variegata* consumed between five and nine sea stars (R = 12–14 cm) when fed ad libitum over 10 weeks (Scheibling 1980b). The triton initially ingested the soft internal tissues of the sea star by penetrating the body wall or mouth with its proboscis, before gradually engulfing the collapsed remains. Given the scarcity of *C. variegata* and its low feeding rate, the triton is unlikely to have an appreciable effect on the abundance of *O. reticulatus* (Scheibling 1980b).

As a consequence of its habit and habitat, *O. reticulatus* is conspicuous and easily accessible to humans, particularly as brightly colored, openly foraging adults. In faunal surveys during the early to mid-1900s, the species was recorded as common in shallow sand and seagrass flats along sheltered coasts throughout the Caribbean (Ummels 1963). H. L. Clark (1933) wrote of *O. reticulatus*: "This is undoubtedly the best known of West Indian sea stars, since it has been taken to many parts of the world as a typical curio and souvenir of the region" (p. 22). Centuries of collection of the sea star as a curio or ornament, and more recently for the aquarium trade, has likely caused severe reductions in populations in some areas of the Caribbean where it was previously common (Scheibling 1979, Guzmán and Guevara 2002, Gasparini et al. 2005). *O. reticulatus* is currently listed as endangered in Florida, where state law prohibits its collection (Metaxas et al. 2008).

Today, the dried test of *O. reticulatus* is still a common item is shell shops throughout its range and beyond, and its use as an ornament extends to displays of commercial products in retail stores, magazines, and on the Internet. Apart from the direct effects of collection by humans, the ongoing alteration and destruction of seagrass and mangrove habitats throughout the Caribbean, by various human activities, may well be limiting recruitment rates and the range of natural populations. There are similar concerns about the impacts of over-extraction and habitat change on other oreasterids in the Indo-Pacific, particularly *Protoreaster nodosus*, which also is widely collected for the ornamental and aquarium trades (Bos et al. 2008). The prolonged and escalating decimation by humans likely has had important ramifications to the ecology of the shallow marine communities that these species inhabit.

Ecological Role

Given the intensity of microphagy by populations of *O. reticulatus* (Scheibling 1980d, 1981b), the sea star likely plays an important role in regulating the diversity, distribution, and abundance of epipsammic and epiphytic microorganisms and meiofauna in sand and seagrass habitats. On sand bottoms in St. Croix, surface sediments were turned over within 1 or 2 weeks, depending on the size structure and density of the sea star population (Scheibling 1979, 1980c). The frequency of such bioturbation may maintain micro- and meiobenthic communities at an early successional state, favoring opportunistic species capable of rapid colonization of cleaned surfaces or species resistant to recurrent decimation. In a large sand patch in St. Croix, chlorophyll content of sediments decreased by 64% in the wake of migratory feeding aggregations of *O. reticulatus* (Scheibling 1980c). The grazing fronts swept through different parts of the sand patch at 3- to 4-week intervals, creating large-scale spatial (10s of meters) and temporal heterogeneity in microbial food resources. Such patchiness may influence the distribution of potential macrofaunal consumers, such as sea urchins, crabs and fish that forage at the sediment surface (Scheibling 1982b). The importance of *O. reticulatus* as an agent of biological disturbance, and its effect on the structure and dynamics of benthic assemblages in shallow sand and seagrass habitats, awaits further experimental study.

Wulff (1995) proposed that predation by *O. reticulatus* restricts the distribution of more palatable sponge species to reef habitats, where predatory fish inhibit foraging by the sea star. In food-choice experiments in Panama, sea stars fed on most sponges that normally grow only in coral reef areas (where *O. reticulatus* is rare) but avoided species that coexist with the sea star in surrounding habitats. She also observed that reef sponges living in a seagrass bed, from which *O. reticulatus* had been absent for at least 4 years, were readily consumed when sea stars migrated into the area. *Tedania ignis* is one of many species of sponge readily consumed by *O. reticulatus*, when swept into seagrass beds from surrounding reefs or mangrove roots. An observation of sponges similar to *T. ignis* in a seagrass bed in Belize, where *O. reticulatus* was abundant, led to the discovery of a cryptic sympatric species *T. klausi* that is rejected by the sea star in feeding trials and differs from *T. ignis* in various ecological, morphological and molecular traits (Wulff 2006). Another sponge readily consumed by *O. reticulatus* is *Lissodendoryx colombiensis*, a large and conspicuous species on coral reefs that has expanded its habitat distribution to include a seagrass bed in Belize. *L. colombiensis* is overgrown by a number of smaller sponge species that are unpalatable to *O. reticulatus* in seagrass beds. In field experiments, Wulff (2008) showed that the associated species can deter sea stars from feeding on *L. colombiensis*. This defensive association, in combination with larval recruitment and rapid regeneration and growth, may enable *L. colombiensis* to coexist with *O. reticulatus* in seagrass habitats.

Intensive microphagy by populations of *O. reticulatus* results in the conversion of particulate organic matter, that occurs at low concentrations on sand and algal/seagrass substrates, to energy-rich gametes that are liberated into the water column. Individual gamete release rates, estimated from GI values (18–150 g yr^{-1} for males, 22–171 g yr^{-1} for females; Scheibling 1981c), suggest thousands of kilograms are released over the broad extent of local populations, particularly those on vast sand bottoms. For example, the population of *O. reticulatus* in Buck Island Channel, St. Croix, spanned an area of sand bottom exceeding 2 km^2 (Scheibling 1980a). Given an estimated gamete release rate of 5 g m^{-2} (eggs and sperm pooled, based on the individual release rate and density of adults; Scheibling 1981c), this population alone spawned over 10^4 kg of gametes annually. Eggs form two-thirds of this mass. Many of these eggs will be fertilized and become larvae that can spend weeks in the water column (Metaxas et al. 2002, 2008). Therefore, by broadcast spawning over large areas of seabed, populations of *O. reticulatus* contribute substantially to suspended particulate organic matter in the water column, ranging in size from microns (sperm) to hundreds of microns (eggs, larvae). These particles are available to a broad range of plantivorous and suspension-feeding invertebrates and fish, suggesting that benthic feeding by *O. reticulatus* can have indirect effects on energy flow in both benthic and pelagic food webs.

Perspective for Future Research

Our knowledge of the biology and ecology of *O. reticulatus* is based largely on field studies of populations that were relatively unaffected by human activity. These studies have yielded detailed information on spatial distribution, abundance and size structure of populations, food and feeding behavior, and spatial and temporal patterns of reproduction. Studies of feeding ecology, in particular, have advanced our understanding of foraging strategy and the spatial organization of microphagous species, both through direct observation and predictive modeling. On the other hand, the growth rate of *O. reticulatus* and extrinsic factors that regulate growth throughout the benthic life history (e.g., availability of microbial, microalgal, and macrofaunal food resources, intraspecific competition) is an aspect that needs further study. Large body size allows external tagging by simple scarification (Scheibling 1980c) but the slow growth rate requires monitoring individuals over long periods. The use of internal wire tags or PIT tags may aid in tracking individuals, particularly juveniles, although these techniques should be carefully tested for efficacy (Lauzon-Guay and Scheibling 2008). The frequency and magnitude of recruitment to adult populations of *O. reticulatus* also remains poorly understood. Ontogenic habitat separation has been well documented but the relative importance of larval behavior at settlement (e.g., substratum selection) versus differences in post-settlement mortality between habitats has not been resolved. Manipulative field experiments that compare settlement or recruitment on artificial collectors or natural substrata in different habitats (e.g., sand bottom vs. dense seagrass bed) and between treatments in which potential predators are excluded or included (e.g., Keller 1976) are needed to address these alternative causal mechanisms. Field studies should be coupled with controlled laboratory experiments aimed at identifying potential predators of early life history stages and at examining the role of biogenic spatial refuges in mediating predation rate.

O. reticulatus appears to exhibit a suite of life history traits that may limit its resilience to human exploitation or habitat change, making it particularly vulnerable to extinction. These include slow growth, large size, delayed reproductive maturity, low recruitment rate, low natural mortality rate, and long life span. With increased exploitation, population size could be reduced below a critical threshold required for successful fertilization (Allee effect), leading to recruitment failure and local extinction. To avoid critical reductions in abundance of *O. reticulatus*, it is necessary to relate changes in population size and spatial structure to fertilization potential. The predictive capacity of coupled advection-fertilization kinetics models is promising in this regard (Metaxas at al. 2002), particularly if existing models are extended to incorporate population size and spatial organization (Lauzon-Guay and Scheibling 2007). Model sensitivity analysis could be used to guide future empirical work by identifying important parameters affecting fertilization rate (e.g., spawning synchrony, population size and spatial dispersion, and hydrodynamic features). Model estimates of zygote production could be coupled with biophysical models to examine larval dispersal potential and connectivity among metapopulations of *O. reticulatus*. Studies of larval behavior are needed to inform these models (Metaxas and Saunders 2009). The occurrence of *O. reticulatus* in shallow habitats with low bottom roughness and well-defined flow facilitates parameterization of the hydrodynamic components of these models.

Studies of *Protoreaster nodosus* in the Indo-Pacific region (Bos et al. 2008, Scheibling and Metaxas 2008) indicate many similarities with *O. reticulatus* (e.g., life history traits, foraging habits, growth and reproduction, general population ecology). This suggests that our knowledge of the biology and ecology of *O. reticulatus* can guide future research on other oreasterids or their ecological equivalents in different marine systems (e.g., deep sea, polar regions). More important, *O. reticulatus* can serve as a model to inform management and conservation strategies aimed at protecting shallow-water tropical species in the face of escalating exploitation and habitat alteration by human activities.

15

Heliaster helianthus

Juan Carlos Castilla,
Sergio A. Navarrete,
Tatiana Manzur, and
Mario Barahona

The Genus *Heliaster* and the Fossil Record

The genus *Heliaster* (GRAY, 1840) includes a remarkable group of seven species that inhabit rocky intertidal and shallow subtidal habitats and are morphologically characterized by a large disc and a large number of rays. Geographically, the genus is restricted along the tropical, subtropical, and temperate coasts of the Eastern Pacific. Four species inhabit mainland littoral sites: *Heliaster helianthus* (LAMARCK, 1816) in the tropical, subtropical and temperate west coasts of South America; *Heliaster polybranchius* H. L. CLARK, 1907, in the tropical coast of Peru; *Heliaster kubiniji* XANTUS, 1860 and *Heliaster microbrachius* XANTUS, 1860, in the west coasts of Mexico and central America. Three species inhabit offshore archipelagos: *Heliaster canopus* PERRIER, 1875, in Chile's Juan Fernández Archipelago; and *Heliaster solaris* A. H. CLARK, 1920, and *Heliaster cumingii* (GRAY, 1840) in the Galápagos Archipelago of Ecuador (see J. E. Gray 1840, H. L. Clark 1907, A. H. Clark 1920, A. M. Clark and Mah 2001).

The fossil record of sea stars is scant, but well-preserved whole-bodied specimens of *H. microbrachius* were discovered in a matrix of Pliocene (2.5–3.5 million years old) calcite-cemented quartz sand in southwest Florida (Jones and Portell 1988). Nowadays, this species is present only along the west coast of Central America (Panama) and Mexico (Acapulco). This provides another example of evidence that a tropical seaway permitted the intermingling of the marine fauna of the eastern Pacific and the western Atlantic during most of the Tertiary, before the emergence of the Panamanian isthmus.

Rays and Regeneration

The South American multiradiate sunstar *H. helianthus* (Forcipulatida, Asteridae, Heliasterinae) is the oldest described species in the genus. It is characterized by having a large disc with 30–40 rays, which are free for ca. 30–50% of their length (Madsen 1956). Adult *H. helianthus* are also distinguished from related species by the evolution of the post-metamorphic ray system and of a specific arm regeneration model

(Hotchkiss 2000, 2009). The plane of autotomy is near the disc at the base of the ray and not at the base of the arm (Lawrence and Gaymer in press). Sánchez (2000) described the pattern by which *H. helianthus* increase the number of rays. He pointed out that the sunstars *Labidiaster*, *Rathbunaster*, and *Pycnopodia* exhibit the same pattern. According to Hotchkiss (see Sánchez 2000), the characteristic color pattern observed in *Labidiaster*, in four distinctive quadrants, may be related to the postmetamorphic pattern of adding rays. This color pattern has not been reported in *Heliaster*. The autotomy of *H. helianthus* rays was first described by Viviani (1978).

Biogeography

According to most literature, the geographic distribution of *H. helianthus* (Fig. 15.1A) is confined to the mainland coast of the southeastern Pacific, from northern Ecuador (ca. 2° S) to Valparaiso, Chile (ca. 33° S). Nevertheless, the species extends south of Valparaíso, to Pichilemu: 34°25′ S (Navarrete and Manzur 2008, Manzur et al. 2010). Although it is uncommon from rocky shores south of Pichilemu, individuals are found again at low density at localities around Concepción and inside the Gulf of Arauco, ca. 37° S (C. Oliver 1944, Viviani 1979). There are recent reports of spotty populations of *H. helianthus* in Puerto Ingles, 33°36′ S and Caleta El Soldado, 36° 41′ S (exposed side of Peninsula Tumbes, reports by J. C. Castilla and R. Otaiza) and Caleta Chome, 36°46′ S (Hualpen Penninsula, reports by R. Otaiza).

Habitat and Co-occurrence with Other Sea Stars

H. helianthus is the most common, conspicuous, and ecologically important rocky intertidal sea star along the mainland coast of the southeastern Pacific. Individuals are found from the mid-upper rocky intertidal zone to shallow (<10 m) subtidal habitats (Olivier 1944, Viviani 1978, Castilla 1981, R. T. Paine et al. 1985, Acosta 1988, Tokeshi 1989, Tokeshi et al. 1989, Gaymer and Himmelman 2008, Manzur et al. 2010, Escobar and Navarrete 2011). In Chile, the species undergoes shifts in habitat use with age: recruits and small juveniles are found under boulders and inside crevices in the mid-upper intertidal zone of relatively sheltered rocky zones and move down to the low tidal and subtidal levels as they grow (Castilla 1981, Acosta 1988, Manzur et al. 2010). Along the range of *H. helianthus* in Chile, there are four species of co-occurring sea stars in the rocky intertidal and shallow subtidal habitats. The five-armed *Stichaster striatus* (MULLER AND TROCHELL, 1840) and the batstar *Patiria chilensis* (LUTKEN, 1859) share the low intertidal zone and shallow subtidal habitats with *H. helianthus*. Like *H. helianthus*, these species are rather gregarious and do not show aggressiveness or predation on other sea star species and are apparently not cannibalistic (but see Viviani 1978). In contrast, the subtidal seastars *Meyenaster gelatinosus* (MEYEN, 1834) and *Luidia magellanica* LEIPOLDT, 1895, are large aggressive species, which not only prey on other echinoderms, including *H. helianthus*, but are also cannibalistic (but for northern Chile, see Viviani 1978). Dayton et al. (1977), Viviani (1978), and Gaymer and Himmelman (2008) described interspecific predation, cannibalism, autotomy, and escape mechanisms among these species. *H. helianthus* is a common prey of *M. gelatinous* in shallow subtidal habitats. Gaymer and Himmelman (2008) suggested that this is one of the factors that largely restrict the distribution of *H. helianthus* to intertidal habitats.

Reproduction, Spawning, Larvae and Ontogenetic Changes

Adult *H. helianthus* reproduce sexually, having separate sexes and external fertilization. On occasions it is possible to observe large aggregations of several dozen spawning individuals at low tide in the lower shore (J. C. Castilla and T. Manzur personal observations), but no studies document the timing and frequency of occurrence of these gregarious reproductive events. However, we have observed patches of bio-foam created around single individuals or aggregations of synchronized spawning individuals, lasting from hours to 2–3 days (Fig. 15.1B). For other species of invertebrates (e.g., the tunicate *Pyura prea-*

Fig. 15.1. (Opposite) A, Group of rocky intertidal *Heliaster helianthus* at the Bay of Antofagasta (23°42′ S), northern Chile. The upper top-right individual is eating the sea urchin *Tetrapygus niger* and snails *Tegula atra*. B, *H. helianthus* spawning, during low tide, at the mid-intertidal rocky shore fringe of Las Cruces, central Chile. C, Juvenile *H. helianthus* with 11 arms, at the upper-intertidal fringe in Las Cruces. D, Adult *H. helianthus* eating the muricid *Concholepas concholepas* ("loco") at Antofagasta. E, Adult *H. helianthus* eating the large balanid *Austromegabalus psittacus*.

putialis), the release of gametes into turbulent, well-aerated seawater of decreased surface tension induces the formation of a buoyant bio-foam, which in turn may increase fertilization rate and retention of early development stages of species with short-lived larvae (see Castilla et al. 2007).

The larval development of *H. helianthus* has not been described in detail. Nevertheless, it is known to have a long-lived planktotrophic larva that does not reach competent stage before at least 2 to 3 months of development (S. A. Navarrete, unpublished data). Although it is not known where the larvae are located in the water column, the apparently long larval development suggests this sunstar has long-distance dispersal potential. Long-distance dispersal in this species is also supported by the consistent lack of correlation between prey availability and demographic parameters of *H. helianthus* populations over spatial scales of tens to hundreds of km (Navarrete and Manzur 2008).

After life in the plankton, larvae settle on rocky intertidal shores. Ontogenetic changes in habitat use patterns, diet composition, and diet breadth of *H. helianthus* have been reported for localities in central Chile near the southern limit of its distribution (Manzur et al. 2010) and in central Peru around the center of *H. helianthus* geographic range (Tokeshi et al. 1989). Young and newly settled individuals (<35 mm maximum diameter) are often found in protected/cryptic microhabitats such as under boulders, inside rock crevices, and in depressions in the high rocky intertidal zone and apparently move down to lower tidal levels as they grow (Acosta 1988, Manzur et al. 2010).

Adults are characteristically found in mid and low intertidal zones, as well as the shallow subtidal of wave-exposed and semi-exposed habitats. They occur only occasionally in the upper-high intertidal zone. It has been suggested that size segregation among tidal levels results from microhabitat selection of the settling larvae, which could in turn be an evolutionary response to reduce post-settlement mortality caused by multiple predators in the lower shore and subtidal habitats (Viviani 1978, Gaymer and Himmelman 2008, Manzur et al. 2010). In central Chile, the ontogenetic changes in habitat use by *H. helianthus* are also accompanied by changes in diet composition. This is mostly reflected in changes in the proportion of different prey items and a widening of the diet (Manzur et al. 2010). These differences in diet breadth between size classes may be due to morphometric or mechanical restrictions of small individuals or to changes in prey availability between microhabitats and, consequently, may not necessarily reflect actual prey preferences (Manzur et al. 2010). This is supported by the fact that many of the prey consumed by adults are also consumed by newly recruited individuals, but in different proportions, suggesting no major physiological changes associated with ontogenetic diet changes. In central Perú, Tokeshi et al. (1989) reported low dietary overlap between individuals <10.9 cm in diameter and larger individuals. Nevertheless, in contrast to the situation in central Chile, they reported that smaller individuals had a wider niche breadth than did larger individuals, suggesting increased specialization in older individuals. The apparent difference between the Chilean and the Peruvian populations of *H. helianthus* may be due to differences in prey availability between the study sites. In general, adult *H. helianthus* show a preference for mussels and barnacles, but at sites with low abundance/recruitment of these types of prey, individuals tend to diversify their diets (Navarrete and Manzur 2008). Therefore, the food specialization of adult *H. helianthus* in central Peru may simply reflect higher abundances of the main prey item, the mussel *Semimytilus algosus*.

Recruitment, Growth, and Population Regulation

Probably due to highly sporadic recruitment events, quantification of *H. helianthus* recruitment has proved difficult, at least in central Chile (Manzur et al. 2010). Monitoring of several types of artificial collectors commonly used to quantify sea star recruitment in other rocky shores species, installed at different tidal levels and microhabitats and at sites with different wave exposure, rendered high temporal and spatial variability and overall very low recruitment rates for *H. helianthus* (Manzur et al. 2010, Navarrete and Manzur, unpublished data). Putting together all long-term (ca. 10 years) records on invertebrate recruitment for multiple sites in central Chile (Navarrete et al. 2008, Caro et al. 2010), suggests that recruitment of *H. helianthus* takes place between austral summer and fall, probably peaking in May.

Reports of the growth rate of *H. helianthus* suggest that recruits and small juveniles grow relatively slowly and that growth rates decrease as individuals increase in size and past the most vulnerable recruit stage. Based on length-frequency analyses of high intertidal *H. helianthus* populations, Manzur et al. (2010) reported that recruits and small juveniles (<35 mm; Fig. 15.1C) grow between 20 and 26 mm/

year. This is in general agreement with growth rates reported by Acosta (1988) for small juveniles fed ad libitum on barnacles under laboratory conditions (approximately 20 mm/year) but are much higher than the growth rates reported by Barrios et al. (2008) for individuals between 20 and 40 mm in diameter (12 mm/year). However, preliminary laboratory studies with small adults (ca. 60–100 mm diameter), fed ad libitum on mussels, shows lower growth rates (2 mm/year; S. A. Navarrete, unpublished data). A similar pattern (initial rapid growth and then slow growth) in morphometric growth has also been recognized for Peruvian *H. helianthus* populations (Tokeshi et. al. 1989). Considering the slow growth rate and the large sizes reached by adults (up to 30–34 cm in diameter in central Chile), *H. helianthus* seems to be a long-lived species.

Predation and Interference on *Heliaster helianthus*

In the intertidal zone, adult *H. helianthus* are virtually immune to most predators, but are occasionally consumed by seagulls *Larus dominicanus* (Castilla 1981). Also, large individuals are occasionally collected by tourists and fishers to be sold as ornaments. In the shallow subtidal, *H. helianthus* is attacked by the starfish *M.gelatinosus* (Dayton et al. 1977, Castilla 1981, Gaymer and Himmelman 2008), the rockfish *Graus nigra* (Fuentes 1982) and the crab *Homalaspis plana* (Castilla 1981). Of these predators, the most important is *M. gelatinosus*. *H. helianthus* can represent up to 15% of the diet of this large subtidal sea star at sites in northern Chile (Viviani 1978). Subtidal predation by *M. gelatinosus* on *H. helianthus* has been studied by Gaymer and Himmelman (2008). These authors document strong effects of *M. gelatinosus* on *H. helianthus* subtidal distribution and foraging behavior by direct lethal and sublethal attacks that cause severe damage by loss of rays. The authors also documented strong escape responses in *H. helianthus*, most probably elicited by chemical cues (see also Dayton et al. 1977). Given this escape response to *M. gelatinosus*, sublethal attacks on adult *H. helianthus* are frequent and cause arm autotomization. This may translate into a reduction of feeding rates and a reduction of energy accumulation in gonads (Barrios et al. 2008). Attacks by *M. gelatinosus* on *H. helianthus* result in higher mortalities of juveniles and adults (Gaymer and Himmelman 2008). In addition, cannibalism by large *H. helianthus* on smaller ones has been reported in northern Chile (Viviani 1978). This

could further segregate small individuals to the upper intertidal zone.

Parasites and Commensals

There are no studies documenting parasites on *H. helianthus*. However, adults, particularly at Las Cruces (33°30′ S; 71°38′ W) in central Chile, are often infested by small idoteid isopods (Isopoda, Valvifera, Idoteidae), which inhabit the oral disc and ambulacral groove areas at densities as high as 199 isopods per sunstar (mean = 24). The relationship is most likely commensal, as there is no evidence of host damage, while the protected nature of the oral surfaces of *H. helianthus* provide the isopods with a safe microhabitat in which to feed and reproduce. The isopod, referred to provisionally as *Edotia* sp. TS1, appears to be an undescribed species of *Edotia* Guérin-Méneville that is probably most closely related to the free-living *Edotia dahli* (MENZIES, 1962) that also occurs in the coastal waters of Chile (T. D. Stebbins, personal communication).

The relationship between *H. helianthus* and *Edotia* represents the first report of an idoteid isopod living commensally on a sea star, with only a few idoteids having been reported as symbionts of other invertebrates. These species include *Colidotea rostrata* (BENEDICT, 1898), which is an obligate commensal of sea urchins, *Strongylocentrotus* spp., living along the coasts of southern California, USA, and northern Baja California, Mexico (see Stebbins 1988a,b, 1989), and *Edotia doellojuradoi* GIAMBIAGI, 1925, and *E. magellanica* CUNNINGHAM, 1871, which have been reported as commensals or parasites of bivalves off the coasts of Chile, Argentina, and the Falkland Islands (e.g., Jaramillo et al. 1981, Gonzalez and Jaramillo 1991, A. P. Gray et al. 1997, Zaixso et al. 2009).

Foraging Behavior

Behavioral observations on *H. helianthus* are limited to foraging sunstar individuals for brief (a few hours) periods of time and movements over periods of days. Tokeshi and Romero (1995) give a detailed description of attack behavior on mussel prey, noting that although *H. helianthus* may forage out of the water, searching activity is concentrated during high tide when animals are immersed, thus avoiding the heat and desiccation stress during low tides. Foraging activity is greatly reduced under conditions of strong waves. Studies in central Chile suggest that foraging activity does not vary greatly among seasons

(Navarrete and Manzur 2008) but does vary among sites, depending on availability of sessile and mobile prey (Barahona and Navarrete 2010). In subtidal habitats, *H. helianthus* tends to adjust its foraging behavior and activity to the presence of its main predator, the sea star *M. gelatinosus*, reducing foraging activity and displaying escape responses (Dayton et al. 1977, Viviani 1978, Gaymer and Himmelman 2008).

Home Range and Movement

Adult *H. helianthus* show rather slow movements with a home range in the scale of few meters over time scales of days to weeks. In Peru, Tokeshi and Romero (1995) reported movements of groups of *H. helianthus* of about 20–30 m over a period of a few weeks. In central Chile, Barahona (2006) tracked marked individuals and reported recapture rates of ca. 70% within areas of ~150 m² after 4–5 days. After several months, ca. 20% of the tagged individuals remained in the study area. Therefore, changes in spatial distribution of *H. helianthus* occur at very slow rates. Daily displacements of *H. helianthus* are among the lowest recorded in asteroids, with average displacements of ca. 1 m per day. Individual displacements >5 m/day are rather unusual. In Peru, Tokeshi and Romero (1995) suggested that, within short time periods, movements of *H. helianthus* appeared as random events. This pattern was confirmed by observations in central Chile by Barahona and Navarrete (2010), who showed that *H. helianthus* does not indicate directional movement or homing behavior. These results suggest *H. helianthus* may have a limited capacity to rapidly aggregate to areas of increased prey availability as shown, for instance, in the sea star *Pisaster ochraceus* (Robles et al. 1995). However, a moderate level of opportunistic aggregation to intertidal prey, such as the large balanid *Austromegabalanus psittacus* (Fig. 15.1E), has been observed to develop over periods of weeks (Navarrete, personal observation). In central Chile, the main factors causing variation in daily displacement of *H. helianthus* among sites and from day to day are the abundance of its main prey, the mussel *Perumytilus purpuratus,* and waves (Barahona and Navarrete 2010). Daily displacements are significantly reduced at a given site with a high cover of *P. purpuratus* or facing strong waves. Thus, this predator appears to adjust displacement rates in response to the availability of food and when confronting high risks associated with strong swell.

Prey Escape from *Heliaster helianthus*

H. helianthus has been described as a generalist predator (Castilla 1981), incorporating in the diet a wide variety of sessile and mobile species such as mussels, limpets, sea urchins, snails, and chitons (Castilla 1981, Navarrete and Manzur 2008). Escape responses induced by *H. helianthus* have been observed in several prey: the limpets *Lotia orbigny, Scurria viridula, S. araucana,* and *S. ceciliana* (Espoz and Castilla 2000); key-hole limpets *Fissurella limbata* and *F. crassa,* chitons, *Acanthopleura echinata,* and *Chiton granosus* (Escobar and Navarrete 2011) and the sea urchin *Tetrapygus niger* (Urriago et al. 2011). In addition to escape responses, other types of responses of prey to *H. helianthus* are (1) increased attachment to the substratum by limpets (Espoz and Castilla 2000), (2) changes in the rate of food consumption by the snail *Acanthina monodon* (Soto et al. 2005), and (3) extension of the mantle over the shell with release of secretions by the snail *Trimusculus peruvianus* (San-Martín et al. 2009).

Community Impact

Several experimental studies and observations conducted in central Chile have demonstrated that *H. helianthus* can play a major role in the entire intertidal community by controlling the abundance and intertidal distribution of its main prey species. The first demonstration of *Heliaster* effects in the mid-intertidal shore was conducted at Las Cruces by R. T. Paine et al. (1985). Continuous removal of *H. helianthus* from a stretch of the coast led to an increase in the intertidal distribution and average cover of the dominant mussel *Perumytilus purpuratus* and a concomitant decrease in the cover of subdominant competitors such as chthamalid barnacles. Since mussels and barnacles occupy a large fraction of the space in this community and since mussels in particular are involved in a myriad of positive and competitive interactions with other species (Navarrete and Castilla 1990, Alvarado and Castilla 1996, Prado and Castilla 2006, Caro et al. 2010), the implication of this early manipulation was that *H. helianthus* is a keystone species (*sensu* Power et al. 1996) in the rocky intertidal community. The great impact of *H. helianthus* on mussel beds is undoubtedly related to its ability to crawl atop the beds and simultaneously remove several juvenile and adult mussel individuals and associated fauna in a single bout (R. T. Paine et al. 1985, Navarrete and Manzur 2008).

Of interest is that later studies in the same study area showed that *H. helianthus* is not the only top predator with large impact on intertidal mussel beds in this community. Indeed, the exclusion of humans from a marine reserve at Las Cruces revealed that the muricid gastropod *Concholepas concholepas* could also exert a keystone role in this community and virtually transform the entire seascape compared with open access areas where humans keep the population under check (Castilla and Durán 1985, Durán and Castilla 1989, Castilla 1999). Experimental quantification of predation rates and the strength of interactions at Las Cruces suggest that the impact of *H. helianthus* and *C. concholepas* on mussel and barnacle beds are similar and much greater than those of any other benthic predator in the system (Navarrete and Castilla 2003). Thus—although there is a diverse guild of carnivore predators (several crab, whelk, and fish species) with ample diet overlap that share the intertidal rocky shore (Soto 1996, Castilla 1981, Castilla and Paine 1987)—the ecological impacts of these predators on the community are greatly different (Navarrete and Castilla 2003). Long-term consequences of predation at this study site have been described by Navarrete et al. (2010). However, we still know little about competitive interactions, interference, or facilitation between *H. helianthus* and *C. concholepas* in this system. Large *H. helianthus* are occasionally observed feeding on recruits and juveniles of *C. concholepas* (Fig. 1D), but it is unclear whether this occasional predation has any effect on the intertidal abundance of *C. concholepas*.

One aspect that make effects of *H. helianthus* particularly distinct than those of any other predator in this intertidal community is the ability of adults to remove and feed on large numbers of mussel recruits (ca. 2 mm) before mussels reach juvenile size. Indeed, large *H. helianthus* individuals commonly sit atop algal turfs (mostly *Gelidium chilensis*) in the low intertidal shore and feed on tens to hundreds of mussel recruits at once (Navarrete and Manzur 2008). As expected, experiments show that these high rates of predation by *H. helianthus* prevent mussels in algal turfs from rapidly overgrowing the algal turf and dominate the low intertidal shore (Wieters 2005). The impact of *H. helianthus* on the low shore through the removal of mussel recruits varies from site to site depending on the cover and frond length of the algal turfs (Wieters 2005).

Despite the well-documented responses of scurrinid limpets, keyhole limpets, chitons, and sea urchins to the presence of *H. helianthus* in the intertidal and shallow subtidal zones (Espoz and Castilla 2000, Escobar and Navarrete 2011, Urriago et al. 2011), there are no studies documenting the impact at the population level of these mobile prey species. Nor do we know whether the striking responses in movement and short-term distribution elicited by the presence of *H. helianthus* affects lower trophic levels in the form of trait-mediated cascading effects (e.g., Trussell et al. 2003).

Across a larger region of the coast in central-northern Chile, experimental manipulations show that the effect of *H. helianthus* and other predators on the mid and low intertidal community varies greatly depending on variation in recruitment of the dominant mussel species, *Perumytilus purpuratus*, which at those scales appears to be driven by large-scale changes in oceanographic regimes (Navarrete et al. 2005). Thus, at sites north of about 32° S, experimental removals of *H. helianthus* from the mid-low intertidal zone does not lead to rapid domination by beds of *P. purpuratus* as shown for most sites south of this latitude. This is not to say that *H. helianthus* does not have an impact on prey populations at northern sites in Chile, but the pattern of predation and community-wide ramifications are expected to be quite different than in central Chile. Indeed, it has been proposed that the absence of the dominant mussel prey species due to limiting recruitment can lead to increased predation pressure and population control of other prey, such as mobile grazers (Navarrete and Manzur 2008).

Much farther north along the Pacific coast, studies conducted on rocky shores of Peru have also shown that *H. helianthus* can be the main species controlling patterns of abundance and zonation of rocky shore communities (Hidalgo et al. in press) In the mid intertidal rocky shore of central Peru, the abundance of the mussel *Semimytilus algosus* is higher than that of *Perumytilus purpuratus* at most wave exposed and semi-exposed shores (Tokeshi 1989, Tokeshi et al. 1989). At these sites, *H. helianthus* consumes *S. algosus* over *P. purpuratus* and appears to actively select them over other prey species (Tokeshi 1989). Experimental manipulations in the mid-low shore showed that removal of sea stars can lead to an expansion of the lower edge and overall cover of the *S. algosus* beds, which is modulated by the presence of recruitment mediators such as filamentous algae (Hidalgo et al. in press). There is evidence suggesting that the role of *H. helianthus* at this locality might change over time according to inter-annual variation in mussel recruitment, which could, during some years, swamp predation effects (Hidalgo et al. 2011). However, it appears that under most conditions *H. helianthus* can play a keystone role in these communities dominated by a different mussel species.

In subtidal environments, accounts of *H. helianthus* feeding on the sea urchin *Tetrapygus niger*, the mussel *Semimytilus algosus*, and the snails *Tegula atra* and *Turritella cingulata* suggest that they can have important effects on prey populations and general community structure of rocky and sediment-covered bottoms (Gaymer and Himmelman 2008). However, no experiments have evaluated whether these apparently high rates are sufficient to control prey abundance of any of these species. Experiments have shown, however, that the foraging activity of *H. helianthus* and, therefore, the impact on local populations in shallow subtidal habitats is regulated by the presence of another, larger starfish, *M. gelatinosus* (Gaymer and Himmelman 2008). Thus, in these environments it appears the controlling top predator in the system is the larger *M. gelatinosus* (Dayton et al. 1977, Viviani 1978, Gaymer and Himmelman 2008).

Acknowledgments

We thank John Lawrence for inviting us to write this chapter. Further, we sincerely thank our friends Patricio Manríquez for providing photographs and Evie Wieters for reviewing the final text. Thanks to T. D. Stebbins for providing information and reviewing the section on parasites and commensals. This is a contribution of the Núcelo Milenio "Centro de Conservación Marina," project financed by ICM, Ministerio de Economía, Fomento y Turismo, Chile.

16

Pisaster ochraceus

Carlos Robles

Walking rocky headlands of the Pacific coast of North America, one finds *Pisaster ochraceus,* scattered throughout the low intertidal zone, as solitary individuals or in groups perhaps numbering into the hundreds, showing vivid colors and comprising the greatest biomass of any resident predator. Whether seen through the eyes of lay people or ecologists, *P. ochraceus* are emblematic of the rich communities of Pacific seashores.

P. ochraceus have drawn the concerted attention of ecologists, because their predatory activities shape the varied structures of the rocky shore community from sheltered bays to wave-beaten promontories. On sheltered shores of the Pacific Northwest, *P. ochraceus* reduce the covers of bay mussels (*Mytilus trossulus*) and acorn barnacles (*Balanus glandula*), producing this effect even at the upper reaches of the tides (Harley 2003). On moderately exposed shores, especially those strewn with boulders, *P. ochraceus* influence the foraging behavior and population structure of the abundant herbivorous snail (*Tegula funebralis*; R. T. Paine 1969b). And on wave-beaten shores throughout the coast, *P. ochraceus* set the lower vertical boundary of sea mussel beds (*Mytilus californianus*), the most conspicuous biological feature in the habitat (R. T. Paine 1974, Blanchette et al. 2005, Robles et al. 2009, Pearse et al. 2010). The lower boundary of the mussel beds usually remain stationary for years at a time, despite substantial spatial and temporal variation in the input (recruitment) of juvenile mussels—variation that would cause explosive shifts in the boundary were they not limited by *P. ochraceus* (Paine 1974, Robles et al. 2009).

This chapter considers how the biology of *P. ochraceus* underpins this archetypal example of stable population limitation. Our current knowledge of life history features (indeterminate growth and reproduction, foraging behavior, prey selection, etc.) of *P. ochraceus* is reviewed, the experimental demonstrations of stable regulation are reexamined in that context, and the life history features are then shown to constitute an integrated predator response that is essential to the capacity of *P. ochraceus* to regulate the mussel populations.

Beyond these issues of boundary formation, however, *P. ochraceus* produces diverse effects in the intertidal community by consuming other sedentary predators and prey and by limiting the mussels, which otherwise would competitively dominate sedentary members of the community. Processes that modify the complex community functions of *P. ochraceus* are reviewed by Menge and Sanford (see Chapter 7).

Biogeography

The latitudinal range of *P. ochraceus* spans the Pacific coastline of North America from Prince William Sound, Alaska, to Cedros Island, Baja California (P. Lambert 2000). Until quite recently, this extensive range was thought to be occupied by two morphologically distinct subspecies: *P. ochraceus ochraceus* north of Point Conception and *P. ochraceus segnis* south of Point Conception (A. M. Clark 1996). However, Lambert (2000) reconsidered the morphological characteristics and proposed a single morphologically variable species. This designation was confirmed by allozyme analysis (Stickle et al. 1992) and mitochondrial haplotypes (Frontana-Uribe et al. 2008; see also Harley et al. 2006). Frontana-Uribe et al. (2008) examined genetic structure of populations on either side of Point Conception and found no significant differentiation. Evidently, the high dispersal of the planktotrophic larvae (see below) causes high gene flow, opposing genetic differentiation (Stickle et al. 1992).

P. ochraceus do show phenotypic variation on local and regional scales. In sheltered bays and estuaries of Vancouver Island, most individuals have relatively thick spines in lower densities on the aboral surface than individuals from wave-exposed shores nearby (Carrion 2008). The ratio of arm width at the base to arm length is also higher, giving them a squat appearance (Fig. 16.1). Such variation may arise from differences in allometric growth caused by differences in food availability, salinity, or temperature regimes that occur between very sheltered and wave-exposed coastlines.

From individual to individual within a local population, the aboral surfaces may show a range of vivid colors: yellow, orange, ocher, maroon, purple, and cobalt blue. The frequencies of the color morphs vary among regions. The outer coast populations through much of the latitudinal range are comprised of 12–28% orange individuals, with hues of maroon predominating (Raimondi et al. 2007). Puget Sound populations have few yellow-orange individuals with some populations comprised almost entirely of deep purple to blue individuals (Harley et al. 2006). Raimondi et al. (2007) found a positive correlation between the frequency of the yellow-orange and size class for sea stars on open coasts. Based on this, they propose a genetically based ontogenic color change to explain the relatively constant frequencies of outer coast populations. The hypothesis should receive further testing, because growth in *P. ochraceus* is indeterminate and strongly influenced by food availability (see below). Therefore, one would not expect a tight

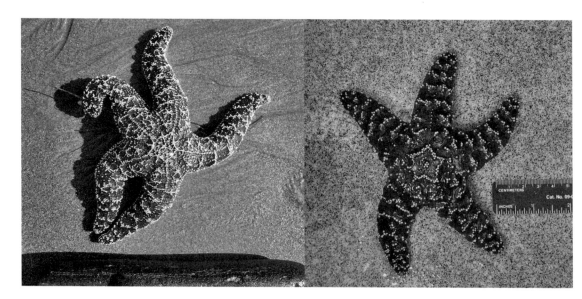

Fig. 16.1. Photographs of *Pisaster ochraceus* from wave-exposed (*left*) and very sheltered (*right*) shores in Barkley Sound, British Columbia. Sea star on left has an arm length of approximately 8.5 cm. Scale on right in centimeters. *Photographs by C. N. Carrion, with permission.*

coupling between size and age across sites. *P. ochraceus* extract carotinoid pigments from their food (Fox and Scheer 1941, Fox and Hopkins 1966), and Harley et al. (2006) note a difference in the composition of prey species from the outer coast of Washington state to Puget Sound, paralleled by a lack of significant genetic differentiation. They propose that color frequencies reflect very plastic phenotypic variation under environmental control and discuss possible effects of spatially varying diet, salinity, and temperature stress.

Growth and Reproductive Development
Indeterminate Growth

P. ochraceus have indeterminate growth (sensu Sebens 1987). Initial growth rates are relatively high, but decline with increasing body size until a maximum size is reached. Initial growth rates and maximum size vary with environmental factors, among the most obvious of which are food availability and temperature. Sanford (2002) showed that small sea stars (mean live mass 130 g) held in aquaria and fed in excess on *Mytilus trossulus* doubled their mass in 2 months. Feder (1970) followed changes in size and live mass in the laboratory and at field sites in Central California. Fed an excess on *Mytilus californianus* in the laboratory, small *P. ochraceus* (approximately 50 g live mass) grew 34.4% per month; large individuals (800–1200 g) grew 7.1% per month. Sea stars starved for >18 months did not die but shrank in proportion to their live masses (27% at approximately 100 g and 40.5% at approximately 475 g). Individuals marked with vital stains were observed at field sites of differing prey abundance. Those on a prey-rich site (mussel beds and abundant barnacles) grew faster and maintained larger maximum sizes than did sea stars on prey-poor sites (sparse barnacles). The size-specific growth rates of the field specimens were bracketed above and below, respectively, by those of the fed and starved laboratory sea stars (Feder 1970). Growth in *P. ochraceus* is also sensitive to temperature, evidently because high temperature increases respiration and maintenance energies (Sanford 2002, Pincebourde et al. 2008; discussion below).

Reproductive Development

P. ochraceus attain sexual maturity at live masses of 70 g (B. Menge 1974) to 150 g (Mauzey 1966). The season of spawning runs April to July for central California (Feder 1980), May to early June for the central Oregon coast (Sanford and Menge 2007), and May through August in the San Juan Islands (Mauzey 1966). Spring spawning places the planktotrophic larvae in the water column when phytoplankton standing stock is usually peaking. The spent gonads remain small until the following fall, when decreasing photoperiod induces gametogensis and renewed gonad development (Greenfield 1959, Nimitz 1976, Pearse and Eernisse 1982, Pearse et al. 1986). In the strongly seasonal environments of the Pacific Northwest, seasonal cycles of gonad mass are inversely related to the mass of the pyloric ceca (Mauzey 1966, Sanford and Menge 2007). Foraging activity and food intake reach a peak in late spring and summer (Mauzey 1966, R. T. Paine 1974, B. Menge et al. 1994), and the assimilated energy is stored in the pyloric ceca, which increase mass to a maximum in fall. This is transferred to the gonads during fall-winter gametogensis, and the mass of the pyloric ceca thus determines the subsequent reproductive output (Sanford and Menge 2007).

The Hypothetical Relationship between Growth and Reproductive Output

The positive relationship among food availability, indeterminate growth, and reproductive output strongly suggests that sites with differing prey availability support differing optimal sizes (Mauzey 1966, R. T. Paine 1976a, Sanford and Menge 2007). Under the hypothesis, juvenile sea stars grow rapidly, allocating energy in excess of maintenance solely to growth. However, as they mature and continue to grow, maintenance energies increase and an increasing proportion of the remainder is allocated to reproduction, until a ceiling is reached. The year's harvest of assimilated energy is apportioned to maintenance and maximal reproduction, with little remaining for additional growth. Sites with relatively high food availability allow growth to larger maximum sizes before rising maintenance and reproduction curtail growth. Furthermore, under the hypothesis, excessively large sea stars in a prey-poor site (i.e., with body sizes above the optimum for the site) must either emigrate to a richer site or decrease in size lest resorption of the pyloric ceca and gonads renders them sterile (Mauzey 1966).

Two studies provide some support for the hypothesis. Sanford and Menge (2007) examined temporal changes in the masses of pyloric ceca and gonads of small sea stars (170–260 g live mass) on a prey-rich site and prey-poor site in central Oregon. Consistent with the hypothesis, they found that small stars on the prey-poor sites had smaller ratios of pyloric ceca to body mass, but apportioned more to reproduction (i.e., produced higher gonad indices and reproductive output) than did small stars from prey-rich sites, although allocations to growth were not measured. R. T. Paine (1976b) collected from a prey-poor site a group of sea stars that had attained a small maximum size and then transplanted them to a prey-rich site. Over the ensuring years, he observed that the transplants grew to the larger maximum size of the resident sea stars. He also attempted the converse experiment, transplanting sea stars of a large maximum size to a prey-poor site. Unlike the first transplants, the large sea stars did not remain, dispersing away from the site in a matter of days. The behavior is consistent with the proposal that prey-poor sites are energetically unsuitable for massive sea stars. However, confirming this interpretation and the underlying hypothesis would require controls for transplanting stress per se and measures of the predicted changes in body mass and stored energy (i.e., the mass of the pyloric ceca and gonads).

The plasticity of growth and size of *P. ochraceus* bears on their capacity to regulate the prey populations, because larger sea stars consume prey at greater rates and incorporate larger prey items in the diet (Landenberger 1968, McClintock and Robnett 1986; discussion below).

Larval Development and Recruitment to the Shore Population

A day or two after broadcast spawning, the drifting eggs develop into a ciliated embryo that within 5 days develops into the bilaterally symmetrical bipinnaria larva approximately 390 μm long (M. Strathmann 1987). The bipinnaria feeds on phytoplankton, gradually increasing in size and transforming into the brachiolaria stage (R. Strathmann 1971). Upon settlement, the brachiolaria metamorphoses into the juvenile, approximately 0.5 mm in diameter with pentaradial symmetry and five arms that appear greatly inflated relative the adult form. The entire developmental sequence has been observed only in culture, and estimates of its duration, from zygote to juvenile, range from 36 days (George 1999) to 228 days (M. Strathmann 1987).

The larval phase shows remarkable developmental plasticity that depends on the food resources of the spawning adults and of the larvae themselves. George (1999) collected adults from a wave-exposed site with high prey abundance and a sheltered site with low prey abundance. Sea stars from the prey-rich site were larger on average and had higher gonad indices than did those from the prey-poor site. Eggs from the prey-rich site were smaller and had lower concentrations of proteins and lipids than did the larger, relatively well-provisioned eggs of the prey-poor sites. These differences evidently determined the morphological development and eventual mortality response of larvae to varying phytoplankton abundance in culture (George 1999). After an initial period of feeding, phytoplankton were either withheld or provided to the two types of larvae. When starved, larvae from the larger eggs developed larger feeding surfaces and grew faster than larvae from the smaller eggs. Most important, greater numbers of large-egg larvae eventually metamorphosed under starvation than did small-egg larvae. This relationship reversed for fed groups, with more small-egg larvae reaching metamorphosis. Assuming that this developmental plasticity occurs in natural conditions, it confers obvious adaptive advantages, allowing copious reproduction of larvae in high productivity areas and rapidly developing starvation-resistant larvae in low productivity areas.

Equally intriguing, bipinnariae of *P. ochraceus* appear to have the capability of cloning in response to high phytoplankton concentrations (Vickery and McClintock 2000b). Larvae in culture were fed varying concentrations of mono-specific or mixed-species phytoplankton, and as they asymptotically approached a maximum size they sometimes reproduced asexually by budding or fission. In treatments with relatively high concentrations of mixed phytoplankton species, larvae grew faster and produced substantially more clones than in the other treatments (Vickery and McClintock 2000b).

Cloning and developmental plasticity of the larval forms are mechanisms by which recruitment of *P. ochraceus* might be linked to prey productivity on the shore. The plausibility of such a link depends on whether higher prey abundances on the shore are often matched, indeed caused by greater nearshore phytoplankton productivity, a relationship for which there is some empirical support (B. Menge 1992, B. Menge et al. 2004). A further linkage would exist if post-settlement survival of juvenile sea stars depends on prey recruitment (i.e., stocks of suitably small prey).

Whether linkages between *P. ochraceus* recruitment and prey production on the shore do indeed occur remains to be seen, because quantitative studies of the sea stars' recruitment are few. Given alongshore currents and duration of the pelagic larval phase of weeks to months, it is reasonable to assume that local mussel beds are open systems, at least in the sense that larvae released from a particular bed seldom, if ever, return there to settle. Wieters et al. (2008) argued the broader proposition that sea star recruitment has no relationship to prey production (recruitment and growth). However, their argument was based solely on a lack of correlation between adult sea star abundances and either prey recruitment or prey cover on sites widely distributed along the Pacific coast. They present no data on sea star recruitment itself. In one of the few studies to examine sea star recruitment, Sewell and Watson (1993) suggest that larvae of *P. ochraceus* are retained within an enclosed embayment in British Columbia, and the retention is responsible for high juvenile recruitment and adult sea stars densities. B. Menge et al. (2004) provide the only data examining the possible correlation between recruitment rates of *P. ochraceus* and mussels. Two sites on the Oregon coast, with relatively high nearshore primary production and high onshore prey production (recruitment and growth), showed significantly higher recruitment of *P. ochraceus* than did two sites with relatively low nearshore and onshore production (see Chapter 7).

Mortality and Morbidity Factors in the Benthic Populations
Physical Stress: Wave Shock

Wave-generated hydrodynamic stresses have long been regarded as a major factor limiting activity and survival of predators and prey in the intertidal zone (Connell 1972, discussion in Denny 1988). There are, however, few quantitative field studies of the effects of wave action on mobile intertidal organisms (but, for examples with other consumers, see B. Menge 1983, Robles et al. 2001). B. Menge et al. (1996) found a negative relationship between the wave exposure of a site and corresponding estimates of per capita predation rates of *P. ochraceus*. Sanford (2002) found an apparent, not quite significant, negative correlation between measured wave force and predation rates of *P. ochraceus*. Although wave stress probably hinders foraging and feeding, *P. ochraceus* are still the dominant predator on many wave-beaten sites (see below).

Physical Stress: Temperature and Desiccation

Feder (1980) observed that *P. ochraceus* seldom occur in shallow sheltered bays or high intertidal pools of central California, presumably because episodes of high temperature and low oxygen greatly reduce survival. Surveys in British Columbia also indicate that, relative to wave-exposed shorelines, densities of *P. ochraceus* are low in very sheltered bays (Robles et al. 1989). However, in this region the scarcity may be the result of meager food availability and episodes of freshwater incursions (see below).

Heating by direct insolation and convection from the air during low tide present an obvious challenge (Helmuth et al. 2006). This is progressively more severe toward higher shore levels, because the lengths of emersion periods increase. This relationship between heat stress and shore level is modified by wave splash and other factors affecting insolation and air temperature. These vary with season, latitude, regional weather patterns, and topographic differences in slope and aspect. Therefore, acute and chronic emergence temperatures may show complex spatial and temporal patterns over a range of scales (Helmuth et al. 2006).

Unlike their sedentary prey, *P. ochraceus* apparently lack the capacity to acclimate to the prolonged high temperature of emergence on upper shore levels (Petes et al. 2008). Sea stars held at high shore levels in cages die following exposures that raise their body temperature above 35°C (Petes et al. 2008, Pincebourde et al. 2008), and their feeding and growth rates are substantially diminished with chronic exposures above 23°C (Petes et al. 2008, Pincebourde et al. 2008). In Barkley Sound, British Columbia, foraging *P. ochraceus* avoid high temperatures by moving with the tides (Robles et al. 1995). They creep upshore with flood tides to the level of concentrated prey and retreat on falling tides to strand at low tide on lower shores levels where prey are scarce. *P. ochraceus* are sensitive to visible light, and in laboratory experiments they avoid this signal of potential heat stress (Burnaford and Vasquez 2008, Garza and Robles 2010). The average intertidal height for foraging sea stars shows semi-monthly variation that suggests the changing vertical ambits are a response to physical stress of emergence (Garza and Robles 2010). As the time of extreme high tide processes over successive days from midnight to midmorning, the average upward excursion decreases, so that on the ebbing tides they strand below midshore levels when low tides fall in the heat of the day. The effect is independent of the height of the flood tide. Prey zonation also plays a role in the height of vertical excursions (see below).

P. ochraceus appear to achieve a measure of accommodation to the potential increase in temperature during emersion by adjusting the volume of coelomic fluid (Pincebourde et al. 2009). Experimental subjects exposed to relatively high temperatures during submersion (16°C) or emersion (26°C) increase the volume of coelomic fluid, and hence their live mass, relative to cooler temperature (10°C) controls. The increase provides greater thermal inertia, which is the extent to which body temperatures lag changes in ambient temperatures (Pincebourde et al. 2009). The mechanisms of thermal inertia are heat capacity and evaporative cooling, both of which can increase with the volume of coelomic fluid. Greater inertia may provide two advantages: limiting the maximum body temperature resulting from short periods of emersion and lowering the cumulative body temperature degrees over the course of a prolonged emersion. The latter would reduce maintenance energy during inactivity, providing more favorable constraints on reproductive energy (discussed above).

Fluid loss by evaporation during emergence results in desiccation, and desiccation at low tide has been considered for some time a significant threat to survival of intertidal invertebrates (e.g., Connell 1972, J. Menge and Menge 1974, Harley 2007). There are, however, few studies of the extent and consequences of desiccation in *P. ochraceus*. Individuals held in aquaria and then exposed in shade or bright sun for 1 h lose 5–10% and 10–20% of live body weight respectively (Landenberger 1969, Pincebourde et al. 2009, appendix D). *P. ochraceus* in the field experience this and greater levels of desiccation but quickly recoup with no apparent damage. While extreme desiccation is surely a threat to survival, the sea stars' vertical foraging ambits suggest this is usually avoided. Thus, the more frequent effect of desiccation may be beneficial, providing evaporative cooling.

Physical Stress: Low Salinity

In the Pacific Northwest, the salinity of near-shore surface waters decreases during periods of high runoff from the land (Thomson 1981). The phenomenon is most pronounced in embayments, where the brackish waters may collect, riding over the relatively dense saline water (Pickard and Stranton 1979). Because they are littoral, the sea stars' exposure to this stress must be substantially greater than subtidal organisms. The absence or scarcity of sea stars in some embayments or shallow seas with chronically hyposaline waters is sometimes attributed to osmotic stress (e.g., Kautsky 1982, Witman and Grange 1998, but see Barker and Russell 2008 for an exception). As with the majority of sea stars (Stickle and Diehl 1987) *P. ochra-*

ceus are osmoconformers and ionoconformers; the coelomic fluid takes on the concentrations of the surrounding water after lags of varying duration (Stickle and Ahokas 1974, Ferguson 1992, Held and Harley 2009). Laboratory experiments examining the effect of lowered salinity on the righting response (the time a sea star takes to turn over after it is placed with the oral surface up on the bottom of the aquarium) show that activity is reduced at salinities below 25 psu and may cease below 15 psu (Held and Harley 2009, Garza and Robles 2010). Individuals exposed to 15 psu may eventually die (Held and Harley 2009). Laboratory experiments matched with field surveys (Garza and Robles 2010) suggest that freshwater incursions following episodes of high rainfall completely suppress vertical foraging excursions; the sea stars remain in crevices at low intertidal and shallow subtidal levels until the brackish incursion disperses. Feeding rates, as measured in laboratory experiments, also decline at low salinities (Held and Harley 2009).

The extent to which activity, feeding rates, and ultimately survival were negatively impacted by lowered salinity in the laboratory trials varied among populations with different histories of hyposaline exposure (Held and Harley 2009). *P. ochraceus* from shores near the mouth of the Fraser River in the Georgia Straight, British Columbia, showed less decrease in performance with declining salinity than did those from Barkley Sound, British Columbia. In the spring-summer period of sea star activity, the Vancouver sites have comparatively low salinities, which are the result of the spring thaw of the snow pack in the watershed of the Fraser River. Thus, it appears that *P. ochraceus* is capable of some acclimation to hyposaline conditions, the extent and mechanisms of which merit further study.

Devastating mortality of littoral invertebrates during periods of extreme physical stress is a conspicuous and well-documented phenomenon (e.g., Brosnan 1994, Carroll and Highsmith 1996, Harley 2007). Yet, as we learn more about the mechanism of stress and adaptation, it appears that temperature, desiccation, and osmotic stresses may exert greatest impact on the population dynamics of *P. ochraceus* not directly through mortality but indirectly by constraining foraging effectiveness and the acquisition of energy for growth and reproduction (Sanford 2002; discussion below).

Predators

P. ochraceus have few predators. Western gulls (*Larus occidentalis*) and sea otters (*Enhydra lutris*) consume

P. ochraceus (VanBlaricom 1987, Reidman and Estes 1990, Pearse et al. 2010). In Barkley Sound, British Columbia, the sunflower sea star, *Pycnopodia helianthoides,* occasionally forages in the intertidal zone. It is known to prey on *Leptsterias hexactis* and *P. giganteus* in the shallow subtidal zone (R. Morris et al. 1980), but whether it also consumes *P. ochraceus* is not known. Substantial effects of predators on the population structure of *P. ochraceus* have been proposed only for sea otters (Pearse et al. 2010; see below).

Disease

Echinoderms seem especially vulnerable to epidemics of microbial diseases, and instances of mass die-offs with the spread of disease are well documented (e.g., Lessios 1988). *P. ochraceus* suffers from a wasting disease, the pathogenic agent of which remains unknown (Eckert et al. 1999, Bates et al. 2009). Bacteria in the genus *Vibrio* have been implicated in wasting diseases of other sea stars (Staehli et al. 2008). In *P. ochraceus,* wasting disease first becomes evident as small white patches on the epithelium of the aboral surface. These quickly enlarge, and regions of the body wall may then swell before the entire organism disintegrates. The time course from initial onset to disintegration may take less than a week (Bates et al. 2009).

The onset of disease is associated with increased temperature. Field-collected specimens transferred to aquaria at 10°C and 14°C showed greater infection rates and pathogenesis at the higher temperature (Bates et al. 2009). In a second experiment, *P. ochraceus* were collected from sheltered and moderately wave-exposed sites and challenged in the aquaria with a 4 day exposure to 20°C. Individuals from the sheltered sites had higher infection rates and greater pathogenesis. Parallel field surveys found the frequency of wasting disease to be higher in sheltered areas, which may stress the sea stars with greater food limitation, pollution, or hyposalinity.

The ciliated protozoan *Orchitophyra stellarum* live in the testis, consume sperm, cause tissue damage, and reduce testis mass in severely infected *P. ochraceus* (Leighton et al. 1991, Stickle et al. 2007, Bates et al. 2010). *Orchitophyra stellarum* infect several species of asteriid sea stars in the Pacific and Atlantic oceans, but it appears that infection of *P. ochraceus* in the Pacific Northwest may be a recent phenomenon. The virulence also appears to be greater than for the other hosts (Leighton et al. 1991, Stickle and Kozloff 2008). Surveys found infected *P. ochraceus* from Cen-

tral California, the southern limit of sampling to Southern Alaska (Stickle and Kozloff 2008). In British Columbia, investigators (Leighton et al. 1991, Stickle and Kozloff 2008, Bates et al. 2010) have repeatedly found sex ratios in the larger size classes biased toward females and infection of males by *O. stellarum.* This suggests the disease may have culled males from the populations. However, other explanations for the sex ratio bias are tenable, and the problem awaits further evaluation.

The prospect of global climate change has drawn attention to the role of elevated temperature in the spread of diseases in marine ecosystems (Harvell et al. 1999, Lafferty et al. 2004). Laboratory studies with *P. ochraceus* (Bates et al. 2010) demonstrated that higher temperatures (14–15°C vs. 10°C) promote greater infection rates and faster pathogenesis of both the wasting disease and infection by *Orchitophyra stellarum.* Field observations showed increases in wasting disease during El Niño events in Southern California (Eckert et al. 1999), higher infection rates in sheltered areas (Bates et al. 2010), and a relative scarcity of infection by *O. stellarum* in cooler waters of southern Alaska populations. This suggests that elevated temperature alone causes the disease and that anomalous temperature spikes might cause catastrophic losses (discussion in Bates et al. 2010). However, the onset of disease may be determined by the interaction of multiple factors affecting the condition of the host (Lafferty et al. 2004). The disease agents themselves appear to be facultative (Stickle et al. 2007), generating pathological effects only in hosts sufficiently stressed. Therefore, possible effects of temperature require examination against the background of varying food availability, pollution, and physical stress.

Feeding and Foraging
Seasonal Foraging

In the southern portion of their range (California and Mexico), *P. ochraceus* remain active and forage year-round (Feder 1959, 1970). In the Pacific Northwest they forage actively in April though October and remain relatively inactive, confining their movement to low in the intertidal and shallow subtidal zone, in November through March (Mauzey 1966, R. T. Paine 1974, Sanford and Menge 2007). The winter period of confinement evidently reduces risk to winter storm waves, freezes during low tides, and brackish water incursions.

Feeding Behavior and Diet

In the turbulence of high tide, waterborne chemical signals seem unlikely to provide reliable information about the location of prey. *P. ochraceus* apparently recognize prey in chance contacts while foraging (Sloan and Campbell 1982, R. Zimmer, personal communication). Small prey, such as acorn barnacles and small mussels, are grasped with adhesive suckers at tips of the numerous tube feet and pulled from the rock, then transferred along the rows of tube feet to the area of the mouth, and the cardiac stomach pressed against the breech in the barnacle test or through the gap in the mussel valves from which the byssus protrudes (Feder 1955, Jangoux 1982a). Small prey are quickly digested, and several may be enveloped by folds of the stomach as others are transferred along the arms while the animal continues to forage (Robles et al. 1995). *P. ochraceus* also use the byssus gap for feeding on large (>5 cm long) *Mytilus californianus*, if the sea star is able to wrench the mussel from the rock. Otherwise the two valves are held by opposing arms, the disks of the tube feet adhere, and a pull is maintained until the valves separate slightly. An opening of less than 1 mm is sufficient to admit the everted stomach. Sea stars feeding in this way take hours to open and consume large mussels (Feder 1959, Jangoux 1982a, McClintock and Robnett 1986).

P. ochraceus are known to consume at least 29 invertebrate species (Feder 1959, Mauzey 1966, Mauzey et al. 1968), but they show clear preferences for a limited number of species. In pair-wise electivity trails, Landenberger (1968) found that *P. ochraceus* prefers, in decreasing order, mussels, *Mytilus galloprovicialis* (as *Mytilus edulis*), *M. californianus*; gastropods, *Tegula funebralis, Nucella ostrina* (as *Thais emarginata*) and *Acanthina spirata*; and chitons *Nutalina californica*. Acorn barnacles and gooseneck barnacles commonly seen in field diets were not tested. The order of preference was fixed and proportions in the diet relative to availability could be altered only slightly, and then only briefly, by past experience (e.g., prior feeding on gastropods alone).

The composition of field diets is best judged by direct observations at high tide, when the sea stars are actively foraging (e.g., Mauzey 1966, Robles et al. 1995). Large prey items, which take longer to consume, may be seen clutched by otherwise inactive sea stars at low tide and thus have a relatively high representation in surveys made at low tide. The high tide diets depend on the availability of preferred prey. When *Mytilus trossulus* and small (<5 cm long) *M. californianus* are present among less preferred prey, they show a disproportionately high tally within the total numbers of prey in the high tide diets (Robles et al. 1995). On the outer coast, the *Mytilus* species, acorn barnacles (*Balanus glandula, Semibalanus cariosus, Chthamalus dalli*), and gooseneck barnacles (*Pollicipes polymerus*) are common and occur in the diet in that order (Robles et al. 1995). In Puget Sound, mussels and gooseneck barnacles are rare, and acorn barnacles comprise the bulk of the diet observed at high tide (Mauzey 1966). Similarly, in sheltered regions of Barkley Sound, mussels are often rare, and the diet then consists of acorn barnacles (*Semibalanus glandula* and *Chthamalus dalli*) and gastropods (e.g., Robles et al. 1995). The high abundance of turban snails on sea benches of some moderately exposed sites is reflected in their greater proportion in the diet; yet here too small mussels predominate when available (R. T. Paine 1969b). Although *P. ochraceus* showed a low preference for chitons in electivity trials, chitons (*Katharina tunicata*) frequently occur in the diet and contribute a major portion of the caloric intake of sea stars in winter in the Pacific Northwest (Mauzey 1966, R. T. Paine 1966). The sea stars are confined to low intertidal and very shallow subtidal levels at this time, and chitons are the only large, relatively abundant prey in this portion of the habitat.

Size-Dependent Predation on Mussels

The size-dependent property of predation is crucial to understanding how *P. ochraceus* limits mussel populations (R. T. Paine 1976a). From minute settlers <0.3 mm long, *Mytilus californianus* may grow to more than 25 cm (Chan 1973) and are thus potentially the largest prey in the diet. However, in laboratory feeding trials (Landenberger 1968, McClintock and Robnett 1986), *P. ochraceus* tend to select relatively small *M. californianus*: the mean size consumed increases with size of the sea stars, but to a limited degree, so that the largest mussels are avoided. In natural populations, the net effect of size selectivity is that risk of predation is much higher for relatively small *M. californianus*; large individuals (≥10 cm long) suffer little predation in most circumstances (Paine 1976a, Robles et al. 2009, Plate 13), but there are exceptions (R. T. Paine 1976b, Robles et al. 2009). *Mytilus trossulus*, restricted to the northern half of the range of *P. ochraceus*, and *Mytilus galloprovincialis*, restricted to the southern half, reach a maximum length of 7 cm, are readily handled at all sizes, and are preferred over *M. californianus* (Landenberger 1968).

McClintock and Robnett (1986) proposed that size selectively increases net energy gain. In their laboratory experiments, *P. ochraceus* were presented with clumps of *M. californianus* comprised of all size classes and in which the gaps in the valves were covered by other members of the clump. Successful prey handling required the same behaviors as field conditions, either dislodging the mussel to expose the byssus gap or pulling the values apart to create a gap. The clumps were continually submerged in aquaria, so that the consumption of predominantly small mussels resulted from true size selection and not a consequence of combined effects of tidal exposure, prey handling times, and vertical segregation of prey sizes as may occur in the field. Assuming a positive relationship between mussel size and the energy required to detach or pull apart the valves, McClintock and Robnett (1986) argue that relatively small sizes yield higher net energy returns. As a sea star grows, prey handling capability increases, raising the optimal size somewhat but not sufficiently to warrant feeding on very large mussels.

Temperature-Dependent Predation on Mussels

Sanford (1999, 2002) has shown that predation on mussels by *P. ochraceus* on the Oregon coast decreases in periods of upwelling, during which surface water temperatures may drop more than 4°C for a few days to several weeks. The intermittent decrease in feeding rate is not accompanied by a decline in growth rate. Evidently, the lower temperatures reduce respiration and hence maintenance energy costs, compensating for the reduced intake rates, and maintaining scope for growth. Air temperatures during low tide emergence do not appear to effect feeding rates, because the sea stars usually avoid prolonged low tide exposures by foraging with the flood and ebb of the tides and by evaporative cooling and thermal inertia that delays the rise in body temperature during exposure. Chronic exposure to high temperature would otherwise suppress feeding and growth (Petes et al. 2008, Pincebourde et al. 2008). Furthermore, additional suppression of feeding rates may occur if temperatures do not co-vary between periods of submergence and emergence (e.g. low tide temperatures suddenly fall as high tide temperatures rise; Pincebourde et al. 2012).

Sanford (2002) argues compellingly that the subtle temperature-energy relationships are a neglected but potentially pervasive influence on the *Mytilus-Pisaster* interaction. Temperature can affect both sides of the interaction. B. Menge (1992) and B. Menge et al. (2004) have shown that greater sea surface productivity is associated with greater mussel growth and recruitment, and greater sea star biomass. Thus, one might expect that repeated cold water upwelling events would suddenly decrease the otherwise high intensity of predation, while boosting mussel growth rates, and the two effects together increasing the likelihood that small mussels would attain large, resistant sizes. El Niño events would produce the opposite effect. Thus, shifting temperature regimes with climate change might affect the age/size structure and abundance of the mussel populations.

Foraging Behavior and the Spatial Array of Mussels

As mentioned in the context of physical stress, *P. ochraceus* are averse to prolonged exposure during low tides, often moving up and down the shore with the advance and retreat of the tides (Robles et al. 1995). However, they show some plasticity in this behavior depending on the structure of the mussel populations. Along a shoreline from wave-exposed to sheltered shores, the lower boundary of the zone of mussels occurs on progressively higher shore levels (Robles et al. 2010). Vertical foraging excursions (as measured either by the shore levels of the highest sea star or by the average height of the entire sample) conform to the alongshore variation in the shore level the prey: sea stars on wave-exposed sites with prey extending to relatively low shores levels may move little over the course of a tide, whereas sea stars on sheltered sites with prey confined to a narrow band high on the shore may move to extreme of high water mark (Robles and Desharnais 2002; Fig. 16.2). Thus, *P. ochraceus* appear to respond to differences in the zonation of their principal prey, albeit a zonation that they shape.

Episodes of massive mussel recruitment (large patches of juvenile and small adult *Mytilus* spp.) on mid- to low-shore levels halt vertical excursions, the sea stars remaining at these prey concentrations, their numbers building as more encounter the site and stay, until the prey are depleted and excursions and alongshore movement resume (Robles et al. 1995, Plate 14). Evidently, the opportunity to remain on a patch of high value prey outweighs the possible costs of somewhat longer exposure. Experimental additions and removals of masses of small mussels confirmed that the shifts in density are a direct response to changes in the spatial array of mussel sizes. Such aggregation and dispersal in *P. ochraceus* is a type of numerical response (*sensu* Holling 1959) and one of several mechanisms in the stable regulation of the local mussel populations (Robles et al. 1995).

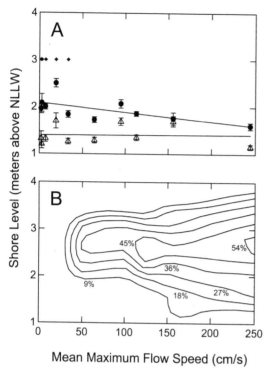

Fig. 16.2. *A,* Relationship between foraging height of *Pisaster ochraceus* and wave exposure within the Barkley Sound, British Columbia. For each of 10 survey sites, the mean height (shore level) in meters above Normal Lower Low Water (NLLW) ±1 SE of foraging by *P. ochraceus* at high tide (*solid circles*) is plotted as a function of mean maximum bottom flow speed, both estimated by divers in the intertidal zone at high tide. Mean height ±1 SE of sea stars at rest at low tide (*open triangles*) is also plotted. Lines are least squares linear fits. Diamonds indicate sites on which the highest sea stars reached 3 m or above at high tide. *B,* Contour lines represent varying mussel abundances over tidal and wave exposure gradients. Percent covers of *Mytilus californianus* in 2,040 quadrats (each 400 cm²) parceled among stratified random quadrat arrays on each of the 10 sites were fitted to a distance weighted least squares surface and the resulting contour lines plotted. *Robles and Desharnais 2002.*

The *Pisaster-Mytilus* Interaction
Stable Prey Population Regulation in Spatially Structured Equilibria

An answer to the question of how the stable regulation of mussel beds occurs lies in the mechanisms of the lower boundary. The shore levels at and below the boundary support relatively high recruitment and growth rates, suggesting the potential for explo-

sive increases in mussel biomass if they were not held in check. Furthermore, changing oceanographic conditions generate substantial interyear variation in the input of juvenile mussels, which might be expected to cause marked fluctuations in boundary location, as well as boundary constituencies showing dominant year classes. Neither is commonly observed. Instead, the lower boundaries are stable, often remaining essentially stationary for years at a time (R. T. Paine 1974, Robles et al. 1995). The upper boundaries of mussel beds occur on high shore levels, where brief immergence times curtail mussel recruitment and growth rates, and the potential for change is much less. Studies of stable population limitation have, therefore, focused on the mechanisms of the lower boundary.

A role for *P. ochraceus* in the formation of lower boundaries was first demonstrated by R. T. Paine (1974, 1976b). He continually removed *P. ochraceus* from the vicinity of mussel beds and compared the changes with control beds where the sea stars were left in place. Marked downward expansions of the lower boundaries, greatly increasing the total area of the beds, occurred over a period of years on the removal sites, while lower boundaries on control sites remained stationary. This outcome, combined with the observation that *P. ochraceus* do not tolerate prolonged emergence on high shore levels, supported the nascent hypothesis of prey refuges (R. T. Paine 1974, 1976a, Connell 1975). Under the hypothesis, the boundary marked the divide between an extirpation zone below and a safe zone above, which the sea stars could not penetrate. The shore level of the divide was set by the gradient of physical stresses imposed by the tidal regime and by adaptive limits of the sea stars, neither of which would be expected to change over the time frames of the observations. Large *M. californianus* sometimes observed below the boundary, either solitary or in clusters of a few, were considered to have been passed over, through the ineptitude of *P. ochraceus*, long enough to attain an invulnerable age/size refuge (Paine 1976a). Spatial and age/size refuges were seen as separate evolutionary strategies assuring the coexistence of the predators and prey (Paine 1976a, see also Connell 1975).

Subsequent studies suggested that additional factors were involved in the stability of boundaries. As explained above, *P. ochraceus* has a surprisingly quick numerical response to episodes of massive recruitment of small *Mytilus* spp., and vertical foraging excursions by *P. ochraceus* vary with spatial and temporal variation in prey zonation. Furthermore, high tide observations revealed that *P. ochraceus* often foraged well above the lower boundary, removing smaller mussels from the matrix of larger mussels (Robles et al. 1995). The flexibility and apparent responsiveness of *P. ochraceus* foraging suggested feedback dynamics that could play a crucial role in prey population regulation.

As an alternative to static spatial refuges, Robles and Desharnais (2002) and Donahue et al. (2011) proposed that mussel bed structure is the result of a complex equilibrium between rates of mussel input (recruitment and growth) and loss to predation. Mussel recruitment, growth, and predation mortality vary over the tidal emersion gradient. Overall rates of predation are also constrained by size-dependent numerical response and prey selection, both of which depend on the spatial arrangements of different prey sizes. Responses of *P. ochraceus* might, therefore, exert spatially differentiated feedback on prey population structure. Under the hypothesis, the lower boundary falls at a shore level at which occurs a phase shift in the equilibrium of input and loss, from sparse covers of very small mussels below, to high covers of large mussels above the boundary line (further discussion in Donahue et al. 2011). Spatially structured equilibria thus explain mussel zonation, and the occurrence of a few large mussels below the boundary is a product of the stochastic property of that unified spatial process, not a separate refuge.

The hypothesis implied two critical predictions that could be evaluated with current, rudimentary experimental methods. First, if the numerical response is countered with density manipulations so that the sea stars are held at artificially low or high densities, the lower boundaries will move: sea star removals will cause downward expansion; sea star additions will cause the upward recession of the boundary. The latter prediction is critical because it contradicts the refuge hypothesis, which posits that the lower boundary is fixed by the regime of tidal emersion. Second, under the hypothesis of spatially structured equilibria, the maintenance of a stationary boundary is functionally linked to size-dependent predation. Countering the numerical response experimentally would shift the size composition of the diet: on removal sites the few remaining sea stars could consume exclusively small mussels, and on addition sites, the dense sea stars would attack relatively large, matrix forming mussels as preferred sizes became scarce. Demonstration of this effect would contradict the proposition of separate spatial and age/size refuges.

The predictions were tested with a large-scale field experiment set up at five locations within a 10 km² area in Barkley Sound, British Columbia (Robles et al. 2009). At each location, three sea benches were chosen for their similarity of topography and wave exposure. One member of the trio was randomly assigned to the removal treatment, another to addition, and the third to unmanipulated control. Over the course of 3 years, thousands of sea stars were transplanted among sites, continually elevating or depressing densities relative to the control sites. Consistent with the first prediction, boundaries expanded downward on the removal sites, receded upward on the addition sites, and remained stationary on the control sites (Figure 16.3a,b).

Divers collected sea stars foraging on the experimental sites at high tide and recorded the sizes and species of prey being consumed. On removal sites, juvenile mussel abundances increased, and diets surveyed at high tide were comprised exclusively of small mussels. Control sites showed a mix of prey species and sizes on both the rock and in the high tide diets. On additional sites, small prey were depleted and large matrix-forming mussels were included in the high tide diet (Figure 16.4). Some mussels consumed were >10 cm long, a size thought to be resistant to predation. Thus, size-dependent predation and maintenance of the mussel bed boundary are not separate mechanisms of alternative types of prey refuges; rather, they are functionally integrated components of an equilibrium process.

The lower boundaries are not always stable under natural conditions. How vertical shifts of the boundaries unfold provides further insight into the regulatory role of *P. ochraceus*. Pearse et al. (2010) report an increase in mussel abundance and marked downward extensions of the mussel beds over a 20-year period on a wave-exposed site in central California. The change was correlated with two other long term trends: reestablishment of sea otter (*Enhydra lutris*) and loss of larger *P. ochraceus* at the site such that overall sea star biomass declined over the period. Pearse et al. (2010) propose that the expansion of the mussel bed resulted from predation by otters preferentially on the larger sea stars. In the resulting near absence of large sea stars, mussels low on the shore grew to sizes that could not be handled readily by the

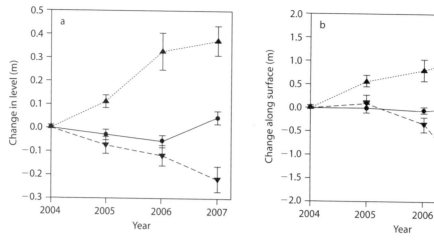

Fig. 16.3. Shifts in the boundaries of adult mussels (*Mytilus californianus*) after three years of altered sea star densities. *a*, Mean change ±1 SE in meters of sea level; *b*, Mean change ±1 SE in run along the sloping rock surface (<0 = meters of seaward movement; >0 = meters of landward movement.). Dotted line = addition; solid line = control; dashed line = removal. *Robles et al. 2009.*

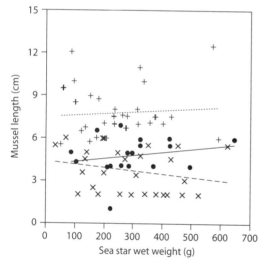

Fig. 16.4. Linear regression plot of largest mussels consumed in a given experimental group. The abscissa represents live wet weights (g) of the sea stars, the ordinate represents the lengths (cm) of the mussels in the upper quartile (largest 25%) of the high tide diet on a given site. Individual records plotted with respect to treatment group. Crosses = addition, dots = control, and Xs = removal. Separate least squares linear regression lines are shown for pooled treatments of the four replicates. Dotted line = addition, solid line = control, and dashed line = removal. *Robles et al. 2009.*

smaller sea stars remaining. Blanchette et al. (2005) followed changes in abundance of *P. ochraceus* and zonation of *M. californianus* at multiple sites of the California Channel Islands. *P. ochraceus* and black abalone (*Haliotis cracherodii*) suffered catastrophic losses to wasting disease during an El Niño event in beginning in 1997. In the ensuing years, sea star abundances increased, especially on south facing sites that also saw the greatest losses of the abalone. Blanchette et al. (2005) propose that loss of the abalone released *P. ochraceus* from competition for resting sites (shaded crevices low on the shore), elevating their densities above historical levels. The persistent increase was accompanied by the upward recession of the boundary. In these examples, it appears that rare natural events altered sea star population structure causing a shift in the boundary.

The Integrated Predator Response: Unifying Role of Reproductive Energetics

The equilibrium process in boundary formation depends on population responses of *P. ochraceus* to varying prey availability. These provide the negative feedback required for stable regulation. The most obvious responses involve flexile foraging behavior: the varying vertical excursions and numerical response. Blocking the numerical response on experimental addition sites (Robles et al. 2009) caused the sea stars to deplete preferred prey on mid to low shore levels and attack large matrix mussels, which then caused the mussel zone to recede to higher shore levels. *P. ochraceus* forced to feed on relatively large mussels apparently take in less energy per unit time or effort (McClintock and Robnett 1986). Feeding at higher shore levels as prey are depleted below incurs additional energy costs, because foraging excursions lengthen and feeding times are curtailed. Therefore, maximization of energy returns, and hence scope for growth and reproduction, appear to drive the foraging behaviors (prey selection, extent of vertical excursions, aggregation and dispersion). Recent work on the role of coelomic fluid volumes in thermal inertia (Pincebourde et al. 2009) further implicates reproductive energetics as a fundamental adaptive constraint on foraging responses.

Although the numerical response plays a central role in maintaining stationary boundaries, in itself it is not likely to be a failsafe mechanism, because alongshore movement of *P. ochraceus* can be hindered naturally by topographic barriers (R. T. Paine 1976a). Changes in prey abundance that are not met by the numerical response may be opposed by changes in body size through indeterminate growth. Although all sea star sizes consume small mussels, the rate of consumption and the maximum size routinely taken increases with sea star size (Landenberger 1968, McClintock and Robnett 1986). Therefore, plastic body size constitutes an alternative predator response to varying prey abundance (the developmental response, *sensu* Murdoch 1971). As explained above, size changes in response to varying prey availability are hypothesized to result from maximization of energy for reproduction (Mauzey 1966, Paine 1976a, Sanford and Menge 2007). Accordingly, R. T. Paine (1976b) and Sebens (1987) propose that among-site differences in sea stars sizes are a product of the long term trends in mussel recruitment, and B. Menge (1992) found a positive relationship between the levels of prey recruitment prevailing on a site, sea star size or biomass, and the intensity of predation. That these long-term trends occur does not imply that size adjustments necessarily require long time courses. The prior laboratory studies confirmed the ability of *P. ochraceus* to grow or shrink with varying food supplies (Feder 1970, Sanford 2002). The size changes are surprisingly rapid, altering a significant percentage of body mass within weeks, a time frame when scaled to the annual addition and loss of mussel recruits seems sufficiently quick to compensate for shortfalls in the numerical response.

Adaptations do not operate as isolated characteristics but rather are functionally integrated parts of a coevolved suite of adaptive characteristics. The varied foraging characteristics and body size plasticity appear to be joined through reproductive energetics in an integrated predator response (discussion in Robles et al. 1995). The relative contributions of each component of the integrated response to prey regulation can only be guessed for any point in the range of environmental conditions over which mussel beds occur, but the relative contributions seem likely to depend on such factors as the time course of elevated production (episodic vs. chronic) and topographic features that moderate alongshore movement or affect temperature regimes. The apparent functional redundancy of integrated predator response may provide a measure of buffering, making the response more robust for its complexity.

There remains one missing piece to the regulation puzzle: the relationship between recruitment of *P. ochraceus* and varying mussel productivity on the shore (see Chapter 7). If prey and predator recruitment are not coupled, as is often assumed for open systems of broadcast spawners (e.g., Wieters et al. 2008), then the resulting mismatches between predator and prey recruitment rates must be compensated for, at least in part, by the integrated response. If, however, the two are linked, as might occur if nearshore productivity supported both greater mussel recruitment and greater production and planktonic survival or cloning of larvae of *P. ochraceus* (George 1999, Vickery and McClintock 2000b), then further redundancy in the regulatory machinery might be expected. The same would apply if early post-settlement mortality of *Pisaster* depended on the abundance of post-metamorphic mussels. Whichever way the recruitment issue is resolved, past experience suggests that further revelations about the life history of this fascinating predator will continue to challenge our preconceptions and expand our imaginations.

17

Asterias amurensis

Maria Byrne,
Timothy D. O'Hara, and
John M. Lawrence

*A*sterias is a major genus of asteroids that live in the cold temperate waters of the Northern Hemisphere and now also occurs in the Southern Hemisphere due to introduction of *Asterias amurensis* to Australia (Fisher 1930, A. M. Clark and Downey 1992, Buttermore et al. 1994). These asteroids have the classic starfish form with five arms radiating out from a small central disc broad at the base and tapering gradually to a pointed tip (Fig. 17.1). Depending on habitat and food availability they can grow to 40–50 cm in diameter. The ventral surface is flat with four rows of suckered tube feet.

Asterias species are among the most extensively studied asteroids (Jangoux 1982a, B. Menge 1982). Several reviews cover aspects of their biology and ecology, particularly for the North Atlantic species (Warner 1979, Sloan 1980a, Jangoux 1982a, Menge 1982, Uthicke et al. 2009). This review focuses on the ecology and biology of *A. amurensis* (Fig. 17.1), which, following its introduction into Australia, has caused major changes to local ecosystems (Ross et al. 2002, 2003a,b, 2004). The ecology and biology of Northern and Southern Hemisphere populations of *A. amurensis* is considered as well as information about other *Asterias* species.

Distribution and Habitat

A. amurensis occurs in the northwest and northeast Pacific Ocean from China and Russia to Alaska. In southern Japan, *A. amurensis var. versicolor* is common (Nojima et al. 1986). *Asterias rubens* is common in the eastern and western regions of the north Atlantic, and *Asterias forbesi* occurs in the western north Atlantic. Western Atlantic populations of *A. rubens* were previously described as *Asterias vulgaris*, a junior synonym (A. M. Clark and Downey 1992).

Asterias species are among the most common shallow-water asteroids. *A. amurensis* occurs in a range of habitats from intertidal rocky shores, off shore reefs, and sheltered unconsolidated habitats (e.g., shell, gravel muddy substrate; Hatanaka and Kosaka 1959, Y. S. Kim 1969, Nojima et al. 1986). These diverse habitats are typical for *Asterias* species (P. Allen 1983, L. Harris et al. 1998, Witman et al. 2003). *A. amu-*

Fig. 17.1. *Asterias amurensis.*

rensis is also known from the continental shelf down to 200 m.

The distribution of *Asterias* species is influenced by a number of abiotic factors, with temperature and salinity being particularly important (Stickle and Diehl 1987, Yakolev 1998). Yakolev (1998) reported an activity temperature range for *A. amurensis* from 6–26°C. Franz et al. (1981) gave a general range of temperature of 3–17°C for *Asterias forbesi*. Wares (2001) attributed that the spread of *Asterias* from the Pacific to the Atlantic in the Miocene and early Pliocene to 5–6°C warmer temperatures in the North Atlantic and Arctic at that time. The temperature requirements of *Asterias* have implications for the consequences of global warming–induced increase in ocean temperatures.

Although all *Asterias* species are found in oceanic salinities, some occur at lower salinities. Stickle and Diehl (1987) cite low salinities of 18% for *A. forbesi* in Long Island Sound and 8% for *A. rubens* in the Baltic Sea. Kashenko (2003) reported the lower salinity limit of *A. amurensis* is 22%. He reported *A. amurensis* migrates to shallow-water oyster beds in Vostok Bay in the summer and can experience mass mortality then when salinity is reduced by monsoon rains.

Food and Feeding Ecology

Asterias species are dominant predators in many north Atlantic and north Pacific ecosystems often playing a keystone role in determining community structure (Fukuyama and Oliver 1985, Himmelman and Dutil 1991, B. Menge and Branch 2001, Witman et al. 2003). This is also true for populations of the introduced *A. amurensis* in Australia (Ross et al. 2002).

As characteristic of *Asterias* species (review Jangoux 1982, B. Menge 1982), *A. amurensis* is reported to prey most often on mollusks, most commonly bivalves (Ino et al. 1955, Hatanaka and Kosaka 1959, Y. S. Kim 1969, Dadaev et al. 1982, Fukuyama and Oliver 1985, Nojima et al. 1986).

The keystone role of *Asterias* species is exemplified by their controlling influence on bivalve distribution, although these asteroids are also subject to predation (see below). Depth limits and abundance of mussels and other bivalves are often determined by predation by these asteroids (Sloan and Alderidge 1981, B. Menge and Branch 2001; see also Chapter 7). Since its introduction, predation by *A. amurensis* in Australia has considerably reduced the abundance of once common bivalves (e.g., *Fulvia tenuicostata*; Ross et al. 2002).

Increase in the density of *A. amurensis* results in marked decreases in commercially important bivalves and newly settled cohorts of young bivalves (Ino et al. 1955, Hatanaka and Kosaka 1959, Ross et al. 2002, 2003a,b, 2004). Periodic outbreaks of *Asterias* species leave fields of opened bivalve shells in their wake (Brun 1968, Sloan 1980b, Sloan and Aldridge 1981, Witman et al. 2003). In soft sediment habitats *A. rubens* occurs in high density in beds of scallops and other bivalves (Brun 1968). *Asterias* species are considered a pest for bivalve fisheries and aquaculture with targeted removal of them from wild harvest and farmed areas by dredging and direct removal (Galtsoff and Loosanoff 1939, D. Johnson 1994, Park et al. 1997), although these measures have not been shown to be effective.

A. amurensis is a generalist opportunistic feeder preying on a wide range of taxa, including bivalves, gastropods, crabs, echinoids, asteroids, and ascidians (Hatanaka and Kosaka 1959, Morrice 1995, Lockhart and Ritz 1998, 2001a,b, Ross et al. 2002, 2003a,b, 2004). It is cannibalistic on small conspecifics (Morrice 1995). In Japan *A. amurensis* utilizes 80% of benthic prey species (Hatanaka and Kosaka 1959). A similar broad diet has been noted for other *Asterias* species (Gulliksen and Skjaeveland 1973, Sloan 1980a, Himmelman and Dutil 1991). In soft sediments *Asterias* species dig for infauna (Arima et al. 1972, Sloan 1980a, Jangoux 1982, P. Allen 1983).

Variation in consumption of prey species such as heart urchins (*Echinocardium cordatum*) and gastropods (*Nassarius nigellus*) by *A. amurensis* is attributed to ease of capture and prey defensive responses (Lockhart and Ritz 1998). Heart urchins may escape predation by burrowing below the 5 cm penetration depth reported for the tube feet of *A. amurensis* (Arima et al. 1972, Fukuyama and Oliver 1985, Lockhart and Ritz 1998). Similarly, the ability of *A. rubens* to capture *Abra alba* varies seasonally depending on the burrowing depth of this bivalve (P. Allen 1983). The bivalve *Nucula turgida* though abundant was not a major component of the diet of *A. rubens* perhaps due to a difficulty in the asteroid's getting a grip on this species (Allen 1983). Diet also varies because *Asterias* species exhibit preference for prey (Sloan 1980, Allen 1983, Barbeau and Scheibling 1994a,b). Barbeau and Scheibling (1994a) found that all sizes (30–150 mm diameter) of *A. forbesi* consumed more small scallops than they did large ones. They concluded this apparent preference resulted from size-related differences in prey vulnerability. *A. forbesi* in the laboratory actively select mussels over scallops (Wong and Barbeau 2005). The energetic content of prey is also reported to influence prey selection of *A. rubens* (Allen 1983).

Asterias species are an important group of benthic scavengers and often aggregate around fish carcasses and animals damaged by fishing. J. S. Oliver et al. (1985) described *A. amurensis* as a widespread scavenger in the Bering Sea. Increase in the density of *A. amurensis* in shipyards following removal of mussels and other fouling organisms from ship hulls attests to the opportunistic nature of feeding in *Asterias* (Fig. 17.2). Increase in the density of *A. rubens* (>100 individual m^{-2}) following dredging and trawling activity is suggested to be due to enhanced scavenging opportunities (Ramsay et al. 1998, Zolotarev 2002). When prey becomes scarce cannibalism and interspecific predation between *Asterias* species occurs, the density-dependent population declines (Anger et al. 1977, B. Menge 1979, L. G. Harris et al. 1998, Witman et al. 2003).

Fig. 17.2. Aggregation of *Asterias amurensis* near a dock area in the Derwent Estuary feeding on mussels cleaned from hulls from boats in dry dock.

The high concentrations of larvae of *A. amurensis* (>150 larvae/m^3) in the plankton of the Derwent Estuary, Tasmania, likely exerts a significant impact on the abundance and diversity of planktonic species (B. Bruce et al. 1995, B. Bruce 1998). These densities are some of the highest recorded for echinoderm larvae (Bruce 1998). Genetic analyses of plankton samples revealed that ca. 80% of the larvae were those of *A. amurensis* (Evans et al. 1998).

Population Ecology and Fluctuations

Asterias populations and their ecological impacts are regulated by a combination of (1) pulses of food, (2) opportunistic migration to food rich areas, (3) cannibalism and emigration when food becomes scarce, and (4) episodic pulses of recruitment (Sloan 1980, Sloan and Aldridge 1981, B. Menge 1982, Nichols and Barker 1984b, Nojima et al. 1986, Mackenzie and Pikanowski 1999, Witman et al. 2003). Uthicke et al. (2009) presented a model for population fluctuation and regulation of *Asterias* species. In this model the lifestyle of *Asterias*—high fecundity and development through a feeding larvae vulnerable to the vagaries of food availability—is considered to be a high risk / high gain strategy resulting in episodic pulses of recruitment.

The phenomenon of periodic high-density aggregations of *A. amurensis* and other *Asterias* species, particularly on shellfish grounds, is reported in many early studies (reviews by Warner 1979, Sloan 1980, Sloan and Aldridge 1981, B. Menge 1982, Jangoux 1982). These asteroids are well known to exhibit boom-and-bust population density fluctuations in association with changes in density of their prey (Menge 1979, Harris et al. 1998, Gaymer and Himmelman 2002, Witman et al. 2003, Uthicke et al. 2009).

Outbreaks of *A. amurensis* (mean density 3.4 individual m^{-2}) occur in response to a superabundance of bivalve prey (Nojima et al. 1986). Nojima et al. (1986) suggested that these outbreaks consist of individuals from 1 year class. Oliver et al. (1985) state that *A. amurensis* is the most abundant and widespread scavenger of the Bering Sea benthos. In Morecome Bay, England the largest intertidal swarm (ca. 10^{-6} asteroids) of *A. rubens* over a 2.5 ha area had a density of 300–400 individual m^{-2} (Dare 1982). The population advanced 5–7 m day^{-1} over 3 months, clearing mussels from 50 ha of shore. Feeding fronts (up to 89 individual m^{-2}) of *A. rubens* are often seen along the shores of Britain, where they aggregate on shallow subtidal and intertidal mussel beds (Sloan and Aldridge 1981,

Dare 1982). The unimodal size distribution of *A. rubens* in these feeding fronts suggests that they represent a single a year class resulting from high recruitment (Sloan and Aldridge 1981). The feeding fronts are usually short lived (ca. 1 yr).

Asterias species respond quickly to pulses of food availability as seen in the increase in density of *Asterias* species following massive settlement of *Mytilus edulis* in the Gulf of Maine (Harris et al. 1998, Witman et al. 2003). This was due to immigration of asteroids to feed on mussels and to recruitment of *Asterias* species (Harris et al. 1998, Witman et al. 2003). Elimination of *M. edulis* as a prey resulted in cannibalism by Asterias, which in turn contributed to a density-dependent population decline in the asteroids (Harris et al. 1998, Witman et al. 2003).

Predators, Parasites, Disease, and Mass Mortality

Known predators of *Asterias* species include conspecifics, other asteroid species, trumpet shells (*Charonia* sp.), and sea gulls (B. Menge 1982, Harris et al. 1998, Parry and Cohen 2001, Witman et al. 2003, Kang and Kim 2004). *A. amurensis* and other asteroid species were the preferred prey of the large gastropod *Charonia* sp. in laboratory trials (Kang and Kim 2004). In Australia the asteroid *Coscinasterias muricata*, giant spider crabs, *Leptomithrax gaimardii,* and fish prey on *A. amurensis* (Parry et al. 2000, Parry and Cohen 2001, Ling and Johnson 2013). Predation by *L. gaimardii* in the field may be related to competition for food between this crab species and *A. amurensis* (Ling and Johnson in press). Observations suggest large and persistent aggregations of *L. gaimardii* may be capable of local displacement of *A. amurensis* (S. D. Ling, personal communication). In the Atlantic *Luidia sarsi* and *Luidia ciliaris* are important predators on *A. rubens* (Eichelbaum 1910, O. D. Hunt 1925). *Asterias* species readily autotomize an arm at the base in response to attack from a predator an effective tactic to avoid mortality (review: Lawrence et al. 1999). Following autotomy *Asterias* species have an impressive capacity to regenerate (Ramsay et al. 1998).

Asterias species are host to a number of parasites (reviews: Jangoux 1987a,b, Goggin and Bouland 1997), the most prominent of which is the castrating parasitic scuticociliate ciliate *Orchitophyra stellarum* (Vevers 1951, Bouland and Jangoux 1988b, Claereboudt and Bouland 1994, Byrne et al. 1997a, 1998, Goggin and Bouland 1997). This ciliate disrupts the germinal layer of the testes and phagocytoses sperm (Bouland and Jangoux 1988b, Byrne et al. 1997a, Goggin and Bouland 1997). Heavy infestations of this parasite cause partial or complete castration of the host. *O. stellarum* has been suggested as a potential biological control agent against populations of *Asterias* on oyster beds in the North America and for invasive populations of *A. amurensis* in Australia (Cépède 1907, Piatt 1935, Goggin and Bouland 1997). Use of *O. stellarum* for biological control of *Asterias* however is not appropriate, because this parasite is not host specific and is likely to parasitize non-target asteroid species (Leighton et al. 1991, Byrne 1996b, Byrne et al. 1997a, 1998, Stickle et al. 2001a,b). *Orchitophyra stellarum* parasitizes six species of Asteriidae (Leighton et al. 1991, Byrne 1996b, Byrne et al. 1997a, 1998, Stickle et al. 2001a,b) and one species in the Asterinidae (Sunday et al. 2008).

Mass mortality of *Asterias* species due to disease and abiotic factors also plays an important role in regulation of populations of these asteroids (B. Menge 1979, Lawrence 1996). Widespread mortality of *A. rubens* due to a necrotic disease occurs in North American populations (Menge 1979). Mass stranding of thousands of *A. rubens* has been reported many times in observations along the coast of England and of *A. forbesi* on the east coast of North America (review by Lawrence 1996). These are associated with heavy seas during winter storms that cast up the asteroids on the beach. In Vostok Bay mass mortality of *A. amurensis* occurs in shallow water occurs due to decreased salinity from monsoon rains (Kashenko 2003). Other causes of mass mortality of *Asterias* include high temperatures, pollution, and physical damage from commercial fishing activity (Kaiser 1996, Ramsay et al. 1998, 2001). These effects of abiotic factors can impact the ecological role of *Asterias* species.

Reproduction and Development

Asterias species have a distinct reproductive cycle with seasonally predictable periods of gametogenesis and spawning (Kim 1968, Jangoux and Vloebergh 1973, Novikova 1978, Barker and Nichols 1983, Nichols and Barker 1984a, Franz 1986, Pearse and Walker 1986, Byrne et al. 1997b). In Japan and Australia *A. amurensis* spawns in winter (Sagara and Ino 1954, Ino et al. 1955, Takashi et al. 1955, Byrne et al. 1997b). Paik et al. (2005) report that *A. amurensis* spawns in the spring in southern Korean waters. According to Kashenko (2005b) *A. amurensis* has a broad spawning season (spring to autumn) in Peter the Great Bay when the temperature is 12–16°C. Kashenko suggests that *A. amurensis* spawns in different seasons in response to differences in water temperature at different locations.

Long-term experiments involving manipulation of light regime in the laboratory showed that gametogenesis in *A. rubens* is controlled by photoperiod with temperature playing a modulatory role (Pearse and Walker 1986, Watts et al. 1990b). The introduction of *A. amurensis* into Tasmania provided an opportunity to test Giese's (1959) hypothesis on photoperiod control of gametogenesis by comparing *Asterias* at similar latitudes (43° N and S) and seasonally out of phase with each other. Comparison of the histology of gametogenesis and gonad index cycles of populations of *A. amurensis* in Japan and Australia revealed that their annual reproductive cycle is identical, with respect to the relationships between the timing of onset of gametogenesis, gamete maturation, and spawning and day length and season (Takashi et al. 1955, Byrne et al. 1997b). Spawning in Tasmanian populations in the Austral winter is 6 months out of phase with that of conspecifics in central Japan, the source location for the introduction of *A. amurensis* into Tasmania (see below). This phase shift supports photoperiodic regulation of gametogenesis in *A. amurensis*. The similar sea temperature regimes at 43° N and S suggests temperature also plays a role in regulating reproduction, perhaps by modulating the onset of spawning activity (Byrne et al. 1997b). The similarity of environmental factors experienced by native and introduced populations would have facilitated reproductive and larval success of *A. amurensis* in Australia. This is likely to have been a major factor in the invasive potential of this species following its introduction.

Asterias species are highly fecund broadcast spawners that have small eggs (ca. 100–150 µm diameter) and development through feeding (planktotrophic) bipinnaria and brachiolaria larvae (Barker and Nichols 1983, B. Bruce et al. 1995, Lee et al. 2004, Kashenko 2005b, Paik et al. 2005; see Chapter 5). *A. amurensis* has an enormous capacity to reproduce, spawning 5–20 million eggs (140 µm diameter). Mass spawning in *A. rubens* (as *vulgaris*) in the Gulf of St. Lawrence involved an estimated 80% of an aggregated population (Himmelman et al. 2008). Spawning individuals crawled to high positions above the seafloor and stood on their arm tips that maximally elevated the disc from the gametes that were released. Spawning coincided with a sharp decrease in sea temperature (Himmelman et al. 2008). Groups of *A. rubens* also migrated toward each other and rap-

idly crawled over one another during the spawning event. These observations indicate that *Asterias* species have behavioral strategies to attain high densities of gametes enhancing fertilization success. Spawning in *A. amurensis* is also correlated with temperature (Ino et al. 1955. Nojima et al. 1986). Aggregations of *A. amurensis* in association with pulses of food availability during the winter spawning period (Fig. 17.2) would facilitate high rates of fertilization and reproductive success (Ling et al. 2012).

Depending on temperature, the larvae are in the plankton for 2–3 months (Hatanaka and Kosaka 1959, Kasyanov 1988, B. Bruce et al. 1995, Lee et al. 2004, Kashenko 2005b, Paik et al. 2005). The optimal temperature for larval development of *A. amurensis* is reported to be 15 °C (Lee et al. 2004). Juvenile *A. amurensis* (R ≤ 5 mm) have been found attached to commercial bivalves in suspended culture, scallop spat, ascidians and on epifauna in port areas (Ino et al. 1955, Morrice 1995). It appears that larval settlement and metamorphosis occurs on a variety of substrates.

Asterias amurensis as an Invasive Pest Species

A. amurensis is one of the few echinoderms identified as an invasive species and is recognized as one of Australia's most significant marine pest species (Ross et al. 2002, 2003a,b). Genetic studies indicate that the source populations of *A. amurensis* were from central Japan (Suruga and Tokyo Bay; Ward and Andrew 1995). Release of larvae in ballast water taken up in a Japanese harbor and discharged at the Hobart wharf is the most likely vector by which *A. amurensis* entered Australia. This is the first known example of an echinoderm being introduced in this way. The subsequent colonization success of *A. amurensis* shows that local conditions were favorable for metamorphic success of the introduced larvae and for the reproductive success of the colonizers.

A. amurensis was first collected in Tasmania in 1986 and was mistaken as a variant of a native asteriid species (*Uniophora granifera*). It was not identified as *A. amurensis* until 1992, by which time it was locally abundant (E. Turner 1992, Zeidler 1992). Because of the concern about an effect on fisheries, "starfish buster" campaigns by divers removed tens of thousands of *A. amurensis* and collections in baited traps (Fig. 17.3 a,b), but these had little impact on the population (Morrice 1995, Grannum et al. 1996). This asteroid underwent a massive increase, reaching densities of up to 24 individuals m^{-2} within 10 yrs (Morrice

1995, Grannum et al. 1996). By the mid-1990s there were ca. 30 million *A. amurensis* present in the Derwent Estuary (Grannum et al. 1996). A model of the population dynamics of *A. amurensis* in Australia indicated that larval supply from existing populations determined location of downstream invasions and that local retention underlies population growth (Dunstan and Bax 2007). Recent surveys (MacLeod and Helidoniotis 2005, Barrett et al. 2012) show that *A. amurensis* remains conspicuous in the Derwent Estuary, especially in areas where food is abundant such as around wharves, jetties, and boat mooring areas where discarded fish carcasses and fouling communities of mussels are a food source (Fig. 17.2). Aggregated populations of *A. amurensis* around wharves with excess food in the Derwent Estuary have higher gonad indices than populations at a distance from such man-made structures (Ling et al. 2012). Enhanced reproductive output by individuals in close proximity would facilitate fertilization success.

Due to their high fecundity and dispersive life history, the high density populations of *A. amurensis* in the Derwent Estuary served as a seed source for establishment of new populations in Tasmania and across Bass Strait to the Australian mainland. This probably resulted from domestic port to port shipping, with larvae transported in ballast water (O'Hara 1995, Parry and Cohen 2001). Juveniles or adults may also have been transported in hull structures. The highest densities of *A. amurensis* larvae in the plankton were recorded in the port areas of Tasmania (B. Bruce et al. 1995) where they could be taken up with ballast water.

The first reports of *A. amurensis* from Port Phillip Bay in 1995 was followed by an exponential increase from 300,000 in 1997 to 165 million individuals in 2000, mirroring the rapid increase seen in Tasmania following its introduction (Parry and Cohen 2001). The total biomass of *A. amurensis* in Port Phillip Bay in 2001 was estimated to be 2000–3000 tons, half that of all demersal fish (Parry et al. 2004). The subsequent

Fig. 17.3. *Left,* The "Starfish Buster" campaign in Hobart, Tasmania, involved divers removing tens of thousands of *Asterias amurensis* from the Derwent Estuary following the introduction of the sea star to Australia. *Right,* The use of baited traps was also tried as a means of removing this asteroid from areas of concern such as commercially important bivalve habitats.

decline in *A. amurensis* to 35.5 million individuals in 2004 may be in response to decreasing food availability (Parry et al. 2004), although recent surveys indicate that this asteroid remains a conspicuous component of the benthic fauna of most parts of Port Phillip Bay (A. Hirst, personal communication).

Why *A. amurensis* has been such a spectacularly successful invader, while other asteroids with planktonic larvae that could also be transported in ballast water have not, is not known. Several factors are likely to have contributed to the invasive success of this species. The similar environmental conditions (temperature, day length) in Australia and Japan were key to successful propagation in its new home (Byrne et al. 1997b). In addition the anthopogenically disturbed condition of the Derwent River may have been a factor. It is well established that disturbance plays an important role in introductions in marine environments (Fields et al. 1993, Stachowicz et al. 2002).

Temperate Australia has a number of endemic predatory asteroids that co-occur with *A. amurensis* including asteriid species (*C. muricata*, *U. granifera*). These asteroids also prey on bivalves (Day et al. 1995). Why endemic asteroids do not reach the high density achieved by *A. amurensis* is not known. They utilize the same prey as the *A. amurensis,* so their densities have not been restricted by prey availability. Asteroids have a great variety in body form (see Chapter 2) that affects their capacity to feed. *Asterias* species with their flexible five-armed bodies may have a higher capacity to feed than species with a more rigid frame such as *Uniophora.*

The high density of *A. amurensis* and the intense level of predation experienced by the benthic ecosystems in the Derwent Estuary and Port Philip Bay are unprecedented. It appears that the benthic communities invaded by *A. amurensis* had not experienced such a predator and so were naive to it. Indeed native scallop species do not recognize *A. amurensis* as a predator (Hutson et al. 2005).

Due to its great abundance, *A. amurensis* has had a major impact on the composition of benthic communities in the Derwent Estuary and Port Philip Bay with drastic reductions, recruitment failure, and local extinction of many prey species (Ross et al. 2002, 2004). Mussel farmers in Port Phillip Bay have had to alter their activities considerably, ensuring that ropes are regularly cleaned of newly settled *A. amurensis* juveniles (Parry et al. 2000). The endangered Australian Spotted Handfish (*Brachionichthys hirsutus*) may succumb to predation by *A. amurensis*. The Handfish lays its eggs at the base of the ascidian *Sycozoa* sp. and *A. amurensis* consumes the ascidians (B. Bruce et al. 1999). Thus *A. amurensis* is implicated to have role in the decline of the Handfish (Bruce et al. 1999).

While it is inevitable that the numbers of *A. amurensis* will decrease as the carrying capacity of the environment reduces with decline in the standing crop of prey (Parry et al. 2004), its ability to survive low food conditions and its opportunistic ability to switch between prey (Hatanaka and Kosaka 1959, Morrice 1995, Lockhart and Ritz 1998, 2001a,b) suggests that *A. amurensis* will remain a conspicuous component of the benthic communities of southeastern Australia.

Acknowledgments
Supported by grants from the Australian Research Council.

18

Leptasterias polaris

Carlos F. Gaymer and
John H. Himmelman

Distribution and Habitat

The northern sea star *Leptasterias polaris* (MÜLLER AND TROSCHEL, 1842) is the best-studied species of the genus *Leptasterias*. *L. polaris* is a common species in the cold-water regions of the northwestern North Atlantic and attains its southern limit in the southern Gulf of St. Lawrence. It also extends through the Canadian Arctic region (Grainger 1966) and into the Beaufort, Chukchi, and Bering Seas (Hoberg et al. 1980, Feder et al. 2005, Bluhm et al. 2009).

In the northern Gulf of St. Lawrence, juveniles (<5 cm in diameter) and small adults (5–20 cm in diameter) are abundant on hard substrata in shallow water and can attain densities of 5 ind. m^{-2} in the first meters of depth (Gaymer et al. 2001a). Large adults (>20 cm in diameter) are mainly found at lower densities, around 0.03 ind. m^{-2}, on soft bottoms at greater depths (Jalbert et al. 1989, Himmelman and Dutil 1991). In the Chukchi Sea, *L. polaris* accounts for ~25% of the echinoderms present at depths of 15 to 60 m (Feder et al. 2005).

Reproduction

Like others of its genus, *L. polaris* is a brooding species that spawns during the winter (Kubo 1951, Chia 1966, 1968a, O'Brien 1976; see Chapter 4). Its gonads develop during the summer and fall and attain peak size at the onset of winter (Boivin et al. 1986). Observations of *L. polaris* maintained in laboratory tanks, supplied with natural seawater pumped in from outside, indicate that individuals aggregate, and even climb over one another, during the 2 months prior to spawning, but curiously contact among individuals decreases when individuals begin to spawn (Hamel and Mercier 1995). Spawning occurs during December and January as temperatures decrease to near the annual minimum (Boivin et al. 1986, Hamel and Mercier 1995).

From laboratory observations, Hamel and Mercier (1995) and Mercier and Hamel (2009) indicate that spawning is initiated by males, probably in response to the decrease in temperature. Females spawn when they detect sperm. Males take the usual asteroid spawning position, with the disc raised and only the arm tips maintaining contact with the bottom as described by Himmelman et al. (2008). The sperm sink and adhere to the bottom. The spermatozoa within a sperm mass on the bottom become inactive and can be viable for up to 6–7 days. Females spawn in a flattened position with the arms extended radially, and the eggs attach to the bottom. Fertilization takes place once spermatozoa become activated in response to contact with the ova.

Following spawning the females curve their arms laterally in one direction to form a disc to cover the embryos that are attached to the bottom (Himmelman et al. 1982; Fig. 18.1a). This "pinwheel" brooding position is maintained through the winter and spring (6–7 months) (Himmelman et al. 1982, Hamel and Mercier 1995, Raymond et al. 2004). The embryological development of *L. polaris* has been described by Emerson (1977) and Hamel and Mercier (1995). Hamel and Mercier (1995) indicate that the brooding mother does not provide nutrients to the embryos, as they observed development occurred at the same rate for non-brooded and brooded embryos in the laboratory. However, the mother probably ventilates the embryos and appears to keep them free of sediment and debris (Himmelman et al. 1982). In addition, protection by the mother from grazers, particularly the sea urchin *Strongylocentrotus droebachiensis*, is likely critical to the survival of the developing sea stars (Himmelman et al. 1982).

Brooding is a genetically fixed reproductive mode for the genus *Leptasterias*, as all species thus far examined follow this pattern (Lieberkind 1920, Chia 1966, R. H. Smith 1971, O'Brien 1976, Worley et al. 1977, Hendler and Franz 1982, Himmelman et al. 1982). The speculation by B. Menge (1975) that brooding in *Leptasterias hexactis* is "an evolutionary response to its competition-induced small size" ignores the evidence that this species comes from a genus of brooders.

Associated with the brooding reproductive mode of *Leptasterias* spp. is (1) the production of a small number of large ova that are produced over a prolonged period (>2 yr) and (2) resorption of excess oocytes (Worley et al. 1977, Boivin et al. 1986). Whereas most species of *Leptasterias* are small (<10 cm in diameter), *L. polaris* is a striking exception as it can attain a size that approaches that of the largest sea star species (50 cm in diameter). The question of why *L. polaris* has evolved such a large size remains to be answered. The capacity to produce gamete increases with size. However, the production of a large number of ova should not be an advantage for female *L. polaris*, because the number of embryos that can be brooded is limited by the available surface area under the mother. Possibly, the "pinwheel" brooding position of *L. polaris*, with the embryos attached to the bottom, allows some increase on the size of its brood. Other *Leptasterias* species, which are small, brood their embryos in the stomach or in a chamber formed by folding the arms inward.

The pattern of energy allocation varies markedly between male and female *L. polaris* (Boivin et al. 1986, Raymond et al. 2004). Males show a substantial increase in gonad size (and gonad energy content) in the 6 months leading up to spawning and a great decrease during spawning. In contrast, females only show a slight gonad increase prior to spawning and a slight decrease during spawning. The massive release of sperm during spawning by males is equivalent to that by broadcast spawners. This expenditure makes up almost the entire reproductive output of males. In contrast, egg production represents only a small portion of reproductive investment in females. Their major reproductive investment is made up by the expenditures for maintenance during the long period of brooding when they do not feed.

In spite of the differences in energy allocation between males and females, the calculations of Raymond et al. (2004), based on the analysis of different tissues of individuals collected prior to and after spawning, and after brooding, indicate similar reproductive investment by the two sexes. However, their analysis was not designed to detect potential decreases in body size (including all body components) that might occur as a result of the long period of starvation during brooding by females. A further reproductive

Fig. 18.1. (Opposite) *A*, A female *Leptasterias polaris* brooding on a boulder in a sea urchin barrens habitat; *B*, *L. polaris* feeding on mussels in the rocky zone; *C*, Aggregation of *L. polaris* and *Asterias rubens* (as *Asterias vulgaris*) foraging on a patch of mussels; *D*, *L. polaris* digging into sediments in search of bivalve prey; *E*, A digging *L. polaris* overturned to show the extended podia and the excavated hole; *F*, *L. polaris* trying to cover a recently dug clam, *Spisula*, as much as possible, and the presence of kleptoparasites, a crab *Cancer irroratus* and three whelks *Buccinum undatum*; *G*, *L. polaris* digesting a large bivalve and whelks *Buccinum undatum* awaiting opportunities to steal prey tissues; *H*, An *A. vulgaris* sneaking in under a feeding *L. polaris* to steal its bivalve prey and nearby whelks *Buccinum undatum* awaiting feeding opportunities. *D,E,H* from Thompson et al. 2005; *G* from Himmelman et al. 2005; *A,B,C,F* unpublished.

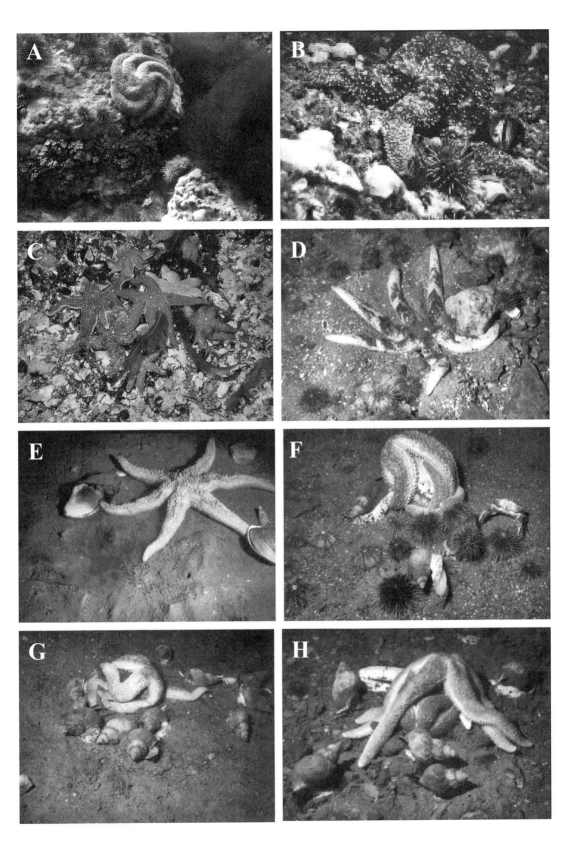

cost for females, also not included in the above analysis of the energetic content of different tissues, is increased mortality during brooding due to predators and ice action. Individuals with scars on the aboral surface, likely from ice abrasion, are common in the spring (Gaymer et al. 2001a). Raymond et al. (2004) showed that the ratio of females to males decreases with increasing body size. This shift could be due to such factors as shrinkage in body size, increased mortality, and decreased time for feeding for females during the prolonged brooding period.

Predators of *Leptasterias polaris*

Three species of sea stars—*Asterias rubens* (as *Asterias vulgaris*), *Crossaster papposus,* and *Solaster endeca*—are reported to prey on *L. polaris* (Himmelman 1991, Gaymer et al. 2004). A study on foraging of sea stars in a soft-bottom community estimated that *A. rubens* consumes ~10 %, and *C. papposus* ~7%, of the biomass of *L. polaris* during a 100-day period in the summer, suggesting that they are the main controllers of populations of *L. polaris* (Gaymer et al. 2004).

Feeding

Leptasterias polaris is a generalist feeder that preys on a large number of benthic invertebrates, including mollusks, annelids, crustaceans, echinoderms, brachiopods, and tunicates. The prey species consumed largely reflect availability (Himmelman and Lavergne 1985, Himmelman 1991, Himmelman and Dutil 1991, Gaymer et al. 2004). Generalist feeding has also been reported for other species of *Leptasterias.*, for example the small *L. hexactis* (B. Menge 1972a) and the holophagous (i.e., internal digestion of the prey) *Leptasterias tenera* (Hendler and Franz 1982). The prey consumed by *L. polaris* in the northern Gulf of St. Lawrence (Mingan Islands) has been reported by studies made over several decades (Himmelman and Lavergne 1985, Himmelman 1991, Himmelman and Dutil 1991, Gaymer et al. 2001a,b, 2004). During diving surveys, to a maximum depth of about 40 m, hundreds of individuals were overturned to identify prey items being consumed. A number of studies were made in a specific zone, either the rocky zone in shallow water or in the sediment zone in deeper water. Given that the prey resources consumed vary markedly between the juveniles and small adults found mainly in the shallow rocky zone and the large adults found mainly in the deeper sediment zone, we will describe feeding and use of prey resources by *L. polaris* for each of these zones separately.

Prey Use in the Rocky Zone

The rocky zone can be divided into two distinct subzones: (1) a shallow subtidal kelp fringe extending to varying depths depending on urchin grazing and (2) an urchin barrens at greater depths where there are only dispersed clumps of algal species that can resist urchin grazing. Dense mussel beds occur mainly in shallow water to about the same depths as the kelp fringe (Himmelman 1991).

In the shallow rocky zone, where mussel beds are usually present, *L. polaris* feeds almost exclusively on the blue mussel, *Mytilus edulis* (Fig. 18.1b). As mussel abundance decreases with depth, *L. polaris* begins to include the crevice-dwelling clam *Hiatella arctica* in its diet. When mussels are absent, *H. arctica* becomes the main prey item (Gaymer et al. 2001a). This prey switching with depth can largely be accounted for by prey availability (Himmelman and Dutil 1991, Gaymer et al. 2001b).

Laboratory Studies of Feeding on Rocky Bottom Prey
PREY SELECTION

Prey selection is generally determined by a combination of factors related to the characteristics of the prey and predator and also to environmental conditions (Emlen 1966, B. Menge 1972a, Barbeau and Scheibling 1994b). Sea stars, particularly *L. polaris*, are not an exception. Laboratory experiments showed that individuals supplied with similar quantities of three common prey species—the mussel *Mytilus edulis*, the clam *Hiatella arctica,* and the limpet *Acmaea testudinalis*—consume the mussel 70% of the time (Gaymer et al. 2001b). However, when the abundance of mussels decreased, *L. polaris* shifted to consuming mainly *H. arctica* and secondarily *A. testudinalis*, thus a similar prey-switching pattern to that observed in the field. Several studies have shown that the prey preferences of *L. polaris* are affected by (1) prey behavior and energetic content, (2) its abilities to capture different prey, and (3) its interactions with *A. rubens* (as *A. vulgaris*), which is both a competitor and potential predator (Gaymer et al. 2001a,b, 2002, 2004).

Size selection of mussels by *L. polaris* is also affected by availability and interactions with *A. rubens* (as *A. vulgaris*; Gaymer et al. 2001a, 2002). In the field *L. polaris* feeds mainly on small mussels that are the most abundant size class (>60%; Gaymer et al. 2001b), however, when offered an abundance of small, medium-sized, and large mussels in the laboratory, it selected medium-sized mussels 50% of the time (Gaymer et al. 2001b). Feeding on medium-sized mussels provides *L. polaris* with a threefold increase in energy intake compared with feeding on small ones (assuming energy is directly related to wet tissue mass). Large mussels are half as abundant as medium-sized mussels and feeding on them provides a similar amount of energy as feeding on medium-sized mussels (Gaymer et al. 2001b). On the west coast of North America, the smaller *L. hexactis* similarly feeds mainly on small prey, which are most abundant in the field (even though larger prey are preferred), a strategy that provides it with a greater energetic intake (J. L. Menge and Menge 1974).

FEEDING RATES

In the laboratory, feeding rates of *L. polaris* on the preferred prey, *M. edulis*, are strongly affected by interactions with both conspecific individuals and the competitor and potential predator, *A. rubens* (as *A. vulgaris*; Gaymer et al. 2002). The presence of conspecific individuals is detected from odors and generally provokes increased feeding (Gaymer et al. 2002). In contrast, the odor of the *A. rubens* reduces feeding rate. This effect is strongest when *A. rubens* is feeding and thus liberating digestive enzymes. The stimulating effect of conspecific individuals would be expected to increase foraging by dense aggregations of *L. polaris* that form on mussels beds in the field (Gaymer et al. 2001a).

It is interesting to note that the feeding rate of *L. polaris* is not affected by temperature over the range of 2–12°C, which covers a large portion of the temperature range found over the geographic distribution of the species. This relationship was shown by a least-squared fitted regression curve based on individual feeding rates at different temperatures recorded during seven laboratory experiments made between 1996 and 1998 (Gaymer et al. 2002). In contrast *A. rubens* (as *A. vulgaris*), which in the northern Gulf of St. Lawrence is near its northern distribution limit, shows a marked increase in feeding rate with increasing temperatures.

Laboratory experiments further demonstrate that *L. polaris* displays a type 2 functional response, which indicates a good capacity to exploit increases in food abundance up to a certain level where the feeding rate levels off (Gaymer et al. 2001b). This type of functional response is characteristic of invertebrates (Hol-

ling 1965) and has also been reported for *A. rubens* (as *A. vulgaris*) feeding on scallops and mussels (Barbeau et al. 1998, Gaymer et al. 2001b).

Prey Use in the Sediment Zone

In the northern Gulf of St. Lawrence, rocky bottoms with bedrock and boulders usually give way to smaller grain materials at ~10–15 m in depth. There is typically a transition with depth within the sediment zone, with cobbles and pebbles gradually being replaced by sand and mud (Himmelman 1991, Gaymer et al. 2004). This sediment zone supports decreased densities of urchins. Sometimes sand dollars are abundant, and large infaunal bivalves are almost always abundant. Mobile carnivores, such as whelks, crabs, and sea stars are also common.

L. polaris found on sediment bottoms, which are mostly large adults (>20 cm in diameter), feed mainly on large mollusks, such as the infaunal bivalves *Mya truncata*, *Spisula polynyma* and *Ensis directus* and the whelk *Buccinum undatum* (Himmelman and Dutil 1991, Gaymer et al. 2004). Although prey use on sediment bottoms is largely determined by prey availability, it is also related to the large size of the *L. polaris* present, because of their energetic need for larger prey (Himmelman and Dutil 1991, Gaymer et al. 2001b). The Yule's V-selectivity index indicates that *L. polaris* selects the large clam *Ensis directus* and large gastropods, *Aporrhais occidentalis* and *B. undatum* (Gaymer et al. 2004).

Studies in the Chukchi Sea exploring the position of *L. polaris* in the food web using the analysis of stable isotopes have shown that this sea star is located among the predators near the highest trophic level in the benthic food web (Iken et al. 2010, Feder et al. 2011). Similar results were also found by Nadon and Himmelman (2010) in the northern Gulf of St. Lawrence. Furthermore, they found that the trophic level of *L. polaris* increased with size.

Role of *Leptasterias polaris* as a Benthic Predator

L. polaris plays a major role in structuring subtidal communities in the northern Gulf of St. Lawrence. As a generalist feeder, it interacts with a great number of invertebrate prey species. The interactions best studied are those with the blue mussel, *M. edulis*, the whelk *B. undatum*, the scallops *Chlamys islandica* and *Placopecten magellanicus,* and several species of soft-bottom clams (Himmelman 1991, Himmelman and Dutil 1991, Nadeau and Cliche 1998, Gaymer and Himmelman 2002, Gaymer et al. 2004).

In the northern Gulf of St. Lawrence, the lower limit of mussel beds in the upper subtidal zone is controlled by foraging aggregations of *L. polaris* and *A. rubens* (as *A. vulgaris*; Himmelman and Lavergne 1985, Himmelman 1991, Himmelman and Dutil 1991, Gaymer et al. 2001a). As a result, mussels are generally limited to shallow-water refuges in the first 1–2 m of the subtidal zone (Gaymer et al. 2004). Entire mussel beds can be eliminated by sea star predation in a 2-year period. The disappearance of the beds leads to changes in the local distribution and abundance of *L. polaris*, as individuals move on in search of other sources of food (Himmelman and Dutil 1991, Gaymer et al. 2001a). Other sea stars are also reported to migrate in search of prey after decimating their food supply (Sloan 1980a, Sloan and Aldridge 1981, Dare 1982, Robles et al. 1995). Distance chemoreception is likely used by *L. polaris* to locate new prey patches (Rochette et al. 1994). Interestingly, there are rare instances of mussel beds in deep (6 m) water in the northern Gulf of St. Lawrence (Fig. 18.1C). We observed dense aggregations of *L. polaris* eliminating such beds in less than a year (Gaymer and Himmelman 2002). The deep mussel beds were mainly composed of large-sized mussels, and the foraging on these mussels by *L. polaris* was likely facilitated by the reduced wave activity in deep water (Gaymer and Himmelman 2002).

Whereas *L. polaris* mainly feeds on sessile prey in the shallow rocky zone, in the sediment zone, it mainly forages on mobile species (e.g., whelks, scallops, clams) that it locates using distance chemodetection (Rochette et al. 1994, M. Thompson et al. 2005). A number of the prey of *L. polaris* deploy strong escape responses to limit attacks (Legault and Himmelman 1993, Rochette et al. 1995, Brokordt et al. 2000).

L. polaris is the main sea star predator of the Iceland scallop *Chlamys islandica,* and its foraging on juveniles likely causes a bottleneck that limits the size of scallop populations (Arsenault and Himmelman 1996). The main strategy of juvenile scallops for avoiding sea star predation is taking refuge in empty shells and crevices where they are less likely detected (Arsenault et al. 1997). Further, the strong escape response deployed by *C. islandica* is a key factor reducing mortality from sea stars, and this is particularly important for adult scallops (Legault and Himmelman 1993, Arsenault and Himmelman 1996). The escape response consists of a series of valve claps that allow the scallop to jump or swim to distance itself

by several meters from *L. polaris* (Brokordt et al. 2000). *L. polaris* is similarly an important predator of the giant scallop *Placopecten magellanicus. L. polaris,* together with *A. rubens* (as *A. vulgaris*) and the crab *Cancer irroratus* are thought to have a major impact on giant scallop populations in eastern Canada (Nadeau and Cliche 1998).

L. polaris is the only predator that can efficiently extract large clams such as *Spisula polynyma* and *M. truncata* from sediment bottoms in the northern Gulf of St. Lawrence (Himmelman 1991, Morissette and Himmelman 2000a). It does this by using its tube feet to transfer sediments from under the central disc to the extremities of the rays, sinking its disc until it comes into contact with the clam and is able to pull the bivalve to the sediment surface (Himmelman et al. 2005, M. Thompson et al. 2005; Fig. 18.1D,E). Compared with *A. rubens* (as *A. vulgaris*), the more rigid body of *L. polaris,* its six (rather than five) arms and a greater number of tube feet per arm appear to provide it with a mechanical advantage in extracting infaunal species (Himmelman and Dutil 1991, Gaymer et al. 2004). The same characteristics may also aid *L. polaris* in extracting the boring clam *H. arctica* from rocky habitats (Gaymer et al. 2001b). The strong preference of *L. polaris* for large mollusks, such as *Ensis directus, Spisula polynyma,* and *B. undatum,* could lead to its limiting the populations of these prey species (Gaymer et al. 2004). Once a clam is brought to the surface, *L. polaris* wraps its arms around the prey, and begins extraoral digestion with the stomach inserted between the clam's valves (Gaymer et al. 2004; Fig. 18.1F).

Himmelman et al. (2005) provided insights into the activity budgets of *L. polaris* and three other sea stars that coexist in the sediment bottom zone by making repeated observations of individuals over extended periods (9 to 23 days). The most frequent activity of *L. polaris* was moving over the bottom (42% of the time), followed by being stationary (26%), digesting prey (17%), and capturing prey (13%). The time spent capturing prey was less for *A. rubens* (as *A. vulgaris;* 5%), because it more frequently fed on epifaunal prey (that did not need to be extracted from the sediments), and substantially less for *C. papposus* and *S. endeca,* which only fed on epifaunal prey. For all four species of sea stars the most frequent change in activities was between moving and being stationary. For *L. polaris,* most 12-h displacements were by <2 m, but displacements of up to 9 m were observed. The distance covered by *L. polaris* was estimated to be 24.1 m in 24 days, and there was no evidence of a migration along the shore. The estimated mean length of a feeding bout (number of days per prey eaten) was 6.6 days for *L. polaris*, more than twice that of *A. rubens* and *C. papposus*, but shorter than that of *S. endeca* (11 days).

The extraoral feeding method of *L. polaris* leads to interesting interactions with a number of other carnivores that are common on sediment bottoms (see below). The interaction between *L. polaris* and the whelk *B. undatum* has been studied in considerable detail (Harvey et al. 1987, M. Thomas and Himmelman 1988, Rochette et al. 1994, 1995, 1996, 1998, Justome et al. 1998). *L. polaris* is the major predator of *B. undatum* (Himmelman 1991), which displays violent foot contortions in response to contact with *L. polaris* (Harvey et al. 1987, Thomas and Himmelman 1988). However, *L. polaris* represents feeding opportunities as well as a predation risk for the whelk, as the whelk may kleptoparasitize (steal) portions of the prey being "sloppily" eaten by *L. polaris* (Fig. 18.1G). The whelk displays complex decision making to balance potential feeding with the risk of being preyed upon, and the bold act of stealing food is only observed when *L. polaris* is actively feeding on clams (Rochette et al. 1995). Laboratory experiments by Rochette et al. (1994) reveal that *L. polaris* displays several behaviors that should increase its success in capturing whelks. These include (1) a preference for strong currents that increases the definition of odor plumes, (2) cross-stream movements to increase the chances of detecting prey, and (3) approaching prey from downstream to avoid being detected. A number of studies have examined in some detail behavioral and physiological aspects of the escape response of the whelk in this predator-prey interaction (Rochette et al. 1995, 1996, 1998, Rochette and Himmelman 1996, Brokordt et al. 2003).

In addition to the whelk, the sea star *A. rubens* (as *A. vulgaris*), the crabs *Cancer irroratus* and *Hyas araneus*, and several small fishes also kleptoparasitize *L. polaris* in the sediment zone (Fig. 18.1F–H). The sequence of interactions during feeding bouts of *L. polaris* on clams, as well as the impact of kleptoparasitism on its food intake, has been evaluated by diving and video filming feeding bouts (Morissette and Himmelman 2000a). The likelihood that *L. polaris* will abandon its prey generally increases with the amount already eaten (the proportion of the prey that has been eaten), thus as its level of hunger decreases. The most important kleptoparasite in terms of loss of food for *L. polaris* is *A. rubens* (Fig. 18.1H). As *A. rubens* also preys on *L. polaris*, *L. polaris* is more likely to abandon its prey once *A. rubens* is drawn to the feeding bout (Morissette and Himmelman 2000b). The "nervous" behavior of *L. polaris* when *A. rubens* is present (holding onto its prey less tightly and being more ready to flee) also provides increased feeding opportunities for the other kleptoparasites (Morissette and Himmelman 2000a). These authors estimated that at least 10% of the prey mass captured by *L. polaris* is lost to kleptoparasites.

As *L. polaris* is abundant in the northern Gulf of St. Lawrence, its feeding activities may contribute substantially to the diet of whelks (Himmelman and Hamel 1993) and other carnivores (Morissette and Himmelman 2000b). Field experiments showed that the diet of *A. rubens* (as *A. vulgaris*) includes *M. truncata* only when it occurs with *L. polaris*. Thus kleptoparitism provides *A. rubens* with access to *M. truncata* (Gaymer et al. 2004).

Spatial and Temporal Variations in Competitive Interactions with *Asterias rubens*

L. polaris and *A. rubens* (as *A. vulgaris*) are major predators in subtidal communities in the northern Gulf of St. Lawrence (Himmelman 1991). Both species overlap in both their geographic and depth distributions (Himmelman 1991). Moreover, they strongly overlap in their use of prey resources (Himmelman 1991, Himmelman and Dutil 1991, Gaymer et al. 2001a). However the degree of overlap changes with depth (Gaymer 2006). Both Morisita's overlap index (C_λ) and Menge's percent overlap index indicate strong overlap in the use of prey species by the two sea star species in the mussel bed but decreased overlap with the disappearance of mussels at greater depths. In the shallow rocky zone, both sea star species feed on *M. edulis* and *H. arctica* and *A. rubens* also consumes the ophiuroid *Ophiopholis aculeata* (Gaymer et al. 2001a). Little overlap is observed in deeper water areas where mussels are absent. On soft bottoms, *L. polaris* feeds mainly on large mollusks, but *A. rubens* feeds on echinoderms (mainly *Strongylocentrotus droebachiensis*) but also consumes large infaunal mollusks that likely become available to it through kleptoparasitism of *L. polaris* (Gaymer et al. 2004, Morissette and Himmelman 2000a).

Initially, the strong overlap in resource use was considered to indicate there were competitive interactions between the two sea star species (Himmelman 1991, Himmelman and Dutil 1991). Studies of different aspects of the ecology of *L. polaris* and *A. rubens* (as *A. vulgaris*) over the past two decades (Himmelman et al. 1982, Himmelman and Lavergne 1985, Boivin et al. 1986, Jalbert et al. 1989, Himmelman and Dutil 1991, Rochette et al. 1994, Hamel and Mercier 1995, Morissette and Himmelman 2000a,b) have provided a better understanding of how the two sea stars interact and of mechanisms that allow their coexistence (Gaymer et al. 2001a,b, 2002, 2004, Gaymer and Himmelman 2002). Laboratory and field experiments demonstrate that these sea stars strongly interfere with each other (encounter competition *sensu* Schoener 1983), causing decreased feeding (rates and percent feeding) and changes in the behavior of L.

polaris (Gaymer et al. 2002). *L. polaris* avoids physical interaction with *A. rubens* and flees from interspecific feeding aggregations when prey become scarce (Fig. 18.2). Initially interference is caused by odors liberated by *A. rubens*, but the effects of interference are strongest when *A. rubens* is feeding (i.e., releasing digestive enzymes) and when there is contact between the two species (Gaymer et al. 2002).

The probability of competition between *L. polaris* and *A. rubens* (as *A. vulgaris*) is most likely in the first meters of depth, where the two sea star species form dense aggregations that feed on mussels. Below the

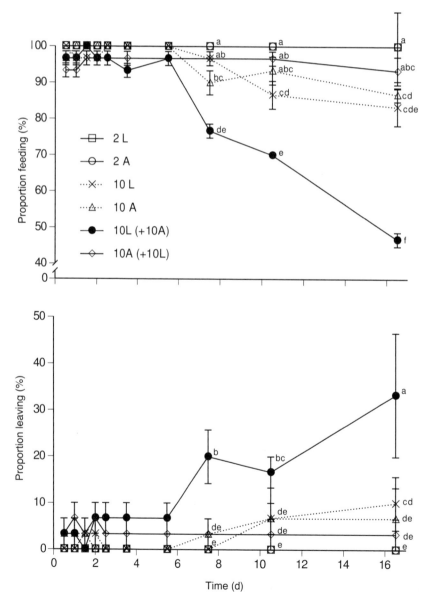

Fig. 18.2. Effect of intra- and inter-specific interactions on the proportion of *Leptasterias polaris* or *Asterias rubens* (as *Asterias vulgaris*) feeding (*above*) and the proportion leaving the experimental area (*below*) during a 16-day field experiment. The treatments were 2 *L. polaris*, 2 *A. rubens*, 10 *L. polaris*, 10 *A. rubens*, and 10 *L. polaris* and 10 *A. rubens*. Values for 2 *L. polaris* and 2 *A. rubens* overlap through the experiment at 100% for proportion feeding and at 0% for proportion leaving. Symbols sharing the same letter on days 7, 10, and 16 are not different (ANOVAs, $p > .05$). None of the treatments differed during the first 6 days. Vertical bars represent standard errors. L represents *L. polaris* and A represents *A. rubens*. Gaymer et al. 2002.

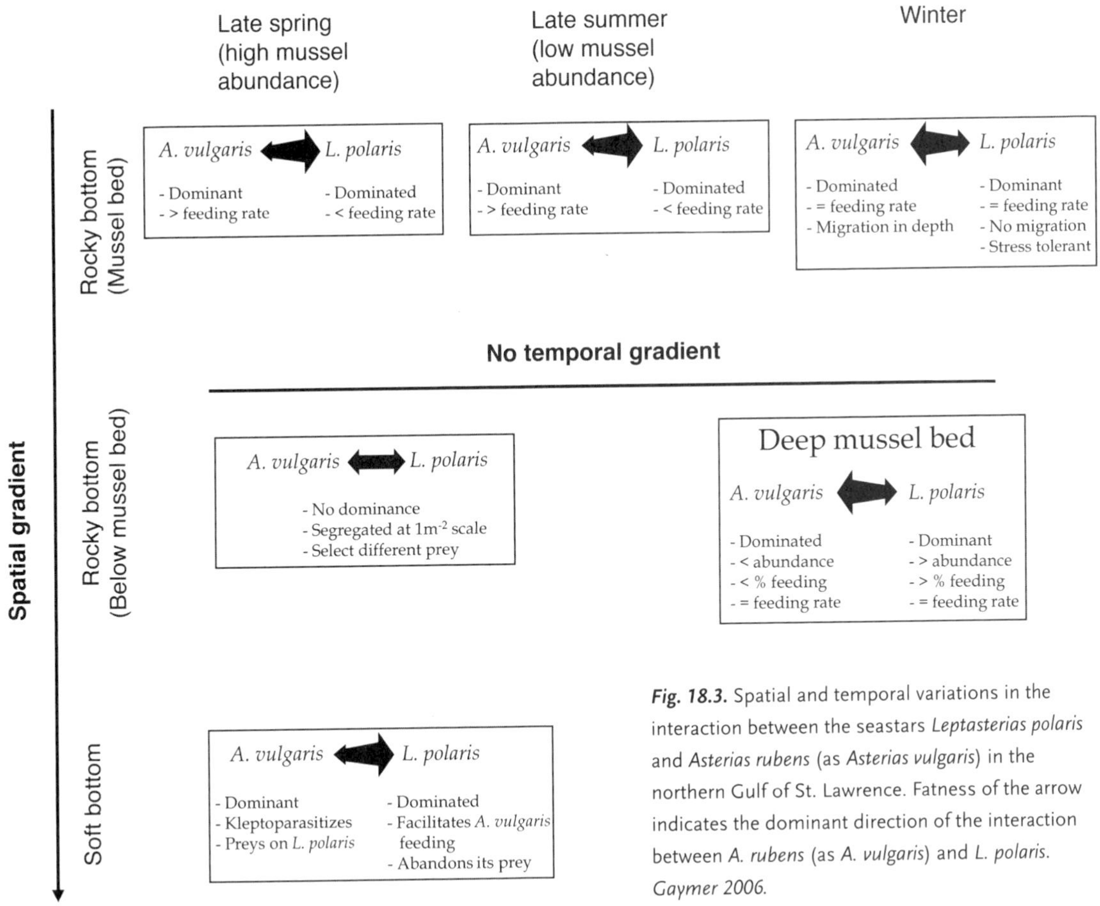

Fig. 18.3. Spatial and temporal variations in the interaction between the seastars *Leptasterias polaris* and *Asterias rubens* (as *Asterias vulgaris*) in the northern Gulf of St. Lawrence. Fatness of the arrow indicates the dominant direction of the interaction between *A. rubens* (as *A. vulgaris*) and *L. polaris*. Gaymer 2006.

mussel bed, interactions are reduced as *L. polaris* and *A. rubens* are segregated at a small spatial scale (1 m⁻²) and select different prey (Fig. 18.3). *L. polaris* selects *H. arctica*, whereas *A. rubens* selects *O. aculeata* (Gaymer et al. 2001a). Although this could be an evolutionary response to past competitive interactions (Connell 1980), it could also simply reflect the abilities of sea stars to feed on different prey (Gaymer et al. 2001b). As explained above, in the sediment zone *L. polaris* facilitates feeding by *A. rubens* as portions of clams it extracts are kleptoparasitized (Fig. 18.3). *L. polaris* often abandons its prey to *A. rubens* (as *A. vulgaris*) because *A. rubens* represents a predatory risk (Morissette and Himmelman 2000a,b). Thus, on soft bottoms *A. rubens* seems to be a dominant competitor. A particular situation is found in the unusual situation when mussel beds occur in deeper water (~6 m in depth). Here the threefold greater numbers of *L. polaris* compared with *A. rubens* at the same depth seem to attenuate or even reverse the competitive dominance of *A. rubens* (Fig. 18.3). The dominance

of *L. polaris* is indicated by three facts: (1) a higher proportion of *L. polaris* feeds compared with *A. rubens*, (2) it does not appear to avoid contacts with *A. rubens*, and (3) it consumes the largest mussels (Gaymer and Himmelman 2002).

Competitive interactions also change in time, both within and between seasons (Fig. 18.3). During the summer, sea star foraging can rapidly decimate mussel beds (Gaymer et al. 2001a, Gaymer and Himmelman 2002) so that by late summer, when mussel abundance is low, competition is increased and the two sea star species begin to partition mussels by size (Gaymer et al. 2002). *A. rubens* (as *A. vulgaris*) selects medium-sized mussels, which are most profitable, whereas *L. polaris* feeds on small mussels, which are most abundant (Gaymer et al. 2001a,b). Partitioning prey by size has also been reported between the sea stars *Pistaster ochraceus* and *L. hexactis* (B. Menge 1972b, J. L. Menge and Menge 1974). *P. ochraceus* selects larger prey than *L. hexactis* in situations where a broad range of prey sizes is available. In the northern

Gulf of St. Lawrence, *L. polaris* appears to be an inferior competitor during the summer, as it selects smaller mussels, shows a decreased feeding rate and avoids *A. rubens* (Gaymer and Himmelman 2002). At the same time the increased feeding rate and selection of medium-sized mussels by *A. rubens* should provide it with a greater energetic intake than that of *L. polaris* (Gaymer et al. 2001b, Gaymer et al. 2002). During the winter, *A. rubens* migrates to deeper water, probably to avoid low temperatures, strong waves, and ice abrasion (Gaymer et al. 2001a). Thus, at that time it cannot feed on its preferred and most profitable prey (i.e., mussels). Further, whereas *A. rubens* feeds at higher rates than *L. polaris* at summer temperatures, the two sea star species likely feed at similar rates when temperatures are low during the winter (Gaymer et al. 2001b). These behavioral changes, together with the ability of *L. polaris* to tolerate reduced temperatures and ice abrasion, suggests that the dominance of *A. rubens* is attenuated and possibly even reversed (i.e., *L. polaris* would become dominant) during the winter (Fig. 18.3).

Thus, the interaction between *L. polaris* and *A. rubens* (as *A. vulgaris*) in the northern Gulf of St. Lawrence varies in both space and time. This implies changes in the dominance of *A. rubens*. During the summer (when temperatures are highest), *A. rubens* (which is near its northern distribution limit) dominates the interaction in competing for resources in shallow mussel beds. However, this dominance is attenuated or disappears during the winter. The dominance of *A. rubens* also decreases in the rocky zone below the shallow mussel beds. There is a clear dominance of *A. rubens* in the sediment zone. Since *A. rubens* is a broadcast spawner, its dominance in the whole system may be attenuated during years of low recruitment. *L. polaris* is a brooder and thus likely has more stable recruitment (Himmelman et al. 1982, Boivin et al. 1986). The coexistence of the two sea star species in the subtidal communities of the northern Gulf of St. Lawrence is likely favored by (1) the periodic attenuation of the dominance of *A. rubens*, (2) the aggregating behavior of both sea star species (especially *A. rubens*; Gaymer et al. 2002), and (3) the patchy distribution of their preferred prey (Atkinson and Shorrocks 1981).

Parasites

L. polaris is parasitized by an ascothoracid parasite *Dendrogaster elegans* (Crustacea: Ascothoracida), both in Alaska and eastern Canada (Grygier 1986). This parasite is highly branched and becomes mixed with the ramifications of the gonads and pyloric ceca. No studies have been made on its effect on *L. polaris*, however, given the effect of *Dendrogaster* spp. on other sea stars (Hamel and Mercier 1994), it likely causes atrophy of the gonad. In the Chukchi Sea and northeastern Bering Sea, the gastropod *Asterophila japonica* is also a parasite of *L. polaris* (Hoberg et al. 1980).

19

Coscinasterias

Michael F. Barker

The genus *Coscinasterias* comprises four species of large forcipulate (Family Asteriidae) starfish common in the intertidal and
in shallow coastal habitats. The genus has a remarkably wide
geographic range in both Northern and Southern Hemispheres, a possible consequence of dispersal, vicariance, or translocation associated
with shipping (Waters and Roy 2003a). Asexual reproduction by fission of the disc is a common feature of all species. The arms are
sharply set off from the disc and are variable in length and number as
a consequence of arm regeneration after fission. Arm number in undi-
vided specimens is six to nine in Atlantic taxa and up to 13 in Pacific
taxa (A. M. Clark and Downey 1992). In this review the greatest detail
will be provided for one species, *Coscinasterias muricata,* for which
many aspects of the biology are well known. Where information for
each section is available for the other species similarities and differences will be noted.

Distribution and Systematic Variation

Coscinasterias muricata (Fig. 19.1) is one of the most widely distributed
and common starfish on the New Zealand coast. It is also widely distributed in Australia from the Abrolhos Islands in Western Australia
to southern Queensland, Tasmania, and Lord Howe and Norfolk Islands (Rowe and Gates 1995). Arm number is variable, ranging from
five to 14, but there are commonly 11. The disc is small, often with several madreporites. Multiple madreporites are generally seen in smaller
specimens (1–20 cm in diameter). This is almost certainly the result of
fission, which is more common in small individuals (see later section
on asexual reproduction). Thus, starfish in this size range frequently
have one to several regenerating arms, while large specimens, greater
than 20 cm in diameter, generally have arms of equal length. Color is
highly variable, often ranging from red or orange to blue, green or
grey, with cream to yellow tube feet.

    Coscinasterias acutispina is widely distributed in the northwestern
Pacific, particularly around the Japanese Archipelago, in coastal regions

Fig. 19.1. *Coscinasterias muricata* collected from the intertidal of Otago Harbour, Dunedin, New Zealand. *a*, A starfish with five long arms, four other arms that have almost reached the same size, and one very small new arm (almost hidden). *b*, A starfish that has divided quite recently with five original and six new small arms.

of southern China (Fujita et al. 2001, Haramoto et al. 2006), and in Taiwan (Waters and Roy 2003a).

Coscinasterias tenuispina is found on the east coast of the USA from North Carolina to Florida; off the coasts of Cuba, Bermuda, eastern Brazil, the Azores, the southern Bay of Biscay, southwest France, the Canary and Cape Verde islands; and in the Mediterranean.

Coscinasterias calamaria was previously synonomized with *C. muricata* but now is regarded as a separate species (Rowe and Gates 1995). It is found in South Africa, Mauritius, Northern Australia, and possibly Madagascar (A. M. Clark and Downey 1992, Waters and Roy 2003a).

Population Genetics

In a wide ranging phylogeographical study (42 samples from 27 locations), Waters and Roy (2003b) sampled *C. muricata* from southern Australia and New Zealand. Analysis of mitochondrial and nuclear

DNA sequences indicated a major phylogenetic split within Australia which was strongly correlated with latitude. Northern haplotypes (latitude less than or equal to 37.6 °S; 15 samples from nine sites) were 7.3–9.4% divergent from southern haplotypes (latitude greater than or equal to 37.6 °S; 27 samples from 19 sites), consistent with late Pliocene separation. Eastern and western representatives of the "northern" clade were 0.5–1.0% divergent, which Waters and Roy (2003b) suggest is a result of Pleistocene isolation. They suggest the "southern" clade of both Australia and New Zealand provides evidence of Pleistocene oceanic dispersal.

Habitat

C. muricata is common in the intertidal, often under rocks, sometimes in pools or puddles. It is not commonly found exposed intertidally, although it is generally found exposed subtidally (Barker, unpublished manuscript). Very small specimens sometimes occur in coralline turf or on algae (pers. obs.). This may be a nursery habitat for recruitment. Generally *C. muricata* does not occur in high energy habitats (e.g., open west coast rocky shores). However, it does occur on the west coast where offshore islands or reefs provide some shelter. It is also common in the shallow subtidal to depths of 30 m+ and can be dredged to depths of 100 m. Other species of *Coscinasterias* are also predominantly found in the intertidal or shallow subtidal (e.g., *C. acutispina;* Yamazi 1950; Fujita and Seto 1998, 2000) and *C. tenuispina* (Yamazi 1950, A. M. Clark and Downey 1992, S. L. S. Alves et al. 2002).

Tolerance to Temperature and Salinity

Although *Coscinasterias muricata* is an open coast species and is seldom found in estuarine situations, it has some tolerance of reduced salinities. Lamare et al. (2009) showed that adults (20–30 mm diameter) had high tolerance of salinities in the laboratory down to 25 PSU but could not tolerate exposure to salinities lower than this. Mortality was 100% within 6 h after exposure to 15 PSU and within 29 h after exposure to 20 PSU.

Coastal temperatures around New Zealand are seldom higher than 22–23°C and then only in the far north. The temperature range in Otago Harbor is 7–18°C. Nevertheless most large *C. muricata* (R = 50 mm) collected from Otago Harbor subjected to a daily temperature variation over a 12-h period of 13.5–25°C) survived in the laboratory for more than 125 days (Sköld et al. 2002). Coastal temperatures in shallow water or the intertidal of Queensland or Western Australia within the geographic range of *C. muricata* would frequently exceed 25°C in summer.

Coscinasterias acutispina inhabiting the intertidal and shallow coastal zone of Toyama Bay (Sea of Japan, central Honshu Island) can tolerate salinities as low as 13 PSU and temperatures as high as 30°C for several days (Fujita et al. 2001).

Salinity and temperature tolerances of *C. tenuispina* and *C. calamaria* do not appear to have been determined. However as both species live in subtropical to tropical habitats, their temperature tolerances might be expected to be higher than those of *C. muricata* or *C. acutispina*.

Associated Species

C. muricata is found with a range of lower intertidal and subtidal species that are common within the geographic range of the starfish. Many of these are potential prey (see below). For example in northern New Zealand the starfish is likely to be associated with mussel beds, particularly the greenshell mussel *Perna canaliculus,* while in the south of the South Island *Mytilus galloprovincialis* is the more common species. In Doubtful Sound on the southwest coast of New Zealand a low salinity layer (LSL) of up to 12 m deep on the surface restricts the upward distribution of many invertebrates (K. R. Grange et al. 1981, Barker and Russell 2008). In this zone, *C. muricata* and the sea urchin *Evechinus chloroticus* are often found in close association, probably because of the special hydrological features of the environment. Aside from this feature and food associations, there appear to be no specific group of invertebrates that occur frequently with *C. muricata*. Associations have not been reported for other species of *Coscinasterias*.

Feeding and Diet

Laboratory observations by Temara et al. (1999) showed that like many asteroids, *C. muricata* locates its prey by chemoreception. Following detection of food cues, the behavioral sequences described by Temara et al. (1999) include arm curling, extension and movement of sensory podia, and orientation toward the prey and are identical to those described for other asteroid species (Sloan 1980b). Like many starfish, *C. muricata* has extraoral feeding, with the cardiac stomach everted over prey and digestive enzymes released. The time for digestion and resorption of prey items is variable with both the size of starfish and the prey. In experiments with individual mussels (*M. galloprovincialis*) of 3.5–5.0 cm shell length attached with Velcro to subtidal vertical rocky walls in Doubtful Sound (southwest coast of New Zealand), consumption of mussels by *C. muricata* took between 5–30 h compared with 6–24 h of similar sized mussels in natural beds (Witman and Grange 1998). In a subsequent field experiment in Doubtful Sound, M. Clarke (2002) observed *C. muricata* consumed two to three similar sized *M. galloprovincialis* held in open cages, approximately one mussel every 8–12 h. The average arm length of *C. muricata* is not given in either of these studies, although Clarke (2002) described the starfish as "large" (400–700 g wet weight). In a laboratory study, Channon (2010) found that *C. muricata,* with an average weight of 249 g and interradial length of 145 mm, consumed an average of 1 *M. galloprovincialis* (mussel size not given) every 2 days over a period of 185 days. Feeding behavior and feeding rates of other species of *Coscinasterias* have not been reported.

Coscinasterias muricata is a voracious carnivore and will feed on a variety of prey organisms, predominantly mollusks. Mussels of a variety of species (*P. canaliculus, M. galloprovincialis, Aulacomyna maoriana,* and *Musculus impactus*) are particularly important general prey in New Zealand (Crump and Barker 1985, M. Clarke 2002, Witman and Grange 1998, Lamare et al. 2009). *C. muricata* on the piles of a pier at Rapid Bay, South Australia, fed on mollusks, crustaceans, and moribund items (Keough and Butler 1979). In a study on foraging behavior of *C. muricata,* Day et al. (1995) report that in Australia abalone may be important prey items in some circumstances. Although the blacklip abalone *Haliotis rubra* was rarely eaten, on many reefs when other prey species were uncommon, aggregations of starfish did feed on *H. rubra.* In the laboratory, small abalone were preferred to larger animals (Day et al. 1995). It appears that abalone produce a chemical deterrent, and, as a result, they are often not eaten unless mussels are not available and starfish are in high densities. In New Zealand's southern fiords, particularly large *C. muricata* (R > 20 cm) can occasionally be found preying on small to medium-sized (test diameter, 80 mm) sea urchins *E. chloroticus* (Grange et al. 1981, pers. obs.).

Other species of *Coscinasterias* appear to consume a similar variety of prey. *C. acutispina* has been reported to be a major predator of released juvenile abalones and turban shells in many parts of Japan (Fujita and Seto 1998) and to feed on barnacles on bare rocks and on small gastropods in crevices or under boulders in Hong Kong (Lam 2002). *C. tenuispina* feeds on mussels at Ponta de Itaipu, Rio de Janeiro, Brazil. Ventura (personal communication) reports that on the Brazilian coast this species is a generalist and eats mussels (*Perna perna*) and gastropods (*Tegula* spp.). In laboratory aquaria it will also consume encrusting calcareous algae.

Movement

The rate of movement of starfish generally, either alongshore or with depth, is poorly understood and little has been published on their movements. As noted earlier, in the southwestern fiords in southern New Zealand, *C. muricata* living below the LSL prey on *M. galloprovincialis.* In this habitat their feeding rates are controlled to some degree by fluctuations in the depth of the fresh water (Witman and Grange 1998, M. Clarke 2002). Using small archival electronic

tags that recorded water temperature and depth every 5 min, Lamare et al. (2009) recorded movement of three individual *Coscinasterias muricata* living immediately below the LSL over periods of 7 to 14 days. Vertical movement was quite variable with starfish altering their depth over a 5 m depth band between 0.39 m above and 4.9 m below mean low water (MLW). It is surprising to note that the movements of starfish were asynchronous, with deeper and shallower depths occupied by each starfish at different times. Over the course of the study period, starfish moved a total vertical distance of between 53.3 and 178.7 m. The average vertical distance moved per day was between 4.86 m and 14.15 m. The average vertical speed of 0.51 to 0.58 m^{-1}h^{-1} was similar for each starfish. Starfish showed little horizontal movement over the course of the study. The vertical movements of *C. muricata* were almost certainly the result of starfish foraging at shallower depth to gain access to the *M. galloprovincialis* at times of high tide, when the increased salinity allowed them access to food. This study clearly shows that, when food is abundant, *C. muricata* probably move little unless forced to do so by the physical environment.

Reproduction
Sexual Reproduction

All species of *Coscinasterias* reproduce sexually and have separate sexes.

GAMETOGENIC CYCLES

In *C. muricata,* sexual reproduction has been described in detail for populations in both the north and south of New Zealand (Crump and Barker 1985, M. Clarke 2002) and in Port Phillip Bay, Victoria, Australia (Georgiades et al. 2006; see Chapter 3). Reproductive cycles were similar in starfish sampled from both the north (Maori Bay, west coast) and south (Otago Harbour east coast; Crump and Barker 1985) and Doubtful Sound (southwestern coast; Clarke 2002) of New Zealand, despite the wide latitudinal difference between these populations. Mean monthly gonad indices show a clearly defined reproductive cycle, synchronous for both males and females, although the Maori Bay population sampled (see later section on asexual reproduction) had few males. Gonad growth commenced in the Austral autumn (April–May) reaching a peak gonad index (GI; gonad weight / eviscerated body weight / 100) of 30–50 in August to December (index dependent on site) with spawning taking place from October through to the end of January. All starfish had completed spawning (GI 1–2) by February or March at all sites. Spring and early summer peaks in the gonad index were noted at both Maori Bay and at two of the three sites in Doubtful Sound (Clarke 2002), suggesting that early spawning is often followed by some gonad regeneration before the main summer spawning occurs.

Histological examination of the gonads showed oocyte formation was initiated in February and maximum oocyte diameter (120–140 μm) was reached from September to November. The cycle of spermatogenesis paralleled oogenesis. Pyloric ceca indices (PCI) also show a clearly defined and parallel pattern in both northern and southern populations, reaching their highest values in May–July (PCI = 20–30). Minimum values were seen in December to January (PCI = 10–12). There were no differences between males and females and the cycle showing an approximately (although slightly later) inverse relationship to the gonad index cycle as has been noted for many other species of starfish. Starfish sampled from two sites in Port Phillip Bay, Australia, showed a markedly similar reproductive cycle with gametogenesis initiated in March and April, the gonad index peaking in August to September. Starfish were spawned out by February. As with the New Zealand populations, there were two spawning peaks, spring and early summer. The pattern of seasonal change in oocyte diameters was also almost identical to that described by Crump and Barker (1985), with an increase in oocyte diameter between March and August, indicating active gametogenesis. Spawning was shown by a decline in oocyte diameter between December 1999 and February 2000. The PCI cycle was also similar to New Zealand populations, peaking in midwinter and declining as gonad growth occurs. Both the gonad (peak value ~14) and pyloric ceca (peak value ~12) indices were however much lower in Port Philip Bay than they were in the New Zealand populations, possibly an indication of lower food availability to these populations.

Georgiades et al. (2006) also monitored seasonal changes in progesterone concentrations in the pyloric ceca of female C. muricata in both populations in Port Philip Bay. Although no seasonal cycle in progesterone concentration was apparent in the pyloric ceca of female starfish, in one population (Governors Reef) there were apparent peaks in progesterone levels during April and October 2000, following the vernal equinox. This may indicate photoperiodic regulation of this steroid as have been described for

other starfish (Pearse and Eernisse 1982, Xu and Barker 1990b,c).

Sexual reproduction is less well described in other species of Coscinasterias. S. L. S. Alves et al. (2002) and Ventura et al. (2004) have described the cycle of sexual reproduction in populations of C. tenuispina sampled over a 13 month period at three sites on the coast of Rio de Janeiro, Brazil. An all male population at Ponta de Itaipu showed a clearly defined seasonal cycle. The GI was low (1–2) during the summer months (December 1997 to May 1998) then increased from May, reaching highest values over the winter and spring (August to November 1997), and decreased because of spawning over November to January (winter and spring). Although the seasonal change in seawater temperature was only 5°C, there was a significant correlation between temperature and GI but no correlation with salinity or photoperiod. The pyloric ceca indices showed an annual variation of only around 6.5%, peaking in October 1997 (20%) with a minimal index in March 1998 (13.5%). The gonadal and pyloric ceca indices showed no clear inverse relationship. Although the other two populations included both males and females, the GI was comparatively low (>5%) for the whole period with no clear seasonal pattern.

Seto et al. (2000) describe the cycle of sexual reproduction of C. acutispina at two shallow subtidal rocky coastal sites (Kurosaki and Uozu) in Toyama Bay on the Sea of Japan side of central Japan. At Kurosaki the populations showed a distinct annual cycle. In males, indices were low from April to October, increased to a peak in December, and gradually decreased to low values in March. In females indices, increased from August, peaked in November, and decreased to very low values by February. In contrast at Uozo indices remained low throughout the study, with small increases in November 1998 and July 1999 (males and females combined). Histology of gonads showed gametogenesis in both testes and ovaries from April to August, maturation with ripe testes and large oocytes (80–160 μm) in November and December, and gradual depletion of gametes until February (females) and March (males).

LARVAL CULTURE AND DEVELOPMENT

Embryonic and larval development has been described in detail for C. muricata (Barker 1978a), and the chronology is outlined in Table 19.1.

Shibata (2010) has given a detailed description of embryogenesis and development for C. acutispina. The oocyte diameter (137 μm) and the overall pattern of embryogenesis and the timing of events up to the

Table 19.1. Chronology of embryonic and larval development of the starfish *Coscinasterias muricata*

Time	Stage	Description
		Embryonic development
0	Fertilization	
1–2 h	First cleavage	
3 h	Second/third cleavages	
5–8 h	Early blastula	64–128 blastomeres. Blastocoel developing
20–22 h	Late blastula	Hatching, invagination of archenteron commencing
30–36 h	Gastrula	Invagination of archenteron well advanced, 2° mesenchyme forming
45–46 h	Late gastrula	Invagination of archenteron nearly complete
45–60 h	Very early bipinnaria (length 0.4 mm)	Formation of the stomodaeum complete, gut present only as a narrow tube, dorsoventral flattening of larva occurring
70–8 h	Early bipinnaria	Embryonic development complete, stomach has expanded to a food-holding area
		Larval development
4 days	Bipinnaria (length 0.4–0.5 mm)	Pre-oral loop and post-oral and lateral ciliated bands complete, larvae actively feeding
8 days	Bipinnaria (0.8–0.9 mm)	Left and rightenterocoels extending anteriorly and posteriorly, ciliated tracts becoming more complex
10–11 days	Bipinnaria (1.2–1.3 mm)	Enterocoels lengthening, posteriorly the ventral horn of left posterior coelom forming
14 days	Mid-late bipinnaria (1.7 mm)	Ciliated larval processes (arms) have elongated, left and right enterocoels have joined anterior and have extended anteriorly into the median dorsal process and posteriorly around the stomach and intestine
18 days	Early brachiolaria (2.1–2.2 mm)	Rudimentary brachiolar arms and adhesive disc have formed, all processes have lengthened with light brown pigmentation at terminal ends
20 days	Brachiolaria (2.5 mm)	Brachiolar arms lengthening and adhesive papillae forming on their tips, adhesive disc now well developed, five hydrocoel lobes forming on left posterior coelom
24 days	Brachiolaria (3.2–3.3 mm)	Larval processes elongate, brachiolar arms terminating in a crown of brown pigmented adhesive papillae and rudimentary spines, ossicles forming on aboral surface of yellow brown adult starfish primordium
27 days	Late brachiolaria (3.6–3.7 mm)	Primordium well developed with pronounced ossicles and spines, larvae swimming near bottom of culture vessel and will attach and undergo metamorphosis on suitable substrata

Source: From Barker 1978a.
Note: Length of bipinnaria and later stage larvae measured from the posterior margin of the larva to the tip of median dorsal process.

early blastula stage is quite similar to that described above for *C. muricata*. However larval development proceeds more slowly with the fully developed brachiolaria not formed until 37 days after fertilization and first metamorphosis at 61 days (27 days in *C. muricata*). Larval morphology (shape and size of larvae) is also almost identical to that described by Barker (1978a) for *C. muricata*.

SETTLEMENT AND METAMORPHOSIS

Barker (1977) has described the process of settlement and metamorphosis of *C. muricata* brachiolaria larvae in detail. On first contact with the substratum the brachiolaria made temporary attachment using the adhesive papillae on the tip of one or two of the three anterior brachiolar arms while the long larval processes were contracted. The larvae then exhibited classic substratum "searching behavior" for periods of a few seconds to 40–50 min. In the presence of adverse stimuli or the absence of positive stimuli, arms were extended and the larvae swam off. When metamorphosis was to occur, cement was secreted from the adhesive disc (Barker 1978b) with permanent attachment 1–2 h after the initial contact with the substratum. The larva remained attached to the substratum for 6 to 7 days during metamorphosis before the five-armed juvenile starfish (disc diameter approximately 0.95 mm) broke away from the attachment stalk and adhesive disc rudiment and assumed free life.

Brachiolaria larvae settled and completed metamorphosis on a variety of substrata. The only requirement appeared to be the presence of a primary surface film. No larvae settled on clean rock surfaces or clean Plexiglas plates. They settled in numbers on surfaces covered with a surface film. Surface topography did not appear to influence settlement, and more larvae settled on the undersides than the upper surfaces of substrata. Shibata et al. (2010) described newly metamorphosed *C. acutispina*. They were 700μm in diameter and had six arms rather than the five in *C. muricata* described by Barker (1978a). Each arm had two pairs of tube feet with an anus recognizable near the central plate and the hydropore opening in an interradial plate on the aboral surface. Three months after completion of metamorphosis, the hydropore was gradually covered with calcareous elements of the interrradial plate, the first formation of the madreporite. A fully formed grooved madreporite did not appear until 11 months after metamorphosis. At this stage the small starfish still had six arms, approximately 50 pairs of tube feet on each arm, and an arm length (R) of ~15mm. At 12 months after the completion of metamorphosis juveniles reared in the laboratory had an R of ~40mm and at that time divided into daughter starfish each with three arms (Shibita et al. 2011)

Asexual Reproduction

Coscinasterias reproduces asexually by dividing the disc, with attached arms, into two parts (fissiparity) and regenerating the missing portion of the disc and new arms. This was first noted in *C. tenuispina* by Steenstrup in 1857 (in Emson and Wilkie 1980) and later observed in the laboratory by Kowalevsky in 1872 (in Emson and Wilkie 1980). Recently split individuals can generally be recognized by the presence of intact and new arms of varying sizes from small buds 1–2mm in length, to both the original and the new arms approaching an even size. Often fission will reoccur before new arms have reached their full length. Generally *Coscinasterias* that have undergone fission at least once will have two madreporites. There may be as many as three or four in *C. muricata*.

Despite the frequency with which it occurs in *Coscinasterias*, the actual process of fission has seldom been described. Crump and Barker (1985, p. 118) observed that *C. muricata* "shows no clearly defined dividing plane across the disc and individuals simply tear themselves in half leaving an open wound across the centre of the disc," a process that took up to 60 min. A gaping wound following fission in *C. acutispina* is also described by Haramoto et al. (2007). Wound healing and production of new arm buds seems to vary with the size of the dividing starfish. In small *C. muricata* (R= 15–25 mm) wound closure and formation of new arm buds is evident a week after fission, but in larger specimens (R= 45–75 mm) more than 3 weeks is required for the closure of the wound and production of new arm buds (Ducati et al. 2004). Wound closure and arm regeneration appears to be more rapid after arm autotomy than after fission. Mazzone et al. (2003) induced arm autotomy in 4–6 cm diameter *C. muricata*. They found the wound was completely healed after 3 days with growth of new tissue in the center of the wound after 5–7 days. The process of fission, wound healing, and regeneration of new tissue has not been described in other species.

INFLUENCE OF SIZE ON FISSION

There is some evidence that fission is more prevalent in small individuals in some species. Crump and Barker (1985) noted that fission was less common in *C. muricata*, where R of the longest arms was more than 120 mm. Yamazi (1950) also found no large specimens (R= 20–50 mm) had undergone fission in *C. acutispina*. However Haramoto et al. (2007) found no relationship of size with fission in samples of this species collected from 23 widely separated different sites around southern Japan. S. L. S. Alves et al. (2002) found split *C. tenuispina* in all sizes of starfish sampled at Ponta de Itaipu on the coast of Rio de Janeiro, Brazil.

SEASONALITY OF FISSION

Fission may continue throughout the year, an individual splitting as soon as or shortly before the new arms reach the size of original arms or may be seasonal, perhaps as a response to changing environmental conditions. Fission in intertidal populations of *C. muricata* in New Zealand seems to be more common in summer, whereas subtidal populations show no clear seasonal trends (Crump and Barker 1985, Barker et al. 1992)

Seto et al. (2000) found asexual reproduction in *C. acutispina* was frequent from early summer to early winter with some differences between populations at Kurosaki and Uozu. *Coscinasterias tenuispina*, however, seems to divide mainly in winter (November to

March) in Bermuda (Crozier 1920) and August to September on the coast of Rio de Janeiro (S. L. S. Alves et al. 2002).

CAUSES OF FISSION

Clear seasonality in fission in the above species raises the question as to the specific physical or biological triggers that induce a starfish to divide. There is some evidence that fission is more common in intertidal than subtidal populations of *C. muricata* (Crump and Barker 1985, Johnston and Threlfall 1987, Barker et al. 1992, Sköld et al. 2002) and *C. acutispina* (Haramoto et al. 2006), although in a subsequent study examining this latter species at a different set of intertidal and subtidal sites, there seemed to be no differences (Haramoto et al. 2007).

Subtidal *C. muricata* are also often larger than intertidal ones, perhaps reflecting better food conditions or the opportunity for animals to feed for longer periods in the subtidal. Crump and Barker (1985) suggested that physical stress and poor food conditions, particularly in the intertidal (e.g., high summer temperatures coinciding with times of low tide) are important stimuli to fission. Sköld et al. (2002) investigated the effects of food supply on fission of *C. muricata* from a fissiparous population in Otago Harbour. Food supply did not affect fission except when the starfish were starving, as indicated by a decrease in size. Gonad development and energy storage under intermediate food supply were similar to measurements of individuals in the field and suggest food limitation for the population studied. Since fissiparity decreases with increasing size in *C. muricata*, and growth rates and gonad development increase with supply of food, Sköld et al. (2002) suggested that the occurrence of fewer fissiparous individuals in habitats where food is abundant is correlated with food supply. Rapid growth would allow individuals to more quickly reach the size threshold at which fission begins to be suppressed. Conversely, where food supply or feeding time is limiting growth, the pattern would be a relative increase in abundance of small fissiparous individuals. However, some populations (e.g., southern fiords and Stewart Island) are comprised of mainly very large (R = 125–200 mm) individuals, all with one madreporite, that do not seem to have ever divided. A possible explanation could be that the asexual mode of reproduction of this species is not triggered in the sheltered subtidal fiord environment. Other possibilities are that natural selection acts against asexual reproduction by fission in the fiord environment or that the ability to split was lost due to founder effects and random genetic drift when the fiords were successively colonized following the rise in sea level after the last ice age.

POPULATION GENETICS AND ASEXUAL REPRODUCTION

Many *Coscinasterias* populations have highly biased sex ratios. Crump and Barker (1985) described a northern (Maori Bay) New Zealand population of *C. muricata* highly biased toward females (sex ratio 4:1) and a southern (Portobello) population that was entirely males, and a third (Wellers Rocks) that had even numbers of both sexes. Sköld et al. (2002) reexamined two of these populations 23 years later (1998) and found that at Maori Bay the sex ratio had changed to be biased more toward males (sex ratio 23:9) but the Portobello starfish were still all male. S. L. S. Alves et al. (2002) describe an all male population of *C. tenuispina* at Ponta de Itaipu on the coast of Rio de Janeiro, Brazil. Of two populations of *C. acutispina* in Toyama Bay Japan, one (Uozu) was all male but a second (Kurosaki) was biased toward females (3:1; Seto et al. 2000). Such highly skewed sex ratios provide strong evidence that recruitment from sexual reproduction may be very limited.

It is possible that some populations of all species of *Coscinasterias* are maintained by asexual rather than sexual reproduction. Within New Zealand, genetic differentiation of *C. muricata* has been examined in several populations. Sköld et al. (2003) used allozymes to investigate the population genetics of subtidal open coast and southern fiord populations. Mean genetic variability measured by Wright's F_{ST} for all the New Zealand populations was 0.061, indicating a moderate level of genetic divergence on a New Zealand wide scale, unusually high for a starfish with widely dispersed planktotrophic larvae over the distances between populations examined (i.e., 1000 to 4000 km). For populations from the fiordland region F_{ST} was generally lower, ranging from 0.024 to 0.028, but it deviated from zero for all the loci examined. These findings suggest that there is likely to be restricted gene flow between the populations of *C. muricata* examined.

Using similar methodology, Johnson and Threlfall (1987) examined genetic variation of *C. muricata* populations near Perth, Western Australia, at 14 sites spanning a geographic range of approximately 50 km, although some sites on Rottnest Island and Garden Island were only 50 m apart. Electrophoretic analysis of six polymorphic loci indicated low genotypic diversity and strong genetic disequilibrium at each site, confirming the highly clonal structure of local populations, even in sites as close as 50 m. In the combined samples from all sites, however, the expected range of multilocus genotypes indicated random mixing of larvae produced sexually, confirming that overall clonal diversity is a result of sexual reproduction.

In a study that also examined genetic differences between populations but using molecular techniques, Perrin et al. (2003, 2004) used a mitochondrial DNA marker to compare genetic differentiation among populations of *C. muricata* from fiordland with others sampled from both the North and South Islands. Over a broad geographical scale (>1000 km), restricted gene flow between the North and South Islands was observed. At a smaller scale (10–200 km), significant population structure was found among fiords and between fiords and the open coast.

The importance of fission and larval recruitment in the maintenance of populations has also been examined in other species of *Coscinasterias*. In a study of *C. acutispina*, Seto et al. (2002) examined isozyme variations at four polymorphic loci in 348 starfish collected from nine sites within Toyama Bay. An excessive number of heterozygotes at each of the nine sites caused significant deviations from a Hardy-Weinberg equilibrium. Eleven genotypes were detected from nine sites but only one or two genotypes were detected in each of seven other sites. Seto et al.

(2002) suggest that the genetic structure of *C. acutispina* populations in Toyama Bay might be maintained mainly by fission rather than by larval recruitment. In a study covering a broader geographic range of sites in the Sea of Japan, including three in Toyama Bay, Haramoto et al. (2006) used a random amplified polymorphic DNA (RAPD) marker to detect clones and to assess gene flow among sites. A simulation approach using RAPD data revealed the presence of clonal individuals at almost all sites, confirming the importance of asexual reproduction in maintaining population size. The result of phylogenetic analysis according to RAPD genotype showed no relationship however between genetic and geographic distances. The authors consider that there is considerable gene flow between sites, due to dispersal of planktonic larvae from sexual reproduction.

Genetic analysis does not necessarily show that all populations of *Coscinasterias* are made up of clones. In two studies of population genetics of *C. tenuispina* (Pazoto et al. 2010, Ventura et al. 2004), allozyme electrophoresis was used to estimate genetic variability measured by Wrights F_{ST} for polymorphic loci of starfish populations at three different sites on the Brazilian coast. None of the three populations departed from the expectations of Hardy-Weinberg equilibrium, leading to the conclusion that none were comprised of clonal individuals.

20

Echinaster

Richard L. Turner

The red Mediterranean sea star *Echinaster sepositus* might well have been Aristotle's ἀστήρ ("aster") in his *Historia Animalium* (Περί Τά Ζώα ατορίαι) and, therefore, the first sea star mentioned in scientific literature, more than 2300 years ago. Although Cresswell (1902) attributed Aristotle's "aster" to *Uraster rubens* (as *Asterias rubens*), this species—probably familiar to the British author—is absent from the Mediterranean Sea. However, Tortonese (1965) chose *E. sepositus* for the cover photograph of his book on echinoderms of Italy, probably because it is so common a seashore animal that it was familiar to amateur naturalists and scientists alike. This species is likely also to have come to Aristotle's attention; it is present in the Adriatic and Aegean Seas and the Saronikos Gulf near Athens (Papadopoulou et al. 1976). Indeed, Ludwig (1897) wrote that *E. sepositus* was familiar to coastal Mediterranean folk since ancient times because of its abundance and color. He claimed that the earliest reference to this sea star in scientific literature was Aldrovandi's (1638) *Stella rubra*. Furthermore, Aristotle's description of digestive physiology of "aster" fits that of an echinasterid as well as *A. rubens*: "The nature of the aster is so hot, that if it is captured immediately after swallowing anything, its food is found digested" (Cresswell 1902). The presumption of swallowing aside, the gut contents indicate an extraoral or particle feeder, the feeding style of *Echinaster* (Jangoux 1982a).

Most literature on the genus *Echinaster* deals with systematics, morphology, and biogeography, but *E. sepositus* was an important laboratory animal for descriptive and experimental research in the middle 1800s to early 1900s, partly because of the availability of this sea star to Mediterranean biologists. Research on *E. sepositus* was especially stimulated by the founding of stations at Roscoff by Henri de Lacaze-Duthiers in 1871 and at Naples by Anton Dohrn in 1872. The latter institution was especially significant in early studies on developmental biology and experimental physiology (Müller 1996). Since those early years, only the research activity on *Echinaster* species from the eastern Gulf of Mexico rivals that on *E. sepositus*. Nevertheless, attention to several species of *Echinaster* worldwide is largely attributable to how common and easily accessible these shallow-water sea stars are within their ranges of distribution. In some places, they are found in enough abundance to be harvested for folk medicine (R. R. N. Alves and Alves 2011) and handicrafts (Mariante et al. 2010), although Ferguson (1969a) otherwise described *Echinaster* as "economically harmless."

Echinaster has been used for descriptive and experimental studies from molecular biology and cytology to organ and whole-animal physiology. The focus of this chapter is on several highly interrelated aspects of the biology of *Echinaster* that give an integrated picture of the biology and ecology of this common and widespread genus.

Morphology

Echinaster is a genus of five-rayed, yellow to red, medium-sized sea stars with tapered cylindrical arms and a small disc composed of a reticulate skeleton beset with small spines and covered by a thick, glandular, chemically defended skin, which probably offsets the absence of pedicellariae. Of six asteroid body forms defined by Blake (1989), his Echinasteroid form (Group 3) includes those asteroids having a small thick disc, thick cylindrical arms with blunt tips, and arm ossicles not strongly differentiated in size or function (Fig. 20.1a,b). Extensive descriptions of morphology are given by Agassiz (1877), Cuénot (1887), Ludwig (1897), Blake (1978, 1980), O'Neill (1989, 1990), A. M. Clark and Downey (1992), and Gale (2011a).

Populations of five-rayed sea stars often have a few individuals with six or more arms. Teratous forms aside (Domantay 1938, Tortonese 1965, Hotchkiss 1979, Watts et al. 1983), the consistent exception to five-rayed *Echinaster* is *E. luzonicus*, a fissiparous sea star for which comet and later stages of regrowth from autotomized arms with five to seven rays can be found (H. L. Clark 1921, Domantay 1938, Yamaguchi 1975, L. M. Marsh 1977). Maximal body sizes based on ray length (R) range from 30 mm in *Echinaster graminicola* to 190 mm in *Echinaster varicolor* (Table 20.1). Most species have fairly slender tapering arms, with ratios of ray length to disc diameter (r) of 4–7. Their body form probably helps in maneuvering in tight spaces of the hard irregular substratum that most species inhabit.

Color and pattern vary widely in *Echinaster* (Table 20.1). The palest species are yellow, and the darkest are reddish brown or deep purple. The paler color of the adoral surface and its seemingly greater sensitivity to light (Cowles 1909) indicate that pigments might play a photoprotective role in this photopositive

shallow-water genus. Little is known about pigments in the genus; *E. sepositus* has a carotenoid astacene (astacin, asterubin), with the formula $C_{40}H_{48}O_4$ (Cuénot 1948). Cuénot (1891) believed the bright colors to be aposematic ("coloration prémonitrice") to forewarn potential predators of toxic secretions of the skin glands in lieu of pedicellariae. Commensal crustaceans closely match their *Echinaster* hosts in color (Humes 1976, A. J. Bruce 1979, Pires 1995). Isopods epizoic on *Echinaster brasiliensis* consume skin secretions of the sea star, apparently without toxic effect, and probably achieve concealing coloration by incorporation of ingested pigment (Pires 1995).

The body wall is fairly rigid, tough, and chemically protected. The thick skin includes microscopic ossicles, adding to its toughness (Ludwig 1897, Gale 2011a). This protective morphology might be strongly associated with food limitation by its microphagous diet, indicating a suite of stress-tolerant adaptations (Lawrence 1990). It offers protection against arm loss by predation (Blake 1989). Arm regeneration is rarely found in the field except for the fissiparous *Echinaster luzonicus*, which reproduces asexually, but regeneration is slow when loss occurs (Lares and Lawrence 1994).

The framework of each ray consists of several longitudinal series of imbricating primary ossicles connected transversely by secondary ossicles (Fig. 20.1f–h). Aborally, the staggered primary series and the transverse series create fairly large open areas in the meshwork where respiratory papulae occur (Fig. 20.1d,f). The two marginal series are more linear, closely spaced, and sometimes separated by intermarginals. Meshes on the sides of the ray are usually smaller and contain fewer papulae. The inferomarginal series abuts the adambulacrals so closely that little open mesh is available for papulae. The presence of respiratory papulae over much of the exposed surface probably reflects the epifaunal habit of *Echinaster*. Ambulacral ossicles form the axis of the ray. Suckered podia with single ampullae extend between them in two series (Fig. 20.1b,e). All series but the ambulacral ossicles bear spines, which can occur on secondary ossicles as well. Spines are small (mostly 1–3 mm). Their morphology, distribution, and number are important in systematics of the genus. In some species, primary ossicles also bear glassy tubercles, exposed patches of expanded stereom bearing few narrow stromal spaces (Fig. 20.1h–j). Their function in sea stars is unknown, but their similarity to purported light-receptive structures of ophiuroids (Hendler and Byrne 1987) begs further study.

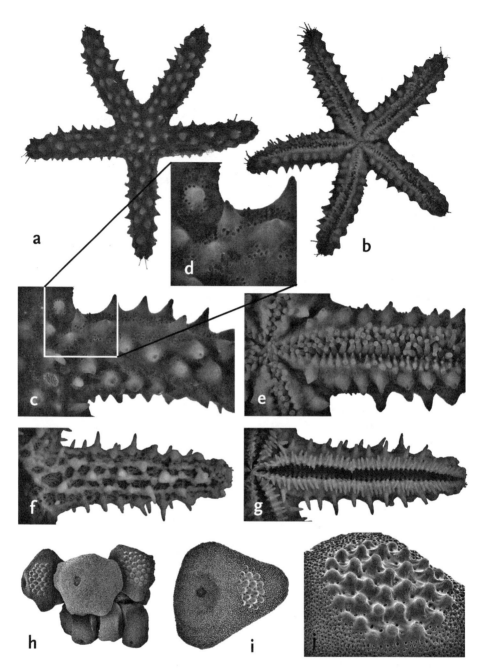

Fig. 20.1. Morphology of *Echinaster echinophorus*, Big Pine Key, Florida (R = 32 mm, R/r = 3.6). *a*, General body form, aboral view, with lighter knobs on darker ground color. *b*, General body form, adoral view, lighter ground color than *a*. *c*, Aboral view, showing distribution of spines covered by thick skin. *d*, magnified view of *c*, with dark papulae in open areas among spines and at bases of spines. *e*, Adoral view, with biserial podia and small adambulacral spines bordered by fewer large inferomarginal spines. *f*, Aboral view, bleached ray, revealing spine-bearing primary ossicles connected longitudinally and transversely by secondary ossicles. *g*, Adoral view, bleached ray, showing narrow ambulacral groove. *h*, Three isolated inferomarginal ossicles (*upper row*) adjacent to three adambulacral ossicles (*lower row*); spine-bearing inferomarginal (spine removed) flanked by two secondary inferomarginals with glassy tubercles. *i*, Isolated spine-bearing aboral primary ossicle (spine removed) with patch of glassy tubercles. *j*, Oblique view of a patch of glassy tubercles. *h*–*j*, scanning electron micrographs.

Table 20.1. Body color, maximal size, and R:r of *Echinaster*

Species	Color	R (mm)	R:r	Reference
E. aculeata		35	4.5	H. L. Clark 1940
E. arcystatus	Deep red or red-brown	172	8	H. L. Clark 1928
	Dull yellow to dark red	150		Shepherd 1968
	Yellow or pink background with bright purple papulae			Rowe and Albertson 1986
E. brasiliensis	Light to reddish chestnut, lighter below	83	4.0–5.5	A. M. Clark and Downey 1992
E. callosus		176	10.4	VandenSpiegel et al. 1998
	Brown to violet, red, and yellow			de Celis 1980
	Yellow with arms annulated with brown or red	130		Guille et al. 1986
E. colemani	Velvet brown with purple papulae	137	4.5–8.4	Rowe and Albertson 1986
			9.3	McKnight 2006
E. cylindricus	Dark blackish brown (whether live or preserved not known)	90	7.5	H. L. Clark 1910
E. echinophorus	Bright red, crimson	70	3.2–5.5	A. M. Clark and Downey 1992
E. farquhari		120		Fell 1959
			4.5–5.8	McKnight 2006
E. glomeratus		100	5	H. L. Clark 1928
	Dark wine red			Shepherd 1968
E. graminicola	Dark red with yellow knobs, blackish-red papularia			Campbell and Turner 1984
	Blackish-red with orange knobs	30	3.3	Clark and Downey 1992
E. guyanensis		67	4.1–4.9	Clark and Downey 1992
	Bright red			Hopkins et al. 2003
E. luzonicus	Rusty red speckled with black	125		H. L. Clark 1921, 1946
	Bright orange, dark green		4.0–8.9	L. M. Marsh 1977
	Orange with brown tips or dark mauve or dark brown			Guille et al. 1986
E. modestus		78	4.0–6.1	Clark and Downey 1992
E. panamensis	(No information available)			
E. parvispinus	Dark purplish, reddish brown or light pink (in alcohol)	53	5.3	A. H. Clark 1916
E. paucispinus	Gray or orange	63	4.5–5.8	Clark and Downey 1992
E. purpureus		140	5.6–9.6	A. M. Clark and Rowe 1971
	Pale creamy pink or brown with fine darker mottling	80	7.1	Rowe and Richmond 2004
	Red			A. J. Bruce 1979
E. reticulatus		75	4.7–5.0	Clark and Downey 1992
E. sentus	Dark red, reddish brown, dark purple, yellowish brown	90	3.8–4.8	Clark and Downey 1992
E. sepositus	Bright red, lighter below	155	5.0–8.0	Clark and Downey 1992
	Bright red, rarely yellow; lighter beneath; pink or yellow podia; red suckers			Tortonese 1965
E. serpentarius		57	4.5–5.3	Clark and Downey 1992
E. smithi	(No information available)			
E. spinulosus	Orange with blue papularia			Campbell and Turner 1984
		90	4.3–5.9	Clark and Downey 1992
E. stereosomus		70		H. L. Clark 1946
		13–18	4.4–5.3	VandenSpiegel et al. 1998
E. superbus	Light yellowish green dorsally with five large blotches along each ray, spines rose in darker areas, otherwise cream-white, oral spines lavender, papulae pale brown	145	6	H. L. Clark 1938
E. tenuispinus		91	6.5	Verrill 1914
E. varicolor		190	5.6–9.5	A. M. Clark and Rowe 1971
	Three morphs: buff spotted with purple, uniformly violet with yellow adorally, or uniformly bright red	190	8–10	H. L. Clark 1938
		150	8.8	Aziz and Jangoux 1984

Note: Colors are given for living animals unless otherwise indicated. Abbreviations: R, ray length; r, radius of the disc; R:r, ratio of the two measurements.

The skeletal meshwork of *Echinaster* is bound internally and externally by orthogonally arranged layers of collagen in twisted cords (O'Neill 1989). Interconnecting fibers bind the cords and layers. Confinement of the layers below and above the skeleton leaves voids for the papular areas, allowing access of the perivisceral coelom to the papulae for gas exchange. The orthogonal arrangement and voids, the mutable nature of the collagen, and the narrow and deep ambulacral groove of a cylindrical ray permit torsion of the ray, an action prevented by geodesic arrangements of fibers in most "worms" (O'Neill 1990). This construction allows *Echinaster* to right itself in tighter spaces than other morphologies of sea star, an advantage in crevices of the rocky or reef habitat or shallow pools typical of the genus (O'Neill 1989).

The body wall and papulae are kept inflated by turgor generated by several factors (Ferguson 1988, 1989). Tension of the body and podial walls creates hydrostatic pressure in the perivisceral coelom (PVC) and the water-vascular system (WVS) that causes a continuous loss of body water by ultrafiltration. The loss is countered by two factors: osmotic pressures of the perivisceral coelomic fluid and water-vascular fluid that are significantly higher (+5.1 and +12.9 mosmol, respectively) than that of ambient seawater; resupply of water through powerful ciliary currents of the stone canal (Ferguson 1988, 1989). Furthermore, water passes down the osmotic gradient from the PVC into the WVS across the podial ampullae. Body volume is further regulated by closure and opening of madreporic pores and by venting of gastric water and feces via the valved anus. In these ways, the animal accommodates changes in body volume from movement (e.g., righting, occupancy of crevices), osmotic challenges (e.g., life in estuaries and on shallow reefs), and feeding (e.g., sudden onset of high-volume ciliary ingestion). Anal venting was first reported by H. L. Clark (1921) for *E. luzonicus* as a stream 25–50 mm high upon handling in air.

Despite the toughness of the body wall, it can change from rigid to soft by mutability of collagen (O'Neill 1989; see Chapter 2). The body wall has a higher protein composition and lower level of ash than in sea stars with less robust arms (McClintock et al. 1990, Lares and Lawrence 1994). Although the dermal skeletal layer is 54% ossicle by volume, the inner fibrous dermal layer is 81% collagen and 9% muscle; and the outer fibrous dermal layer and epithelium are 63% collagen and 20% epidermis by volume (O'Neill 1989). The skin is thick and glandular, hiding much of the skeleton. The skin glands, studied long ago

(Cuénot 1887, 1948, Ludwig 1897, Barthels 1906, Ferguson 1967b) in *Echinaster* and *Henricia*, are numerous and clustered in the papular areas. Sometimes called "dermal" glands, they are acinar epidermal aggregations of secretory cells with a duct to the surface (Fig. 20.2). They are 0.6–0.8 mm in diameter in *E. sepositus* and release copious mucus upon irritation. The glands of *E. graminicola* absorb free amino acids from the surrounding seawater (Ferguson 1967b) and from the PVC (Ferguson 1970). The short residence time of labeled free amino acids indicates that the glands release contents also continuously at a low rate in addition to a high rate upon stimulation (Ferguson 1967b).

All spinulosid sea stars lack pedicellariae (Blake 1981), which have multiple functions in echinoderms that have them. Their absence seems to be a secondary loss (Lafay et al. 1995). Cuénot (1887), puzzled by their absence, wondered how *E. sepositus* defends itself and predicted that the skin glands play the same physiological role as pedicellariae do in other sea stars. Barthels (1906) also thought the glands to play a role in defense. Their secretions are claimed to be toxic, and asterosaponins and other secondary metabolites from *Echinaster* species are known to have bacteriostatic, antifouling, and predator-deterrent properties (Iorizzi et al. 2001; see Chapter 8). Among these active chemicals are the cyclic steroidal glycosides, found among echinoderms only in a few species of *Echinaster* and *Henricia* (Palagiano et al. 1996). Nevertheless, the contents of skin glands have not been analyzed, and secondary metabolites have not been localized in the bodies of sea stars. Chemical defense, body morphology (Blake 1989), and, in some cases,

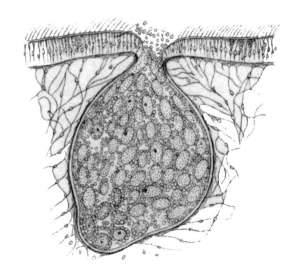

Fig. 20.2. Skin gland of *Echinaster sepositus*. After Cuénot 1887.

occurrence in predator-free reef locations (Yamaguchi 1975) might explain the rarity of predators of *Echinaster* (Bentivegna et al. 1989, Villamor and Becerro 2010). Lucien Cuénot would not be so puzzled today.

Systematics

Echinaster Müller and Troschel, 1840, one of eight genera in the family Echinasteridae Verrill, 1870, is a family trimmed of several genera in recent decades (Blake 1981, A. M. Clark and Downey 1992). The genus includes 27 species in two subgenera and six additional species of doubtful nomenclatural status (World Asteroidea Database 2012). Aside from the possible first reference 2300 years ago by Aristotle (Cresswell 1902), knowledge of the genus originated in 1783 with Retzius's description of *Asterias seposita*, the type species (now *Echinaster* [*Echinaster*] *sepositus*). With the application of molecular genetics to phylogeny, *Echinaster* and *Henricia* as representative echinasterids now are important in understanding asteroid relationships (Lafay et al. 1995, Knott and Wray 2000, Janies et al. 2011).

Echinaster is a Recent genus. Its absence from the fossil record might hold important significance for our understanding of recent faunal changes: based on their abundance, the robustness of their ossicles, and the tough fibrous connective tissue that binds the skeleton, one would expect to find fossils (Blake and Reid 1998). Echinasterids are a rapidly evolving group, with mutation rates of 28S rRNA among the highest for echinoids and asteroids (Lafay et al. 1995). Perhaps rapid speciation during the Quaternary Period explains their puzzling absence from the fossil record and the frustration experienced by systematists who have struggled with *Echinaster* and *Henricia* since the 1800s (A. M. Clark and Downey 1992).

Two of several characters traditionally used to distinguish *Echinaster* from *Henricia* and species within *Echinaster* are the presence and distribution of papulae and glassy tubercles. Since the 1800s, the presence of actinal papulae was taken to be diagnostic for *Henricia*. Confusion of skin glands and papulae in museum specimens gave rise to numerous misidentifications at the generic level (A. M. Clark and Downey 1992). Distribution of papulae is no longer considered a reliable character. Tortonese and Downey (1977) resurrected *Othilia* as a subgenus of *Echinaster* based on the presence of glassy tubercles on primary plates of the arms, and they placed most Atlantic species in *Echinaster* (*Othilia*). Since then, reduced and sunken tubercles were found in some *Echinaster* (*Echinaster*) *modes-*

tus and *E.* (*E.*) *sepositus* (Clark and Downey 1992). Their presence in most Pacific species has not been evaluated. The taxonomic value of glassy tubercles has yet to be convincingly demonstrated.

Mayr (1954) cited Tortonese's (1950) suspicion that *E. luzonicus* and *Echinaster purpureus* are subspecies and his statements about variation in Atlantic *Echinasters* as evidence of geographic speciation in asteroids. The *Echinaster* complex in the eastern Gulf of Mexico has been especially vexing, and attempts to separate them morphologically and molecularly have met with variable success (Tuttle and Lindahl 1980, Campbell and Turner 1984, A. M. Clark and Downey 1992, Fontanella and Hopkins 2003). Two species, *E. graminicola* and *Echinaster spinulosus*, are excellent examples of the problem (Table 20.2). The first study clearly done on *E. graminicola* (as *E. spinulosus*) was that of Ferguson (1966), who collected his material from "shallow grass flats." Uncertainty about the identity of the two forms near Tampa Bay was raised by Atwood (1973), who called them *Echinaster echinophorus* type 1 and type 2 and distinguished them by spination, color, and size of adults and by color, size, buoyancy, and adhesiveness of eggs (Table 20.2). Atwood (1973) gave no details on how adults differed in arrangement of spines. Accumulating data eventually raised suspicion that the two forms were separate species (Ferguson 1976, Scheibling and Lawrence 1982). Morphological distinctions (spination, plate arrangements, madreporite morphology; Table 20.2) useful for identifying museum specimens were found by Campbell and Turner (1984) when they described Atwood's (1973) "type 2" as *E. graminicola*. The distinctions were, however, very weak, although Clark and Downey (1992) accepted the validity of *E. graminicola*. Most evidence for separation of *E. graminicola* and *E. spinulosus* is ecological and physiological (Table 20.2). In the field, the two species are best recognized by habitat and body color. However, the distinctions between them in reproduction, life history, and organic composition are many and might reflect a recent history of divergence from differential adaptation to stress as in other species- and genus-level complexes of echinoderms (Lawrence and Herrera 2000). Continued studies of molecular genetics might help resolve this and other cases of sibling species in *Echinaster*.

Nutrition

Species of *Echinaster* are epifaunal particle feeders, using ciliary currents to ingest microscopic organisms

Table 20.2. Distinctions between two sibling species of *Echinaster* in the eastern Gulf of Mexico

Character	E. graminicola	E. spinulosus	Reference
Morphology			
Aboral disc spines	Tapered, subacute	Cylindrical, obtuse	Campbell and Turner 1984
Mouth plate spines	2–3	3–4	Campbell and Turner 1984
Spines on primary ray ossicles	0–1	1–2	Campbell and Turner 1984
Furrow spines	1	2	Campbell and Turner 1984
Intercalary carinals	Usually present	Usually absent	Campbell and Turner 1984
Madreporic gyri	Meandering	Radiating	Campbell and Turner 1984
Gyral madreporic spines	0–10	3–34	Campbell and Turner 1984
Peripheral madreporic spines	2–16	12–37	Campbell and Turner 1984
Color			
General body color	Bright red and black	Pale orange	Atwood 1973
	Dark red with yellow knobs	Orange	Ferguson 1975
			Campbell and Turner 1984
Color of papulariae	Blackish red	Bluish white	Ferguson 1975
			Campbell and Turner 1984
Eggs			
Color	Orange	Brown	Atwood 1973
			Turner and Lawrence 1979
Shape	Oblate spheroid	Prolate spheroid	Turner and Lawrence 1979
Diameter	0.84–0.88 mm	1.0–1.3 mm	Atwood 1973
Volume	0.258 µl	0.199 µl	Turner and Lawrence 1979
Lipid:protein ratio	1:1	5:2	Turner and Lawrence 1979
	1.34	2.49–3.66	George et al. 1997
Organic density	595 µg/µl	439 µg/µl	Turner and Lawrence 1979
Energy content	4.7 Joules	3.5 Joules	McEdward and Morgan 2001
	1.7 cal	1.0 cal	Scheibling and Lawrence 1982
Buoyancy upon spawning	Sinks	Floats	Atwood 1973
			Turner and Lawrence 1979
Adhesion upon spawning	Sticky	Not sticky	Atwood 1973
Reproduction and growth			
Size at sexual maturity	R = 14–15 mm	R ~ 30 mm	Scheibling and Lawrence 1982
Ovary color	Bright orange	Dark green	Scheibling and Lawrence 1982
Testis protein level	76–79%	66–75%	Scheibling and Lawrence 1982
Testis lipid level	9–10%	14–15%	Scheibling and Lawrence 1982
Ovary protein level	50–56%	39–47%	Scheibling and Lawrence 1982
Ovary lipid level	36–42%	53–56%	Scheibling and Lawrence 1982
Spawning	Late April to early May	Late May to early June	Scheibling and Lawrence 1982
Larva	Demersal	Pelagic	Atwood 1973
Larval thermal tolerance	Greater	Less	Watts et al. 1982
Age at mouth opening	12 d	14 d	Watts et al. 1982
Growth rate first year	10 mm	22 mm	Scheibling 1982c
Maturity	Later age, smaller size	Earlier age, larger size	Scheibling and Lawrence 1982
Reproductive effort	1.3–2.0×	1×	Scheibling and Lawrence 1982
Fecundity	1×	3–5×	Scheibling and Lawrence 1982
Reproductive output	1×	2–3×	Scheibling and Lawrence 1982
Caloric investment per egg	1.7×	1×	Scheibling and Lawrence 1982

Table 20.2. continued

Character	E. graminicola	E. spinulosus	Reference
		Adult Ecology	
Body size	6 cm "diameter"	≥11 cm "diameter"	Atwood 1973
	Smaller, R = 16–20 mm	Larger, R = 35–70 mm	Campbell and Turner 1984
			Scheibling and Lawrence 1982
Body wall index	70–80% of total dry weight	80–90% of total dry weight	Scheibling and Lawrence 1982
Body wall protein level	35–41%	27–34%	Scheibling and Lawrence 1982
Body wall ash level	54–59%	61–70%	Scheibling and Lawrence 1982
Peak pyloric index	December to January	January to February	Scheibling and Lawrence 1982
Fatty acid profile	Similar to sponges	Unlike *E. graminicola* or sponges	Ferguson 1976
Habitat	Seagrass, oyster	Sand	Atwood 1973
			Ferguson 1975
			Scheibling 1982c
Depth	1–2 m	3–5 m	Atwood 1973
	Shallower	Deeper	Ferguson 1975
			Scheibling 1982c
	1–2 m	2–3 m	Scheibling and Lawrence 1982
Location on seawalls	High	Low	Scheibling and Lawrence 1982
Geographic range	475 km; eastern Gulf of Mexico	2200 km; Cape Hatteras (35.5° N) to west Florida (85.5° W)	A. M. Clark and Downey 1992
Allozymic heterozygosity (H)	0.255	0.255–0.260	Tuttle and Lindahl 1980

and detrital material for internal digestion or to consume tissues of larger organisms partly digested extraorally. Their body form (Blake 1989) probably does not allow intraoral feeding or burrowing. Their feeding style does not, however, prevent carnivory on large animals, at least in aquaria, or scavenging (Siddall 1979, Ferguson 1984). There is no evidence that *Echinaster* is a suspension feeder like *Henricia* (Ferguson and Walker 1993). Fatty acid profiles for *E. graminicola* and *E. spinulosus* differ from those of mixed plankton (Ferguson 1976).

The small cardiac stomach, typical of microphagous asteroids (Jangoux 1982b), is everted through the mouth as a small button over the substratum or prey (Vasserot 1961, Ferguson 1969a). Eversion is stimulated by nutrient-rich seawater (Ferguson 1984). Powerful ciliary currents of the Tiedemann's pouches draw water and particles into the pyloric stomach and into the pyloric ceca for storage (Ferguson 1969b, 1974). Absorption in the gut is accompanied by levels of alkaline phosphatase two orders of magnitude higher than levels in carnivorous sea stars (Jangoux and Van Impe 1971). Nutrients absorbed by the pyloric ceca, cardiac stomach, and rectal ceca are translocated to the rest of the body, including gonads, via the PVC

and hemal system without much involvement of amoebocytes (Ferguson 1984). Distribution within the PVC is aided by peritoneal cilia and by muscular movements of the pyloric ceca (Ferguson 1966). Metabolic needs of the epidermis are met by transepidermal absorption of dissolved organic matter from seawater without translocation to or from deeper tissues (Ferguson 1980a,b). Ferguson's groundbreaking research on nutritional and reproductive physiology of sea stars using *E. graminicola* appeared almost annually over 19 years (Ferguson 1966, 1984) and should be consulted for details.

Anecdotal field and aquarium observations indicate that *Echinaster* feeds on a variety of encrusting invertebrates and microbial films and scavenges if carrion is encountered (Sloan 1980a, Jangoux 1982a, Scheibling 1982c, Ferguson 1984, Brooks and Gwaltney 1993). For example, Ferguson (1969a) found 33% of 1136 *E. graminicola* in the field with stomachs everted while on sponges and 9% with stomachs everted while on sand, algae, or ascidians. The remaining 58% did not have stomachs everted and were usually moving over the seafloor. Experimental studies that established *Echinaster* as highly specialized spongivores were based on choice of substratum by the sea stars but included no data on consumption (Vasserot 1961, Waddell and Pawlik 2000). No study has demonstrated that *Echinaster* consumes sponges in the field. Ferguson (1969a) found that *E. graminicola* with stomachs everted on sponges in the field caused little damage to the sponges, and he suggested that the sea stars absorb substances leached from the sponges. The experimental work by Maldonado and Uriz (1998) gave little evidence that *E. sepositus* is spongivorous. In fact, Ferrat and Escoubet (1987) and Waddell and Pawlik (2000) found that some sponges are chemically defended from *E. sepositus*. A descriptive and experimental study by Villamor and Becerro (2010) on *E. sepositus* over a 150-km coastal region in northeastern Spain revealed that its distribution was positively correlated with percent cover of coralline algae and not other benthos, fish, or several abiotic factors. Their experimental study found no impact of seastar density on community structure over a 4-mo period. Definitive studies on the diet of *Echinaster* species are much needed, but evidence so far does not support the idea that it is a genus of sponge specialists.

Reproduction

Most species of *Echinaster* for which information is available are gonochoric and reproduce annually by freely spawned gametes, which include large, pigmented, yolky eggs and typical asteroid spermatozoa. Their uniformly ciliated modified brachiolariae are non-feeding, short-lived, and pelagic. Studies over the last 150 yr on *E. sepositus* and the comparative work since the 1970s on *E. graminicola* and *E. spinulosus* (Table 20.2) have contributed much to our knowledge of sexual reproduction in the genus. Asexual reproduction occurs only in *E. luzonicus*.

Gonads of *Echinaster* are paired in each ray with a gonoduct to the interbrachium (Cuénot 1948). Minimal size for sexual maturity is reported for three species: *E. graminicola* and *E. spinulosus* (Table 20.2), although the authors did not specify the criterion for maturity; *Echinaster guyanensis*, in which all with R > 36 mm have high gonad indices (Mariante et al. 2010). Most asteroids have ratios of ovary index to testis index of 1.0 or greater (Lawrence 1976, Mariante et al. 2010). The ratio is 0.4–0.5 in *E. graminicola* (as *E. echinophorus* and *Echinaster* sp.; Lawrence and Lane 1982), which has demersal adherent eggs. The proportionately greater size of testes in *E. graminicola* indicates that its mating system differs from other *Echinaster* species and might involve sperm competition. The sex ratio is 1:1 in *E. graminicola*, *E. guyanensis*, and

E. spinulosus (Ferguson 1974, Scheibling and Lawrence 1982, Mariante et al. 2010). Hermaphroditism is rarely reported (Delavault 1960, Scheibling and Lawrence 1982). Ghiselin's (1969) observation that sea stars with abbreviated development are hermaphroditic does not hold for *Echinaster*.

Eggs are large and are spherical, elongate, or flattened and have the pigment of the adult (Table 20.3). Fecundity is low (100–1000 eggs per female; Table 20.3). Gametes are released in warmer months, as early as March and as late as October in various species and locations in the Northern Hemisphere. Spawning periods for the sympatric *E. graminicola* and *E. spinulosus* are short and separated by 2–3 weeks, although viable hybrids can be produced in vitro (Atwood 1973, Scheibling 1982c). Spawning follows distinct annual cycles of gonad indices (Ferguson 1974, 1975, Mariante et al. 2010). Eggs usually float, but those of *E. graminicola* and *Echinaster sentus* are adhesive and demersal (Atwood 1973, Siddall 1979), sticking to the interbrachium at first (Turner, unpublished) and eventually to seagrass, rocks, and oysters (Atwood 1973) in *E. graminicola*. Agassiz's (1877) claim of adhesive eggs and broodcare in *E. sentus* has not been confirmed.

Development is lecithotrophic, modified, and short. This abbreviated "*Echinaster* mode" (Agassiz 1877)—with the evolutionary loss of a feeding larva (Emlet 1994)—has interested echinoderm biologists since its discovery in the 1850s (Löhner 1913). The bipinnaria is absent, and the modified brachiolaria is uniformly ciliated, a condition necessary for effective locomotion by large, pelagic, buoyant larvae (Emlet 1994). The brachiolaria has no mouth or gut and has three or four arms, among which the attachment disc appears. For those few species studied, embryogenesis lasts only 1–2 days, reaching the brachiolaria in 2–6 days, settlement or attachment in 3–7 days, and opening of the juvenile mouth in 12–23 days (Table 20.4); this program is much shorter than for species with planktotrophic larvae (Hoegh-Guldberg and Pearse 1995). The juvenile reaches R = 1 mm in 12–29 days (Table 20.4).

Asexual reproduction occurs only in *E. luzonicus* by autotomy of rays, unlike *Henricia sexradiata*, which divides across the disc (A. M. Clark and Downey 1992). The animal passes through the comet stage as it regrows the disc and rays, with five to seven rays, one or two madreporites, and one or two anuses (H. L. Clark 1921, Domantay 1938, Yamaguchi 1975, L. M. Marsh 1977, Hotchkiss 1979). Based on field and laboratory observations, Clark (1921) and Domantay (1938) concluded that this species is fissiparous, but

Table 20.3. Characteristics of reproduction and eggs of *Echinaster* species

Species	Fecundity (number of eggs)	Season of spawning, maximal gonad index, or ripe gonads	Egg diameter (mm)	Egg shape	Egg color	Egg buoyancy	Reference
E. brasiliensis			1	Round/oval	Dark brown		Nobre and Campos 2004
E. echinophorus	100–300	March			Dirty brown	Float	Kempf 1966
E. graminicola	485	Late April to early May					Scheibling 1982c
							Scheibling and Lawrence 1982
			0.84–0.88		Orange	Sink, adhesive	Atwood 1973
E. guyanensis		March to May		Oblate spheroid			Turner and Lawrence 1979
							Mariante et al. 2010
E. purpureus		July	Large		Almost black	Float	Mortensen 1938
E. sentus	200–500	May to June	0.7–0.8		Bright orange	Sink, adhesive	Siddall 1979
						Brooded	Agassiz 1877
E. sepositus		Fall (Turkey)					Artüz 1967
		Summer (Adriatic)					Schapiro 1914
	1000	August to late September (Banyuls)	1.2–1.3				Picard et al. 1985
		Late August to late September at night (Adriatic)	1	Spherical	Bright red	Float free	Nachtsheim 1914
		May to October (Adriatic)	1		Rosso mattone	Float free	Mitic 1992
		August (Adriatic)	1				Lo Bianco 1909
							Löhner 1913
E. spinulosus		late May to early June	1.0–1.3		Brownish-black	Float	Scheibling 1982c
							Atwood 1973
				Prolate spheroid			Turner and Lawrence 1979

Table 20.4. Chronology of embryos, larvae, and juveniles of *Echinaster* species

Species	Age at gastrulation	Age at brachiolaria	Age at settlement	Age at mouth formation	Post-metamorphic growth (R)	Reference
E. brasiliensis		4 days	7 days	23 days		Nobre and Campos 2004
E. echinophorus	2 days		3 days		1.05 mm by age 12 days	Kempf 1966
E. graminicola	1 day			14 days	0.82–0.98 mm by age 28 days	Atwood 1973
				14 days		Watts et al. 1982
					10 mm by age 1 yr	Scheibling 1982c
E. purpureus		<6 days				Mortensen 1938
E. sentus		2 days		14 days		Siddall 1979
E. sepositus				20 days		Artüz 1967
	1 day	5 days	5–6 days	14 days	1.9 mm by age 7.5 months	Nachtsheim 1914
					1.2 mm by age 29 days	Löhner 1913
E. spinulosus		2 days	3 days			Atwood 1973
				12 days		Watts et al. 1982
					22 mm by age 1 yr	Scheibling 1982c

Note: Consult the references for greater detail on embryogenesis, larvigenesis, and post-metamorphic growth.

there is no evidence that either author induced fragmentation and followed regrowth in individuals. Yamaguchi (1975) attributed the local abundance of *E. luzonicus* to its ability to reproduce asexually. There are no reports in the literature of sexual reproduction by *E. luzonicus*.

Biogeography

Echinaster generally is a eurytopic, circumtropical, shallow-water genus of sea star (Table 20.5, Fig. 20.3). The present distribution of the genus seems to be the outcome of its encounter with vicariances and corridors and several aspects of its biology: preference for hard substrata, suckered podia, ability to contort rays, rapid speciation, intolerance for low salinity and soft substrata, short-lived brachiolaria. Two species from the western Atlantic with contrasting distributions are the sibling species *E. graminicola* and *E. spinulosus* (Table 20.2). The narrower distribution of *E. graminicola* might be explained by its need for firm substrata with crevices for attachment of its adhesive eggs and for protection of its demersal short-lived larva (Atwood 1973). Other regions of the world have interesting species pairs of *Echinaster* that might prove instructive with further study (e.g., Lower California, Red Sea, South Australia).

Echinaster is one of about 140 asteroid genera in the Atlantic Ocean and includes 11 species (A. M. Clark and Downey 1992). The only species in the eastern Atlantic is *E. (Echinaster) sepositus*, which extends from Roscoff, France (48.7° N) to the Democratic Republic of the Congo (6° S) and throughout the Mediterranean Sea (Fig. 20.3). Its greatest affinity is with species of the Indo-West Pacific, from which it has been isolated since closure of the Tethys Seaway 19–12 Ma. It must have reinvaded the Mediterranean from the Atlantic after the Messinian Salinity Crisis about 5 Ma. Its recent extension to Roscoff might have been only since the last glacial maximum (21–17 ka), when the Atlantic and Mediterranean populations were disjunct at Gibraltar. Its southern limit might be influenced by outflow from the Congo River. Although present in the Cape Verde Islands, its absence from the Azores, Madeiras, Canaries, Ascension, and St. Helena is explained by its short larval period.

The western Atlantic has nine species, most of which occupy the Gulf of Mexico and Caribbean Sea. All belong to subgenus *Echinaster (Othilia)* except the deep-water *E. (Echinaster) modestus* (depth 67–470 m). The northernmost, *E. spinulosus*, extends to Cape Hatteras, USA (35.5° N), and the southernmost, *E. brasiliensis*, to Gulf of San Matias, Argentina (42° S). None occurs in Bermuda, Ascension Island, or St. Helena.

Table 20.5. Distribution of *Echinaster*

Species	Depth range (m)	Substratum	Distribution	Reference
E. aculeata	26	Fine sand	Pacific coast of Nicaragua and Costa Rica	H. L. Clark 1940
E. arcystatus	10–30	Rock	Western Australia to South Australia	Shepherd 1968
	0–46		Western Australia to South Australia	Rowe and Gates 1995
E. brasiliensis	0–60		Brazil and Argentina	A.M. Clark and Downey 1992
(as E. nudus)		Rock	Brazil	Gondim et al. 2008
E. callosus	0–62		Southeast Asia and Philippines	Lane et al. 2000
	5–20	Shelly sand, rarely on coral or muddy sand with algae		Guille et al. 1986
			Red Sea and Indian Ocean to Japan and South Pacific Islands	A. M. Clark and Rowe 1971
	0–14		Northeastern Australia	Rowe and Gates 1995
E. colemani	17–40		Southeastern Australia and Norfolk Island, New Zealand	McKnight 2006
	5–110	Rock, reef, sand, rubble, sponge, coral	Southeastern Australia and Norfolk Island, New Zealand	Rowe and Albertson 1986
E. cylindricus			Callas, Peru	Maluf 1988
E. echinophorus	0–73	Rock, coral	Florida Keys, Caribbean, and Brazil	A.M. Clark and Downey 1992
E. farquhari	33–711		New Zealand	McKnight 2006
	531–604	Fine green sand and mud	New Zealand	Fell 1960
E. glomeratus	20–30	Rock	South Australia	Shepherd 1968
	0–64		Western Australia to South Australia	Rowe and Gates 1995
E. graminicola	0–2	Seagrass, oyster, seawalls, bulkheads	Eastern Gulf of Mexico	Clark and Downey 1992
E. guyanensis	13–106	Coarse calcareous sand, shell, gravel, muddy sand	Caribbean	Clark and Downey 1992
		Rock	Espirito Santo, Brazil	Mariante et al. 2010
E. luzonicus	0–73		Southeast Asia	Lane et al. 2000
		Muddy flats, seagrass	Malay Peninsula to Japan and South Pacific Islands	H. L. Clark 1921
		Mixed sand and rock, coral patches	Philippines	de Celis 1980
		Lightly shelly or muddy sand, sponge	New Caledonia	Guille et al. 1986
E. modestus	67–470	Sandy mud, calcareous sand	Southeastern Gulf of Mexico and Caribbean	Clark and Downey 1992
E. panamensis			Pearl Islands, Panama	Maluf 1988
E. parvispinus	18–29	Gray sand, broken shell	Baja California, Mexico	Maluf 1988
E. paucispinus	12–85	Sponge with coral, shell, bryozoans, algae, sand, rock	Eastern Gulf of Mexico	Clark and Downey 1992
E. purpureus	Low water		Gulf of Kachchh, India	Soota and Sastry 1977
	<20		Red Sea and Indian Ocean	A. M. Clark and Rowe 1971
E. reticulatus	33–320	Sand, rock	South Africa	Clark and Downey 1992

continued

Table 20.5. continued

Species	Depth range (m)	Substratum	Distribution	Reference
E. sentus	0–68	Coral reef, rock, sand, seagrass	Florida Keys, Bahamas, and northern Caribbean	Clark and Downey 1992
E. sepositus	2–250	Sand, rock, shingle, algae, seagrass	Eastern Atlantic and Mediterranean	Clark and Downey 1992
		Crustose coralline algae	Mediterranean, northeastern Spain	Villamor and Becerro 2010
		Stones	Cape Verde Islands	Entrambasaguas et al. 2008
E. serpentarius	< a few	Coral reef	Vera Cruz, Mexico	Clark and Downey 1992
E. smithi	450		Bellingshausen Sea, Antarctica	Fisher 1940
E. spinulosus	0–55	Sand, shell, sponge, coral, algae	Southeastern United States	Clark and Downey 1992
E. stereosomus	55–146		Northern Australia	Rowe and Gates 1995
	14–757		Southeast Asia and Philippines	Lane et al. 2000
E. superbus		Rock	Northwestern Australia	H. L. Clark 1938
	18–62		Northwestern Australia	Rowe and Gates 1995
E. tenuispinus	0–73	Bolders, rock, sand, shale, shell	Baja California, Mexico	Maluf 1988
E. varicolor		Sand	Northwestern Australia	H. L. Clark 1938
	0–52		Western Australia	Rowe and Gates 1995
	30–35		Java Sea	Aziz and Jangoux 1984
			New Caledonia	Jangoux 1984

Low salinity probably accounts for the disjunctions of *E. echinophorus* and *E. guyanensis* near the Amazon River and the absence of *E. graminicola*, *Echinaster paucispinus*, and *E. spinulosus* west of Florida, USA (Fig. 20.3); the latter probably have reinvaded the Gulf of Mexico since the Wisconsin Glacial Epoch (100–14 ka) and the postglacial flow of meltwater down the Mississippi River from the Laurentide ice sheet. Reinvasion and rapid speciation might account for some of the current systematic problems in the region.

Echinaster (*Echinaster*) *reticulatus* barely extends into the Atlantic off South Africa (30–35° S), where a Mediterranean climate prevails. Its greatest affinities probably are with species of the Indian Ocean, into which this species also barely extends.

Echinaster is one of 85 genera and includes five of 185 species of sea star in the central eastern Pacific Ocean (35° N to 12° S; Maluf 1988). The southernmost species, *Echinaster cylindricus*, is known only by the type specimen, from Peru. *Echinaster panamensis* inhabits the Pearl Islands off Panama, and *Echinaster* (*Othilia*) *aculeata* ranges along the coasts of Costa Rica and Nicaragua. The distributions of *Echinaster parvispinus* and *Echinaster* (*Othilia*) *tenuispinus* completely overlap in the Gulf of California (to 32° N);

along the Pacific coast of Baja California, *E.* (*O.*) *tenuispinus* extends farther north than *E. parvispinus* to Vizcaino Bay (Dawson 1952). Of the two species, *E.* (*O.*) *tenuispinus* has the greater range of depth and substratum (Table 20.5). There are no data on depth and substratum for the other three species. The presence of two species in the subgenus *Othilia* indicates an affinity to the western Atlantic *Echinaster* fauna and a minimum age of 3 million years for the subgenus. The occurrence of *Acanthaster* and *Mithrodia* in this region argues for the potential influence from the Indo-West Pacific fauna, but the life history of *Echinaster* and the absence of the genus from the Hawaiian Islands make a Caribbean origin more likely. Life history pattern also explains its absence from the Galápagos Archipelago (Hickman 1998).

Despite its richness in other biota, the tropical Indo-West Pacific has only six species of *Echinaster* among its 230 species in 59 genera of asteroids (A. M. Clark and Rowe 1971). Among them, only *Echinaster callosus* extends throughout the region, from tropical east Africa (30° N to 27° S) to Japan (35° N) and the South Pacific Islands in the east, south to east central Australia (26° S, Fig. 20.3). None is present in the Hawaiian Islands. Distributions of the other species are

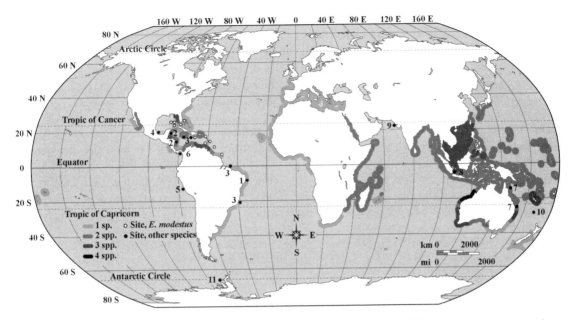

Fig. 20.3. Biogeography of *Echinaster*. Distributional data are taken from references listed in Table 20.5. Extensive gray areas along coastlines indicate general presence based on records in the literature and do not account for possible small gaps in distribution. Extensive gray areas away from coastlines serve only to enclose presence of the species among island groups except for the deep-water *E. farquhari*. White dots are individual collection sites for *E. modestus*. Numbered black dots indicate single locations for other species: 1, *E. brasiliensis*; 2, *E. echinophorus*; 3, *E. guyanensis*; 4, *E. serpentarius*; 5, *E. cylindricus*; 6, *E. panamensis*; 7, *E. callosus*; 8, *E. varicolor*; 9, *E. purpureus*; 10, *E. colemani*; 11, *E. smithi*. Intensity of gray indicates number of species present in overlapping regions of distribution.

divided at the Malay Peninsula, where an extensive land bridge formed at the last glacial maximum. The region from this divide eastward includes *Echinaster luzonicus* and, to a lesser geographic extent, *Echinaster stereosomus*. The occurrence of *E. callosus* and *E. luzonicus* in many of the island groups in the south Pacific is puzzling, considering their life histories. Yamaguchi (1975) noted the limited distribution and rarity of other asteroids with lecithotrophic larvae in the region. Five species extend to or are confined (*Echinaster superbus*) to northern Australia. In addition to *E. callosus* and *E. (E.) reticulatus*, only *Echinaster purpureus* occupies the Indian Ocean. It meets *E. luzonicus* at the Malay Peninsula; Tortonese (1950) thought the two to be subspecies. The near absence of *E. callosus* and *E. purpureus* from much of the Arabian Peninsula through the west coast of India might be related to a lack of hard substratum.

The Australian asteroid fauna comprises 230 species in 111 genera (Rowe and Gates 1995). Among them, *Echinaster* has probably been invading Australia as the continent slowly moved northward over recent epochs. During the northward drift, warm southward-moving currents allowed invasion of and

speciation in southern Australia (Rowe and Vail 1982, O'Hara and Poore 2000). The distributions of the three southern species overlap little with northern Australian species. A land bridge across the Bass Strait during the last glacial maximum (O'Hara and Poore 2000) might have contributed to the present separation of *Echinaster arcystatus* and *Echinaster glomeratus* to the west and *Echinaster colemani* to the east and north of Tasmania (Fig. 20.3). *E. colemani* and *E. arcystatus* probably diverged during the glacial period (Rowe and Albertson 1986). Rowe and Albertson (1986) speculated that *E. colemani* reached the distant Norfolk Island, New Zealand, east of Australia by dispersal of a planktonic larva. Its life history is, however, unknown. Perhaps a better explanation of its dispersal is rafting of adults by west-wind-drift dispersal (Fell 1962). This might also explain the presence of *Echinaster farquhari* in New Zealand (McKnight 2006). This endemic species of New Zealand is the southernmost *Echinaster* in the Pacific Ocean at 50.7° S in the Auckland Islands.

Echinaster smithi was described from one specimen collected in the late 1890s from the Bellingshausen Sea near Palmer Peninsula, Antarctica, from a depth of 450 m. It has not been collected again. Although currently accepted as a valid name (World Asteroidea Database 2012), Fisher (1940) thought it belonged in the genus *Rhopiella*. The status of *E. smithi* should be reevaluated, for its location and depth are problematic for the genus.

Acknowledgments
My resources for writing this chapter greatly benefited from gifts of systematic literature in recent years from John Lawrence and the late John Dearborn as they trimmed their own libraries during retirement. I thank Christopher Kovalik for his assistance with a literature search on *Echinaster* and the interlibrary loan staff of Evans Library, Florida Institute of Technology, for acquisition of literature. Translations of literature by Marina Dell'Utri (Italian), John Lawrence (French), Kathe Jensen (Danish), and Munevver Mine Subasi (Turkish) are greatly appreciated.

Dedication
This chapter is dedicated to John Carruthers Ferguson (Professor Emeritus, Eckerd College, St. Petersburg, Florida), who probably has studied and written more about the biology of *Echinaster* than any other author.

References

Achituv, Y., and E. Sher. 1991. Sexual reproduction and fission in the sea star *Asterina burtoni* from the Mediterranean coast of Israel. *Bulletin of Marine Science* 48:670–678.

Acosta, M. G. 1988. Aspectos autoecológicos de juveniles de *Heliaster helianthus* (Asteroidea: Forcipulatida) en el intermareal de Las Cruces, Chile Central. Tesis de Magister, Facultad de Ciencias, Universidad de Chile, Santiago, Chile.

Agassiz, A. 1877. North American starfishes. *Memoirs of the Museum of Comparative Zoology at Harvard College* 5:1–136.

Aldrovandi, U. 1638. *De animalibus insectis libri septem.* C. Ferronium, Bologna.

Allen, E. J. 1899. On the fauna and bottom-deposits near the thirty-fathom line from the Eddystone Grounds to Start Point. *Journal of the Marine Biological Association of the United Kingdom* 5:365–542.

Allen, J. D., and B. Pernet. 2007. Intermediate modes of larval development: bridging the gap between planktotrophy and lecithotrophy. *Evolution and Development* 9:643–653.

Allen, P. L. 1983. Feeding behaviour of *Asterias rubens* (L.) on soft bottom bivalves: a study in selective predation. *Journal of Experimental Marine Biology and Ecology* 70:79–90.

Allison, G. W. 1994. Effects of temporary starvation on larvae of the sea star *Asterina miniata*. *Marine Biology* 118:255–261.

Alvarado, J. L., and J. C. Castilla. 1996. Tridimensional matrices of mussels *Perumytilus purpuratus* on intertidal platforms with varying wave forces in Central Chile. *Marine Ecology Progress Series* 133:135–141.

Alves, R. R. N., and H. N. Alves. 2011. The faunal drugstore: animal-based remedies used in traditional medicines in Latin America. *Journal of Ethnobiology and Ethnomedicine* 7(9):1–43.

Alves, S. L. S., A. D. Pereira, and C. R. R. Ventura. 2002. Sexual and asexual reproduction of *Coscinasterias tenuispina* (Echinodermata: Asteroidea) from Rio de Janeiro, Brazil. *Marine Biology* 140:95–101.

Ameye, L., R. Hermann, P. Dubois, and P. Flammang. 2000. Ultrastructure of the echinoderm cuticle after fast-freezing/freeze substitution and conventional chemical fixations. *Microscopy Research and Technique* 48:385–393.

Amsler, C. D., K. B. Iken, J. B. McClintock, and B. J. Baker. 2001. Secondary metabolites from Antarctic marine organisms and their ecological implications. Pp. 263–300. In *Marine Chemical Ecology* (J. B. McClintock and B. J. Baker, eds.). CRC Press, Boca Raton, Florida.

Amsler, C. D., K. B. Iken, J. B. McClintock, M. O. Amsler, K. J. Peters, J. M. Hubbard, F. B. Furrow, and B. J. Baker. 2005. A comprehensive evaluation of the palatability and chemical defenses of subtidal macroalgae from the Antarctic Peninsula. *Marine Ecology Progress Series* 294:141–159.

Anderson, J. M. 1960. Histological studies on the digestive system of a starfish, *Henricia*, with notes on Tiedemann's pouches in starfishes. *Biological Bulletin* 119:371–398.

Anderson, J. M. 1978. Studies of the functional morphology in the digestive system of *Oreaster reticulatus* (L.) (Asteroidea). *Biological Bulletin* 154:1–14.

Anderson, J. M. 1979. Histological studies on the pyloric stomach and its appendages in *Oreaster reticulatus* (L.) (Asteroidea). *Biological Bulletin* 156:1–19.

Anger K., U. Rogal, G. Scheiever, and C. Valentin. 1977. In situ investigations on the echinoderm *Asterias rubens* as a predator of soft-bottom communities in the western Baltic Sea. *Helgoländer Wissenschaftliche Meeresunters* 29:439–459.

Araki, G., and A. C. Giese. 1970. Carbohydrases in sea stars. *Physiological Zoology* 43:296–305.

Arima, K., S. Hamaya, and Y. Miyakawa. 1972. Feeding behaviour of the starfishes to the bivalves. *Scientific Report of the Hokkaido Fisheries Experimental Station* 4:63–69.

Arnaud, P. M. 1970. Frequency and ecological significance of necrophagy among the benthic species of Antarctic coastal waters. Pp. 259–267. In *Antarctic Ecology*, vol. 1 (M. W. Holdgate, ed.). Academic Press, London.

Arnaud, P. M. 1974. Contribution à la bionomie marine benthique des regions antarctiques et subantarctiques. *Téthys* 6:467–653.

Aronson, R. B., S. Thatje, J. B. McClintock, and K. A. Hughes. 2011. Anthropogenic impacts on marine ecosystems in Antarctica. *Annals of the New York Academy of Science* 1223:82–107.

Arsenault, D. J., and Himmelman, J. H. 1996. Ontogenic habitat shifts of the Iceland scallop, Chlamys *islandica*

(Müller, 1776), in the northern Gulf of St. Lawrence. *Canadian Journal of Fisheries and Aquatic Sciences* 53:884–895.

Arsenault, D. J., Girard, P., and Himmelman, J. H. 1997. Field evaluation of the effects of refuge use and current velocity on the growth of juvenile Iceland scallops, *Chlamys islandica* (O. F. Müller, 1776). *Journal of Experimental Marine Biology and Ecology* 217:31–45.

Artüz, M. İ. 1967. Türkiye denizlerinde rastlanan denizyildizlari (Asteroidea). *Türk Biyoloji Dergisi* 17:11–24.

Atkinson, W. D., and B. Shorrocks. 1981. Competition on a divided and ephemeral resource: a stimulation model. *Journal of Animal Ecology* 50:461–471.

Atwood, D. G. 1973. Larval development in the asteroid *Echinaster echinophorus*. *Biological Bulletin* 144:1–11.

Atwood, D. G., and J. L. Simon. 1971. Correlation of gamete shedding with presence of neuro secretory granules in *Echinaster* and *Patiria* (Echinodermata: Asteroidea). *American Zoologist* 11:701–702.

Avila, C., K. Iken, A. Fontana, and G. Cimino. 2000. Chemical ecology of the Antarctic nudibranch *Bathydoris hodgsoni* Eliot, 1907: defensive role and origin of its natural products. *Journal of Experimental Marine Biology and Ecology* 252:27–44.

Ayre, D. J., T. E. Minchinton, and C. Perrin. 2009. Does life history predict past and current connectivity for rocky intertidal invertebrates across a marine biogeographic barrier? *Molecular Ecology* 18:1887–1903.

Ayukai, T. 1994. Ingestion of ultraplankton by the planktonic larvae of the crown-of-thorns starfish, *Acanthaster planci*. *Biological Bulletin* 186:90–100.

Aziz, A., and M. Jangoux. 1984. Les astéries (echinodermes) du Plateau de la Sonde (Indonesia). *Indo-Malayan Zoology* 1:127–140.

Babcock, R. C., and C. N. Mundy. 1992. Reproductive biology, spawning and field fertilisation rates of *Acanthaster planci*. *Australian Journal of Marine and Freshwater Research* 43:525–534.

Babcock, R., C. N. Mundy, J. Keesing, and J. Oliver. 1992. Predictable and unpredictable spawning events: *in situ* behavioural data from free-spawning coral reef invertebrates. *Invertebrate Reproduction & Development* 22:213–227.

Babcock, R. C., C. N. Mundy, U. Engelhardt, and B. Lassig. 1993. Seasonal changes in fertility and fecundity in *Acanthaster planci*. Great Barrier Reef Marine Park Authority, Townsville, Queensland, Australia.

Babcock R., C. N. Mundy, and D. Whitehead. 1994. Sperm diffusion models and in situ confirmation of long-distance fertilization in the free-spawning asteroid *Acanthaster planci*. *Biological Bulletin* 186:17–28.

Babcock, R., E. Franke, and N. Barr. 2000. Does spawning depth affect fertilization rates? Experimental data from the sea star *Coscinasterias muricata*. *Marine and Freshwater Research* 51:55–61.

Balch, T., and R. E. Scheibling. 2000. Temporal and spatial variability in settlement and recruitment of echinoderms in kelp beds and barrens in Nova Scotia. *Marine Ecology Progress Series* 205:139–154.

Balch, T., and R. E. Scheibling. 2001. Larval supply, settlement and recruitment in echinoderms. Pp. 1–83. In *Echinoderm Studies,* vol. 6 (M. Jangoux and J. M. Lawrence, eds.). Balkema, Lisse, the Netherlands.

Barahona, M. 2006. Patrones de desplazamiento del sol de mar *Heliaster heliathus* (Lamarck 1816) (Echinodermata: Asteroidea) en sitios con distinta productividad de especies presa en la costa de Chile central. Thesis, Facultad de Ciencias del Mar y Recursos Naturales, Universidad de Valparaíso, Chile.

Barahona, M., and S. A. Navarrete. 2010. Movement patterns of the seastar *Heliaster helianthus* in central Chile: relationship with environmental conditions and prey availability. *Marine Biology* 157:647–661.

Barbaglio, A., D. Mozzi, M. Sugni, P. Tremolada, F. Bonasoro, R. Lavado, C. Porte, and M. D. C. Carnevali. 2006. Effects of exposure to ED contaminants (TPT-Cl and Fenarimol) on crinoid echinoderms: comparative analysis of regenerative development and correlated steroid levels. *Marine Biology* 149:65–77.

Barbeau, M. A., and Scheibling, R. E. 1994a. Behavioral mechanisms of prey size selection by sea stars (*Asterias vulgaris* Verrill) and crabs (*Cancer irroratus* Say) preying on juvenile sea scallops (*Placopecten magellanicus* Gmelin). *Journal of Experimental Marine Biology and Ecology* 180:103–136.

Barbeau, M. A., and R. E. Scheibling. 1994b. Temperature effects on predation of juvenile sea scallops *Placopecten magellanicus* (Gmelin) by seastars (*Asterias vulgaris* Verrill) and crabs (*Cancer irroratus* Say). *Journal of Experimental Marine Biology and Ecology* 182:27–47.

Barbeau, M. A., Scheibling, R. E., and Hatcher, B. G. 1998. Behavioural responses of predatory crabs and sea stars to varying densities of juvenile sea scallops. *Aquaculture* 169:87–98.

Barbosa, S. S. 2012. Population genetic structure of two sympatric asterinids *Meridiastra calcar* and *Parvulastra exigua* along the southeast coast of Australia. PhD Thesis, University of Sydney, Australia.

Barbosa, S. S., O. S. Klanten, H. Jones, and M. Byrne. 2012. Selfing in *Parvulastra exigua*: an asterinid sea star with benthic development. *Marine Biology* 159:1071–1077.

Bargmann, W., and B. Behrens. 1964. Uber die Tiedemannschen Organe des Seesterns (*Asterias rubens* L.). *Zeitschrift fur Zellforschung* 63:120–133.

Barker, M. F. 1977. Observations on the settlement of the brachiolaria larvae of *Stichaster australis* (Verrill) and *Coscinasterias calamaria* (Gray) (Echinodermata: Asteroidea) in the laboratory and on the shore. *Journal of Experimental Marine Biology and Ecology* 30:95–108.

Barker, M. F. 1978a. Descriptions of the larvae of *Stichaster australis* (Verrill) and *Coscinasterias calamaria* (Gray) (Echinodermata: Asteroidea) from New Zealand, obtained from laboratory culture. *Biological Bulletin* 154:32–46.

Barker, M. F. 1978b. Structure of the organs of attachment of brachiolaria larvae of *Stichaster australis* (Verrill) and *Coscinasterias calamaria* (Gray) (Echinodermata: Asteroidea). *Journal of Experimental Marine Biology and Ecology* 33: 1–36.

Barker, M. F. 1979. Breeding and recruitment in a population of the New Zealand starfish *Stichaster australis* (Verrill). *Journal of Experimental Marine Biology and Ecology* 41:195–211.

Barker, M. in press. Class Asteroidea, phylum Echinodermata. In *New Zealand Coastal Marine Invertebrates*, vol. 2 (de Cook, S., ed.). Canterbury University Press, Christchurch, New Zealand.

Barker, M. F., and D. Nichols. 1983. Reproduction, recruitment and juvenile ecology of the starfish, *Asterias rubens* and *Marthasterias glacialis*. *Journal of the Marine Biological Association of the United Kingdom* 63:745–765.

Barker, M. F., and M. P. Russell. 2008. The distribution and behaviour of *Patiriella mortenseni* and *P. regularis* in the extreme hyposaline conditions of the Southern New Zealand Fiords. *Journal of Experimental Marine Biology and Ecology* 355:76–84.

Barker, M. F., and R. E. Scheibling. 2008. Rates of fission, somatic growth and gonadal development of a fissiparous sea star, *Allostichaster insignis*, in New Zealand. *Marine Biology* 153:815–824.

Barker, M. F., and R. A. Xu. 1991a. Population differences in gonad and pyloric caeca cycles of the New Zealand sea star *Sclerasterias mollis* (Echinodermata: Asteroidea). *Marine Biology* 108:97–103.

Barker, M. F., and R. A. Xu. 1991b. Seasonal changes in biochemical composition of body walls, gonads and pyloric caeca in two populations of *Sclerasterias mollis* (Echinodermata: Asteroidea) during the annual reproductive cycle. *Marine Biology* 109:27–34.

Barker, M. F., and R. A. Xu. 1993. Effects of estrogens on gametogenesis and steroid levels in the ovaries and pyloric caeca of *Sclerasterias mollis* (Echinodermata: Asteroidea). *Invertebrate Reproduction & Development* 24:53–58.

Barker M. F., R. Scheibling, and P. V. Mladenov. 1992. Population biology of the sympatric fissiparous sea star *Allostichaster insignis* (Farquhar) and *Coscinasterias calamaria* (Gray). Pp. 191–195. In *Echinoderm Research 1991* (L. Scalera-Liaci and C. Canicatti, eds.). Balkema, Rotterdam, the Netherlands.

Barnes, D. K. A., and K. E. Conlan. 2007. Disturbance, colonization and development of Antarctic benthic communities. *Philosophical Transactions of the Royal Society B* 362:11–38.

Barnes, H. 1975. Reproductive rhythms and some marine invertebrates: an introduction. *Pubblicazioni della Stazione Zoologica di Napoli* 39:8–25.

Barrett, N., G. Edgar, C. J. Zagal, and E. Oh. In press. Surveys of intertidal and subtidal biota of the Derwent Estuary. Department of Parks, Water and Environment, Hobart, Tasmania.

Barrios, J. V., C. F. Gaymer, J. A. Vásquez, and K. B. Brokordt. 2008. Effect of the degree of autotomy on feeding, growth, and reproductive capacity in the multi-armed sea star *Heliaster helianthus*. *Journal of Experimental Marine Biology and Ecology* 361:21–27.

Barthels, P. 1906. The large dermal glands of the species of *Echinaster*. *Annals and Magazine of Natural History* 17:511–512.

Bates, A. E., B. J. Hilton, and C. D. G. Harley. 2009. Effects of temperature, season, and locality on wasting disease in the keystone predatory sea star *Pisaster ochraceus*. *Diseases of Aquatic Organisms* 86:245–251.

Bates, A. E., W. B. Stickle, and C. D. G. Harley. 2010. Impact of temperature on an emerging parasitic association between a sperm-feeding scuticociliate and Northeast Pacific sea stars. *Journal of Experimental Marine Biology and Ecology* 384:44–50.

Beach, D. H., N. J. Hanscomb, and R. F. G. Ormond. 1975. Spawning pheromone in crown-of-thorns starfish. *Nature* 254:135–136.

Beddingfield, S. D., and J. B. McClintock. 1993. Feeding behavior of the sea star *Astropecten articulatus* (Echinodermata: Asteroidea): an evaluation of energy-efficient foraging in a soft-bottom predator. *Marine Biology* 115:669–676.

Benitez-Villalobos, F., P. A. Tyler, and C. M. Young. 2006. Temperature and pressure tolerance of embryos and larvae of the Atlantic seastars *Asterias rubens* and *Marthasterias glacialis* (Echinodermata: Asteroidea): potential for deep sea invasion. *Marine Ecology Progress Series* 314:109–117.

Benitez-Villalobos, F., J. P. Diaz-Martínez, and P. A. Tyler. 2007. Reproductive biology of the deep-sea asteroid *Henricia abyssicola* from the NE Atlantic Ocean. *Ciencias Marinas* 33:49–58.

Bentivegna, F., P. Cirino, and A. Toscano. 1989. Expériences, en aquarium, sur le comportement alimentaire de *Charonia* (*Charonia*) *rubicunda* (Perry 1811) (Mollusca, Gastropoda, Cymatiidae). *Bulletin de l'Institut Océanographique, Monaco* 5:133–139.

Benzie, J. A. H. 1992. Review of the genetics, dispersal and recruitment of crown-of-thorns starfish (*Acanthaster planci*). *Australian Journal of Marine and Freshwater Research* 43:597–610.

Benzie, J. A. H. 1999. Major genetic differences between crown of-thorns starfish (*Acanthaster planci*) populations in the Indian and Pacific Oceans. *Evolution* 53:1782–1795.

Berger, V. Y., and A. D. Naumov. 1996. Influence of salinity on the ability of starfishes *Asterias rubens* L. to attach to substrate. *Biologiya Morya* 22:99–101.

Bergmann, W. 1962. Sterols: their structure and distribution. Pp. 103–162. In *Comparative Biochemistry*, vol. 3 (M. Florkin and H. S. Mason, eds.). Academic Press, New York.

Berlow, E. L., J. A. Dunne, N. D. Martinez, P. B. Stark, R. J. Williams, and U. Brose. 2009. Simple prediction of interaction strengths in complex food webs. *Proceedings of the National Academy of Sciences USA* 106:187–191.

Binyon, J. 1976. The permeability of the asteroid podial wall to water and potassium ions. *Journal of the Marine Biological Association of the United Kingdom* 56:639–647.

Birkeland, C. 1974. Interactions between a sea pen and seven of its predators. *Ecological Monographs* 44:211–232.

Birkeland, C. 1982. Terrestrial runoff as a cause of outbreaks of *Acanthaster planci* (Echinodermata: Asteroidea). *Marine Biology* 69:175–185.

Birkeland, C. 1989. The influence of echinoderms on coral reef communities. Pp.1–79. In *Echinoderm Studies*, vol. 3

(M. Jangoux and J. M. Lawrence, eds.). Balkema, Rotterdam, Netherlands.

Birkeland, C., and J. S. Lucas. 1990. Acanthaster planci: *Major Management Problem of Coral Reefs*. CRC Press, Boca Raton, Florida.

Birkeland, C., F.-S. Chia, and R. R. Strathmann. 1971. Development, substratum selection, delay of metamorphosis and growth in the seastar, *Mediaster aequalis* Stimpson. *Biological Bulletin* 141:99–108.

Birkeland, C., P. K. Dayton, and N. A. Engstrom. 1982. A stable system of predation on a holothurian by four asteroids and their top predator. *The Australian Museum Memoir* 16:175–189.

Black, K. P., and P. J. Moran. 1991. Influence of hydrodynamics on the passive dispersal and initial recruitment of larvae of *Acanthaster planci* (Echinodermata: Asteroidea) on the Great Barrier Reef. *Marine Ecology Progress Series* 69:55–65.

Black, K., P. Moran, D. Burrage, and G. De'ath. 1995. Association of low-frequency currents and crown-of-thorns starfish outbreaks. *Marine Ecology Progress Series* 125:185–94.

Blake, D. B. 1972. Sea Star *Platasterias*: ossicle morphology and taxonomic position. *Science* 176:306–307.

Blake, D. B. 1973. Ossicle morphology of some recent asteroids and description of some west American fossil asteroids. *University of California Publication in Geological Sciences* 104:1–59.

Blake, D. B. 1978. The taxonomic position of the modern sea-star *Cistina* Gray, 1840. *Proceedings of the Biological Society of Washington* 91:234–241.

Blake, D. B. 1980. On the affinities of three small sea-star families. *Journal of Natural History* 14:163–182.

Blake, D. B. 1981. A reassessment of the sea-star orders Valvatida and Spinulosida. *Journal of Natural History* 15:375–394.

Blake, D. B. 1982. Somasteroidea, Asteroidea, and the affinities of *Luidia (Platasterias) latiradiata*. *Palaeontology* 25:167–191.

Blake, D. B. 1983. Some biological controls on the distribution of shallow water sea stars. *Bulletin of Marine Science* 33:703–712.

Blake, D. B. 1987. A classification and phylogeny of post-Palaeozoic sea stars (Asteroidea: Echinodermata). *Journal of Natural History* 21:481–528.

Blake, D. B. 1988. Paxillosidans are not primitive asteroids: a hypothesis based on functional considerations. Pp. 309–314. In *Echinoderm Biology* (R. D. Burke, P. V. Mladenov, P. Lambert and R. L. Parsley, eds). Balkema, Rotterdam, the Netherlands.

Blake, D. B. 1989. Asteroidea: functional morphology, classification and phylogeny. Pp. 197–223. In *Echinoderm Studies*, vol. 3 (M. Jangoux and J. M. Lawrence, eds.). Balkema, Rotterdam.

Blake, D. B. 1990a. Adaptive zones of the class Asteroidea (Echinodermata). *Bulletin of Marine Science* 46:701–718.

Blake, D. B. 1990b. Hettangian Asteriidae (Echinodermata: Asteroidea) from southern Germany: taxonomy, phylogeny and life habits. *Paläontologische Zeitschrift* 64:103–123.

Blake, D. B. 1996. Redescription and interpretation of the asteroid species *Tropidaster pectinatus* from the Jurassic of England. *Palaeontologoy* 39:179–188.

Blake, D. B. 2010. Crown-group asteroid phylogeny: an enduring quandary. Pp. 131–134. In *Echinoderms: Durham* (L. G. Harris, S. A. Bottger, C. W. Walker, and M. P. Lesser, eds.) Taylor & Francis Group, London.

Blake, D. B., and D. R. Elliot. 2003. Ossicular homologies, systematics, and phylogenetic implications of certain North American Carboniferous asteroids (Echinodermata). *Journal of Paleontology* 77:476–489.

Blake, D. B., and T. E. Guensburg. 1988. The water vascular system and functional morphology of Paleozoic asteroids. *Lethaia* 21:189–206.

Blake, D. B., and T. E. Guensburg. 1989a. Two new multi-armed Paleozoic (Mississippian) asteroids (Echinodermata) and some paleobiologic implications. *Journal of Paleontology* 63:331–340.

Blake, D. B., and T. E. Guensburg. 1989b. *Illusioluidia teneryi* gen and sp. (Asteroidea: Echinodermata) from the Pennsylvanian of Texas and its homeomorphy with the extant genus *Luidia* Forbes. *Journal of Paleontology* 63:662–668.

Blake, D. B., and H. Hagdorn. 2003. The Asteroidea (Echinodermata) of the Muschelkalk (Middle Triassic of Germany). *Paläontologische Zeitschrift* 77:23–58.

Blake, D. B., and F. H. C. Hotchkiss. 2004. Recognition of the asteroid (Echinodermata) crown group: implication of the ventral skeleton, *Journal of Paleontology* 78:359–370.

Blake, D. B., and R. Reid III. 1998. Some Albian (Cretaceous) asteroids (Echinodermata) from Texas and their paleobiological implications. *Journal of Paleontology* 72:512–532.

Blanchette, C. A., et al. 2005. Regimes shifts, community change and population booms of keystone predators at the Channel Islands. *Proceedings of the 6th California Channel Islands Symposium* 6:435–441.

Blanchette, C. A., et al. 2009. Trophic structure and diversity in rocky intertidal upwelling ecosystems: a comparison of community patterns across California, Chile, South Africa, and New Zealand. *Progress in Oceanography* 83:107–116.

Blankley, W. O., and G. M. Branch. 1984. Co-operative prey capture and unusual brooding habits of *Anasterias rupicola* (Verrill) (Asteroidea) at sub-Antarctic Marion Island. *Marine Ecology Progress Series* 20:171–176.

Bloch, K. 1965. The biological synthesis of cholesterol. *Science* 150(692):19–28.

Bluhm, B. A., K. Iken, S. Mincks Hardy, B. I. Sirenko, and B. A. Holladay. 2009. Community structure of epibenthic megafauna in the Chukchi Sea. *Aquatic Biology* 7:269–293.

Blumer, M. 1969. Oil pollution in the ocean. Pp. 5–13. In *Oil on the Sea* (D. P. Hoult, ed.). Plenum Press, New York.

Blunt, J. W., et al. 2011. Marine natural products. *Natural Products Report*, 28:196–268.

Boivin, Y., Larrivée, D., and J. H. Himmelman. 1986. Reproductive cycle of the subarctic brooding asteroid *Leptasterias polaris*. *Marine Biology* 92:329–337.

Bonasoro, F., M. D. Candia Carnevali, C. Moss, and M. C. Thorndyke. 1998. Epimorphic versus morphallactic mechanisms in arm regeneration of crinoids and asteroids. Pp. 13–18. In *Echinoderms: San Francisco* (R. Mooi and M. Telford, eds.). Balkema, Rotterdam, the Netherlands.

Bos, A. R., G. S. Gumanao, J. C. E. Alipoyo, and L. T. Cardona. 2008. Population dynamics, reproduction and growth of the Indo-Pacific horned sea star, *Protoreaster nodosus* (Echinodermata; Asteroidea). *Marine Biology* 156:55–63.

Bos, A. R., B. Mueller, and G. S. Gumanao. 2011. Feeding biology and symbiotic relationships of the corallimorpharian *Paracorynactis hoplites* (Anthozoa: Hexacorallia). *The Raffles Bulletin of Zoology* 59:245–250.

Bosch, I. 1989. Contrasting modes of reproduction in two Antarctic asteroids of the genus *Porania*, with a description of unusual feeding and non-feeding larval types. *Biological Bulletin* 177:77–82.

Bosch, I. 1992. Symbiosis between bacteria and oceanic clonal sea star larvae in the western North Atlantic Ocean. *Marine Biology* 114:495–502.

Bosch, I., and J. S. Pearse. 1990. Developmental types of shallow-water asteroids of McMurdo Sound, Antarctica. *Marine Biology* 104:41–46.

Bosch, I., and M. Slattery. 1999. Costs of extended brood protection in the Antarctic sea star, *Neosmilaster georgianus*. *Marine Biology* 134:449–459.

Bosch, I., R. B. Rivkin, and S. P. Alexander. 1989. Asexual reproduction by oceanic planktotrophic echinoderm larvae. *Nature* 337:169–170.

Bosch I., S. J. Colwell, J. S. Pearse, and V. B. Pearse. 1991. Nutritional flexibility in yolk-rich planktotrophic larvae of an Antarctic echinoderm. *Antarctic Journal of the U. S.* 26:168–170.

Botticelli, C. R., F. L. Hisaw and H. H. Wotiz. 1960. Estradiol-17 β and progesterone in ovaries of starfish (*Pisaster ochraceous*). *Proceedings of Society of Experimental Biology* 103:875–877.

Bouland, C., and M. Jangoux. 1988a. Investigation of the gonadal cycle of the asteroid *Asterias rubens* under static condition. Pp. 169–176. In *Echinoderm Biology* (R. D. Burke, P. V. Mladenov, P. Lambert, and R. L. Parsley, eds.). Balkema, Rotterdam, the Netherlands.

Bouland, C., and M. Jangoux. 1988b. Infestation of *Asterias rubens* (Echinodermata) by the ciliate *Orchitophyra stellarum*: effect on gonads and host reaction. *Diseases of Aquatic Organisms* 5:239–242.

Bowden, D. A., A. Clarke, and L. S. Peck. 2009. Seasonal variation in the diversity and abundance of pelagic larvae of Antarctic marine invertebrates. *Marine Biology* 156:2033–2047.

Branch, G. M., C. L. Griffiths, M. L. Branch, and L. E. Beckley. 2007. *Two Oceans: A Guide to the Marine Life of Southern Africa*. Struik Publishers, Cape Town, South Africa.

Branch, M. L., M. Jangoux, V. Alva, and C. Massin. 1993. The Echinodermata of subantarctic Marion and Prince Edward Islands. *South Africa Journal of Antarctic Research*. 23:37–70.

Brey, T., J. Pearse, L. Basch, J. McClintock, and M. Slattery. 1995. Growth and production of *Sterechinus neumayeri* (Echinoidea: Echinodermata) in McMurdo Sound, Antarctica. *Marine Biology* 124:279–292.

Brodie, J. E. 1992. Enhancement of larval and juvenile survival and recruitment in *Acanthaster planci* from the effects of terrestrial runoff: a review. *Australian Journal of Marine and Freshwater Research* 43:539–554.

Brodie, J. E., K. E. Fabricius, A. G. De'ath, and K. Okaji. 2005. Are increased nutrient inputs responsible for more outbreaks of crown-of-thorns starfish? An appraisal of the evidence. *Marine Pollution Bulletin* 51:266–278.

Brogger, M. I., and P. E. Penchaszadeh. 2008. Infaunal mollusks as main prey for two sand bottoms sea stars off Puerto Quequén (Argentina). *Revista de Biología Tropical* 56:238–334.

Brogger, M. I., D. G. Gil, T. Rubilar, M. I. Martinez, M. E. Díaz de Vivar, M. Escolar, L. Epherra, A. Pérez, and A. Tablado. 2012. Echinoderms from Argentina: biodiversity, distribution and current state of knowledge. Pp. 359–403. In *Echinoderm Research and Diversity in Latin America* (J. J. Alvarado and F. A. Solís-Marín, eds). Springer, Berlin.

Brokordt, K. B., J. H. Himmelman, and H. E. Guderley. 2000. Effect of reproduction on escape responses and muscle metabolic capacities in the scallop *Chlamys islandica* Müller 1776. *Journal of Experimental Marine Biology and Ecology* 251:205–225.

Brokordt, K. B., H. E. Guderley, M., Guay, C. F. Gaymer, and J. H. Himmelman. 2003. Sex differences in reproductive investment: maternal care reduces escape response capacity in the whelk *Buccinum undatum*. *Journal of Experimental Marine Biology and Ecology* 291:161–180.

Brooks, W. R., and C. L. Gwaltney. 1993. Protection of symbiotic cnidarians by their hermit crab hosts: evidence for mutualism. *Symbiosis* 15:1–13.

Brosnan, D. M. 1994. Environmental Factors and Plant-Animal Interactions on Rocky Shores Along the Oregon Coast. Ph.D. Dissertation. Oregon State University, Corvallis.

Bruce, A. J. 1979. A report on a small collection of pontoniine shrimps from Eniwetok Atoll. *Crustaceana, Supplement* 5:209–230.

Bruce, B. 1998. A summary of CSIRO studies on the larval ecology of *Asterias amurensis*. Pp. 36–41. In *Proceedings of a Meeting on the Biology and Management of the Introduced Seastar* Asterias amurensis *in Australian Waters* (C. L. Goggin, ed.). Centre for Research on Introduced Marine Pests Technical Report Number 15, CSIRO Division of Fisheries, Hobart, Tasmania, Australia.

Bruce, B. D., and M. A. Green. 1998. *The Spotted Handfish 1999–2001 Recovery Plan*. www.environment.gov.au /biodiversity/threatened/publications/recovery/spotted -handfish/.

Brueggeman, P. 1999. *Underwater Field Guide to Ross Island & McMurdo Sound, Antarctica. Echinodermata-Asteroidea: Seastars*. Office of Polar Programs, National Science Foundation. Available at www.peterbrueggeman.com /nsf/fguide/index.html.

Brun, E. 1968. Extreme population density of the starfish *Asterias rubens* L. on a bed of Iceland scallop, *Chlamys islandica* (O.F. Müller). *Astarte* 32:1–4.

Brun, E. 1972. Food and feeding habits of *Luidia ciliaris* Echinodermata: Asteroidea. *Journal of the Marine Biological Association of the United Kingdom* 52:225–236.

Bruno, J., and E. Selig. 2007. Regional decline of coral cover in the Indo-Pacific: timing, extent, and subregional comparisons. *Public Library of Science ONE 2, e711.* doi:710.1371/journal.pone.0000711.

Bryan, P. J. 2004. Energetic cost of development through metamorphosis for the seastar *Mediaster aequalis* (Stimpson). *Marine Biology* 145:293–302.

Bryan, P. J., D. Rittschof, and J. B. McClintock. 1996. Bioactivity of echinoderm ethanolic body-wall extracts: an assessment of marine bacterial settlement and macro-invertebrate larval settlement. *Journal of Experimental Marine Biology and Ecology* 196:79–96.

Bryan, P. J., J. B. McClintock, and T. S. Hopkins. 1997. Structural and chemical defenses of echinoderms from the northern Gulf of Mexico. *Journal of Experimental Marine Biology and Ecology* 210:173–186.

Bullard, S. G., N. L. Lindquist, and M. E. Hay. 1999. Susceptibility of invertebrate larvae to predators: how common are post-capture larval defenses? *Marine Ecology Progress Series* 191:153–161.

Burke, R. D. 1989. Echinoderm metamorphosis: comparative aspects of the change in form. Pp. 81–108. In *Echinoderm Studies,* vol. 3 (M. Jangoux and J. M. Lawrence, eds.). Balkema, Rotterdam, the Netherlands.

Burla, H., V. Ferlin, G. Pabst, and G. Ribi, G. 1972. Notes on the ecology of *Astropecten aranciacus. Marine Biology* 14:235–241.

Burnaford, J. L., and M. Vasquez. 2008. Solar radiation plays a role in habitat selection by the sea star *Pisaster ochraceus. Marine Ecology Progress Series* 368:177–187.

Burnell, D. J., J. W. ApSimon, and M. W. Gilgan. 1982. Seasonal and geographic variations of the sterols from the starfish *Asterias vulgaris* (Verrill). *Steroids* 39(4): 357–369.

Burns, B. G., V. H. Logan, J. Burnell, and J. W. ApSimon. 1977. Estimation of a steroid released from the crude saponins of the starfish *Asterias vulgaris* by solvolysis: seasonal and geographic abundance. *Analytical Biochemistry* 81:196–208.

Bustamante, R. H., and G. M. Branch. 1996. Large scale patterns and trophic structure of southern African rocky shores: the roles of geographic variation and wave exposure. *Journal of Biogeography* 23:339–351.

Bustamante, R. H., G. M. Branch, S. Eekhout, B. Robertson, P. Zoutendyk, M. Schleyer, A. Dye, N. Hanekom, D. Keats, M. Jurd, and C. McQuaid. 1995. Gradients of intertidal primary productivity around the coast of South Africa and their relationships with consumer biomass. *Oecologia* 102:189–201.

Buttermore, R. E., E. Turner, and M. G. Morrice. 1994. The introduced northern Pacific seastar *Asterias amurensis* in Tasmania. *Memoirs of the Queensland Museum* 36:21–25.

Byrne, M. 1992. Reproduction of sympatric populations of *Patiriella gunnii, P. calcar* and *P. exigua* in New South Wales, asterinid seastars with direct development. *Marine Biology* 114:297–316.

Byrne, M. 1995. Change in larval morphology in the evolution of benthic development by *Patiriella exigua* (Asteroidea: Asterinidae), a comparison with the larvae of *Patiriella* species with planktonic development. *Biological Bulletin* 188:293–305.

Byrne, M. 1996a. Viviparity and intragonadal cannibalism in the diminutive asterinid sea stars *Patiriella vivipara* and *P. parvivipara. Marine Biology* 125:551–567.

Byrne, M. 1996b. Starfish wanted, dead or alive. *New Scientist* 2052:53.

Byrne, M. 1999. Echinodermata. Pp. 940–954. In *Encyclopedia of Reproduction* (E. Knobil and J. Neill, eds.). Academic Press, New York.

Byrne, M. 2005. Viviparity in the sea star *Cryptasterina hystera* (Asterinidae): conserved and modified features in reproduction and development. *Biological Bulletin* 208:81–91.

Byrne, M. 2006. Life history diversity and evolution in the Asterinidae. *Integrative and Comparative Biology* 46:243–254.

Byrne, M., and M. F. Barker. 1991. Embryogenesis and larval development of the asteroid *Patiriella regularis* viewed by light and scanning electron microscopy. *Biological Bulletin* 180:332–345.

Byrne, M., and A. Cerra. 1996. Evolution of intragonadal development in the diminutive asterinid sea stars *Patiriella vivipara* and *P. parvivipara* with an overview of development in the Asterinidae. *Biological Bulletin* 191:17–26.

Byrne, M., and A. Cerra. 2000. Lipid dynamics in the embryos of *Patiriella* species with divergent modes of development. *Development, Growth & Differentiation* 42:79–86.

Byrne, M., A. Cerra, T. Nishigaki, and M. Hoshi. 1997a. Infestation of the testes of the Japanese asteroid *Asterias amurensis* by the ciliate *Orchitophyra stellarum*: a caution against the use of this ciliate for biological control. *Diseases of Aquatic Organisms* 28:235–239.

Byrne, M., M. G. Morrice, and B. Wolf. 1997b. Introduction of the northern Pacific asteroid *Asterias amurensis* to Tasmania: reproduction and current distribution. *Marine Biology* 127:673–685.

Byrne, M., A. Cerra, T. Nishigaki, and M. Hoshi. 1998. Male infertility: a new phenomenon affecting Japanese populations of the sea star *Asterias amurensis* (Asteroidea) due to introduction of the parasitic ciliate *Orchitophyra stellarum* to Japan. Pp. 203–207. In *Echinoderms: San Francisco* (R. Mooi and M. Telford, eds). Balkema, Rotterdam, the Netherlands.

Byrne, M., A. Cerra, and J. T. Villinski. 1999. Oogenic strategies in the evolution of development in *Patiriella* (Asteroidea). *Invertebrate Reproduction Development* 36:195–202.

Byrne, M., M. W. Hart, A. Cerra, and P. Cisternas. 2003. Reproduction and larval morphology of broadcasting and viviparous species in the *Cryptasterina* species complex. *Biological Bulletin* 205:285–294.

Cameron, A. M., and R. Endean. 1981. Renewed population outbreaks of a rare and specialized carnivore (the starfish *Acanthaster planci*) in a complex high-diversity system (the Great Barrier Reef). Pp. 593–96. In *Fourth International*

Coral Reef Symposium (E. D. Gomez, C. E. Birkeland, R. W. Buddemeier, R. E. Johannes, J. A. Marsh Jr., and R. T. Tsuda, eds). Marine Sciences Center, University of the Philippines, Manila, Philippines.

Campbell, D. B., and R. L.Turner. 1984. *Echinaster graminicola*, a new species of spinulosid sea star (Echinodermata: Asteroidea) from the west coast of Florida. *Proceedings of the Biological Society of Washington* 97:167–178.

Caregnato, F. F., F. Wiggers, J. C. Tarasconi, J.C., and I. L. Veitenheimer-Mendes. 2009. Taxonomic composition of mollusks collected from the stomach content of *Astropecten brasiliensis* (Echinodermata: Asteroidea) in Santa Catarina, Brazil. *Brazilian Journal of Biosciences* 7:252–259.

Carey, J. G., Jr. 1972. Food sources of sublittoral, bathyal and abyssal asteroids in the northeast Pacific Ocean. *Ophielia* 10:35–47.

Carlier, A., P. Riera, J.-M. Amouroux, J.-Y. Bodiou, and A. Grémare. 2007. Benthic trophic network in the Bay of Banyuls-sur-Mer (northwest Mediterranean, France): an assessment based on stable carbon and nitrogen isotopes analysis. *Estuarine, Coastal and Shelf Science* 72:1–15.

Caro, A. U., S. A. Navarrete, and J. C. Castilla. 2010. Ecological convergence in a rocky intertidal shore metacommunity despite high spatial variability in recruitment regimes. *Proceedings of the National Academy of Science.* 107:18528–18532.

Carrion, C. N. 2008. Habitat-specific growth and morphology of the sea star *Pisaster ochraceus*. M.S. Thesis, California State University, Los Angeles.

Carroll, M. C., and R. C. Highsmith. 1996. Role of catastrophic disturbance in mediating *Nucella-Mytilus* interactions in the Alaskan intertidal. *Marine Ecology Progress Series* 138:125–133.

Carvalho, A. L. P. S., and C. R. R. Ventura. 2002. The reproductive cycle of *Asterina stellifera* (Möbius) (Echinodermata: Asteroidea) in the Cabo Frio region, southeastern Brazil. *Marine Biology* 141:947–954.

Castilla, J. C. 1981. Perspectivas de investigación en estructura y dinámica de comunidades intermareales rocosas de Chile central. II. Depredadores de alto nivel trófico. *Medio Ambiente* 5:190–215.

Castilla, J. C. 1999. Coastal marine communities: trends and perspectives from human-exclusion experiments. *Trends in Ecology and Evolution* 14:280–283.

Castilla, J. C., and R. L. Durán. 1985. Human exclusion from the rocky intertidal zone of Central Chile: the effects on *Concholepas concholepas* (Gastropoda), *Oikos* 45:391–399.

Castilla, J. C., and R. T. Paine. 1987. Predation and community organization on Eastern Pacific, temperate zone, rocky intertidal shores. *Revista Chilena de Historia Natural* 60: 131–151.

Castilla, J. C., P. H. Manríquez, A. Delgado, L. Gargallo, A. Leiva, and D. Radic. 2007. Bio-foam enhances larval retention in a free-spawning marine tunicate. *Proceedings of the National Academy of Sciences* 104:18120–18122.

Cavey, M. J. 1998. Neuromyoepithelial relationships in the starfish ambulacrum and excitation-contraction coupling among the podial retractor cells. Pp. 215–219. In

Echinoderms: San Francisco (R. Mooi and M. Telford, eds.). Balkema, Rotterdam, the Netherlands.

Cépède, C. 1907. La castration parasitaire des étoiles de mer mâles par un nouvel infusoire astome: *Orchitophyra stellarum* n.g., n.sp. *Comptes rendus de l'Académie des Sciences Paris* 145:1305–1306.

Cerra, A., and M. Byrne. 1995. Cellular events of wrinkled blastula formation and the influence of the fertilization envelope on wrinkling in the sea star *Patiriella exigua*. *Acta Zoologica* 76:155–165.

Chaet, A. B. 1964. A mechanism for obtaining mature gametes from starfish. *Biological Bulletin* 126:8.

Chaet, A. B. 1965. Invertebrates adhering surfaces: secretions of the starfish, *Asterias forbesi*, and the coelenterate, *Hydra pirardi*. *Annals of the New York Academy of Sciences* 18:921–929.

Chaet, A. B. 1966. Neurochemical control of gamete release in starfish. *Biological Bulletin* 130:43–58.

Chaet, A. B. 1967. Gamete release and shedding substance of sea-stars. *Symposia of the Zoological Society of London* 20:13–24.

Chaet, A. B., and R. A. McConnaughy. 1959. Physiologic activity of nerve extracts. *Biologcal Bulletin* 117:407–408.

Chaet, A. B., and D. E. Philpott. 1964. A new subcellular particle secreted by the starfish. *Journal of Ultrastructure Research* 11:354–362.

Chan, G. L. 1973. Subtidal mussel beds in Baja California with a new record size for *Mytilus californianus*. *Veliger* 16:239–240.

Chanley, J. D., R. Ledeen, J. Wax, R. F. Nigrelli, and H. Sobotka. 1959. Holothurin. I. The isolation, properties, and sugar components of holothurin A. *Journal of the American Chemical Society* 81:5180–5183.

Channon, T. 2010. Archival tagging and feeding of the New Zealand sea star, *Coscinasterias muricata* (Echinodermata: Asteroidea), in Doubtful Sound. M.S.Thesis, University of Otago, Dunedin, New Zealand.

Chao, S.-M. 1999. A revision of the Family Astropectinidae (Echinodermata: Asteroidea) from Taiwan, with description of five new records. *Zoological Studies* 38:257–267.

Chen, B.-Y., and C.-P. Chen. 1992. Reproductive cycle, larval development, juvenile growth and population dynamics of *Patiriella pseudoexigua* (Echinodermata: Asteroidea) in Taiwan. *Marine Biology* 113:271–280.

Chia, F.-S. 1966. Brooding behavior of a six-rayed starfish *Leptasterias hexactis*. *Biological Bulletin* 130:304–315.

Chia, F.-S. 1968a. The embryology of a brooding starfish *Leptasterias hexactis* (Stimpson). *Acta Zoologica* 49:321–354.

Chia, F.-S. 1968b. Some observations on the development and cyclic changes of the oocytes in a brooding starfish, *Leptasterias hexactis*. *Journal of Zoology* 154:453–461.

Chia, F.-S., and H. Amerongen. 1975. On the prey-catching pedicellariae of a starfish, *Stylasterias forreri* (de Loriol). *Canadian Journal of Zoology* 53:748–755.

Chia F.-S., and R. D. Burke. 1978. Echinoderm metamorphosis: fate of larval structures. Pp. 219–234. In *Settlement and Metamorphosis of Marine Invertebrate Larvae* (F. S. Chia and M. E. Rice, eds.). Elsevier, New York.

Chia, F.-S., and C. W. Walker. 1991. Echinodermata: Aste-roidea. Pp. 301–353. In *Reproduction of Marine Inver-tebrates,* vol. 6, *Echinoderms and Lophophorates* (A. C. Giese, A.C., J. S. Pearse, and V. B. Pearse, eds.). Boxwood Press, Pacific Grove, California.

Chia, F.-S., C. M. Young, and F. S. McEuen. 1984. The role of larval settlement behavior in controlling patterns of abundance in echinoderms. Pp. 409–424. In *Advances in Invertebrate Reproduction 3* (W. Engels et al., eds.). Elsevier, New York.

Chia, F.-S., C. Oguro, and M. Komatsu. 1993. Sea-star (asteroid) development. *Oceanography and Marine Biology: An Annual Review* 31:223–257.

Chiantore, M., R. Cattaneo-Vietti, L. Elia, M. Guidetti, and M. Antonini. 2002. Reproduction and condition of the scallop *Adamussium colbecki* (Smith 1902), the sea urchin *Sterechinus neumayeri* (Meissner 1900), and the sea-star *Odontaster validus* (Koehler 1911) at Terra Nova Bay (Ross Sea): different strategies related to inter-annual variations in food availability. *Polar Biology* 25:251–255.

Chiu, S. T., V. W. W. Lam, and P. K. S. Shin. 1990. Further observations on the feeding biology of *Luidia* spp. in Hong Kong. Pp. 907–933. In *The Marine Flora and Fauna of Hong Kong and Southern China II* (B. Morton, ed.). Hong Kong University Press, Hong Kong.

Christensen, A. M. 1957. The feeding behaviour of the sea-star *Evasterias troschelii* Stimpson, *Limnology and Oceanography* 2:l80–197.

Christensen, A. M. 1970. Feeding biology of the sea-star *Astropecten irregularis* Pennant. *Ophelia* 8:1–134.

Claereboudt, M. R., and C. Bouland 1994. The effect of parasitic castration by a ciliate on a population of *Asterias vulgaris. Journal of Invertebrate Pathology* 63:172–177.

Clark, A. H. 1916. Six new starfishes from the Gulf of California and adjacent waters. *Proceedings of the Biological Society of Washington* 29:51–62.

Clark, A. H. 1920. A new name for *Heliaster multiradiatus* (Gray). *Proceedings of the Biological Society of Washington* 33:183.

Clark, A. M. 1977. *Starfishes and Related Echinoderms.* Trustees of the British Museum (Natural History), London.

Clark, A. M. 1982. Echinoderms of Hong Kong. Pp. 485–501. In *The Marine Flora and Fauna of Hong Kong and Southern China* (B. S. Morton and C. K. Tseng, eds.). Hong Kong University Press, Hong Kong.

Clark, A. M. 1993. An index of names of recent Asteroidea. Part 2: Valvatida. Pp. 187–366. In *Echinoderm Biology,* vol. 4 (M. Jangoux and J. M. Lawrence, eds.) Balkema, Rotterdam, the Netherlands.

Clark, A. M. 1996. An index of names of recent Asteroidea. III. Velatida and Spinulosida. Pp. 183–250. In *Echinoderm Studies,* vol. 5 (M. Jangoux and J. M. Lawrence, eds.). Balkema, Rotterdam, the Netherlands.

Clark A. M., and M. E. Downey. 1992. *Starfishes of the Atlantic.* Chapman and Hall, London.

Clark, A. M. and C. Mah. 2001. An index of names of recent Asteroidea. IV. Forcipulatida and Brisingida. Pp. 229–347. In *Echinoderm Studies,* vol. 6 (M. Jangoux and J. M. Lawrence, eds.). Balkema, Lisse, the Netherlands.

Clark, A. M., and F. W. E. Rowe. 1971. *Monograph of the Shallow-water Indo-West Pacific Echinoderms.* British Museum (Natural History), London.

Clark, H. E. S. 1963. The Fauna of the Ross Sea. III. Asteroi-dea. *New Zealand Department of Science and Industry Research Bulletin* 151:1–84.

Clark, H. E. S., and D. G. McKnight. 2001. The marine fauna of New Zealand: Echinodermata; Asteroidea (sea-stars). *National Institute of Water and Atmospheric Research (NIWA) Biodiversity Memoir* 117:1–269.

Clark, H. L. 1907. The starfishes of the genus *Heliaster. Bulletin of the Museum of Comparative Zoology at Harvard College* 51:25–76.

Clark, H. L. 1910. The echinoderms of Peru. *Bulletin of the Museum of Comparative Zoology at Harvard College* 52:321–358.

Clark, H. L. 1921. *The Echinoderm Fauna of Torres Strait: Its Composition and its Origin.* Carnegie Institution of Washington, Washington, D.C.

Clark, H. L. 1928. The sea-lilies, sea-stars, brittle-stars and sea-urchins of the South Australian Museum. *Records of the South Australian Museum* 3:361–482.

Clark, H. L. 1933. A handbook of the littoral echinoderms of Porto Rico and the other West Indian Islands. *Scientific Survey of Puerto Rico* 16:1–147.

Clark, H. L. 1938. Echinoderms from Australia. *Memoirs of the Museum of Comparative Zoology at Harvard College* 55:1–596.

Clark, H. L. 1940. Eastern Pacific expeditions of the New York Zoological Society. XXI. Notes on echinoderms from the west coast of Central America. *Zoologica* 25:331–352.

Clark, H. L. 1946. *The Echinoderm Fauna of Australia:Its Composition and its Origin.* Carnegie Institution of Washington, Washington, D.C.

Clark, M. S., K. P. P. Fraser, and L. S. Peck. 2008. Lack of an HSP70 heat shock response in two Antarctic marine invertebrates. *Polar Biology* 31:1059–1065.

Clarke, A. 2008. Ecological stoichiometry in six species of Antarctic marine benthos. *Marine Ecology Progress Series* 369: 25–37.

Clarke, A., and N. M. Johnston. 2003. Antarctic marine benthic diversity. *Oceanography and Marine Biology: An Annual Review* 41:47–114.

Clarke, A., D. K. A. Barnes, and D. A. Hodgson. 2005. How isolated is Antarctica? *Trends in Ecology and Evolution.* 20:1–3.

Clarke, M. 2002. The effect of salinity on distribution, reproduction and feeding of the starfish *Coscinasterias muricata* (Echinodermata:Asteroidea) in a rocky subtidal community of a New Zealand fiord. M.S.Thesis, University of Otago, Dunedin, New Zealand.

Cloud, J., and A. W. Schuetz. 1973. Spontaneous maturation of starfish oocytes: role of follicle cells and calcium ions. *Experimental Cell Research* 79:446–450.

Cobb, J. C., J. M. Lawrence, and T. Talbot-Oliver. In press. Long-term stability (1971–1972 to 2006–2009) of the size-frequency distribution of *Luidia clathrata* (Echinoder-mata: Asteroidea) in Old Tampa Bay. *Florida Scientist.*

Cobb, J. L. S. 1987. Neurobiology of the Echinodermata. Pp. 483–525. In *Nervous Systems in Invertebrates* (M. A. Ali, ed.). Plenum Press, New York.

Colgan, D. J., M. Byrne, E. Rickard, and L. R. Castro. 2005. Limited nucleotide divergence over large spatial scales in the asterinid sea star *Patiriella exigua. Marine Biology* 146:263–270.

Colgan, M. W. 1987. Coral reef recovery on Guam (Micronesia) after catastrophic predation by *Acanthaster planci. Ecology* 68:1592–1605.

Collins, L. S., A. F. Budd, and A. G. Coates. 1996. Earliest evolution associated with closure of the Tropical American Seaway. *Proceedings of the National Academy of Science USA* 93:6069–6072.

Colombo, L., and P. Belvedere. 1976. Gonadal steroidogenesis in echinoderms. *General and Comparative Endocrinology* 29: 255–256.

Connell, J. H. 1972. Community interactions on the marine rocky intertidal shores. *Annual Review of Ecology and Systematics* 3:169–192.

Connell, J. H. 1975. Some mechanisms producing structure in natural communities. Pp. 460–490. In *Ecology and Evolution of Communities* (M. L. Cody and J. Diamond, eds.). Belknap, Cambridge, Massachusetts.

Connell, J. H. 1980. Diversity and the coevolution of competitors, or the ghost of competition past. *Oikos* 35:131–138.

Coppard, S. E., A. Kroh, A., and A. B. Smith. 2010. The evolution of pedicellariae in echinoids: an arms race against pests and parasites. *Acta Zoologica* doi:10.1111/j.1463-6395.2010.00487.x.

Costello, D. P., and C. Henley. 1971. *Methods for obtaining and handling marine eggs and embryos.* Marine Biological Laboratory, Woods Hole, Massachusetts.

Cowden, C., C. M. Young, and F.-S. Chia. 1984. Differential predation on marine invertebrate larvae by two benthic predators. *Marine Ecology Progress Series* 14:145–149.

Cowles, R. P. 1909. Preliminary report on the behavior of echinoderms. *Carnegie Institution of Washington Year Book* 8:128–129.

Crawford, B. J., and D. Jackson. 2002. Effect of microgravity on the swimming behaviour of larvae of the starfish *Pisaster ochraceus. Canadian Journal of Zoology* 80:2218–2225.

Cresswell, R. [translator]. 1902. *Aristotle's History of Animals in Ten Books.* George Bell & Sons, London.

Crozier, W. J. 1920. Notes on some problems of adaptation. II. On the temporal relations of asexual propagation and gametic reproduction in *Coscinasterias tenuispina*: with a note on the direction of progression and on the significance of the madrepores. *Biological Bulletin* 39:116–129.

Crump, R. G. 1971. Annual reproductive cycles in three geographically separated populations of *Patiriella regularis* (Verrill), a common New Zealand asteroid. *Journal of Experimental Marine Biology and Ecology* 7:137–162.

Crump, R. G., and M. F. Barker. 1985. Sexual and asexual reproduction in geographically separated populations of the fissiparous asteroid *Coscinasterias calamaria* (Gray). *Journal of Experimental Marine Biology and Ecology* 88:109–127.

Crump R. G., and R. H. Emson. 1983. The natural history, life history and ecology of the two British species of *Asterina. Field Studies* 5:867–882.

Cuénot, L. 1887. *Contribution a l'Étude Anatomique des Astérides.* Poitiers, Paris.

Cuénot, L. 1891. Étude morphologiques sur les Echinodermes. *Archives de Biologie* 11:313–680.

Cuénot, L. 1948. Anatomie, éthologie et systématique des Échinodermes. Pp. 3–272. In *Traité de Zoologie: Anatomie, Systématique, Biologie,* vol. 11 (P.-P. Grassé, ed.). Masson et Cie, Paris.

Dadaev, A. A., V. S. Levin, and A. M. Murahvry. 1982. Changes in distribution of the scallop *Patinopecten yessoensis* and sea stars after introducing scallops on the bottom in the Vityaz Bay (Sea or Japan). *Marine Biology* (Vlad.) No. 4:37–43.

Daigle, R., and A. Metaxas. 2011. Vertical distribution of marine invertebrate larvae in response to thermal stratification in the laboratory. *Journal of Experimental Marine Biology and Ecology* doi:10.1016/j.jembe.2011.08.008.

Dare, P. J., 1982. Notes on the swarming behaviour and population density of *Asterias rubens* L. (Echinodermata: Asteroidea) feeding on the mussel, *Mytilus edulis* L. *Journal du Conseil International pour l'Exploration de la Mer* 40:112–118.

Dartnall, A. J. 1971. Australian sea stars of the genus *Patiriella* (Asteroidea, Asterinidae). *Proceedings of the Linnean Society of New South Wales* 96: 39–51.

Dartnall A. J., M. Byrne, J. Collins, and M. W. Hart. 2003. A new viviparous species of asterinid (Echinodermata, Asteroidea, Asterinidae) and a new genus to accommodate the species of pan-tropical exiguoid sea stars. *Zootaxa* 359:1–14.

Das, P. K., R. L. Watts, D. C. Watts, and E. J. Dimelo. 1971. Distribution, specificity and function of some proteases, general esterases and cholinesterase from several species of starfish. *Comparative Biochemistry and Physiology* 39B:979–997.

Dautov, S. S. 2000. Distribution, species composition, and dynamics of echinoderm larvae in an area of the Tumen River Estuary and of the Far East State Marine Reserve. *Russian Journal of Marine Biology* 26:12–17.

Dautov, S. S. 2006. Distribution of echinoderms and their larvae in the southwestern Peter the Great Bay, Sea of Japan. *Russian Journal of Marine Biology* 32:259–264.

Dautov, S. S., and M. S. Selina. 2009. Foraging conditions of planktotrophic larvae of echinoderms in the southwest part of Peter the Great Bay of the Sea of Japan. *Russian Journal of Marine Biology* 35:25–33.

Dawson, E. Y. 1952. Circulation within Bahia Vizcaino, Baja California, and its effects on marine vegetation. *American Journal of Botany* 39:425–432.

Dawydoff, C. 1948. Embryologie des Echinodermes. Pp. 277–363. In *Traité de Zoologie,* vol. 11 (P. P. Grassé, ed.). Masson et Cie, Paris.

Day, R. W., A. Dowell, G. Sant, J. Klemke, and C. Shaw. 1995. Patchy predation: foraging behaviour of *Coscinasterias calamaria* and escape responses of *Haliotis rubra. Marine and Freshwater Behaviour and Physiology* 26:11–33.

Dayton, P. K. 1971. Competition, disturbance, and community organization: the provision and subsequent utilization of space in a rocky intertidal community. *Ecological Monographs* 41:351–389.

Dayton, P. K. 1972. Toward an understanding of community resilience and the potential effects of enrichment to the benthos at McMurdo Sound, Antarctica. Pp. 81–95. In *Proceedings of the colloquium on Conservation Problems in Antarctica* (B. C. Parker, ed.). Allen Press, Lawrence, Kansas.

Dayton, P. K. 1989. Interdecadal variation in the Antarctic sponge and its predators from oceanographic climate shifts. *Science* 245:1484–1486.

Dayton, P. K., and J. S. Oliver. 1977. Antarctic soft-bottom benthos in oligotrophic and eutrophic environments. *Science* 197:55–58.

Dayton, P. K., and M. J. Tegner. 1984. The importance of scale in community ecology: a kelp forest example with terrestrial analogs. Pp. 457–481. In *A New Ecology: Novel Approaches to Interactive Systems* (P. W. Price, C. M. Slobodchikoff, and W. S. Gaud, eds.). Wiley, New York.

Dayton, P. K., G. A. Robilliard, and A. L. DeVries. 1969. Anchor ice formation in McMurdo Sound, Antarctica, and its biological effects. *Science* 163:273–274.

Dayton, P. K., G. A. Robilliard, and R. T. Paine. 1970. Benthic faunal zonation as a result of anchor ice at McMurdo Sound, Antarctica. Pp. 244–258. In *Antarctic Ecology*, vol. 1 (M. W. Holdgate, ed.). Academic Press, London.

Dayton, P. K., G. A. Robilliard, R. T. Paine, and L. B. Dayton. 1974. Biological accommodation in the benthic community at McMurdo Sound, Antarctica. *Ecological Monographs* 44:105–128.

Dayton, P. K., R. J. Rosenthal, L. C. Mahen and T. Antezana. 1977. Population structure and foraging biology of the predaceous Chilean asteroid *Meyenaster gelatinosus* and the escape response of its prey. *Marine Biology* 39:361–370.

Dearborn, J. H. 1965. Ecological and faunistic investigations of the marine benthos at McMurdo Sound, Antarctica. Ph.D. dissertation, Stanford University, Stanford, California.

Dearborn, J. H. 1977. Foods and feeding characteristics of Antarctic asteroids and ophiuroids. Pp. 293–326. In *Adaptations within Antarctic Ecosystems* (G. A. Llano, ed). Gulf Publishing, Houston.

Dearborn, J. H., K. C. Edwards, and D. B. Fratt. 1991. Diet, feeding behavior, and surface morphology of the multi-armed Antarctic sea star *Labidiaster annulataus* (Echinodermata: Asteroidea). *Marine Ecology Progress Series* 77:65–84.

De'ath, G. and P. J. Moran. 1998a. Factors affecting the behaviour of crown-of-thorns starfish (*Acanthaster planci* L.) on the Great Barrier Reef. I. Patterns of activity. *Journal of Experimental Marine Biology and Ecology* 220:83–106.

De'ath, G., and P. J. Moran. 1998b. Factors affecting the behaviour of crown-of-thorns starfish (*Acanthaster planci* L.) on the Great Barrier Reef. II. Feeding preferences. *Journal of Experimental Marine Biology and Ecology* 220:107–26.

de Celis, A. K. 1980. The asteroids of Marinduque Island. *Acta Manilana, Natural and Applied Sciences* 19A:20–74.

Defretin, R. 1952. Etude histochimique des mucocytes des pieds ambulacraires de quelques échinodermes. *Recueil des Travaux de la Station Marine d'Endoume.*

Dehn, P. F. 1980a. The annual reproductive cycles of two populations of *Luidia clathrata* (Asteroidea). I. Organ indices and occurrence of larvae. Pp. 361–367. In *Echinoderms: Past and Present* (M. Jangoux, ed.). Balkema, Rotterdam, the Netherlands.

Dehn, P. F. 1980b. Growth and reproduction in *Luidia clahrata* (Say) (Echinodermata: Asteroidea). Thesis. University of South Florida, Tampa, Florida.

Delavault, R. 1960. La sexualité chez *Echinaster sepositus* Gray du Golfe de Naples. *Pubblicazioni della Stazione Zoologica di Napoli* 32:41–55.

De Marino, S., E. Palagiano, F. Zollo, L. Minale, and M. Iorizzi. 1997. A novel sulphated steroid with a 7-membered 5-oxalactone B-ring from an Antarctic starfish of the family Asteriidae. *Tetrahedron* 53:8625–8628.

De Marino, S., N. Borbone, M. Iorizzi, G. Esposito, J. B. McClintock, and F. Zollo. 2003. Bioactive asterosaponins from the starfish *Luidia quinaria* and *Psilaster cassiope*. *Journal of Natural Products.* 66:515–519.

Denny, M. W. 1988. *Biology and the Mechanics of Wave-swept Environments.* Princeton University Press, Princeton, New Jersey.

De Riccardis, F., M. Iorizzi, L. Minale, and R. Riccio. 1992. The first occurrence of polyhydroxylated steroids with phosphate conjugation from the starfish *Tremaster novaecaledoniae. Tetrahedron Letters* 33:1097–1100.

De Simone, F., A. Dini, L. Minale, R. Riccio and F. Zollo. 1980. The sterols of the asteroid *Echinaster sepositus. Comparative Biochemistry and Physiology* 66B:351–357.

De Vantier, L. M., and G. Deacon. 1990. Distribution of *Acanthaster planci* at Lord Howe Island, the southern-most Indo-Pacific reef. *Coral Reefs* 9:145–148.

DeVantier, L., and T. Done. 2007. Inferring past outbreaks of the crown-of-thorns seastar from scar patterns on coral heads. Pp. 85–125. In *Geological Approaches to Coral Reef Ecology* (R. Aronson, ed.). Springer, Berlin.

De Waal, M., J. Poortman, and P. A Voogt. 1982. Steroid receptors in invertebrates. A specific 17β-estradiol binding protein in a sea-star. *Marine Biology Letters* 3:317–323.

Diamond, J. 1986. Overview: laboratory experiments, field experiments, and natural experiments. Pp. 3–22. In *Community Ecology* (J. Diamond and T. J. Case, eds.). Harper and Row, New York.

Diaz-Guisado, D., C. F. Gaymer, K. B. Brokordt, and J. M. Lawrence. 2006. Autotomy reduces feeding, energy storage and growth of the sea star *Stichaster striatus. Journal of Experimental Marine Biology and Ecology* 338:73–80.

Diehl, W. J., III, and J. M. Lawrence. 1979. Effect of nutrition on the excretion rate of soluble nitrogenous products of *Luidia clathrata* (Say) (Echinodermata: Asteroidea). *Comparative Biochemistry and Physiology* 62A:801–806.

Diehl, W. J., and J. M. Lawrence. 1982. Effect of decreased salinity on levels of glycine in the pyloric caeca and tube feet of *Luidia clathrata* (Say) (Asteroidea). Pp. 365–370. In *Echinoderms: Proceedings of the International Conference, Tampa Bay* (J. M. Lawrence, ed.). Balkema, Rotterdam, the Netherlands.

Diehl, W. J., III, and J. M. Lawrence. 1984. The effect of salinity on coelomic fluid osmolyte concentrations and intracellular water content in *Luidia clathrata* (Say) (Echinodermata: Asteroidea). *Comparative Biochemistry and Physiology A* 79:119–126.

Diehl, W. J., and J. M. Lawrence. 1985. Effect of salinity on the intracellular osmolytes in the pyloric caeca and tube feet of *Luidia clathrata* (Say) (Echinodermata: Asteroidea). *Comparative Biochemistry and Physiology A* 82:559–566.

Diehl, W.J., III, L. McEdward, E. Proffitt, V. Rosenberg, and J. M. Lawrence. 1979. The response of *Luidia clathrata* (Echinodermata: Asteroidea) to hypoxia. *Comparative Biochemistry and Physiology A* 62:669–671.

Dieleman, S. J., and H. J. Schoenmakers. 1979. Radioimmunoassays to determine the presence of progesterone and estrone in the starfish *Asterias rubens. General and Comparative Endocrinology* 39(4): 534–542.

Dight, I. J., L. Bode, and M. K. James. 1990a. Modelling the larval dispersal of *Acanthaster planci* I. Large scale hydrodynamics, Cairns Section, Great Barrier Reef Marine Park. *Coral Reefs* 9:115–123.

Dight, I. J., M. K. James, and L. Bode. 1990. Modelling the larval dispersal of *Acanthaster planci*. II.. Patterns of reef connectivity. *Coral Reefs* 9:125–134.

Döderlein, L. 1917. Die Asteriden der Siboga-Expedition. I. Die Gattung *Astropecten* und ihre Stammesgeschichte. In *Siboga-Expeditie. Uitkomsten op zoölogisch, botanisch, ozeanographisch en geologisch gebied verzameld in Nederlandsch Oost-Indie 1899–1900 aan boord H.M. "Siboga"* (E. J. Brill, ed.) 46 (a), Leiden, the Netherlands.

Döderlein, L. 1935. Die Asteriden der Siboga-Expedition. III. Oreasteridae. In *Siboga-Expeditie. Uitkomsten op zoölogisch, botanisch, ozeanographisch en geologisch gebied verzameld in Nederlandsch Oost-Indie 1899–1900 aan boord H.M. "Siboga"* (E. J. Brill, ed.) 46 (a), Leiden, the Netherlands.

Doi, T. 1976. Some aspects of feeding ecology of the sea stars, genus *Astropecten. Publications from the Amakusa Marine Biological Laboratory, Kyushu University* 4:1–19.

Domanski, P. A. 1984. Giant larvae: prolonged planktonic larval phase in the asteroid *Luidia sarsi. Marine Biology* 80:189–195.

Domantay, J. S. 1938. An unusual bud due to heteromorphosis in *Echinaster luzonicus* (Gray). *Philippine Journal of Science* 64:281–285.

Donahue, M. J., R. A. Desharnais, C. D. Robles, and P. Arriola. 2011. Mussel bed boundaries as dynamic equilibria: thresholds, phase shifts, and alternative states. *American Naturalist* 178:612–625.

Done, T., L. DeVantier, E., Turak, D., Fisk, M., Wakeford, and R. Van Woesik. 2010. Coral growth on three reefs: development of recovery benchmarks using a space for time approach. *Coral Reefs* 29:815–33.

Dong, G., T. Xu, B. Yang, X. Lin, X. Zhou, X. Yang, and Y. Liu. 2011. Chemical constituents and bioactivities of starfish. *Chemistry & Biodiversity* 8(5):740–791.

Dorée, C. 1909. The occurrence and distribution of cholesterol and allied bodies in the animal kingdom. *Biochemistry Journal* 4(1–2): 72–106.

Downey, M. E. 1970. Zorocallida, new order, and *Doraster constellatus*, new genus and species, with notes on the Zoroasteridae (Echinodermata, Asteroidea). *Smithsonian Contributions to Zoology* 64.

Downey, M.E., and G. M. Wellington. 1978. Rediscovery of the giant sea-star *Luidia superb* A. H. Clark in the Galapagos Islands. *Bulletin of Marine Science* 28:375–376.

Ducati, C. C., M. D. Candia Carnevali, and M. F. Barker. 2004. Regenerative potential and fissiparity in the stafish *Coscinasterias muricata*. Pp. 113–118. In *Echinoderms: München* (T. Heinzeller and J. H. Nebelsick, eds.). Balkema, Leiden, the Netherlands.

Dulvy, N., R. Freckleton, and N. Polunin. 2004. Coral reef cascades and the indirect effects of predator removal by exploitation. *Ecology Letters* 7:410–16.

Dunstan, P. K., and N. J. Bax. 2007. How far can marine species go? Influence of population biology and larval movement on future range movements. *Marine Ecology Progress Series* 344:15–28.

Dupont, S., B. Lundve, and M. Thorndyke. 2010a. Near future ocean acidification increases growth rate of lecithotrophic larvae and juveniles of the sea star *Crossaster papposus. Journal of Experimental Zoology (Molecular and Developmental Evolution)* 314B:382–389.

Dupont, S., N. Dorey, and M. Thorndyke. 2010b. What meta-analysis can tell us about vulnerability of marine biodiversity to ocean acidification? *Estuarine, Coastal and Shelf Science* 89:182–185.

Durako, M. J., R. H. Goddard, W. Hoffman, and J. M. Lawrence.1979. Malate and lactate dehydrogenase activities in the pyloric caeca of *Luidia clathrata* (Echinodermata: Asteroidea). *Comparative Biochemistry and Physiology B* 62:127–128.

Duran, L. R., and J. C. Castilla. 1989. Variation and persistence of the middle rocky intertidal community of central Chile, with and without human harvesting. *Marine Biology* 103:555–562.

Ebert, T. A. 1983. Recruitment in echinoderms. Pp. 169–203. In *Echinoderm Studies*, vol. 1 (M. Jangoux and J. M. Lawrence, eds.). Balkema, Rotterdam, the Netherlands.

Eckert, G. L., J. M. Engle, and D. J. Kushner. 1999. Sea star disease and population declines at the Channel Islands. *Proceedings of the 5th California Channel Islands Symposium* 5:390–393.

Eichelbaum, E. 1910. Uber Nährung und Ernährungsorgane von Echinoderment. *Wissenschaften Meeresunters. Kiel* 11:187–274.

Ellington, W. R. 1982. Intermediary metabolism. Pp. 395–415. In *Echinoderm Nutrition* (M. Jangoux and J. M. Lawrence, eds.). Balkema, Rotterdam, the Netherlands.

Ellington, W. R., and J. M. Lawrence. 1974. Coelomic fluid volume regulation and isosmotic intracellular regulation by *Luidia clathrata* (Echinodermata: Asteroidea) in response to hyposmotic stress. *Biological Bulletin* 16:20–31.

Elliott, J. K., D. M. Ross, C. Pathirana, S. Miao, R. J. Andersen, P. Singer, W. C. M. C. Kokke, and W. A. Ayer. 1989. Induction of swimming in *Stomphia* (Anthozoa: Actiniaria) by imbricatine, a metabolite of the asteroid *Dermasterias imbricata*. *Biological Bulletin* 176:73–78.

Emerson, C. J. 1977. Larval development of the sea star, *Leptasterias polaris*, with particular reference to the optic cushion and ocelli. *Scanning Electron Microscopy* 2:631–638.

Emlen, J. M. 1966. The role of time and energy in food preference. *American Naturalist* 100:611–617.

Emlet, R. B. 1994. Body form and patterns of ciliation in nonfeeding larvae of echinoderms: functional solutions to swimming in the plankton? *American Zoologist* 34:570–585.

Emlet R. B., L. R. McEdward, and M. F. Strathmann. 1987. Echinoderm larval ecology viewed from the egg. Pp. 55–136. In *Echinoderm Studies*, vol. 2 (M. Jangoux and J. M. Lawrence, eds). Balkema, Rotterdam, the Netherlands.

Emson, R. H., and Crump, R. G. 1979. Description of a new species of Asterina (Asteroidea) with an account of its ecology. *Journal of the Marine Biological Association of the United Kingdom* 59:77–94.

Emson, R. H., and I. C. Wilkie. 1980. Fission and autotomy in echinoderms. *Oceanography and Marine Biology: An Annual Review* 18:155–250.

Emson, R. H., and C. M. Young. 1994. Feeding mechanism of the brisingid starfish *Novodinia antillensis*. *Marine Biology* 118:433–442.

Endean, R. 1974. *Acanthaster planci* on the Great Barrier Reef. Pp. 563–576. In *Proceedings of the Second International Symposium on Coral Reefs* (A. M. Cameron, B. M. Campbell, A. B. Cribb, R. Endean, J. S. Jell, O. A. Jones, P. Mather, and F. H. Talbot, eds.). The Great Barrier Reef Committee, Brisbane, Australia.

Endean, R., and W. Stablum. 1973. A study of some aspects of the crown-of-thorns starfish (*Acanthaster planci*) infestation of reefs of Australia's Great Barrier Reef. *Atoll Research Bulletin* 167.

Engster, M. S., and S. C. Brown. 1972. Histology and ultrastructure of the tube foot epithelium in the phanerozonian starfish, *Astropecten*. *Tissue & Cell* 4:503–518.

Entrambasaguas, L., A. Pérez-Ruzafa, J. A. García-Charton, B. Stobart, and J. J. Bacallado. 2008. Abundance, spatial distribution and habitat relationships of echinoderms in the Cabo Verde Archipelago (eastern Atlantic). *Marine and Freshwater Research* 59:477–488.

Ericsson, J., and H. G. Hansson. 1973. Observations on the feeding biology of *Porania pulvillus* (O.F. Müller), (Asteroidea), from the Swedish west coast. *Ophelia* 12:53–58.

Escobar, J., and S. A. Navarrete. 2011. Risk recognition and variability in escape responses among intertidal molluskan grazers to the sun star *Heliaster helianthus*. *Marine Ecology Progress Series* 421:151–161.

Espoz, C., and J. C. Castilla. 2000. Escape responses of four Chilean intertidal limpets to seastars. *Marine Biology* 137:887–892.

Estes, J. A., and D. O. Duggins. 1995. Sea otters and kelp forests in Alaska: generality and variation in a community ecological paradigm. *Ecological Monographs* 65:75–100.

Evans, B. S., R. W. G. White, and R. D. Ward. 1998. Genetic identification of asteroid larvae from Tasmania, Australia, by PCR-RFLP. *Molecular Ecology* 7:1077–1082.

Eylers, J. P. 1976. Aspects of skeletal mechanics of the starfish *Asterias forbesi*. *Journal of Morphology* 149:353–367.

Fabricius, K. E., and F. H. Fabricius. 1992. Re-assessment of ossicle frequency patterns in sediment cores: rate of sedimentation related to *Acanthaster planci*. *Coral Reefs* 11:109–14.

Fabricius, K. E., K. Okaji, and G. De'ath. 2010. Three lines of evidence to link outbreaks of the crown-of-thorns seastar *Acanthaster planci* to the release of larval food limitation. *Coral Reefs* 29:593–605.

Fagerlund, U. H. M., 1969. Lipid metabolism. Pp. 123–134. In *Chemical Zoology*, vol. 3, *Echinodermata, Nematoda, and Acanthocephala* (M. Florkin and B. T. Scheer, eds.). Academic Press, New York.

Fagerlund, U. H. M., and D. R. Idler. 1960. Marine sterols. VI. Sterol biosynthesis in molluscs and echinoderms. *Canadian Journal of Biochemistry and Physiology* 38: 997–1002.

Fairweather, P. G. 1985. Differential predation on alternative prey, and the survival of rocky intertidal organisms in New South Wales. *Journal of Experimental Marine Biology and Ecology* 89:135–156.

Fairweather, P. G. 1988. Consequences of supply-side ecology: manipulating the recruitment of intertidal barnacles affects the intensity of predation upon them. *Biological Bulletin* 175:349–354.

Fairweather, P. G., and A. J. Underwood. 1991. Experimental removals of a rocky intertidal predator: variations within two habitats in the effects of the prey. *Journal of Experimental Marine Biology and Ecology* 154:29–75.

Fairweather, P. G., A. J. Underwood, and M. J. Moran. 1984. Preliminary investigations of predation by the whelk *Morula marginalba*. *Marine Ecology Progress Series* 17:143–156.

Farmanfarmaian, A., A. C. Giese, R. A. Boolootian, and J. Bennett. 1958. Annual reproductive cycles of four species of west coast starfishes. *Journal of Experimental Zoology* 138:355–367.

Feder, H. 1955. On the methods used by the starfish *Pisaster ochraceus* in opening three types of bivalve mollusk. *Ecology* 36:764–767.

Feder, H. M. 1959. The food of the star fish *Pisaster ochraceus* along the California coast. *Ecology* 40:721–724.

Feder, H. M. 1970. Growth and predation by the ochre sea star, *Pisaster ochraceus* (Brandt) in Monterey Bay, California. *Ophelia* 8:161–185.

Feder, H. M. 1980. Asteroidea. Pp. 416–448. In *Intertidal Invertebrates of California* (R. H. Morris, D. P. Abbott, and E. C. Haderlie, eds.). Stanford University Press, Stanford, California.

Feder, H. M., and A. M. Christensen. 1966. Aspects of asteroid biology. Pp. 88–127. In *Physiology of Echinodermata* (R. Boolootian, ed.). Wiley, New York.

Feder, H. M., Jewett, S. C., and Blanchard, A. 2005. South-eastern Chukchi Sea (Alaska) epibenthos. *Polar Biology* 28:402–421.

Feder, H. M., Iken, K., Jewett, S. C., Blanchard, A., and Schonberg, S. 2011. Benthic food-web structure in the southeastern Chukchi Sea: an assessment using δ^{13}C and δ^{15}N analyses. *Polar Biology* 34:521–532.

Fell, H. B. 1954. New Zealand fossil Asterozoa. 3. *Odontaster priscus* sp. nov. from the Jurassic. *Transactions of the Royal Society of New Zealand* 82:817–819.

Fell, H. B. 1959. Starfishes of New Zealand. *Tuatara* 7:127–142.

Fell, H. B. 1960. Biological results of the Chatham Islands 1954 expedition. II. Archibenthal and littoral echino-derms. *New Zealand Department of Scientific and Industrial Research Bulletin* 139:47–98.

Fell, H. B. 1962. West-wind drift dispersal of echinoderms in the southern hemisphere. *Nature* 193:759–761.

Fell, H. B. 1963. The phylogeny of sea-stars. *Philosophical Transactions of the Royal Society Series B.* 246:381–435.

Fenchel, T. 1965. Feeding biology of the sea-star *Luidia sarsi* Düben & Koren. *Ophelia.* 2:223–236.

Ferguson, J. C. 1966. Mechanical responses of isolated starfish digestive glands to metabolic drugs, inhibitors and nutrients. *Comparative Biochemistry and Physiology* 19:259–266.

Ferguson, J. C. 1967a. Utilization of dissolved exogenous nutrients by the starfishes, *Asterias forbesi* and *Henricia sanguinolenta*. *Biological Bulletin* 132:161–173.

Ferguson, J. C. 1967b. An autoradiographic study of the utilization of free exogenous amino acids by starfishes. *Biological Bulletin* 133:317–329.

Ferguson, J. C. 1969a. Feeding activity in *Echinaster* and its induction with dissolved nutrients. *Biological Bulletin* 136:374–384.

Ferguson, J. C. 1969b. Feeding, digestion, and nutrition in Echinodermata. Pp. 71–100. In *Chemical Zoology,* vol. 3, *Echinodermata, Nematoda, and Acanthocephala* (M. Florkin and B. T. Scheer, eds.). Academic Press, New York.

Ferguson, J. C. 1970. An autoradiographic study of the translocation and utlization of amino acids by starfish. *Biological Bulletin* 138:14–25.

Ferguson, J. C. 1974. Growth and reproduction of *Echinaster echinophorus*. *Florida Scientist* 37:57–60.

Ferguson, J. C. 1975. The role of free amino acids in nitrogen storage during the annual cycle of a starfish. *Comparative Biochemistry and Physiology A* 51:341–350.

Ferguson, J. C. 1976. The annual cycle of fatty acid composi-tion in a starfish. *Comparative Biochemistry and Physiology B* 54:249–252.

Ferguson, J. C. 1980a. Fluxes of dissolved amino acids between sea water and *Echinaster*. *Comparative Biochemis-try and Physiology A* 65:291–295.

Ferguson, J. C. 1980b. The non-dependency of a starfish on epidermal uptake of dissolved organic matter. *Comparative Biochemistry and Physiology A* 66:461–465.

Ferguson, J. C. 1984. Translocative functions of the enigmatic organs of starfish—the axial organ, hemal vessels, Tiedemann's bodies, and rectal caeca: an autoradio-graphic study. *Biological Bulletin* 166:140–155.

Ferguson, J. C. 1988. Madreporite and anus function in fluid volume regulation of a starfish (*Echinaster graminicola*). Pp. 603–609. In *Echinoderm Biology* (R. D. Burke, P. V. Mladenov, P. Lambert, and R. L. Parsley, eds.). Balkema, Rotterdam, the Netherlands.

Ferguson, J. C. 1989. Rate of water admission through the madreporite of a starfish. *Journal of Experimental Biology* 145:147–156.

Ferguson, J. C. 1990. Hyperosmotic properties of the fluids of the perivisceral coelom and water vascular system of starfish kept under stable conditions. *Comparative Biochemistry and Physiology A* 9:245–248.

Ferguson, J. C. 1992. The function of the madriporite in the body fluid volume maintenance by an intertidal starfish, *Pisaster ochraceus*. *Biological Bulletin* 183:482–489.

Ferguson, J. C., and C. W. Walker. 1993. Adhesion seams in Tiedemann's diverticula of the starfish *Henricia sanguino-lenta*. *Transactions of the American Microscopical Society* 112:158–167.

Ferrat, D., and P. Escoubet. 1987. Recherche d'une toxicité potentielle des éponges vis-à-vis de quelques organismes marins. *Revue Internationale d'Océanographie Médicale* 85–86:162–164.

Fields, P. A., J. B, Graham, R. H. Rosenblatt, and G. N. Somero. 1993. Effects of expected global climate change on marine faunas. *Trends in Ecology and Evolution* 8:361–367.

Findlay, J. A., Z. Q. He, and M. Jaseja. 1989. Forbeside E: a novel sulfated sterol glycoside from *Asterias forbesi*. *Canadian Journal of Chemistry* 67:2078–2080.

Fisher, W. K. 1906. *The Starfishes of the Hawaiian Islands.* U.S. Commission of Fish and Fisheries. Government Printing Office, Washington, D.C.

Fisher, W. K. 1911. Asteroidea of the North Pacific and adjacent waters. I. Phanerozonia and Spinulosa. *Smithsonian Institution. United States National Museum. Bulletin* 76:1–420.

Fisher, W. K. 1919. Starfishes of the Phillipine seas and adjacent waters. *Smithsonian Institution. United States National Museum, Bulletin* 3(100):1–547, plates 1–156.

Fisher, W. K. 1930. Asteroidea of the North Pacific and adjacent waters. III. Forcipulata (concluded), *Smithsonian Institution. United States National Museum, Bulletin* 76:1–356.

Fisher, W. K. 1940. Asteroidea. *Discovery Reports* 20:69–305.

Fisk, D. A. 1992. Recruitment of *Acanthaster planci* over a five-year period at Green Island Reef. *Australian Journal of Marine and Freshwater Research* 43:629–633.

Flammang, P. 1995. Fine structure of the podia in three species of paxillosid asteroids of the genus *Luidia* (Echinodermata). *Belgian Journal of Zoology* 125:125–134.

Flammang, P. 1996. Adhesion in echinoderms. Pp. 1–60. In *Echinoderm Studies*, vol. 5 (M. Jangoux and J. M. Lawrence, eds.). Balkema, Rotterdam, the Netherlands.

Flammang, P. 2006. Adhesive secretions in echinoderms: an overview. Pp. 183–206. In *Biological Adhesives* (A. M. Smith and J. A. Callow, eds.). Springer-Verlag, Berlin and Heidelberg.

Flammang, P., and G. Walker. 1997. Measurement of the adhesion of the podia in the asteroid *Asterias rubens*

(Echinodermata). *Journal of the Marine Biological Association of the United Kingdom* 77:1251–1254.

Flammang, P., S. Demeulenaere, and M. Jangoux. 1994. The role of podial secretions in adhesion in two species of sea stars (Echinodermata). *Biological Bulletin* 187:35–47.

Flammang, P., A. Van Cauwenberge, H. Alexandre, and M. Jangoux. 1998. A study of the temporary adhesion of the podia in the sea star *Asterias rubens* (Echinodermata, Asteroidea) through their footprints. *Journal of Experimental Biology* 201:2383–2395.

Flammang, P., R. Santos, and D. Haesaerts. 2005. Echinoderm adhesive secretions: from experimental characterization to biotechnological applications. Pp 201–220. In *Marine Molecular Biotechnology: Echinodermata* (V. Matranga, ed.). Springer-Verlag, Berlin.

Florey, E., and M. A. Cahill. 1977. Ultrastructure of sea urchin tube feet: evidence for connective tissue involvement in motor control. *Cell and Tissue Research* 177:195–214.

Fontaine, A. R., and F.-S. Chia. 1968. Echinoderms: an autoradiographic study of assimilation of dissolved organic molecules. *Science* 161:1153–1155.

Fontanella, F. M., and T. S. Hopkins. 2003. Preliminary phylogeny of *Echinaster* (*Othilia*) from the Gulf of Mexico based on morphological characters (Echinodermata: Asteroidea). Pp. 91–95. In *Echinoderm Research 2001* (J.-P. Féral and B. David, eds.). Balkema, Rotterdam, the Netherlands.

Forbes, E. 1839. On the Asteriadae of the Irish Sea. *Memoirs of the Wernerian Society, Edinburgh* 8:114–129.

Forbes, E. 1841. *A History of British Starfish and Other Animals of the Class Echinodermata.* John Van Voorst, London.

Forbes, E. 1848. On the Asteridae found fossil in British strata. *Memoirs of the Geological Survey of Great Britain, British Organic Remains.*, Decade 2:457–482.

Forcucci, D., and J. M. Lawrence. 1986. Effect of low salinity on the activity, feeding, growth and absorption efficiency of *Luidia clathrata* (Echinodermata: Asteroidea). *Marine Biology* 92:315–321.

Fox, D. L., and T. S. Hopkins. 1966. The comparative biochemistry of pigments. Pp. 277–300. In *Physiology of Echinodermata* (R. A. Boolootian, ed.). Interscience Publishers, New York.

Fox, D. L., and B. T. Scheer. 1941. Comparative studies of the pigments of some Pacific coast echinoderms. *Biological Bulletin* 80:441–455.

Franz, D. R. 1986. Seasonal changes in pyloric caecum and gonad indices during the annual reproductive cycle in the seastar *Asterias forbesi. Marine Biology* 91:553–560.

Franz, D. R., and E. K. Worley. 1982. Seasonal variability of prey in the stomachs of *Astropecten americanus* (Echinodermata: Asteroidea) from off Southern New England, U.S.A. *Estuarine, Coastal and Shelf Science* 14:355–368.

Franz, D. R., E. K. Worley, and A. S. Merrill. 1981. Distribution patterns of common seastars of the Middle Atlantic continental shelf of the Northwest Atlantic (Gulf of Maine to Cape Hatteras). *Biological Bulletin* 160:394–418.

Freeman, S. M., C. A. Richardson, and R. Seed. 1999. Seasonal abundance, prey selection and locomotory activity patterns of *Astropecten irregularis* (Echinodermata: Asteroidea). Pp. 459–464. In *Echinoderm Research 1998* (M. D. Candia Carnevali and F. Bonasuro, eds). Balkema, Rotterdam, the Netherlands.

Freeman, S. M., C. A. Richardson, and R. Seed. 2001. Seasonal abundance, spatial distribution, spawning and growth of *Astropecten irregularis* (Echinodermata: Asteroidea). *Estuarine, Coastal and Shelf Science* 53:39–49.

Freire, A. S., T. M. Absher, A. C. Cruz-Kaled, Y. Kern, and K. L. Elbers. 2006. Seasonal variation of pelagic invertebrate larvae in the shallow Antarctic waters of Admiralty Bay (King George Island). *Polar Biology* 29:294–302.

Frontana-Uribe, S., J. De la Rosa-Velez, L. Enriquez-Paredes, L. B. Ladah, and L. Sanvicente-Anorve. 2008. Lack of genetic evidence for the subspeciation of *Pisaster ochraceus* (Echinodermata: Asteroidea) in the north-eastern Pacific Ocean. *Journal of the Marine Biological Association of the United Kingdom* 88:395–400.

Fuentes, H. R. 1982. Feeding habitats of *Graus nigra* (Labridae) in coastal waters of Iquique in northern Chile. *Japanese Journal of Ichthyology* 29:95–98.

Fujita, D., and Y. Seto. 1998. On a sea star *Coscinasterias acutispina* (Asteroidea, Asteriidae) in Toyama Bay (Preliminary report). *Bulletin Toyama Prefectural Fisheries Research Institute* 10:53–64.

Fujita, D., and Y. Seto. 2000. Distribution of a fissiparous multi-armed sea star, *Coscinasterias acutispina*, in Japan. *Bulletin Toyama Prefectural Fisheries Research Institute* 12:19–31.

Fujita, D., Y. Seto, Y. Moriyama, and M. Komatsu. 2001. *Coscinasterias acutispina*: distribution and ecology in Toyama Bay. Pp. 169–174. In *Echinoderms 2000* (M. F. Barker, ed.). Balkema, Lisse, the Netherlands.

Fukuyama, A. K., and J. S. Oliver 1985. Sea star and walrus predation on bivalves in Norton Sound, Bering Sea, Alaska. *Ophelia* 24:17–36.

Gaffney, J., and L. J. Goad. 1974. Progesterone metabolism by the echinoderms *Asterias rubens* and *Marthasterias glacialis. Biochemistrical Journal* 138(2): 309–311.

Gale, A. S. 1987. Phylogeny and classification of the Asteroidea (Echinodermata). *Zoological Journal of the Linnean Society* 89:107–132.

Gale, A. S. 2005. *Chrispaulia*, a new genus of mud star (Asteroidea, Goniopectinidae) from the Cretaceous of England. *Geological Journal* 40:383–397.

Gale, A. S. 2011a. Phylogeny of the post-Palaeozoic Asteroidea (Neoasteroida, Echinodermata). *Special Papers in Palaeontology* 85:5–112.

Gale, A. S. 2011b. The upper Oxfordian (Jurassic) asteroid fauna of Savigna, Départment of Jura, France. *Swiss Journal of Palaeontology* 130:69–89.

Gale, A. S., and L. Villier. In press. Mass mortality of fossil asteriid asteroids from the Maastrictian (Late Cretaceous) of Morocco. *Palaeontology.*

Galtsoff, P. S., and V. L. Loosanoff. 1939. Natural history and method of controlling the starfish (*Asterias forbesi*, Desor). *Bulletin Bureau Fisheries Washington* 49:75–132.

Ganmanee, M., T. Narita, S. Ida, and H. Sekiguchi. 2003. Feeding habits of asteroids, *Luidia quinaria* and *Astropecten scoparius*, in Ise Bay, central Japan. *Fisheries Science* 69:1121–1134.

Garza, C., and C. D. Robles. 2010. Effects of brackish water incursions and diel phasing of the tides on vertical excursions of the keystone predator *Pisaster ochraceus*. *Marine Biology* 157:673–682.

Gasparini, J. L., S. R. Floeter, C. E. L. Ferreira, and I. Sazima. 2005. Marine ornamental trade in Brazil. *Biodiversity and Conservation* 14:2883–2899.

Gaymer, C. F. 2006. Interference between the sea stars *Leptasterias polaris* and *Asterias vulgaris*: the importance of spatial and temporal variations in species interactions. Pp. 25–27. In *Atelier de travail "Prédation du pétoncle et gestion des ensemencements" Compte rendu nº 24* (M. M. Nadeau and G. Tita, eds.). Ministére de l'Agriculture, des Pêcheries et de l'Alimentation du Québec. Gaspé, Québec.

Gaymer, C. F., and J. H. Himmelman. 2002. Mussel beds in deeper water provide an unusual situation for competitive interactions between the seastars *Leptasterias polaris* and *Asterias forbesi*. *Journal of Experimental Marine Biology and Ecology* 277:13–24.

Gaymer, C. F., and J. H. Himmelman. 2008. A keystone predatory sea star in the intertidal zone is controlled by a higher-order predatory sea star in the subtidal zone. *Marine Ecology Progress Series* 370:143–153.

Gaymer, C. F., Himmelman, J. H., and Johnson, L. E. 2001a. Distribution and feeding ecology of the seastars *Leptasterias polaris* and *Asterias vulgaris* in the northern Gulf of St. Lawrence, eastern Canada. *Journal of the Marine Biological Association of the United Kingdom* 81:827–843.

Gaymer, C. F., Himmelman, J. H., and Johnson, L. E. 2001b. Use of prey resources by the seastars *Leptasterias polaris* and *Asterias vulgaris*: a comparison between field observations and laboratory experiments. *Journal of Experimental Marine Biology and Ecology* 262:13–30.

Gaymer, C. F., Himmelman, J. H., and Johnson, L. E. 2002. Effect of intra- and inter-specific interactions on the feeding behavior of two subtidal seastars. *Marine Ecology Progress Series* 232:149–162.

Gaymer C. F., Dutil, C., and Himmelman, J. H. 2004. Prey selection and predatory impact of four major seastars on a soft bottom subtidal community. *Journal of Experimental Marine Biology and Ecology* 313:353–374.

Gaymer, C. F., A. T. Palma, J. M. A. Vega, C. J. Monaco, and L. A. Henríquez. 2010. Effects of La Niña on recruitment and abundance of juveniles and adults of benthic community-structuring species in northern Chile. *Marine and Freshwater Research* 61:1185–1196.

Gehman, A.-L. M., and B. L. Bingham. 2010. Maternal diet and juvenile quality in the sea star *Leptasterias aequalis*. *Journal of Experimental Marine Biology and Ecology* 386:86–93.

Gemmill, J. F. 1915. On the ciliation of asterids, and on the question of ciliary nutrition in certain species. *Proceedings of the Zoological Society of London*, 1–19.

George, S. B. 1994. Phenotypic plasticity in the larvae of *Luidia foliolata* (Echinodermata: Asteroidea). Pp. 297–307 In *Echinoderms through Time* (B. David, A. Guille, J.-P. Féral, and M. Roux, eds.) Balkema, Rotterdam, the Netherlands.

George, S. B. 1994. Population differences in maternal size and offspring quality for *Lepasterias epichlora* (Brandt) (Echinodermata: Asteroidea). *Journal of Experimental Marine Biology and Ecology* 175:121–131.

George, S. B. 1999. Egg quality, larval growth, and phenotypic plasticity in a forciculate sea star. *Journal of Experimental Marine Biology and Ecology* 27:203–224.

George, S., J. Lawrence, and L. Fenaux. 1990. Effect of starvation and the time to first feeding on larvae of *Luidia clathrata* (Say) (Echinodermata: Asteroidea). Pp. 91–98. In *Echinoderm Research 1990* (C. De Ridder, P.Dubois, M.-C. Lahaye, and M. Jangoux, eds.). Balkema, Rotterdam, the Netherlands.

George, S. B., J. M. Lawrence, and L. Fenaux. 1991. The effect of food ration on the quality of eggs of *Luidia clathrata* (Say) (Echinodermata: Asteroidea). *Invertebrate Reproduction and Development* 20:237–242.

George, S. B., C. M. Young, and L. Fenaux. 1997. Proximate composition of eggs and larvae of the sand dollar *Encope michelini* (Agassiz): the advantage of higher investment in planktotrophic eggs. *Invertebrate Reproduction and Development* 32:11–19.

Georgiades, E. T., A. Temara, and D. A. Holdway. 2006. The reproductive cycle of the asteroid *Coscinasterias muricata* in Port Phillip Bay, Victoria, Australia. *Journal of Experimental Marine Biology and Ecology* 332:188–197.

Ghiselin, M. T. 1969. The evolution of hermaphroditism among animals. *Quarterly Review of Biology* 44:189–208.

Giese, A. C. 1959. Comparative physiology: annual reproductive cycles of marine invertebrates. *Annual Review of Physiology* 21:547–576.

Giese, A. C., and J. S. Pearse. 1974. Introduction: general principle. Pp. 1–49. In *Reproduction of Marine Invertebrates*, vol. 1 (A. C. Giese, and J. S. Pearse, eds.). Academic Press, New York.

Giese, A. C., J. S. Pearse, and V. B. Pearse (Eds.). 1991. *Reproduction of marine invertebrates*. Boxwood Press, Pacific Grove, California, USA.

Ginsburg, D. W., and D. T. Manahan. 2009. Developmental physiology of Antarctic asteroids with different life-history modes. *Marine Biology* 156:2391–2402.

Glynn, P. W., and D. A. Krupp. 1986. Feeding biology of a Hawaiian sea star corallivore, *Culcita novaeguineae* Muller & Troschel. *Journal of Experimental Marine Biology and Ecology* 96:75–96.

Goad, L. J., I. Rubinstein, and A. G. Smith. 1972. The sterols of echinoderms. *Proceedings of the Royal Society of London Series B, Biological Sciences* 180(1059):223–246.

Goggin, C. L., and C. Bouland. 1997. The ciliate *Orchitophrya* cf. *stellarum* and other parasites and commensals of the northern Pacific sea star *Asterias amurensis* from Japan. *International Journal of Parasitology* 27:1415–1418.

Gondim, A. I., P. Lacouth, C. Alonso, and C. L. C. Manso. 2008. Echinodermata da Praia do Cabo Branco, João Pessoa, Paraíba, Brasil. *Biota Neotropica* 8:151–159.

Gondolf, A. L. 2000. Light and scanning electron microscopic observations of the developmental biology of the common starfish *Asterias rubens* Linné (Echinodermata: Asteroidea). *Ophelia* 52:153–170.

Gonzalez, M., and E. Jaramillo. 1991. The association between *Mulinia edulis* (Mollusca, Bivalvia) and *Edotia magellanica* (Crustacea, Isopoda) in southern Chile. *Revsta Chilena de Historia Natural* 64:37–51.

Goss, R. J. 1969. *Principles of Regeneration.* Academic Press, New York.

Gosselin, L. A., and P.-Y. Qian. 1997. Juvenile mortality in benthic marine invertebrates. *Marine Ecology Progress Series* 146:265–282.

Grabe, S. A., D. J. Karlen, C. M. Holden, B. Goetting, T. Dix, and S. Markham. 2003. *Tampa Bay Benthic Monitoring Program: Status of Old Tampa Bay: 1993–1998.* Environmental Protection Commission of Hillsborough County, Tampa, Florida.

Grainger, E. H. 1966. Sea stars (Echinodermata: Asteroidea) of Arctic North America. *Bulletin of the Fisheries Research Board of Canada* 152.

Grange, K. R., R. J. Singleton, J. R. Richardson, P. J. Hill, and W. D. 1981. Shallow rock-wall biological associations of some southern fiords of New Zealand. *New Zealand Journal of Zoology* 8:209–227.

Grange, L. J., P. A. Tyler, and L. S. Peck. 2007. Multi-year observations on the gametogenic ecology of the Antarctic seastar *Odontaster validus. Marine Biology* 153:15–23.

Grannum, R. K., N. B. Murfet, D. A. Ritz, and E. Turner. 1996. The distribution and impact of the exotic seastar, *Asterias amurensis* (Lütken), in Tasmania. Pp. 53–138. In *The Introduced Northern Pacific Seastar,* Asterias amurensis *(Lütken), in Tasmania* (R. K. Grannum, ed.). Australian Nature Conservation Agency, Canberra.

Grant, A., and P. A. Tyler. 1986. An analysis of the reproductive pattern in the seastar *Astropecten irregularis* (Pennant) from the Bristol Channel. *International Journal of Invertebrate Reproduction and Development* 9:345–361.

Gray, A. P., C. A. Richardson, and R. Seed. 1997. Ecological relationships between the valviferan isopod *Edotia doellojuradoi* Giambiagi, 1925, and its host *Mytilus edulis chilensis* in the Falkland Islands. *Estuarine Coastal and Shelf Science* 44:231–239.

Gray, J. E. 1840. A synopsis of the genera and species of the class Hypostoma (Asterias Linn.). *Annals and Magazine of Natural History* 6:175–184, 275–290.

Greenfield, L. J. 1959. Biochemical and environmental factors involved in the reproductive cycle of the sea star *Pisaster ochraceus* (Brandt). Ph.D. dissertation. Stanford University, Stanford, Callifornia.

Greenfield, L., A. C. Giese, A. Farmanfarmaian, and R. A. Boolootian. 1958. Cyclic biochemical changes in several echinoderms. *Journal of Experimental Zoology* 139:507–524.

Greer, S. P., K. B. Iken, J. B. McClintock and C. D. Amsler. 2003. Individual and coupled effects of echinoderm extracts and surface hydrophobicity on spore settlement and germination in the brown alga *Hincksia irregularis. Biofouling* 19:315–326.

Greer, S. P., K. Iken, J. B. McClintock, and C. D. Amsler. 2006. Bioassay-guided fractionation of antifouling compounds using computer-assisted motion analysis of brown algal spore swimming. *Biofouling* 22:125–132.

Griffiths, H. J. 2010. Antarctic marine biodiversity: what do we know about the distribution of life in the Southern Ocean? *PLoS ONE* 5(8): e11683. doi:10.1371/journal .pone.0011683.

Grime, J. P. 1977. Evidence for the existence of three primary strategies in plants and its relevance to ecological and evolutionary thought. *American Naturalist* 11:1196–1194.

Grosberg, R. K., and C. W. Cunningham, 2000. Genetic structure in the sea: from populations to communities. Pp. 61–84. In *Marine Community Ecology* (M. D. Bertness, S. D. Gaines, and M. E. Hay, eds.). Sinauer Associates, Sunderland, Massachusetts.

Grygier, M. J. 1981. A representative of the genus *Dendrogaster* (Cirripedia: Ascothoracida) parasitic in an Antarctic starfish. Pp. 1–16. In *Antarctic Research Series,* vol. 32, *Biology of the Antarctic Seas X* (L. S. Kornicker, ed.). American Geophysical Union, Washington, D.C.

Grygier, M. J. 1986. *Dendrogaster* (Crustacea: Ascothoracida) parasitic in Alaskan and eastern Canadian *Leptasterias* (Asteroidea). *Canadian Journal of Zoology* 64:1249–1253.

Grygier, M. J. 1987. Antarctic records of asteroid-infesting Ascothoracida (Crustacea), including a new genus of Ctenosculidae. *Proceedings of the Biological Society of Washington* 100:700–712.

Guille, A., P. Laboute, and J.-L. Menou. 1986. *Guide des Étoiles de Mer, Oursins et Autres Échinodermes du Lagon de Nouvelle-Calédonie.* ORSTOM, Paris.

Guillou, M. 1990. Biotic interactions between predators and super-predators in the Bay of Douarnenez, Brittany. Pp. 141–156. In *Trophic Relationships in the Marine Environment* (M. Barnes and R. N. Gibson, eds.). Aberdeen University Press, Aberdeen.

Guillou, M. 1996. Biotic and abiotic interactions controlling starfish outbreaks in the Bay of Douarnenez, Brittany, France. *Oceanologica Acta* 19:415–420.

Gulliksen, B., and S. H. Skaeveland. 1973. The sea star, *Asterias rubens* L., as a predator on the ascidian, *Ciona intestinalis* (L.). in Borgenfjorden, North-Trondelag, Norway. *Sarsia* 52:15–20.

Gupta, K. C. and P. J. Scheuer. 1968. Echinoderm sterols. *Tetrahedron* 24(17): 5831–5837.

Gutt, J., and A. Starmans. 1998. Structure and biodiversity of megabenthos in the Weddell and Lazarev Seas (Antarctica): ecological role of physical parameters and biological interactions. *Polar Biology* 20:229–247.

Guzmán, H. M., and C. A. Guevara. 2002. Annual reproductive cycle, spatial distribution, abundance, and size structure of *Oreaster reticulatus* (Echinodermata: Asteroidea) in Bocas del Toro, Panama. *Marine Biology* 141:1077–1084.

Haesaerts, D., M. Jangoux, and P. Flammang. 2005. The attachment complex of the brachiolaria larvae of the starfish *Asterias rubens* (Echinodermata): an ultrastructural and immunocytochemical study. *Zoomorphology* 124:67–77.

Haesaerts, D., M. Jangoux and P. Flammang. 2006. Adaptations to benthic development: functional morphology of the attachment complex of the brachiolaria larva in the sea star *Asterina gibbosa*. *Biological Bulletin* 211:172–182.

Hagerman, D. D. 1956. Invertebrate metabolism *in vitro* not affected by estradiol. *Biological Bulletin* 111:318–319.

Hagerman, D. D., F. M. Wellington, and C. A. Villee. 1957. Estrogens in marine invertebrates. *Biological Bulletin* 112:180–183.

Halpern, J. A. 1970. Growth rate of the tropical sea star *Luidia senegalensis* (Lamarck). *Bulletin of Marine Science.* 20:626–633.

Hamel, J. F., and A. Mercier. 1994. New distribution and host record for the starfish parasite *Dendrogaster* (Crustacea: Ascothoracica). *Journal of the Marine Biological Association of the United Kingdom* 74:419–425.

Hamel, J.-F., and A. Mercier. 1995. Prespawning behavior, spawning, and development of the brooding starfish *Leptasterias polaris*. *Biological Bulletin* 188:32–45.

Hancock, D. A. 1958. Notes on starfish on an Essex oyster bed. *Journal of the Marine Biological Association of the United Kingdom* 37:565–589.

Haramoto, S., M. Komatsu, and Y. Yamazaki. 2006. Population genetic structures of the fissiparous seastar *Coscinasterias acutispina* in the Sea of Japan. *Marine Biology.* 149:813–820.

Haramoto, S., M. Komatsu, and Y. Yamazaki. 2007. Patterns of asexual reproduction in the fissiparous seastar *Coscinasterias acutispina* (Asteroidea: Echinodermata) in Japan. *Zoological Science* 24:1075–1081.

Hardege, D., M. Bentley, M. Beckmann, and C. Müller. 1996. Sex pheromones in marine polychaetes: volatile organic substances (VOS) isolated from *Arenicola marina*. *Marine Ecology Progress Series* 139:157–166.

Harley, C. D. G. 2003. Species importance and context: spatial and temporal variation in species interactions. Pp. 44–68. In *The Importance of Species: Perspectives on Expendability and Triage* (P. Kareiva and S. A. Levin, eds.). Princeton University Press, Princeton, New Jersey.

Harley, C. D. G. 2007. Zonation. Pp. 647–653. In *Encyclopedia of Tidepools and Rocky Shores* (M. Denny and S. D. Gaines, eds.). University of California Press, Berkeley, California.

Harley, C. D. G., M. S. Pankey, J. P. Wares, R. K. Grosberg, and M. J. Wonham. 2006. Color polymorphism and genetic structure in the sea star *Pisaster ochraceus*. *Biological Bulletin.* 211:248–262.

Harris, L. G., M. Tyrell, and C. M. Chester. 1998. Changing patterns for two sea stars in the Gulf of Maine. Pp. 243–248. In *Echinoderms: San Francisco* (R. Mooi and M. Telford, eds.). Balkema, Rotterdam, the Netherlands.

Harris, P., and G. Shaw. 1984. Intermediate filaments, microtubules and microfilaments in epidermis of sea urchin tube foot. *Cell and Tissue Research* 236:27–33.

Harrison, G., and D. Philpott. 1966. Subcellular particles in echinoderm tube feet. I. Class Asteroidea. *Journal of Ultrastructure Research* 16:537–547.

Harrold, C., and J. S. Pearse. 1987. The ecological role of echinoderms in kelp forests. Pp. 137–233. In *Echinoderm Studies*, vol. 2 (M. Jangoux and J. M. Lawrence, eds.). Balkema, Rotterdam, the Netherlands.

Hart, M. W., and P. B. Marko. 2010. It's about time: divergence, demography, and the evolution of developmental modes in marine invertebrates. *Integrative and Comparative Biology* 50:643–661.

Hart, M. W., M. Byrne, and M. J. Smith. 1997. Molecular phylogenetic analysis of life history evolution in asterinid sea stars. *Evolution* 51:1848–1861.

Hart, M. W., M. Byrne, and S. L. Johnson. 2003. *Patiriella pseudoexigua* (Asteroidea: Asterinidae): a cryptic species complex revealed by molecular and embryological analyses. *Journal of the Marine Biological Association of the United Kingdom* 83:1109–1116.

Hart, M. W., S. L. Johnson, J. A. Addison, and M. Byrne. 2004. Strong character incongruence and character choice in phylogeny of sea stars of the Asterinidae. *Invertebrate Biology* 123:343–356.

Hart, M. W., C. Keever, A. J. Dartnall, and M. Byrne. 2006. Morphological and genetic variation indicate cryptic species within Lamarck's little sea star, *Parvulastra* (=*Patirella*) *exigua*. *Biological Bulletin* 210:158–167.

Harvell, C. D., K. Kim, J. M. Burkholder, R. R. Colwell, P. R. Epstein, D. J. Grimes, E. E. Hofmann, E. K. Lipp, A. D. M. E. Osterhaus, R. M. Overstreet, J. W. Porter, G. W. Smith, and G. R. Vasta. 1999. Emerging marine diseases: climate links and anthropogenic factors. *Science* 285:1505–1510.

Harvey, C., F.-X. Garneau, and J. H. Himmelman. 1987. Chemodetection of the predatory seastar *Leptasterias polaris* by the whelk *Buccinum undatum*. *Marine Ecology Progress Series* 40:79–86.

Hatanaka, M., and M. Kosaka. 1959. Biological studies on the population of the starfish *Asterias amurensis* in Sendai Bay. *Tohoku Journal of Agricultural Research* 9:159–173.

Heddle, D. 1967. Versatility of movement and the origin of the asteroids. Pp. 125–141. In *Echinoderm Biology* (N. Millott, ed.). Symposia of the Zoological Society of London 20.

Heddle, D. 1995. The descent of the Asteroidea and the reaffirmation of paxillosidan primitiveness. Pp 179–183. In *Echinoderm Research 1995* (R. Emson, A. B. Smith, and A. Campbell, eds). Balkema, Amsterdam/Brookfield, the Netherlands.

Held, M. B. E., and C. D. G. Harley. 2009. Responses to low salinity by the sea star *Pisaster ochraceus* from high- and low-salinity populations. *Invertebrate Biology* 128:381–390.

Helmuth, B., B. R. Broitman, C. A. Blanchette, S. Gilman, P. Halpin, C. D. G. Harley, M. J. O'Donnell, G. E. Hofmann, B. Menge, and D. Strickland. 2006. Mosaic pattern in thermal stress in the rocky intertidal zone: implications for climate change. *Ecological Monographs* 76:461–479.

Henderson, J. A., and J. S. Lucas. 1971. Larval development and metamorphosis of *Acanthaster planci* (Asteroidea). *Nature* 232:655–657.

Hendler, G., and M. Byrne. 1987. Fine structure of the dorsal arm plate of *Ophiocoma wendti*: evidence for a photoreceptor system (Echinodermata, Ophiuroidea). *Zoomorphology* 107:261–272.

Hendler, G., and D. R. Franz. 1982. The biology of a brooding seastar, *Leptasterias tenera*, in Block Island Sound. *Biological Bulletin* 162:273–289.

Hendler, G., J. E. Miller, D. L. Pawson, and P. M. Kier. 1995. *Echinoderms of Florida and the Caribbean: Sea Stars, Sea Urchins and Allies.* Smithsonian Institution Press, Washington, D.C.

Hennebert, E. 2010. Adhesion mechanisms developed by sea stars: a review of the ultrastructure and composition of tube feet and their secretion. Pp. 99–109. In *Biological Adhesive Systems: From Nature to Technical and Medical Application* (J. von Byern and I. Grunwald, eds.). Springer, Vienna.

Hennebert, E., P. Viville, R. Lazzaroni, and P. Flammang. 2008. Micro- and nanostructure of the adhesive material secreted by the tube feet of the sea star *Asterias rubens*. *Journal of Structural Biology* 164:108–118.

Hennebert, E., D. Haesaerts, P. Dubois, and P. Flammang. 2010. Evaluation of the different forces brought into play during tube foot activities in sea stars. *Journal of Experimental Biology* 213:1162–1174.

Hennebert, E., R. Wattiez, and P. Flammang. 2011. Characterization of the carbohydrate fraction of the temporary adhesive material secreted by the tube feet of the sea star *Asterias rubens*. *Marine Biotechnology* 13:484–495.

Hennebert, E., R. Wattiez, J. H. Waite, and P. Flammang. 2012. Characterization of the protein fraction of the temporary adhesive secreted by the tube feet of the sea star *Asterias rubens*. *Biofouling* 28:289–303.

Hennebert, E., R. Santos, and P. Flammang. In press. Echinoderms don't suck: evidence against the involvement of suction in tube foot attachment. In *Göttingen: Echinoderm Research 2010.* (M. Reich, A. Kroh, and J. Reitner, eds.). Springer, Heidelberg, Germany.

Hermans, C. O. 1983. The duo-gland adhesive system. *Oceanography and Marine Biology: An Annual Review* 21:281–339.

Herringshaw, L. G., A. T. Thomas, and M. P. Smith. 2007a. Starfish diversity in the Wenlock of England. *Palaeontology* 50:1211–1229.

Herringshaw, L. G., M. P. Smith, and A. T. Thomas. 2007b. Evolutionary and ecological significance of *Lepidaster grayi*, the earliest multiradiate starfish, *Zoological Journal of the Linnean Society* 150:743–754.

Hess, H. 1972. Eine Echinodermen-Fauna aus dem mittleren Dogger des Aargauer Juras. *Schweizerische Paläontologische Abhandlungen* 92.

Hickman, C. P., Jr. 1998. *A Field Guide to Sea Stars and Other Echinoderms of Galápagos.* Sugar Spring Press, Lexington, Virginia.

Hidalgo, F. J., F. N. Firstater, B. J. Lomovasky, and O. O. Iribarne. 2011. The influence of the predatory starfish *Heliaster helianthus* on colonization by a dominant mussel at central Peruvian coast. *Marine Ecology Progress Series* 432:103–114.

Hill, R. B. 1993. Comparative physiology of echinoderm muscle. Pp. 81–109. In *Echinoderm Studies,* vol. 4 (M. Jangoux and J. M. Lawrence, eds.). Balkema, Rotterdam.

Hill, R. B. 2004. Active state in echinoderm muscle. Pp. 351–352. In *Echinoderms: München* (T. Heinzeller and J. H. Nebelsick, eds.). Balkema, Leiden, the Netherlands.

Himmelman, J. H. 1991. Diving observations of subtidal communities in the northern Gulf of St. Lawrence. Pp. 319–332. In *The Gulf of St. Lawrence: Small ocean or big estuary?* (J.-C. Therriault, ed.). Canadian Special Publication of Fisheries and Aquatic Sciences 113, Fisheries and Oceans, Ottawa, Ontario, Canada.

Himmelman, J. H. 1999. Spawning, marine invertebrates. Pp. 524–533. In *Encyclopedia of reproduction* (E. Knobil and J. D. Neill, eds.). Academic Press, New York.

Himmelman. J. H., and C. Dutil. 1991. Distribution, population size-structure and feeding of subtidal seastars in the northern Gulf of St. Lawrence. *Marine Ecology Progress Series* 76:61–72.

Himmelman, J. H., and J. R. Hamel. 1993. Diet, behaviour and reproduction of the whelk *Buccinum undatum* in the northern Gulf of St. Lawrence, eastern Canada. *Marine Biology* 116:423–430.

Himmelman, J. H., and Y. Lavergne. 1985. Organization of rocky subtidal communities in the St. Lawrence Estuary. *Naturaliste Canadien* 112:143–154.

Himmelman, J. H., Y. Lavergne, A. Cardinal, G. Martel, and P. Jalbert. 1982. Brooding behaviour of the Northern sea star *Leptasterias polaris*. *Marine Biology* 68: 235–240.

Himmelman, J. H. C. Dutil, and C. F. Gaymer. 2005. Foraging behavior and activity budgets of sea stars on a subtidal sediment bottom community. *Journal of Experimental Marine Biology and Ecology* 322:153–165.

Himmelman, J. H., C. P. Dumont, C. F. Gaymer, C. Vallières, and D. Drolet. 2008. Spawning synchrony and aggregative behaviour of cold-water echinoderms during multi-species mass spawnings. *Marine Ecology Progress Series* 361:161–168.

Hines, G. A., S. A. Watts, S. A. Sower, and C. W. Walker 1992a. Sex steroid levels in the testes, ovaries, and pyloric caeca during gametogenesis in the sea star *Asterias vulgaris. General and Comparative Endocrinology* 87(3):451–460.

Hines, G. A., S. A. Watts, C. W. Walker, and P. A. Voogt. 1992b. Androgen metabolism in somatic and germinal tissues of the sea star *Asterias vulgaris. Comparative Biochemistry and Physiology* 102B(3):521–526.

Hintz, J. L., and J. M. Lawrence. 1994. Acclimation of gametes to reduced salinity prior to spawning in *Luidia clathrata* (Echinodermata: Asteroidea). *Marine Biology* 12:443–440.

Hirai, S., K. Chida, and H. Kanatani. 1973. Role of follicle cells in maturation of starfish oocytes. *Development, Growth and Differentiation* 15:21–31.

Hoberg, M. K., H. M. Feder, and S. C. Jewett. 1980. Some aspects of the biology of the parasitic gastropod, *Asterophila japonica* Randall and Heath (Prosobranchia: Melanellidae), from southeastern Chukchi Sea and northeastern Bering Sea, Alaska. *Ophelia* 19:73–77.

Hoegh-Guldberg, O. 1994. Uptake of dissolved organic matter by larval stage of the crown-of-thorns starfish *Acanthaster planci. Marine Biology* 120:55–63.

Hoegh-Guldberg, O., and J. S. Pearse. 1995. Temperature, food availability, and the development of marine invertebrate larvae. *American Zoologist* 35:415–425.

Holland, N. D. 1984. Echinodermata: Epidermal cells. Pp. 756–774. In *Biology of the Integument*, vol. 1, *Invertebrates* (J. Bereiter-Hahn, A. G. Matoltsy, and K.S. Richards, eds.). Springer Verlag, Berlin.

Holland, N. D., and K. H. Nealson. 1978. The fine structure of the echinoderm cuticle and the subcuticular bacteria of echinoderms. *Acta Zoologica (Stockholm)* 59:169–185.

Holling, C. S. 1959. The components of predation as revealed by a study of small mammal predation of the European pine saw fly. *Canadian Entomologist* 91:293–320.

Holling, C. S. 1965. The functional response of predators to prey density and its role in mimicry and population regulation. *Memoirs of the Entomological Society of Canada* 45:1–60.

Holme, N. A. 1984. Fluctuations of *Ophiothrix fragilis* in the western English Channel. *Journal of the Marine Biological Association of the United Kingdom* 64:351–378.

Hopkins, T. S., and K. E. Knott. 2010. The establishment of a neotype for *Luidia clathrata* (Say, 1825) and a new species within the genus *Luidia* (Asteroidea: Paxillosida: Luidiidae). Pp. 207–212. In *Echinoderms: Durham* (L. G. Harris, S. A. Boetger, C. W. Walker, and M. P. Lesser, eds.). Taylor & Francis, London.

Hopkins, T. S., F. M. Fontanella, and C. R. R. Ventura. 2003. Morphological diagnosis of three Brazilian sea stars of the genus *Echinaster* (sub-genus *Othilia*). Pp. 97–103. In *Echinoderm Research 2001* (J.-P. Féral and B. David, eds.) Balkema, Rotterdam, the Netherlands.

Hotchkiss, F. H. C. 1979. Case studies in the teratology of starfish. *Proceedings of the Academy of Natural Sciences of Philadelphia* 131:139–157.

Hotchkiss, F. H. C. 2000. On the number of rays in starfish. *American Zoologist* 40:340–354.

Hotchkiss, F. H. C. 2009. Arm stumps and regeneration models in Asteroidea. *Proceedings of the Biological Society of Washington* 122:342–354.

Houk, P., and J. Raubani. 2010. *Acanthaster planci* outbreaks in Vanuatu coincide with ocean productivity, furthering trends throughout the Pacific Ocean. *Journal of Oceanography* 66:435–438.

Houk, P., S. Bograd, and R. van Woesik. 2007. The transition zone chlorophyll front can trigger *Acanthaster planci* outbreaks in the Pacific Ocean: historical confirmation. *Journal of Oceanography* 63:149–154.

Howarth, R., F, Chan, D. J. Conley, J. Garnier, S. C. Doney, R. Marino, and G. Billen. 2011. Coupled biogeochemical cycles: eutrophication and hypoxia in temperate estuaries and coastal marine ecosystems. *Frontiers in Ecology and the Environment* 9:18–26.

Hulings, N. C., and D. W. Hemlay. 1963. An investigation of the feeding habits of two species of sea stars. *Bulletin of Marine Science of the Gulf and Caribbean* 13:354–359.

Humes, A. G. 1976. Cyclopoid copepods associated with asteroid echinoderms in New Caledonia. *Smithsonian Contributions to Zoology* 217:1–19.

Hunt, A. 1993. Effects of contrasting patterns of larval dispersal on the genetic connectedness of local populations of two intertidal starfish *Patiriella calcar* and *Patiriella exigua*. *Marine Ecology Progress Series* 92:179–186.

Hunt, H. L., and R. E. Scheibling. 1997. Role of early post-settlement mortality in recruitment of benthic marine invertebrates. *Marine Ecology Progress Series* 155:269–301.

Hunt, O. D. 1925. The food of the bottom fauna of the Plymouth fishing grounds. *Journal of the Marine Biological Association of the United Kingdom* 13:560–599.

Hutchinson, G. E. 1959. Homage to Santa Rosalia or why are there so many kinds of animals? *American Naturalist* 93:145–159.

Hutson, K. S., D. J. Ross, R. W. Day, and J. J. Ahern. 2005. Australian scallops do not recognize the introduced predatory seastar *Asterias amurensis*. *Marine Ecology Progress Series* 298:305–309.

Hyman, L. H. 1955. *The Invertebrates: Echinodermata*. McGraw-Hill, New York.

Ikegami, S. 1976. Role of asterosaponin A in starfish spawning induced by gonad-stimulating substance and 1-methyladenine. *Journal of Experimental Zoology* 198:359–366.

Ikegami, S., H. Shirai, and H. Kanatani. 1971. On the occurrence of progesterone in ovary of the starfish. *Zoological Magazine* 80:26–28.

Ikegami, S., Y. Kamiya, and S. Tamura. 1972. Isolation and characterization of spawning inhibitors in the ovary of the starfish, *Asterias amurensis*. *Agricultural and Biological Chemistry* 36:2005–2011.

Iken, K., S. P. Greer, C. D. Amsler, and J. B. McClintock. 2003. A new antifouling bioassay monitoring brown algal spore swimming behavior in the presence of echinoderm extracts. *Biofouling* 19:327–334.

Iken, K., B. Bluhm, and K. Dunton. 2010. Benthic food-web structure under differing water mass properties in the southern Chukchi Sea. *Deep-Sea Research II* 57:71–85.

Inagaki, M., R. Isobe, Y. Kawano, T. Miyamoto, T. Komori, and R. Higuchi. 1998. Isolation and structure of three new ceramides from the starfish *Acanthaster planci*. *European Journal of Organic Chemistry* 129–131.

Inagaki, M., T. Miyamoto, R. Isobe, and R. Higuchi. 2005. Biologically active glycosides from Asteroidea, 43. Isolation and structure of a new neuritogenic-active ganglioside molecular species from the starfish *Linckia laevigata*. *Chemistry and Pharmaceutical Bulletin* 53:1551–1554.

Inagaki, M., Y. Ikeda, S. Kawatake, K. Nakamura, M. Tanaka, E. Misawa, M. Yamada, and R. Higuchi. 2006. Isolation and structure of four new ceramides from the starfish *Luidia maculata*. *Chemistry and Pharmaceutical Bulletin* 54:1647–1649.

Ino, T., J. Sagara, S. Hamada, and M. Tamasawa. 1955. On the spawning season of the starfish *Asterias amurensis* in Tokyo Bay. *Bulletin of the Japanese Society of Scientific Fisheries* 21:32–36.

Iorizzi, M., P. Bryan, J. McClintock, L. Minale, E. Palagiano, S. Maurelli, R. Riccio, and F. Zollo. 1995. Chemical and biological investigation of the polar constituents of the starfish *Luidia clathrata*, collected in the Gulf of Mexico. *Journal of Natural Products* 58:653–671.

Iorizzi, M., S. De Marino, and F. Zollo. 2001. Steroidal oligoglycosides from the Asteroidea. *Current Organic Chemistry* 5:951–973.

Ishii, T., T. Okino, and Y. Mino. 2006. A ceramide and cerebroside from the starfish *Asterias amurensis* Lutken and their plant-growth promotion activities. *Journal of Natural Products* 69:1080–1082.

Iyengar, E. V., and C. D. Harvell. 2001. Predator deterrence of early developmental stages of temperate lecithotrophic asteroids and holothuroids. *Journal of Experimental Marine Biology and Ecology* 264:171–188.

Jaeckle, W. B. 1994. Multiple modes of asexual reproduction by tropical and subtropical sea star larvae: an unusual adaptation for genet dispersal and survival. *Biological Bulletin* 186:62–71.

Jaeckle W. B. 1995. Variation in the size, energy content, and biochemical composition of invertebrate eggs: correlates to the mode of larval development. Pp 49–77. In *Ecology of Marine Invertebrate Larvae* (L. R. McEdward, ed.). CRC Press, Boca Raton, Florida.

Jalbert, P., J. H. Himmelman, P. Beland, and B. Thomas. 1989. Whelks (*Buccinum undatum*) and other subtidal predators in the nothern Gulf of St. Lawrence. *Naturaliste Canadien* 116:1–15.

James, M. K., and J. P. Scandol. 1992. Larval dispersal simulations: correlation with the crown-of-thorns starfish outbreaks database. *Australian Journal of Marine and Freshwater Research* 43:569–82.

James, M. K., P. R. Armsworth, L. B. Mason, and L. Bode. 2002. The structure of reef fish metapopulations: modelling larval dispersal and retention patterns. *Proceedings of the Royal Society of London, Series B: Biological Sciences* 269:2079–2086.

Jangoux, M. 1982a. Food and feeding mechanisms: Asteroidea. Pp. 117–159. In *Echinoderm Nutrition* (M. Jangoux and J. M. Lawrence, eds.). Balkema, Rotterdam, the Netherlands.

Jangoux, M. 1982b. Digestive systems: Asteroidea. Pp. 235–279. In *Echinoderm Nutrition* (M. Jangoux and J. M. Lawrence, eds.), Balkema, Rotterdam, the Netherlands.

Jangoux, M. 1982c. On *Tremaster* Verrill, 1879, an odd genus of recent starfish (Echinodermata: Asteroidea). Pp. 155–163. In *Echinoderms: Proceedings of the International Conference, Tampa Bay* (J. M. Lawrence, ed.). Balkema, Rotterdam, the Netherlands.

Jangoux, M. 1984. Les astérides littoraux de Nouvelle-Calédonie. *Bulletin Muséum National d'Histoire Naturelle Paris, 4e Série* 6:279–293.

Jangoux, M., and A. Lambert, A. 1987. Étude comparative des pédicellaires des astérides (échinodermes). *Bulletin de la Société Scientifique et Naturaliste de l'Ouest de France, supplément. H.S.*:47–56.

Jangoux, M., and E. Van Impe. 1971. Étude comparative des activités phosphomonoestérasiques alcalines du tube digestif de plusiers espèces d'Astéroïdes (Échinodermes) précédée d'une note anatomique. *Cahiers de Biologie Marine* 12:405–418.

Jangoux, M., and E. Van Impe. 1977. The annual pyloric cycle of *Asterias rubens* L. (Echinodermata: Asteroidea). *Journal of Experimental Marine Biology and Ecology* 30:165–184.

Jangoux, M., and M. Vloebergh. 1973. Contribution a l'étude du cycle annuel de reproduction d'une population

d'*Asterias rubens* (Echinodermata, Asteroidea) du littoral Belge. *Netherlands Journal of Sea Research* 6:389–408.

Janies, D. A. 2004. Evolution of asterozoan echinoderms and their development. Pp. 586–587. In *Echinoderms: München* (T. Heinzeller and J. H. Nebelsick, eds.). Balkema, Leiden, the Netherlands.

Janies, D. A., J. R. Voight, and M. Daly. 2011. Echinoderm phylogeny including *Xyloplax*, a progenetic asteroid. *Systematic Biology* 60:420–438.

Janosik, A. M. and K. M. Halanych. 2010. Unrecognized Antarctic biodiversity: a case study of the genus *Odontaster* (Odontasteridae; Asteroidea). *Integrative and Comparative Biology* 50:981–992.

Janosik, A. M., A. R. Mahon, and K. M. Halanych. 2011. Evolutionary history of Southern Ocean *Odontaster* sea star species (Odontasteridae; Asteroidea). *Polar Biology* 34:575–586.

Janssen, H. H. 1991. Surprising findings on the sea star *Archaster typicus*. *Philippine Scientist* 28:89–98.

Jaramillo, E., J. Navarro, and J. Winter. 1981. The associations between *Mytilus chilensis* Hupe (Bivalvia, Mytilidae) and *Edotia magellanica* Cunningham (Isopoda, Valvifera) in southern Chile. *Biological Bulletin* 160:107–113.

Jeffery, C. H., R. B. Emlet, and D. T. J. Littlewood. 2003. Phylogeny and evolution of developmental mode in temnopleurid echinoids. *Molecular Phylogenetics and Evolution* 28:99–118.

Jeffreys, R. M., G. A. Wolff, and R. J. Murty. 2009. The trophic ecology of key megafaunal species at the Pakistan Margin: evidence from stable isotopes and lipid biomarkers. *Deep-Sea Research I* 56:1816–1833.

Jennings, L. B., and H. L. Hunt. 2010. Settlement, recruitment and potential predators and competitors of juvenile echinoderms in the rocky subtidal zone. *Marine Biology* 157:307–316.

Johnson, C. 1992a. Settlement and recruitment of *Acanthaster planci* on the Great Barrier Reef: questions of process and scale. *Australian Journal of Marine and Freshwater Research* 43:611–627.

Johnson, C. 1992b. Reproduction, recruitment and hydrodynamics in the crown-of-thorns phenomenon on the Great Barrier Reef: introduction and synthesis. *Australian Journal of Marine and Freshwater Research* 43:517–523.

Johnson, C. R., and D. C. Sutton. 1994. Bacteria on the surface of crustose coralline algae induce metamorphosis of the crown-of-thorns starfish *Acanthaster planci*. *Marine Biology* 120:305–10.

Johnson, C. R., D. C. Sutton, R. R. Olson, and R. Giddins. 1991. Settlement of crown-of-thorns starfish: role of bacteria on surfaces of coralline algae and a hypothesis for deepwater recruitment. *Marine Ecology Progress Series* 71:143–162.

Johnson, D. 1994. Seastar fight gains momentum. *Australian Fisheries* 53:25–27.

Johnson, M. S., and T. J. Threlfall. 1987. Fissiparity and population genetics of *Coscinasterias calamaria*. *Marine Biology* 93:517–525.

Jones, D. S., and R. W. Portell. 1988. Occurrence and biogeographic significance of *Heliaster* (Echinodermata:

Asteroidea) from the Pliocene of southwest Florida. *Journal of Paleontology* 62:126–132.

Justome, B., Rochette, R., and Himmelman, J. H. 1998. Investigation of the influence of exposure to predation risk on the development of defensive behaviors in a marine gastropod. *Veliger* 41:172–179.

Kaack, K. E., and C. M. Pomory. 2011. Salinity effects on arm regeneration in *Luidia clathrata* (Echinodermata: Asteroidea). *Marine and Freshwater Behaviour and Physiology* 44:359–374.

Kaiser, M. J. 1996. Starfish damage as an indicator of trawling damage. *Marine Ecology Progress Series* 134:303–307.

Kanatani, H. 1964. Spawning of starfish: action of gamete-shedding substance obtained from radial nerves. *Science* 146:1177–1179.

Kanatani, H. 1967. Mechanism of starfish spawning with special reference to gonad-stimulating substance (GSS) of nerve and meiosis-inducing substance (MIS) of gonad. *Japanese Journal of Experimental Morphology* 21:61–78.

Kanatani, H. 1969. Induction of spawning and oocyte maturation by 1-methyladenine in starfishes. *Experimental Cell Research* 57:333–337.

Kanatani, H. 1973. Maturation-inducing substance in starfishes. *International Review of Cytology* 35:253–298.

Kanatani, H. 1979. Hormones in echinoderms. Pp. 273–307. In *Hormones and Evolution*, vol. 1 (E. J. W. Barrington, ed.). Academic Press, New York.

Kanatani, H., and Y. Nagahama. 1983. Echinodermata. Pp. 611–654. In *Reproductive Biology of Invertebrates*, vol. 1, *Oogenesis, Oviposition, and Oosorption* (K. G. Adiyodi and R. G. Adiyodi, eds.). Wiley, Chichester.

Kanatani, H., and M. Ohguri. 1966. Mechanism of starfish spawning. I. Distribution of active substance responsible for maturation of oocytes and shedding of gametes. *Biological Bulletin* 131:104–114.

Kanatani, H., and H. Shirai. 1969. Mechanism of starfish spawning. II. Some aspects of action of a neural substance obtained from radial nerves. *Biological Bulletin* 137:297–311.

Kanatani, H., and H. Shirai. 1970. Mechanism of starfish spawning. III. Properties and action of meiosis-inducing substance produced in gonad under influence of gonad-stimulating substance. *Development Growth & Differentiation* 12:119–140.

Kanatani, H., H. Shirai, K. Nakanishi, and T. Kurokawa. 1969. Isolation and indentification on meiosis inducing substance in starfish *Asterias amurensis*. *Nature* 221:273–274.

Kanatani, H., S. Ikegami, and H. Shirai. 1971. On the occurrence of progesterone in ovary of the starfish, *Asterias amurensis*. *Zoological Magazine* 80:26–28.

Kanazawa, A., S. I. Teshima, and S. Tomita. 1974. Sterol biosynthesis in some coelenterates and echinoderms. *Bulletin of the Japanese Society for the Science of Fish* 40(12):1257–1260.

Kanazawa, A., S. Teshima, T. Ando, and S. Tomita. 1976. Sterols in coral-reef animals. *Marine Biology* 34:53–57.

Kang, K. H., and J. M. Kim. 2004. The predation of trumpet shell, *Charonia* sp., on eight different marine invertebrate species. *Aquaculture Research* 35:1202–1206.

Kano,Y. T., M. Komatsu, and C. Ogouro, 1974. Notes on the development of the sea-star *Leptasterias ochotensis similispinis*, with special reference to skeletal system. *Proceedings of the Japanese Society for Systematic Zoology* 10:45–53.

Karako, S., Y. Achituv, R. Perl-Treves, and D. Katcoff. 2002. *Asterina burtoni* (Asteroidea; Echinodermata) in the Mediterranean and the Red Sea: does asexual reproduction facilitate colonization? *Marine Ecology Progress Series* 234:139–145.

Kashenko, S. D. 2003. The reaction of the starfish *Asterias amurensis* and *Patria pectinifera* (Asteroidea) from Vostok Bay (Sea or Japan) to a salinity decrease. *Russian Journal of Marine Biology* 29:110–114.

Kashenko, S. D. 2005a. Development of the starfish *Asterias amurensis* under laboratory conditions. *Russian Journal of Marine Biology* 31:36–42.

Kashenko, S. D. 2005b. Responses of embryos and larvae of the starfish *Asterias amurensis* to changes in temperature and salinity. *Russian Journal of Marine Biology* 31:294–302.

Kasyanov V. L. 1988. Reproductive strategies of seastars in the Sea of Japan. Pp. 205–209. In *Echinoderm Biology* (R. D. Burke, P. V. Mladenov, P. Lambert, and R. L. Parsley, eds.). Balkema, Rotterdam, the Netherlands.

Kautsky, N. 1982. Growth and size structure in a baltic *Mytilus edulis* population. *Marine Biology* 68:117–133.

Keesing, J. K., and A. R. Halford. 1992a. Field measurement of survival rates of juvenile *Acanthaster planci*: techniques and preliminary results. *Marine Ecology Progress Series* 85:107–14.

Keesing, J. K., and A. R. Halford. 1992b. Importance of post settlement processes for the population dynamics of *Acanthaster planci* (L.). *Australian Journal of Marine and Freshwater Research* 43:635–651.

Keesing, J. K., C. M. Cartwright, and K. C. Hall. 1993. Measuring settlement intensity of echinoderms on coral reefs. *Marine Biology* 117:399–407.

Keesing, J. K., W. L.Wiedermeyer, K. Okaji, A. R. Halford, K. C. Hall, and C. M. Cartwright. 1996. Mortality rates of juvenile starfish *Acanthaster planci* and *Nardoa* spp. measured on the Great Barrier Reef, Australia and in Okinawa, Japan. *Oceanologica Acta* 19:441–48.

Keesing, J., F. Graham, T. Irvine, and R. Crossing. 2011. Synchronous aggregated pseudo-copulation of the sea star *Archaster angulatus* Müller & Troschel, 1842 (Echinodermata: Asteroidea) and its reproductive cycle in south-western Australia. *Marine Biology* 158:1163–1173.

Keever, C. C., J. Sunday, J. B. Puritz, J. A. Addison, R. J. Toonen, R. K. Grosberg, and M. W. Hart. 2009. Discordant distribution of populations and genetic variation in a sea star with high dispersal potential. *Evolution* 63:3214–3227.

Keller, B. D. 1976. Sea urchin abundance patterns in seagrass meadows: the effects of predation and competitive interactions. Ph.D. thesis, Johns Hopkins University, Baltimore, Maryland.

Kelman, D., and R. B. Emlet. 1999. Swimming and buoyancy in ontogenetic stages of the cushion star *Pteraster tesselatus* (Echinodermata: Asteroidea) and their implications for distribution and movement. *Biological Bulletin* 197:309–314.

Kempf, M. 1966. On the development of *Echinaster echinophorus* (Lmk). *Anais da Academia Brasileira de Ciências* 38:505–507.

Kenchington, R. A. 1977. Growth and recruitment of *Acanthaster planci* (L.) on the Great Barrier Reef. *Biological Conservation* 11:103–118.

Keough, M. J., and A. J. Butler. 1979. The role of asteroid predators in the organization of a sessile community on pier pilings. *Marine Biology* 51:167–177.

Kerkut, G. A. 1953. The forces exerted by the tube feet of the starfish during locomotion. *Journal of Experimental Biology* 30:575–583.

Kesling, R. V. 1969. Three Permian starfish from Western Australia and their bearing on the revision of the Asteroidea. *Contribultions from the Museum of Paleontology, University of Michigan* 22:361–376.

Kesling, R. V., and H. L. Strimple. 1966. *Calliasterella americana*, a new starfish from the Pennsylvanian of Illinois. *Journal of Paleontology* 40:1157–1166.

Kicha, A. A., N. V. Ivanchina, I. A. Gorshkova, L. P. Ponomarenko, G. N. Likhatskaya, and V. A. Stonik. 2001. The distribution of free sterols, polyhydroxysteroids and steroid glycosides in various body components of the starfish *Patiria* (=*Asterina*) *pectinifera*. *Comparative Biochemistry and Physiology* 128B(1):43–52.

Kicha, A. A., N. V. Ivanchina, A. I. Kalinovsky, P. S. Dmitrenok, and V.A. Stonik. 2003a. Alkaloidosteroids from the starfish *Lethasterias nanimensis chelifera*. *Tetrahedron Letters* 44:1935–1937.

Kicha, A. A., N. V. Ivanchina, and V. A. Stonik. 2003b. Seasonal variations in the levels of polyhydroxysteroids and related glycosides in the digestive tissues of the starfish *Patiria* (*Asterina*) *pectinifera*. *Comparative Biochemistry and Physiology* 136B(4):897–903.

Kidawa, A. 2001. Antarctic starfish, *Odontaster validus*, distinguish between fed and starved conspecifics. *Polar Biology* 24:408–410.

Kidawa, A. 2005a. The role of amino acids in phagostimulation in the shallow-water omnivorous Antarctic sea star *Odontaster validus*. *Polar Biology* 28:147–155.

Kidawa, A. 2005b. Behavioural and metabolic responses of the Antarctic sea star *Odontaster validus* to food stimuli of different concentrations. *Polar Biology* 28:449–455.

Kidawa, A. 2009. Food selection of the Antarctic sea star *Odontaster validus* (Koehler): Laboratory experiments with food quality and size. *Polish Journal of Ecology* 57:139–147.

Kidawa, A., K. Stepanowska, M. Markowska, and S. Rakusa-Suszezewski. 2008. Fish blood as a chemical signal for Antarctic marine invertebrates. *Polar Biology* 31:519–525.

Kidawa, A., M. Potocka, and T. Janecki. 2010. The effects of temperature on the behaviour of the Antarctic sea star *Odontaster validus*. *Polish Polar Research* 31:273–284.

Kier, P. M., and R. E. Grant. 1965. Echinoid distribution and habits, Key Largo Reef Preserve, Florida. *Smithsonian Miscellaneous Collections* 149:1–68.

Kim, S., and A. Thurber. 2007. Comparison of seastar (Asteroidea) fauna across island groups of the Scotia Arc. *Polar Biology* 30:415–425.

Kim, S. L, A. Thurber, K. Hammerstom, and K. Conlan. 2007. Seastar response to organic enrichment in an oligotrophic polar habitat. *Journal of Experimental Marine Biology and Ecology* 346:66–75.

Kim, Y. S. 1968. Histological observations of the annual change in the gonad of the starfish, *Asterias amurensis* Lütken. *Bulletin of the Faculty of Fisheries Hokkaido University* 19:97–108.

Kim, Y. S. 1969. Selective feeding on several bivalve molluscs by the starfish, *Asterias amurensis* Lütken. *Bulletin of the Faculty of Fisheries Hokkaido University* 19:244–249.

Kingsford, M. J., J. M. Leis, A. Shanks, K. C. Lindeman, S. G. Morgan, and J. Pineda. 2002. Sensory environments, larval abilities and local self-recruitment. *Bulletin of Marine Science* 70:309–340.

Kitagawa, I., M. Kobayashi, T. Sugawara, and I. Yosioka. 1975. Thonrasterols A and B, two genuine sapogenols from the starfish *Acanthaster planci*. *Tetrahedron Letters* 967–970.

Klinger, T. S. 1979. A study of sediment preference and its effect on distribution in *Luidia clathrata* Say (Echinodermata: Asteroidea). Thesis, University of South Florida, Tampa, Florida.

Knott, E. K., and G. A. Wray. 2000. Controversy and consensus in asteroid systematics: new insights to ordinal and familial relationships. *American Zoologist* 40:382–392.

Knott, K. E., E. J. Balser, W. B. Jaeckle, and G. A. Wray. 2003. Identification of asteroid genera with species capable of larval cloning. *Biological Bulletin* 204:246–255.

Koehler, R. 1906. Echinodermes (Stéllérides, Ophiures et Echinides). Pp. 1–41. In *Expédition Antarctique Française (1903–1905) commandée par le Dr. Jean Charcot*. Sciences Naturelles: Documents Scientifiques. Masson et Cie, Paris.

Komatsu, M. 1975. On the development of the sea-star, *Astropecten latespinosus* Meissner. *Biological Bulletin* 148:49–59.

Komatsu, M. 1983. Development of the sea star *Archaster typicus* with a note on male on female superposition. *Annotationes Zoologicae Japonenses* 56:187–195.

Komatsu, M., and S. Nojima. 1985. Development of the seastar, *Astropecten gisselbrechti* Döderlein. *Pacific Science* 39:274–282.

Komatsu, M., Y. T. Kano, H. Yoshizara, S. Akabane, and C. Oguro. 1979. Reproduction and development of the hermaphroditic sea-star *Asterina minor* Hayashi. *Biological Bulletin* 157:258–274.

Komatsu, M., C. Oguro, and Y. S. Kano. 1982. Development of the sea-star, *Luidia quinaria* von Martens. Pp. 497–503. In *International Echinoderm Conference, Tampa Bay* (J. M. Lawrence, ed.). Balkema, Rotterdam, the Netherlands.

Komatsu, M., C. Oguro, and J. M. Lawrence. 1991. A comparison of development in three species of the genus *Luidia*, (Echinodermata: Asteroidea) from Florida. Pp. 489–498. In *Biology of Echinodermata* (T. Yanagisawa, I. Yasumasu, C. Oguro, N. Suzuki, and T. Motokawa, eds.). Balkema, Rotterdam, the Netherlands.

Komatsu, M., M. Kawai, S. Nojima, and C. Oguro. 1994. Development of the multiarmed seastar, *Luidia maculata* Müller & Troschel. Pp. 327–333. In *Echinoderms through Time* (B. David, A. Guille, J.-P. Féral, and M. Roux, eds.). Balkema, Rotterdam, the Netherlands.

Komatsu, M., M. Sewell, S. F. Carson, and F.-S. Chia. 2000. Larval development and metamorphosis of the sea star *Luidia foliolata* (Echinodermata: Asteroidea). *Species Diversity* 5:155–162.

Komatsu, M., P. M. O'Loughlin, B. Bruce, H. Yoshizawa, K. Tanaka, and C. Murakami. 2006. A gastric-brooding asteroid, *Smilasterias multipara*. *Zoological Science* 23:699–705.

Kong, F., M. K. Harper, and D. J. Faulkner. 1992. Fuscusine, a tetrahydroisoquinoline alkaloid from the sea star *Perknaster fuscus antarcticus*. *Natural Product Letters* 1:71–74.

Kossel, A., and S. Edlbacher. 1915. Articles on the chemical issue of echinoderms. *Hoppe-Seylers Zeitschrift Fur Physiologische Chemie* 94(4): 264–283.

Kroon, F. J., S. N. Wilkinson, A. Kinsey-Henderson, B. Abbott, J. E. Brodie, and R. D. R. Turner. 2011. River loads of suspended solids, nitrogen, phosphorus and herbicides delivered to the Great Barrier Reef lagoon. *Marine Pollution Bulletin*. DOI:10.1016/j.marpolbul.2011.10.018.

Kubo, K. 1951. Some observations of the development of the sea-star *Leptasterias ochotensis smilispinus* (Clark). *Journal of the Faculty of Science, Hokkaido University, Series 6* 10:97–105.

Lafay, B., A. B. Smith, and R. Christen, 1995. A combined morphological and molecular approach to the phylogeny of asteroids (Asteroidea: Echinodermata). *Systematic Biology* 44:190–208.

Lafferty, K. D., J. W. Porter, and S. E. Ford. 2004. Are diseases increasing in the oceans? *Annual Review of Ecology and Systematics* 35:31–54.

Lam, K. K. Y. 2002. Escape responses of intertidal gastropods on a subtropical rocky shore in Hong Kong. *Journal of Molluscan Studies* 68:297 306.

Lamare, M. D., T. Channon, C. Cornelisen, and M. Clarke. 2009. Archival electronic tagging of a predatory sea star: testing a new technique to study movement at the individual level. *Journal of Experimental Marine Biology and Ecology* 373:1–10.

Lambert, A., L. De Vos, and M. Jangoux, M. 1984. Functional morphology of the pedicellariae of the asteroid *Marthasterias glacialis* (Echinodermata). *Zoomorphology* 104:122–130.

Lambert, P. 2000. *Sea Stars of British Columbia, Southeast Alaska, and Puget Sound,* 2nd ed. Royal British Columbia Museum Handbook, Victoria, British Columbia.

Landenberger, D. E. 1968. Studies on selective feeding in the Pacific starfish *Pisaster* in Southern California. *Ecology* 49:1062–1075.

Landenberger, D. E. 1969. The effects of exposure to air on Pacific starfish and its relationship to distribution. *Physiological Zoology* 42:220–230.

Lane, D. J. W., L. M. Marsh, D. VandenSpiegel, and F. W. E. Rowe. 2000. Echinoderm fauna of the South China Sea: an inventory and analysis of distribution patterns. *Raffles Bulletin of Zoology, Supplement* 8:459–493.

Lares, M. T., and J. M. Lawrence. 1994. Nutrient and energy allocation during arm regeneration in *Echinaster paucispinus* (Clark) (Echinodermata; Asteroidea). *Journal of Experimental Marine Biology and Ecology* 180:49–58.

Lauerman, L. M. L. 1998. Diet and feeding behavior of the deep-water sea star *Rathbunaster californicus* (Fisher) in the Monterey submarine canyon. *Bulletin of Marine Science* 63:523–530.

Lauzon-Guay, J.-S., and R. E. Scheibling. 2007. Importance of spatial population characteristics on the fertilization rates of sea urchins. *Biological Bulletin* 212:195–205.

Lauzon-Guay J.-S., and R. E. Scheibling. 2008. Experimental evaluation of PIT tags in field and laboratory studies of sea urchins: caution advised. *Aquatic Biology* 2:105–112.

Lauzon-Guay, J.-S, R. E. Scheibling, and M. A. Barbeau. 2008. Formation and propagation of feeding fronts in benthic marine invertebrates: a modelling approach using kinesis. *Ecology* 89:3150–3162.

Lavado, R., A. Barbaglio, M. D. C. Carnevali, and C. Porte. 2006. Steroid levels in crinoid echinoderms are altered by exposure to model endocrine disruptors. *Steroids* 71:489–497.

Lavoie, M. E. 1956. How sea stars open bivalves. *Biological Bulletin* 111:114–122.

Lawrence, J. M. 1973. Level, content, and caloric equivalents of the lipid, carbohydrate, and protein in the body components of *Luidia clathrata* (Echinodermata: Asteroidea: Playtasterida) in Tampa Bay. *Journal of Experimental Marine Biology and Ecology* 11:263–274.

Lawrence, J. M. 1976. Patterns of lipid storage in post-metamorphic marine invertebrates. *American Zoologist* 16:747–762.

Lawrence, J. M. 1982. Digestion. Pp. 283–316. In *Echinoderm Nutrition* (M. Jangoux and J. M. Lawrence, eds.). Balkema, Rotterdam, the Netherlands.

Lawrence, J. M. 1985. The energetic echinoderm. Pp. 47–67. In *Echinodermata* (B. F. Keegan, and B. D. S. O'Connor, eds.). Balkema, Rotterdam, the Netherlands.

Lawrence, J. M. 1987. *A Functional Biology of Echinoderms.* Johns Hopkins University Press, Baltimore, Maryland.

Lawrence, J. 1988. Functional consequences of the multiarmed condition in asteroids, Pp. 597–602. In *Echinoderm Biology* (R. D. Burke, P. V. Mladenov, P. Lambert, and R. L. Parsely, eds.). Balkema, Rotterdam, the Netherlands.

Lawrence, J. M. 1990. The effect of stress and disturbance on echinoderms. *Zoological Science* 7:17–28.

Lawrence, J. M. 1991. A chemical alarm response in *Pycnopodia helianthoides* (Echinodermata: Asteroidea). *Marine Behavior and Physiology* 19:39–44.

Lawrence, J. M. 1992. Arm loss and regeneration in Asteroidea (Echinodermata). Pp. 39–52. In *Echinoderm Research 1991* (L. Scalera-Liaci and C. Canacatti, eds.). Balkema, Rotterdam, the Netherlands.

Lawrence, J. M. 1996. Mass mortality of echinoderms from abiotic factors. Pp 103–137. In *Echinoderm Studies,* vol. 5 (M. Jangoux and J. M. Lawrence, eds.). Balkema, Rotterdam, the Netherlands.

Lawrence, J. M. 2010. Energetic costs of loss and regeneration of arms in stellate echinoderms, *Integrative and Comparative Biology* 50:506–514.

Lawrence, J. M. In press a. Form, function, food and feeding in stellate echinoderms. In *Göttingen: Echinoderm Research 2010* (M. Reich, A. Kroh, and J. Reitner, eds.). Springer, Heidelberg, Germany.

Lawrence, J. M. In press b. Arm loss and regeneration in stellate echinoderms: an organismal view. In *Echinoderms in a Changing World* (C. Johnson, ed.). Balkema, Rotterdam, the Netherlands.

Lawrence, J. M., and W. Avery. 2010. Bilateral symmetry of the rays of small *Luidia senegalensis* (Echinodermata: Asteroidea), *Florida Scientist* 73:27–30.

Lawrence, J. M., and P. F. Dehn. 1979. Biological characteristics of *Luidia clathrata* (Echinodermata:Asteroidea) from Tampa Bay and the shallow waters of the Gulf of Mexico. *Florida Scientist* 42:9–13.

Lawrence, J. M., and A. Ellwood. 1991. Simultaneous allocation of resources to arm regeneration and to somatic and gonadal production in *Luidia clathrata* (Say) (Echinodermata: Asteroidea). Pp. 543–548. In *Biology of Echinodermata* (T. Yanaagisawa, I. Yasumasu, C. Oguro, N. Suzuki, and T. Motokawa, eds.). Balkema, Rotterdam, the Netherlands.

Lawrence, J. M., and C. F. Gaymer. In press. Autotomy of rays of *Heliaster helianthus* (Asteroidea: Echinodermata). In *Göttingen: Echinoderm Research 2010* (M. Reich, A. Kroh, and J. Reitner, eds.). Springer, Heidelberg, Germany.

Lawrence, J. M., and J. Herrera. 2000. Stress and deviant reproduction in echinoderms. *Zoological Studies* 39:151–171.

Lawrence, J. M., and M. Komatsu. 1990. Mode of arm development in multiarmed species of asteroids, Pp. 269–275. In *Echinoderm Research 1990* (C. De Ridder, P. Dubois, M.-C. Lahaye, and M. Jangoux, eds.). Balkema, Rotterdam, the Netherlands.

Lawrence, J. M., and J. M. Lane. 1982. The utilization of nutrients by post-metamorphic echinoderms. Pp. 331–371. In *Echinoderm Nutrition* (M. Jangoux, and J. M. Lawrence, eds.). Balkema, Rotterdam, the Netherlands.

Lawrence, J. M., and A. Larrain. 1994. The cost of arm autotomy in the starfish *Stichaster striatus. Marine Ecology Progress Series* 109:311–313.

Lawrence, J. M., and P. Moran. 1992. Proximate composition and allocation of energy to body components in *Acanthaster planci* (Linnaeus) (Echinodermata: Asteroidea), *Zoological Science* 9:321–328.

Lawrence, J. M., and C. M. Pomory. 2008. Position of arm loss and rate of arm regeneration by *Luidia clathrata* (Echinodermata: Asteroidea). *Cahiers de Biologie Marine* 49:369–373.

Lawrence, J., R. Erwin, and R. Turner. 1974. Stomach contents of *Luidia clathrata* (Asteroidea). *Florida Scientist* Supplement 1:8.

Lawrence, J. M., T. S. Klinger, J. B. McClintock, S. A. Watts, C.-P. Chen, A. Marsh, and L. Smith. 1986. Allocation of resources to body components by regenerating *Luidia clathrata* (Say) (Echinodermata: Asteroidea). *Journal of Experimental Marine Biology and Ecology* 102:47–53.

Lawrence, J. M., M. Byrne, L. Harris, B. Keegan, S. Freeman, and B. C. Cowell. 1999. Sublethal predation in *Asterias amurensis* (eastern Bering Sea and southwestern Pacific), *Asterias rubens* (northeastern Atlantic Ocean) and *Asterias vulgaris* and *Asterias forbesi* (northwestern Atlantic Ocean) (Echinodermata: Asteroidea). *Vie et Milieu* 49:69–73.

Lawrence, J. M., J. K. Keesing, and T. R. Irvine. 2011. Population characteristics and biology of two species of *Archaster angulosus* (Echinodermata: Asteroidea) in different habitats off the central-western Australian coast. *Journal of the Marine Biological Association of the United Kingdom* 91:1577–1585.

Lawrence, J. M. 2012. Density and dispersion of *Luidia claathrata* (Echinodermata: Asteroidea) in Old Tampa Bay. *Marine and Freshwater Behaviour and Physiology* 45:101–109.

Lee, C. H., T. K. Ryu, and J. W. Choi. 2004. Effects of water temperature on embryonic development in the northern Pacific asteroid, *Asterias amurensis*, from the southern coast of Korea. *Invertebrate Reproduction and Development* 45:109–116.

Legault, C., and Himmelman, J. H. 1993. Relation between escape behaviour of benthic marine invertebrates and the risk of predation. *Journal of Experimental Marine Biology and Ecology* 170:55–74.

Leighton, B. J., J. D. C. Boom, C. Bouland, E. B. Hartwick, and M. J. Smith. 1991. Castration and mortality in *Pisaster ochraceus* parasitised by *Orchitophyra stellarum* (Ciliophora). *Diseases of Aquatic Organisms* 10:71–73.

Lemmens, J. W. T. J., P. W Arnold, and R. A. Birtles, 1995. Distribution patterns and selective feeding in two *Astropecten* species (Asteroidea: Echinodermata) from Cleveland Bay, Northern Queensland. *Marine and Freshwater Research* 46:447–455.

Lessios, H. A. 1988. Mass mortality of *Diadema antillarum* in the Caribbean: what have we learned? *Annual Review of Ecology and Systematics* 19:371–393.

Leverone, J. R., C. A. Luer, and J. M. Lawrence. 1991. The effect of cations on the specific activities of pyruvate kinase and glucose-6-phosphate dehydrogenase of *Luidia clathrata* (Say) (Echinodermata: Asteroidea). *Comparative Biochemistry and Physiology B* 99:259–264.

Levin, S. A. 1992. The problem of pattern and scale in ecology. *Ecology* 73:1943–1967.

Levitan, D. R. 1988. Asynchronous spawning and aggregative behavior in the sea urchin *Diadema antillarum* (Philippi). Pp. 181–186. In *Echinoderm Biology* (R. D. Burke, P. V. Mladenov, P. Lambert, and R. L. Parsley, eds.). Balkema, Rotterdam, the Netherlands.

Levitan, D. R. 1995. The ecology of fertilization in free-spawning invertebrates. Pp. 123–156. In *Ecology of Marine Invertebrate Larvae* (L. McEdward, ed.). CRC Press, Boca Raton, Florida.

Levitan, D. R. 2004. Density-dependent sexual selection in external fertilizers: variances in male and female fertilization success along the continuum from sperm limitation to sexual conflict in the sea urchin *Strongylocentrotus franciscanus. American Naturalist* 164:298–309.

Lieberkind, I. 1920. On a starfish (*Asterias grønlandica*) which hatches its young in its stomach. *Vidensk Meddr Dansk Naturh Foren Ser 8* 72:121–126.

Lieberkind, I. 1926. *Ctenodiscus australis* Lütken. A brood-protecting asteroid. *Videnskabelige Meddelelser fra Dansk Naturhistorisk, Forening I Khobenhavn* 82:183–196.

Lima-Verde, J. S., and H. R. Matthews. 1969. On the feeding habits of the sea star *Luidia senegalensis* (Lamarck) in the state of Ceará (Brazil). *Arquivo Ciências do Mar* 9:173–175.

Ling, S. D., and C. R. Johnson. 2013. Native spider crab causes high incidence of sub-lethal injury to the introduced seastar *Astrias amuurensis*. In *Echinoderms in a Changing World* (C. R. Johnson, ed.). Balkema, Rotterdam, the Netherlands.

Ling S. D., C. R. Johnson, C. N. Mundy, A. Morris, and D. J. Ross. 2012. Hotspots of exotic free-spawnng sex: man-made environment facilitates success in an invasive seastar. *Journal of Applied Ecology* 49:733–741.

Lo Bianco, S. 1909. Notizie biologiche riguardanti special-mente il periodo di maturità sessuale degli animali del golfo di Napoli. *Mittheilungen aus der zoologischen Station zu Neapel* 19:513–761.

Lockhart, S. J., and D. A. Ritz. 1998. Feeding rates of the introduced sea star, *Asterias amurensis* (Lütken), in Tasmania. Pp. 267–272. In *Echinoderms: San Francisco* (R. Mooi and M. Telford, eds.). Balkema, Rotterdam, the Netherlands.

Lockhart, S. J., and D. A. Ritz. 2001a. Preliminary observa-tions of the feeding periodicity and selectivity of the introduced seastar, *Asterias amurensis* (Lütken), in Tasmania, Australia. *Papers and Proceedings of the Royal Society of Tasmania* 135:25–33.

Lockhart, S. J., and D. A. Ritz. 2001b. Size selectivity and energy maximisation of the introduced seastar, *Asterias amurensis* (Lütken), in Tasmania, Australia. *Papers and Proceedings of the Royal Society of Tasmania* 135:35–40.

Löhner, L. R. 1913. Zur Entwicklungsgeschichte von *Echinaster sepositus* (Gray). *Zoologischer Anzeiger* 41:181–186.

Loosanoff, V. L. 1964. Variations in time and intensity of setting of the starfish, *Asterias forbesi*, in Long Island Sound during a twenty-five-year period. *Biological Bulletin* 126:423–439.

Lourey, M. J., D. A. J. Ryan, and I. R. Miller. 2000. Rates of decline and recovery of coral cover on reefs impacted by, recovering from and unaffected by crown-of-thorns starfish *Acanthaster planci*: a regional perspective of the Great Barrier Reef. *Marine Ecology Progress Series* 196:179–186.

Lubchenco, J., and B. A. Menge. 1978. Community develop-ment and persistence in a low rocky intertidal zone. *Ecological Monographs* 48:67–94.

Lucas, J. S. 1973. Reproductive and larval biology of *Acanthaster planci* (L.) in Great Barrier Reef waters. *Micronesica* 9:197–203.

Lucas, J. S. 1982. Quantitative studies of feeding and nutrition during larval development of the coral reef asteroid *Acanthaster planci* (L.). *Journal of Experimental Marine Biology and Ecology* 65:173–193.

Lucas, J. S., R. J. Hart, M. E. Howden, and R. Salathe. 1979. Saponins in eggs and larvae of *Acanthaster planci* (L.) (Asteroidea) as chemical defenses against planktivorous fish. *Journal of Experimental Marine Biology and Ecology* 40:155–165.

Ludwig, H. 1882. Entwicklungsgeschichte der *Asterina gibbosa* Forbes. *Zeitschrift für wissenschaftliche Zoologie* 37:1–98.

Ludwig, H. 1897. *Die Seesterne des Mittelmeeres*. R. Friedländer & Sohn, Berlin.

Lynn, D. H., and M. Strüder-Kypke. 2005. Scuticociliate endosymbionts of echinoids (Echinodermata): phylogene-tic relationships among species in the genera *Entodiscus, Plagiopyliella, Thyrophylas,* and *Entorhipidium* (Phylum Ciliophora). *Journal of Parasitology* 91:1190–1199.

MacBride, E. W. 1921. Echinoderm larvae and their bearing on evolution. *Nature* 108:529–530.

MacBride, E. W. 1923a. Echinoderm larvae and their bearing on classification. *Nature* 111:47.

MacBride, E. W. 1923b. (Response to T. Mortensen, 1923). *Nature* 111:323–324.

Mackenzie, C. L., Jr., and R. Pikanowski. 1999. A decline in starfish, *Asterias forbesi*, abundance and a concurrent increase in Northern Quahog, *Mercenaria mercenaria*, abundance and landings in the Northeastern United States. *Marine Fisheries Review* 61:66–71.

Mackie, A. M. 1970. Avoidance reactions of marine inverte-brates to either steroid glycosides of starfish or synthetic surface-active agents. *Journal of Experimental Marine Biology and Ecology* 5:63–69.

Mackie, A. M., and A. B. Turner. 1970. Partial characteriza-tion of a biologically active steroid glycoside isolated from the starfish *Marthasterias galcialis*. *Biochemical Journal* 117:543–550.

Mackie, A. M., H. T. Singh, and J. M. Owen. 1977. Studies on the distribution, biosynthesis and function of steroidal saponins in echinoderms. *Comparative Biochemistry and Physiology B* 56:9–14.

MacLeod, C., and F. Helidoniotis. 2005. *Ecological status of the Derwent and Huon estuaries*. Final report for Natural Heritage Trust–National Action Plan, Canberra project number 46928. Marine Research Laboratories, Hobart, Australia.

Madsen, F. J. 1956. Reports of the Lund University Chile Expedition 1948–1949. 24. Asteroidea: With a survey of the Asteroidea of the Chilean shelf. *Lunds Universitets Arsskrift* 52(2):1–53.

Madsen, F. J. 1966. The Recent sea-star *Platasterias* and the fossil Somasteroidea. *Nature* 209:1367.

Mah, C. L. and D. B. Blake. 2012. Global diversity and phylogeny of the Asteroidea (Echinodermata). *PloS ONE* 7(4): e35644. doi:10.1371/Journal.pone—35644.

Mah, C. L., and D. Foltz. 2011a. Molecular phylogeny of the Valvatacea (Asteroidea: Echinodermata). *Zoological Journal of the Linnean Society* 161:769–788.

Mah, C. L., and D. Foltz. 2011b. Molecular phylogeny of the Forcipulatacea (Asteroidea: Echinodermata): systematics and biogeography. *Zoological Journal of the Linnean Society* 162(3):646–660.

Mah, C., and H. Hansson. 2012a. *Luidia* Forbes. In World Asteroidea database, http://www.marinespecies.org/aphia.php?p-taxdetails&id-213245.

Mah, C., and H. Hansson. 2012b. Odontasteridae. In World Asteroidea database, http://www.marinespecies.org/asteroidea/aphia.php?p=taxdetails&id=123136.

Mahon, A. R., et al. 2002. Chemo-tactile predator avoidance responses of the common Antarctic limpet *Nacella concinna*. *Polar Biology* 25:469–473.

Mahon, A. R., et al. 2003. Tissue-specific palatability and chemical defenses against macropredators and pathogens in the common articulate brachiopod *Liothyrella uva* from the Antarctic Peninsula. *Journal of Experimental Marine Biology and Ecology*. 290:197–210.

Maldonado, M., and M. J. Uriz. 1998. Microrefuge exploitation by subtidal encrusting sponges: patterns of settlement and post-settlement survival. *Marine Ecology Progress Series* 174:141–150.

Maluf, L. Y. 1988. Composition and distribution of the central eastern Pacific echinoderms. *Natural History Museum of Los Angeles County Technical Reports* 2:1–242.

Manzur, T., M. Barahona, and S. A. Navarrete. 2010. Ontogenetic changes in habitat use and diet of the sea-star *Heliaster helianthus* on the coast of central Chile. *Journal of the Marine Biological Association of the United Kingdom* 90:537–546.

Mariante, F. L. F., G. B. Lemos, F. J. Eutrópio, R. R. L. Castro, and L. C. Gomes. 2010. Reproductive biology in the starfish *Echinaster* (*Othilia*) *guyanensis* (Echinodermata: Asteroidea) in southeastern Brazil. *Zoologia* 27:897–901.

Marion, K. R.. S. A. Watts, J. B. McClintock, G. Schinner, and T. S. Hopkins. 1998. Seasonal gonad maturation in *Astropecten articulatus* from the northern Gulf of Mexico. Pp. 278–284. In *Echinoderms: San Francisco* (R. Mooi, and M. Telford, eds.). Balkema, Rotterdam, the Netherlands.

Marko, P. B. 2002. Fossil calibration of molecular clocks and the divergence times of geminate species pairs separated by the Isthmus of Panama. *Molecular Biology and Evolution* 19:2005–2021.

Marrs, J., et al. 2000. Size-related aspects of arm damage, tissue mechanics and autotomy in the starfish *Asterias rubens*. *Marine Biology* 137:59–70.

Marsh, A., and J. M. Lawrence. 1985. The effects of cations on the activity of citrate synthase (EC 4.1.3.7) in *Luidia clathrata* (Say) (Echinodermata: Asteroidea). *Comparative Biochemistry and Physiology* 31B:767–770.

Marsh, A. G., and C. W. Walker. 1995. Effect of estradiol and progesterone on c-myc expression in the sea star testis and the seasonal regulation of spermatogenesis. *Molecular Reproduction and Development* 40(1):62–68.

Marsh, L. M. 1977. Coral reef asteroids of Palau, Caroline Islands. *Micronesica* 13:251–281.

Marshall, D., et al. 2012. The biogeography of marine invertebrate life histories. *Annual Review of Ecology, Evolution, and Systematics* 43:97–114.

Maruta, T., et al. 2005. Biologically active glycosides from asteroidea, 41. Isolation and structure determination of glucocerebrosides from the starfish *Linckia laevigata*. *Chemical and Pharmaceutical Bulletin* 53:1255–1258.

Matsubara, M., M. Komatsu, and H. Wada. 2004. Close relationship between *Asterina* and Solasteridae (Asteroidea) supported by both nuclear and mitochondrial gene molecular phylogenies. *Zoological Science* 21:785–793.

Matsubara, M., M. Komatsu, T. Araki, T., and S. Asakawa. 2005. The phylogenetic status of Paxillosida (Asteroidea) based on complete mitochondrial DNA sequences. *Molecular Phylogenetics and Evolution* 36:598–605.

Mauzey, K. 1966. Feeding behavior and reproductive cycles in *Pisaster ochraceus*. *Biological Bulletin*, 131:127–144.

Mauzey, K. P., C. Birkeland, and P. K. Dayton. 1968. Feeding behavior of asteroids and escape responses of their prey in the Puget Sound region. *Ecology* 49:603–619.

Mayo, P., and A. M. Mackie. 1976. Studies of avoidance reactions in several species of predatory British seastars (Echinodermata: Asteroidea). *Marine Biology* 38:41–49.

Mayr, E. 1954. Geographic speciation in tropical echinoids. *Evolution* 8:1–18.

Mazzone, F., et al. 2003. Arm autotomy and regeneration in the seastar *Coscinasterias muricata*. Pp. 209–213. In *Echinoderm Research 2001* (J. P. Féral and B. David, eds.). Swets & Zeitlinger, Lisse, the Netherlands.

McCallum, H. 1992. Completing the circle: stock-recruitment relationships and *Acanthaster*. *Australian Journal of Marine and Freshwater Research* 43:653–662.

McClary, D. J., and P. V. Mladenov. 1988. Brood and broadcast: a novel mode of reproduction in the sea star *Pteraster militaris*. Pp. 163–168. In *Echinoderm Biology* (R. D. Burke, P. V. Mladenov, P. Lambert, and R. L. Parsley, eds.). Balkema, Rotterdam, the Netherlands.

McClintock, J. B. 1987. Investigation of the relationship between invertebrate predation and biochemical composition, energy content, spicule armament and toxicity of benthic sponges at McMurdo Sound, Antarctica. *Marine Biology* 94:479–487.

McClintock, J. B. 1994. Trophic biology of antarctic shallow-water echinoderms. *Marine Ecology Progress Series* 111:191–202.

McClintock, J. B., and B. J. Baker. 1997a. A review of the chemical ecology of Antarctic marine invertebrates. *American Zoologist* 37:329–342.

McClintock, J. B., and B. J. Baker. 1997b. Palatability and chemical defense of eggs, embryos and larvae of shallow-water Antarctic marine invertebrates. *Marine Ecology Progress Series* 154:121–131.

McClintock, J. B., and J. M. Lawrence.1981. An optimization study on the feeding behavior of *Luidia clathrata* Say (Echinodermata: Asteroidea). *Marine Behaviour and Physiology* 78:263–275.

McClintock, J. B., and J. M. Lawrence. 1982. Photoresponse and associative learning in *Luidia clathrata* Say (Echinodermata: Asteroidea). *Marine Behaviour and Physiology* 9:13–21.

McClintock, J. B., and J. M. Lawrence. 1984. Ingestive conditioning in *Luidia clathrata* (Say) (Echinodermata: Asteroidea): effect of nutritional condition on selectivity, teloreception, and rates of ingestion. *Marine Behaviour and Physiology* 10:167–181.

McClintock, J. B., and J. M. Lawrence. 1985a. Characteristics of foraging in the soft-bottom benthic starfish *Luidia clathrata* (Echinodermata: Asteroidea): prey selectivity, switching behavior, functional responses and movement patterns. *Oecologia* 66:291–298.

McClintock, J. B., and J. M. Lawrence. 1985b. Size selectivity by *Luidia clathrata* (Say) (Echinodermata: Asteroidea): effect of nutritive condition and age. Pp. 533–539. In *Echinodermata* (B. F. Keegan and B. D. S. O'Connor, eds.). Balkema, Rotterdam, the Netherlands.

McClintock, J. B., and T. J. Robnett, Jr. 1986. Size selective predation by the asteroid *Pisaster ochraceus* on the bivalve *Mytilus californianus*: a cost benefit analysis. *Marine Ecology* 7:321–332.

McClintock, J. B., and J. D. Vernon. 1990. Chemical defense in the eggs and embryos of antarctic sea stars (Echinodermata). *Marine Biology* 105:491–495.

McClintock, J. B., T. S. Klinger, and J. M. Lawrence. 1983. Extraoral feeding in *Luidia clathrata* (Say) (Echinodermata: Asteroidea). *Bulletin of Marine Science* 33:171–172.

McClintock, J. B., T. S. Klinger, and J. M. Lawrence. 1984. Chemoreception in *Luidia clathrata* (Echinodermata: Asteroidea). *Marine Biology* 84:47–52.

McClintock, J. B., J. S. Pearse, and I. Bosch. 1988. Population structure and energetics of the shallow-water Antarctic sea star *Odontaster validus* in contrasting habitats. *Marine Biology* 99:235–246.

McClintock, J. B., T. Hopkins, S. A. Watts, and K. Marion. 1990. The biochemical and energetic composition of somatic body components of echinoderms from the northern Gulf of Mexico. *Comparative Biochemistry and Physiology* 95A:529–532.

McClintock, J. B., B. J. Baker, M. Hamman, M. Slattery, R. W. Kopitzke, and J. Heine. 1994. Tube-foot chemotactic responses of the spongivorous sea star *Perknaster fuscus* to organic extracts of antarctic sponges. *Journal of Chemical Ecology* 20:859–870.

McClintock, J. B., S. A. Watts, K. R. Marion, and T. S. Hopkins. 1995. Gonadal cycle, gametogenesis and energy allocation in two sympatric mid shelf sea stars with contrasting modes of reproduction. *Bulletin of Marine Science.* 57:442–452.

McClintock, J. B., B. J. Baker, C. D. Amsler, and T. L. Barlow. 2000a. Chemotactic tube-foot responses of the spongivorous sea star *Perknaster fuscus* to organic extracts of sponges from McMurdo Sound, Antarctica. *Antarctic Science* 12:41–46.

McClintock, J. B., B. J. Baker, and D. K. Steinberg. 2000b. The chemical ecology of invertebrate meroplankton and holoplankton. Pp. 195–226. In *Marine Chemical Ecology* (J. B. McClintock and B. J. Baker, eds.). CRC Press, Boca Raton, Florida.

McClintock, J. B., A. R. Mahon, K. J. Peters, C. D. Amsler, and B. J. Baker. 2003. Chemical defenses in embryos and juveniles of two common antarctic sea stars and an isopod. *Antarctic Science* 15:339–344.

McClintock, J. B., M. O. Amsler, C. D. Amsler, and B. J. Baker. 2006. The biochemical composition, energy content and chemical antifeedant defenses of the common Antarctic Peninsular sea stars *Granaster nutrix* and *Neosmilaster georgianus*. *Polar Biology* 29:615–623.

McClintock, J. B., R. A. Angus, C. P. Ho, C. D. Amsler, and B. J. Baker. 2008a. A laboratory study of behavioral interactions of the Antarctic keystone sea star *Odontaster validus* with three sympatric predatory sea stars. *Marine Biology* 154:1077–1084.

McClintock, J. B., R. A. Angus, C. P. Ho, C. D. Amsler, and B. J. Baker. 2008b. Intraspecific agonistic arm-fencing behavior in the Antarctic keystone sea star *Odontaster validus* influences prey acquisition. *Marine Ecology Progress Series* 371:297–300.

McClintock, J. B., C. D. Amsler, and B. J. Baker. 2010. Overview of the chemical ecology of benthic marine invertebrates along the western Antarctic Peninsula. *Integrative and Comparative Biology* 50:967–980.

McCurley, S. R., and W. M. Kier. 1995. The functional morphology of starfish tube feet: the role of a crossed-fiber helical array in movement. *Biological Bulletin* 188:197–209.

McEdward, L. R. 1997. Reproductive strategies of marine benthic invertebrates revisited: facultative feeding by planktotrophic larvae. *American Naturalist* 150:48–72.

McEdward, L. R., and D. A. Janies. 1993. Life cycle evolution in asteroids: what is a larva? *Biological Bulletin* 184:255–268.

McEdward, L. R., and D. A. Janies. 1997. Relationships among development, ecology, and morphology in the evolution of echinoderm larvae and life cycles. *Biological Journal of the Linnean Society* 60:1–40.

McEdward, L. R. and B. G. Miner. 2001. Larval and life cycle patterns in echinoderms. *Canadian Journal of Zoology* 79:1125–1170.

McEdward, L. R., and K. H. Morgan. 2001. Interspecific relationships between egg size and the level of parental investment per offspring in echinoderms. *Biological Bulletin* 200:33–50.

McEdward, L. R., W. B. Jaeckle, and M. Komatsu. 2002. Phylum Echinodermata: Asteroidea. Pp. 499–502. In *Atlas of Marine Invertebrate Larvae* (C. M. Young and M. E. Sewell, eds.). Academic Press, London.

McKnight, D. G. 2006. The marine fauna of New Zealand. Echinodermata: Asteroidea (sea-stars). 3. Orders Velatida, Spinulosida, Forcipulatida, Brisingida with addenda to Paxillosida, Valvatida. *NIWA Biodiversity Memoir* 120:1–187.

Mendonca, V. M., M. M. Al Jabri, I. Al Ajmi, M. Al Muharrami, M. Al Areimi, and H. A. Al Aghbari. 2010. Persistent and expanding population outbreaks of the corallivorous starfish *Acanthaster planci* in the Northwestern Indian Ocean: are they really a consequence of unsustainable starfish predator removal through overfishing in coral reefs, or a response to a changing environment? *Zoological Studies* 49:108–123.

Menge, B. A. 1972a. Foraging strategy of a starfish in relation to actual prey availability and environmental predictability. *Ecological Monographs* 42:25–50.

Menge, B. A. 1972b. Competition for food between two intertidal starfish species and its effect on body size and feeding. *Ecology* 53:635–644.

Menge, B. A. 1974. The effect of wave action on brooding and reproductive effort in the sea star, *Leptasterias hexactis*. *Ecology* 55:84–93.

Menge, B. A. 1975. Brood or broadcast? The adaptive significance of different reproductive strategies in the two intertidal sea stars *Leptasterias hexactis* and *Pisaster ochraceus*. *Marine Biology* 31:87–100.

Menge, B. A. 1976. Organization of the New England rocky intertidal community: role of predation, competition and environmental heterogeneity. *Ecological Monographs* 46:355–393.

Menge, B. A. 1979. Coexistence between the seastars *Asterias vulgaris* and *A. forbesi* in a heterogeneous environment: a non-equilibrium explanation. *Oecologia* 41:245–272.

Menge, B. A. 1982. Effect of feeding on the environment: Asteroidea. Pp. 521–551. In *Echinoderm Nutrition* (M. Jangoux and J. M. Lawrence, eds.). Balkema, Rotterdam, the Netherlands.

Menge, B. A. 1983. Components of predation intensity in the low zone of the New England rocky intertidal community. *Oecologia* 58:141–155.

Menge, B. A. 1991. Generalizing from experiments: is predation strong or weak in New England rocky intertidal? *Oecologia* 88:1–8.

Menge, B. A. 1992. Community regulation: under what conditions are bottom-up factors important on rocky shores? *Ecology* 73:755–765.

Menge, B. A., and G. Branch. 2001. Rocky intertidal communities. Pp. 221–252. In *Marine Community Ecology* (M. D. Bertness, S. D. Gaines, and M. E. Hay, eds.). Sinauer Associates, Sunderland, Massachusetts.

Menge, B. A., and T. L. Freidenburg. 2001. Keystone species. Pp. 613–631. In *Encyclopedia of Biodiversity* (S. A. Levin, ed.). Academic Press, New York.

Menge, B. A., and J. P. Sutherland. 1987. Community regulation: variation in disturbance, competition, and predation in relation to environmental stress and recruitment. *American Naturalist* 130:730–757.

Menge, B. A., E. I. Berlow, C. A. Blanchette, S. A. Navarrete, and S. B. Yamada. 1994. The keystone species concept: variation in interaction strength in a rocky intertidal habitat. *Ecological Monographs* 64:249–286.

Menge, B. A., B. A. Daley, and P. A. Wheeler. 1996. Control of interaction strength in marine benthic communities. Pp. 258–274. In *Food Webs: Integration of Patterns and Dynamics* (G. A. Polis and K. O. Winemiller, eds.). Chapman and Hall, New York.

Menge, B. A., B. Daley, P. A. Wheeler, E. Dahlhoff, E. Sanford, and P. T. Strub. 1997. Benthic-pelagic links in rocky intertidal communities: evidence for bottom-up effects on top-down control. *Proceedings of the National Academy of Sciences USA* 94:14530–14535.

Menge, B. A., B. A. Daley, J. Lubchenco, E. Sanford, E. Dahlhoff, P. M. Halpin, G. Hudson, and J. L. Burnaford. 1999. Top-down and bottom-up regulation of New Zealand rocky intertidal communities. *Ecological Monographs* 69:297–330.

Menge, B. A., E. Sanford, B. A. Daley, T. L. Freidenburg, G. Hudson, and J. Lubchenco. 2002. An inter-hemispheric comparison of bottom-up effects on community structure: insights revealed using the comparative-experimental approach. *Ecological Research* 17:1–16.

Menge B. A., J. Lubchenco, M. E. S. Bracken, F. Chan, M. M. Foley, T. L. Freidenburg, S. D. Gaines, G. Hudson, C. Krenz, H. Leslie, D. M. L. Menge, R. Russell, and M. S. Webster. 2003. Coastal oceanography sets the pace of rocky intertidal community dynamics. *Proceedings of the National Academy of Sciences USA* 100:12229–12234.

Menge, B. A., C. Blanchette, P. Raimondi, T. Freidenburg, S. Gaines, J. Lubchenco, D. Lohse, G. Hudson, M. Foley, and J. Pamplin. 2004. Species interaction strength: testing model predictions along an upwelling gradient. *Ecological Monographs* 74:663–684.

Menge, B. A., F. Chan, K. J. Nielsen, E. Di Lorenzo, and J. Lubchenco. 2009. Climatic variation alters supply-side ecology: impact of climate patterns on mussel recruitment. *Ecological Monographs* 79:379–395.

Menge, J. L., and B. A. Menge. 1974. Role of resource allocation, aggression, and spatial heterogeneity in coexistence of two competing intertidal starfish. *Ecological Monographs* 44:189–209.

Mercier, A., and J.-F. Hamel. 2008. Depth-related shift in life history strategies of a brooding and broadcasting deep-sea asteroid. *Marine Biology* 156:205–223.

Mercier, A., and J.-F. Hamel. 2009. Endogenous and exogenous control of gametogenesis and spawning in echinoderms. *Advances in Marine Biology* 55:1–302.

Mercier, A., and J.-F. Hamel. 2010. Synchronized breeding events in sympatric marine invertebrates: role of behavior and fine temporal windows in maintaining reproductive isolation. *Behavioral Ecology and Sociobiology* 64:1749–1765.

Mercier, A., Z. Sun, S. Baillon, and J.-F. Hamel. 2011. Lunar rhythms in the deep sea: evidence from the reproductive periodicity of several marine invertebrates. *Journal of Biological Rhythms* 26:82–86.

Metaxas, A. 2001. Behaviour in flow: perspectives on the distribution and dispersion of meroplanktonic larvae in the water column. *Canadian Journal of Fisheries and Aquatic Sciences* 58:86–98.

Metaxas, A., and M. Saunders. 2009. Quantifying the "bio-" components in biophysical models of larval transport in marine benthic invertebrates: advances and pitfalls. *Biological Bulletin* 216:257–272.

Metaxas, A., R. E. Scheibling, and C. M. Young. 2002. Estimating fertilization success in marine benthic invertebrates: a case study with the tropical sea star *Oreaster reticulatus*. *Marine Ecology Progress Series* 226:87–101.

Metaxas, A., R. E. Scheibling, M. C. Robinson, and C. M. Young. 2008. Larval development, settlement, and early post-settlement behavior of the tropical sea star *Oreaster reticulatus*. *Bulletin of Marine Science* 83:471–480.

Meyer, D. L., and D. B. Macurda, Jr. 1977. Adaptive radiation of comatulid crinoids. *Paleobiology* 3:74–82.

Miller, R. L. 1989. Evidence for the presence of sexual pheromones in free-spawning starfish. *Journal of Experimental Marine Biology and Ecology* 130:205–222.

Miller, S. R., and J. M. Lawrence. 1999. Gonad and pyloric caeca production in the nine-armed starfish *Luidia senegalensis* off the southwest Florida gulf coast during the annual reproductive cycle. *Bulletin of Marine Science.* 65:175–184.

Minale, L., R. Riccio, F. De Simone, A. Dini, and C. Pizza. 1979. Starfish saponins. II. 22;23-epoxysteroids; minor genins from the starfish *Echinaster sepositus*. *Tetrahedron Letters* 640–643.

Minchin, D. 1987. Sea-water temperature and spawning behavior in the sea star *Marthasterias glacialis*. *Marine Biology* 95:139–144.

Miner, B. G., L. A. McEdward, and L. R. McEdward. 2005. The relationship between egg size and the duration of the facultative feeding period in marine invertebrate larvae. *Journal of Experimental Marine Biology and Ecology* 321:135–144.

Mita, M. 1993. 1-methyladenine production by ovarian follicle cells responsible for spawning in the starfish *Asterina pectinifera*. *Invertebrate Reproduction & Development* 24:237–242.

Mita, M., M. Yoshikuni, K. Ohno, Y. Shibata, B. Paul-Prasanth, S. Pitchayawasin, M. Isobe, and Y. Nagahama. 2009. A relaxin-like peptide purified from radial nerves induces oocyte maturation and ovulation in the starfish, *Asterina pectinifera*. *Proceedings of the National Academy of Sciences* 106:9507–9512.

Mitic, M. 1992. Reproduction of the Mediterranean red-starfish. *Freshwater and Marine Aquarist* 15(10):40–41.

Monteiro, A. M. G., and E. V. Pardo. 1994. Dieta alimentar de *Astropecten marginatus* e *Luidia senegalensis* (Echinodermata-Asteroidea). *Revista Brasileira de Biologia* 54:49–54.

Moore, M. and D. T. Manahan. 2007. Variation among females in egg lipid content and developmental success of echinoderms from McMurdo Sound, Antarctica. *Polar Biology* 30:1245–1252.

Moran, P. J. 1986. The *Acanthaster* phenomenon. *Oceanography and Marine Biology Annual Review* 24: 379–480.

Moran, P. J., and G. De'ath. 1992. Estimates of the abundance of the crown-of-thorns starfish *Acanthaster planci* in outbreaking and non-outbreaking populations on reefs within the Great Barrier Reef. *Marine Biology* 113:509–515.

Moran, P. J., R. H. Bradbury, and R. E. Reichelt. 1985. Mesoscale studies of the crown-of-thorns/coral interaction: A case history from the Great Barrier Reef. Pp. 321–326. In *Proceedings of the 5th International Coral Reef Congress, Tahiti, 1985* (V. M. Harmelin and B. Salvat, eds.). Antenne Museum–EPHE, Moorea, French Polynesia.

Moran, P. J., G. De'ath, V. J. Baker, D. K. Bass, C. A. Christie, I. R. Miller, B. A. Miller-Smith, and A. A. Thompson. 1992. Pattern of outbreaks of crown-of-thorns starfish (*Acanthaster planci* (L.)) along the Great Barrier Reef since 1966. *Australian Journal of Marine Freshwater Resources* 43:555–67.

Morissette, S., and J. H. Himmelman. 2000a. Subtotal food thieves: interaction of four invertebrates kleptoparasites with the seastar *Leptasterias polaris*. *Animal Behavior* 60:531–543.

Morissette, S., and J. H. Himmelman. 2000b. Decision of the asteroid *Leptasterias polaris* to abandon its prey when confronted with the predatory asteroid *Asterias vulgaris*. *Journal of Experimental Marine Biology and Ecology* 252:151–157.

Morrice, M. G. 1995. The distribution and ecology of the introduced northern Pacific seastar, *Asterias amurensis* (Lütken), in Tasmania. Final Report. *Australian Nature Conservation Agency Feral Pests Program 35*.

Morris, R. H., D. P. Abbott, and E. C. Haderlie (Eds.). 1980. *Intertidal Invertebrates of California*. Stanford University Press, Stanford, California.

Morris, V. B., P. Selvakumaraswamy, R. Whan, and M. Byrne. 2009. Development of the five primary podia from the coeloms of a sea star larva: homology with the echinoid echinoderms and other deuterostomes. *Proceedings of the Royal Society B* 276:1277–1284.

Morris, V. B., P.Selvakumaraswamy, R. Whan, and M. Byrne. 2011. The coeloms in a late brachiolarian larva of the asterinid sea star *Parvulastra exigua*: deriving an asteroid coelomic model. *Acta Zoologica* 92: 266–275.

Mortensen, T. 1913. On the development of some British echinoderms. *Journal of the Marine Biological Association of the United Kingdom* 10:1–18.

Mortensen, T. 1921. *Studies of the development and larval forms of echinodems*. G. E. C. Gad, Copenhagen.

Mortensen, T. 1922. Echinoderm larvae and their bearing on classification. *Nature* 110:806–807.

Mortensen, T. 1923. Echinoderm larvae and their bearing on classification. *Nature* 111: 322–323.

Mortensen, T. 1933. Papers from Dr. Th. Mortensen's Pacific Expedition 1914–18. LXVI. The echinoderms of St. Helena (other than crinoids). *Videnskabelige Meddelelser fra Dansk naturhistorisk Forening i Kobenhavn* 93:401–473.

Mortensen, T. 1938. Contributions to the study of the development and larval forms of echinoderms, IV. *Mémoires de l'Académie Royale des Sciences et des Lettres de Danemark, Copenhague, Section des Sciences, 9me Série* 7(3):1–59.

Morton, J. E., and M. C. Miller. 1968. *The New Zealand Sea Shore*. Collins, London, United Kingdom.

Motokawa, T. 1982. Rapid changes in mechanical properties of echinoderm connective tissues caused by coelomic fluid, *Comparative Biochemistry and Physiology* 73C:223–229.

Motokawa, T., and S. A. Wainwright. 1991. Stiffness of starfish arm and involvement of catch connective tissue in the stiffness change. *Comparative Biochemistry and Physiology* 100A:393–397.

Mravec, B. 2006. Salsolinol, a derivate of dopamine, is a possible modulator of catecholaminergic transmission: a review of recent developments. *Physiological Research* 55:353–364.

Müller, I. 1996. The impact of the Zoological Station in Naples on developmental physiology. *International Journal of Developmental Biology* 40:103–111.

Murdoch, W. M. 1971. The developmental responses of predators to changes in prey density. *Ecology* 52:132–137.

Nachtsheim, H. 1914. Über die Entwicklung von *Echinaster sepositus* (Gray). *Zoologischer Anzeiger* 43:600–606.

Nadeau, M., and Cliche, G. 1998. Predation of juvenile sea scallops (*Placopecten magellanicus*), by crabs (*Cancer irroratus*) and starfish (*Asterias vulgaris, Leptasterias polaris* and *Crossaster papposus*). *Journal of Shellfish Research* 17:905–910.

Nadon, M.-O., and J. H. Himmelman. 2010. The structure of subtidal food webs in the northern Gulf of St. Lawrence, Canada, as revealed by the analysis of stable isotopes. *Aquatic Living Resources* 23:167–176.

Nance, J. M., and L. F. Braithwaite. 1972. The function of mucus secretions in the cushion star *Pteraster tesselatus* Ives. *Journal of Experimental Marine Biology and Ecology* 40:259–266.

Navarrete, S. A., and J. C. Castilla. 1990. Barnacles walls as mediators of intertidal mussel recruitment: effects of patch size on the utilization of space. *Marine Ecology Progress Series* 68:113–119.

Navarrete, S. A., and J. C. Castilla. 2003. Experimental determination of predation intensity in an intertidal predator guild: dominant versus subordinate prey. *Oikos* 100:251–262.

Navarrete, S. A., and T. Manzur. 2008. Individual- and population-level responses of a keystone predator to geographic variation in prey. *Ecology* 89:2005–2018.

Navarrete, S. A., and B. A. Menge. 1996. Keystone predation and interaction strength: interactive effects of predators on their main prey. *Ecological Monographs* 66:409–429.

Navarrete, S. A., E. A. Wieters, B. R. Broitman, and J. C. Castilla. 2005. Scales of benthic-pelagic coupling and the intensity of species interactions: from recruitment limitation to top down control. *Proceedings of the National Academy of Sciences USA* 102:18046–18051.

Navarrete, S. A., B. R. Broitman, and B. A. Menge. 2008. Interhemispheric comparison of recruitment to intertidal communities: pattern persistence and scales of variation. *Ecology* 89:1308–1322.

Navarrete, S. A., S. Gelcich, and J. C. Castilla. 2010. Long-term monitoring of coastal ecosystems at Las Cruces, Chile: defining baselines to build ecological literacy in a world of change. *Revista Chilena de Historia Natural* 83:143–157.

Naylor, E. 1999. Marine animal behaviour in relation to lunar phase. *Earth, Moon, and Planets* 85–86:291–302.

Netto, L. F., V. F. Hadel, and C. G. Tiago. 2005. Echinodermata from São Sebastião Channel (São Paulo, Brazil). *Revista Biología Tropicale* 53 (Suppl. 3):207–218.

Nichols, D. 1966. Functional morphology of the water vascular system. Pp. 219–244. In *Physiology of Echinodermata* (R. A. Boolootian, ed.). Interscience, New York.

Nichols, D. 1969. *Echinoderms*. Hutchison University Library, London.

Nichols, D. 1972. The water-vascular system in living and fossil echinoderms. *Palaeontology* 15:519–536.

Nichols, D., and M. F. Barker. 1984a. A comparative study of reproductive and nutritional periodicities in two populations of *Asterias rubens* (Echinodermata: Asteroidea) from the English channel. *Journal of the Marine Biological Association of the United Kingdom* 64:471–484.

Nichols, D., and M. F. Barker. 1984b. Growth of juvenile *Asterias rubens* L. (Echinodermata: Asteroidea) on an intertidal reef in southwestern Britain. *Journal of Experimental Marine Biology and Ecology* 78:157–165.

Nimitz, M. A. 1976. Histochemical changes in gonadal nutrient reserves correlated with nutrition in the sea stars *Pisaster ochraceus* and *Pateria miniata. Biological Bulletin* 151:357–369.

Nishihira, M. 1987. Natural and human interference with the coral reef and coastal environments in Okinawa. *Galaxea* 6:311–21.

Nobre, C. C., and L. S. Campos. 2004. Effect of salinity on the larval development of *Echinaster brasiliensis* Müller & Troschel, 1842 (Echinodermata: Asteroidea). Pp. 45–51. In *Echinoderms: München* (T. Heinzeller and J. H. Nebelsick, eds.). Taylor & Francis, London.

Nojima, S. 1982. Ecological studies of the sea star *Astropecten latespinosus* Meissner. II Growth rate and differences in growth pattern of immature and mature sea stars. *Publications of the Amakusa Marine Biological Laboratory* 6:65–84.

Nojima, S. 1983. Ecological studies on the sea star *Astropecten latespinosus* (M.) V. Pattern of spatial distribution and seasonal migration with special reference to spawning aggregation. *Publications of the Amakusa Marine Biological Laboratory* 7:1–16.

Nojima, S. 1988. Stomach contents and feeding habits of four sympatric sea stars, genus *Astropecten* (Echinodermata: Asteroidea), from northern Kyushu, Japan. *Publications of the Amakusa Marine Biological Laboratory* 9:67–76.

Nojima, S. 1989. Ecological studies on the sea star *Astropecten latespinosus* (M.). VI. Seasonal changes in stomach contents preference of food items size preference and two kinds of switching in feeding habits. *Publications of the Amakusa Marine Biological Laboratory* 10:17–40.

Nojima, S., F. El-Sayed Soliman, Y. Kondo, Y. Kuwano, K. Nasu, and C. Kitajima. 1986. Some notes on the outbreak of the sea star, *Asterias amurensis versicolor* Sladen, in the Ariake Sea, western Kyushu. *Publications of the Amakusa Marine Biological Laboratory* 8:89–112.

Norberg, J., and M. Tedengren. 1995. Attack behaviour and predatory success of *Asterias rubens* L. related to differences in size and morphology of the prey mussel *Mytilus edulis* L. *Journal of Experimental Marine Biology and Ecology* 186:207–220.

Noumura, T., and H. Kanatani. 1962. Induction of spawning by radial nerve extracts in some starfishes. *Journal of the Faculty of Science, Hokkaido University, Series 6* 9:397–402.

Novikova, G. P. 1978. Reproductive cycles of seastars *Asterias amurensis* and *Patiria pectinifera* from Peter the Great Bay (Sea of Japan). *Biologia Morya* 6:33–40.

O'Brien, F. X. 1976. Some adaptations of the seastar, *Leptasterias litoralis* (Stimpsom) to life in intertidal zone. *Thalassia jugoslavica* 12:237–243.

O'Donoghue, C. H. 1926. On the summer migration of certain starfish in Departure Bay, B. C. *Contributions to Canadian Biology and Fisheries* 1:455–472.

Oguro, C. 1989. Evolution of development and larval types in asteroids. *Zoological Science* 6:199–210.

Oguro, C., M. Komatsu, and Y. T. Kano. 1976. Development and metamorphosis of the sea-star, *Astropecten scoparius* Valenciennes. *Biological Bulletin* 151:560–573.

O'Hara T. 1995. Northern Pacific seastar. *The Victorian Naturalist* 112: 261.

O'Hara, T. D., and G. C. B. Poore. 2000. Patterns of distribution for southern Australian marine echinoderms and decapods. *Journal of Biogeography* 27:1321–1335.

Okaji, K. 1996. Feeding ecology in the early life stages of the crown-of-thorns starfish, *Acanthaster planci* (L.). Ph.D. Thesis, James Cook University, Townsville, Australia.

Okaji, K., T. Ayukai, and J. S. Lucas. 1997. Selective feeding by larvae of the crown-of-thorns starfish, *Acanthaster planci* (L.). *Coral Reefs* 16:47–50.

O'Loughlin, P. M., and J. M. Waters. 2004. A molecular and morphological revision of genera of Asterinidae (Echinodermata: Asteroidea). *Memoirs of the Museum of Victoria* 61:1–40.

Oliver, C. 1944. Sobre la distribución geográfica del *Heliaster helianthus*. *Revista Universitaria* 1:75–76.

Oliver, J. S., R. G. Kvitek, and P. N. Slattery. 1985. Walrus feeding disturbance: scavenging habits and recolonization of the Bering Sea benthos. *Journal of Experimental Marine Biology and Ecology* 91:233–246.

O'Loughlin, P. M., and T. D. O'Hara. 1990. A review of the genus *Smilasterias* (Echinodermata, Asteroidea), with descriptions of two new species from south-eastern Australia, one a gastric brooder, and a new species from Macquarie Island. *Memoirs of the Museum of Victoria* 50:307–323.

Olson, R. R. 1985. *In situ* culturing of larvae of the crown-of-thorns starfish *Acanthaster planci*. *Marine Ecology Progress Series* 25:207–210.

Olson, R. R. 1987. *In situ* culturing as a test of the larval starvation hypothesis for the crown-of-thorns starfish, *Acanthaster planci*. *Limnology and Oceanography* 32:895–904.

Olson, R. R., and M. H. Olson. 1989. Food limitation of plankotrophic marine invertebrate larvae: does it control recruitment success? *Annual Review of Ecology and Systematics* 20:225–247.

Olson, R. R., I. Bosch, and J. S. Pearse. 1987. The hypothesis of Antarctic larval starvation examined for the asteroid *Odontaster validus*. *Limnology and Oceanography* 32:686–690.

Omori, K. 1995. The adaptive significance of a lunar or semi-lunar reproductive cycle in marine animals. *Ecological Modelling* 82:41–49.

O'Neill, P. 1989. Structure and mechanics of starfish body wall. *Journal of Experimental Biology* 147:53–89.

O'Neill, P. L. 1990. Torsion in the asteroid ray. *Journal of Morphology* 203:141–149.

Ormond, R. F. G., A. C. Campbell, S. H. Head, R. J. Moore, P. R. Rainbow, and A. P. Saunders. 1973. Formation and breakdown of aggregations of the crown-of-thorns starfish, *Acanthster planci*. *Nature* 246:167–168.

Ormond, R., R. Bradbury, S. Bainbridge, K. Fabricius, J. Keesing, L. DeVantier, P. Medlay, and A. Steven. 1990. Test of a model of regulation of crown-of-thorns starfish by fish predators. Pp. 189–207. In *Lecture Notes in Biometrics* (R. Bradbury, ed.). Springer Verlag, Berlin.

Osborne, K., A. M. Dolman, S. C. Burgess, and K. A. Johns. 2011. Disturbance and the dynamics of coral cover on the Great Barrier Reef (1995–2009). *PLoS One* 6., e17516. doi:17510.11371/journal.pone.0017516.

Osborne, S. W. 1979. The seasonal distribution of *Luidia clathrata* (Say) in Charlotte Harbor with reference to various physical-chemical parameters. Thesis. Florida State University, Tallahassee, Florida.

Ottesen, P. O., and J. S. Lucas. 1982. Divide or broadcast: interrelation of asexual and sexual reproduction in a population of the fissiparous hermaphroditic seastar *Nepanthia belcheri* (Asteroidea: Asterinidae). *Marine Biology* 69:223–233.

Pabst, B., and H. Vicentini. 1978. Dislocation experiments in the migrating sea star *Astropecten jonstoni*. *Marine Biology* 48:271–278.

Pace, D. A., and D. T. Manahan. 2007. Cost of protein synthesis and energy allocation during development of Antarctic sea urchin embryos and larvae. *Biological Bulletin* 212:115–129.

Paik, S.-G., H.-S. Park, S. K. Yi, and S. G. Yun. 2005. Developmental duration and morphology of the sea star *Asterias amurensis* in Tongyeong, Korea. *Ocean Science Journal* 40:1–6.

Pain, S. L., P. A. Tyler, and J. D. Gage. 1982. The reproductive biology of *Hymenaster membranaceus* from the Rockall Trough, north-east Atlantic Ocean, with notes on *H. gennaeus*. *Marine Biology* 70:41–50.

Paine, R. T. 1966. Food web complexity and species diversity. *American Naturalist* 100:65–75.

Paine, R. T. 1969a. A note on trophic complexity and community stability. *American Naturalist* 103:91–93.

Paine, R. T. 1969b. The *Pisaster-Tegula* interaction: prey patches, predator food preference, and intertidal community structure. *Ecology* 50:949.

Paine, R. T. 1971. A short-term experimental investigation of resource partitioning in a New Zealand rocky intertidal habitat. *Ecology* 52:1096–1106.

Paine, R. T. 1974. Intertidal community structure: experimental studies on the relationship between a dominant competitor and its principal predator. *Oecologia* 15:93–120.

Paine, R. T. 1976a. Size limited predation: an observational and experimental approach with the *Mytilus-Pisaster* interaction. *Ecology* 57:858–873.

Paine, R. T. 1976b. Biological observations on a subtidal *Mytilus californianus* bed. *Veliger* 19:125–130.

Paine, R.T. 1980. Food webs: linkage, interaction strength and community infrastructure. *Journal of Animal Ecology* 49:667–685.

Paine, R. T. 1984. Ecological determinism in the competition for space. *Ecology* 65:1339–1348.

Paine, R. T. 1994. *Marine Rocky Shores and Community Ecology: An Experimentalist's Perspective*. Ecology Institute, Oldendorf/Luhe, Germany.

Paine, R. T. 1995. A conversation on refining the concept of keystone species. *Conservation Biology* 9:962–964.

Paine, R. T. 2010. Macroecology: does it ignore or can it encourage further ecological syntheses based on spatially local experimental manipulations? *American Naturalist* 176:385–393.

Paine R. T., J. C. Castilla and J. Cancino. 1985. Perturbation and recovery patterns of starfish-dominated intertidal assemblages in Chile, New Zealand, and Washington State. *American Naturalist* 125:679–691.

Paine, V. L. 1926. Adhesion of the tube feet in starfishes. *Journal of Experimental Zoology* 45:361–366.

Painter, S. D. 1992. Coordination of reproductive activity in *Aplysia*: peptide neurohormones, neurotransmitters, and pheromones encoded by the egg-laying hormone family of genes. *Biological Bulletin* 183:165–172.

Palagiano, E., F. Zollo, L. Minale, M. Iorizzi, P. Bryan, J. McClintock, and T. Hopkins. 1996. Isolation of 20 glycosides from the starfish *Henricia downeyae*, collected in the Gulf of Mexico. *Journal of Natural Products* 59:348–354.

Palma, A. T., E. Poulin, M. G. Silva, R. B. San Martín, C. A. Muñoz, and A. D. Díaz. 2007. Antarctic shallow subtidal echinoderms: is the ecological success of broadcasters related to ice disturbance? *Polar Biology* 30:343–350.

Papadopoulou, C., G. D. Kanias, and E. Moraitopoulou-Kassimati. 1976. Stable elements of radioecological importance in certain echinoderm species. *Marine Pollution Bulletin* 7:143–144.

Park, S.-W., T.-H. Kim, and J.-K. Oh. 1997. A study on the development of the extermination gear for starfish, *Asterias amurensis* and its efficiency. *Bulletin Korean Society of Fishery Technology* 33:166–172.

Parry, G. D., and B. F. Cohen. 2001. The distribution, abundance and population dynamics of the exotic seastar *Asterias amurensis* during the first three years of its invasion of Port Phillip Bay: status at May 1999. Marine and Freshwater Resources Institute Report No. 33. Marine and Freshwater Resources Institute, Queenscliff, Victoria, Australia.

Parry, G., B. F. Cohen B. F., M. A. McArthur, and N. J. Hickman. 2000. *Asterias amurensis* incursion in Port Phillip Bay: status at May 1999. Marine and Freshwater Resources Institute Report No. 19. Marine and Freshwater Resources Institute, Queenscliff, Victoria, Australia.

Parry G., S. Heislers, and G. Werner. 2004. Changes in distribution and abundance of *Asterias amurensis* in Port Phillip Bay 1999–2003. Department of Primary Industries, Queenscliff, Victoria, Australia.

Pastor de Ward, C., T. Rubilar, M. Díaz-de-Vivar, X. Gonzalez-Pisani, E. Zarate, M. Kroeck, and E. Morsan. 2007. Reproductive biology of *Cosmasterias lurida* (Echinodermata: Asteroidea) an anthropogenically influenced substratum from Golfo Nuevo, Northern Patagonia (Argentina). *Marine Biology* 151:205–217.

Pathirana, C., and R. J. Andersen. 1986. Imbricatine, an unusual benzyltetrahydroisoquinoline alkaloid isolated from the starfish *Dermasterias imbricata*. *Journal of the American Chemical Society* 108:8288–8289.

Pazoto, C. E. M., C. R. R. Ventura, and E. P. Silva. 2010. (Echinodermata: Asteroidea) in Rio De Janeiro, Brazil. Pp. 473–478. In *Echinoderms: Durham* (L. G. Harris, S. A. Böttger, C. W. Walker, and M.P. Lesser, eds.). CRC Press, Boca Raton, Florida.

Pearse, J. S. 1965. Reproductive periodicities in several contrasting populations of *Odontaster validus* Koehler, a common Antarctic asteroid. *Antarctic Research Series* 5:39–85.

Pearse, J. S. 1966. Antarctic asteroid *Odontaster validus*: constancy of reproductive periodicities. *Science* 152:1763–1764.

Pearse, J. S. 1967. Coelomic water volume control in the Antarctic star-fish *Odontaster validus*. *Nature* 216:1118–1119.

Pearse, J. S. 1969a. Antarctic sea star. *Australian Natural History* September :234–238.

Pearse, J. S. 1969b. Slow developing demersal embryos and larvae of the Antarctic seastar *Odontaster validus*. *Marine Biology* 3:110–116.

Pearse, J. S. 1994. Cold-water echinoderms break "Thorson's rule." Pp. 26–39. In *Reproduction, Larval Biology, and Recruitment in the Deep-sea Benthos* (K. J. Eckelbarger and C. M. Young, eds.). Columbia University Press, New York.

Pearse, J. S., and K. A. Beauchamp. 1986. Photoperiodic regulation of feeding and reproduction in a brooding sea star from central California USA. *Invertebrate Reproduction & Development* 9:289–298.

Pearse, J. S., and I. Bosch. 1986. Are the feeding larvae of the commonest Antarctic asteroid really demersal? *Bulletin of Marine Science* 39:477–484.

Pearse, J. S., and I. Bosch. 2002. Photoperiodic regulation of gametogenesis in the Antarctic sea star *Odontaster validus* Koehler: evidence for a circannual rhythm modulated by light. *Invertebrate Reproduction and Development* 41:73–81.

Pearse, J. S. and D. J. Eernisse. 1982. Photoperiodic regulation of gametogenisis and gonadal growth in the sea star *Pisaster ochraceus*. *Marine Biology* 67:121–125.

Pearse, J. S., and C. W. Walker 1986. Photoperiodic regulation of gametogenesis in a North Atlantic sea star *Asterias vulgaris*. *Invertebrate Reproduction and Development* 9:71–77.

Pearse, J. S., D. E. Eernisse, V. B. Pearse, and K. A. Beauchamp. 1986. Photoperiodic regulation of gametogenesis in sea stars, with evidence for an annual calendar independent of fixed daylength. *American Zoologist* 26:417–431.

Pearse, J. S., J. B. McClintock, and I. Bosch. 1991a. Reproduction of Antarctic benthic marine invertebrates: tempos, modes, and timing. *American Zoologist* 31:65–80.

Pearse, J. S., I. Bosch, V. B. Pearse, and L. V. Basch. 1991b. Differences in feeding on algae and bacteria by temperate and Antarctic sea star larvae. *Antarctic Journal United States* 26:170–172.

Pearse J. S., R. Mooi, S. J. Lockhart, and A. Brandt. 2009. Brooding and species diversity in the Southern Ocean: selection for brooders or speciation within brooding clades? Pp. 181–196. In *Smithsonian at the Poles: Contributions to International Polar Year Science* (I. Krupnik, M. A. Lang, and S. E. Miller, eds.). Smithsonian Institution, Washington, D.C.

Pearse, J. S., J. B. McClintock, K. E. Vicknair, and H. M. Feder. 2010. Long-term population changes in sea stars at three contrasting sites Pp. 633–640. In *Echinoderms: Durham* (L. G. Harris, S. A. Boettger, C. W. Walker, and M. P. Lesser, eds.). CRC Press, Boca Raton, Florida.

Pechenik, J.A. 2006. Larval experience and latent effects: metamorphosis is not a new beginning. *Integrative and Comparative Biology* 46:323–333.

Peck, L. S. and E. Prothero-Thomas. 2002. Temperature effects on the metabolism of larvae of the Antarctic starfish *Odontaster validus*, using a novel micro-respirometry method. *Marine Biology* 141:271–276.

Peck, L. S., K. E. Webb, and D. M. Bailey. 2004. Extreme sensitivity of biological function to temperature in Antarctic marine species. *Functional Ecology* 18:625–630.

Peck, L. S., K. E. Webb, A. Miller, M. S. Clark, and T. Hill. 2008. Temperature limits to activity, feeding and metabolism in the Antarctic starfish *Odontaster validus*. *Marine Ecology Progress Series* 358:181–189.

Peck, L. S., M. S. Clark, S. A. Morley, A. Massey, and H. Rossetti. 2009. Animal temperature limits and ecological relevance: effects of size, activity and rates of change. *Functional Ecology* 23:248–256.

Peckham, V. 1964. Year-round scuba diving in the Antarctic. *Polar Record* 12:143–146.

Penchaszadeh, P. E., and M. E. Lera. 1983. Alimentación de tres species tropicales de *Luidia* (Echinodermata, Asteroidea) en Golfo Triste, Venezuela. *Caribbean Journal of Science* 19:1–5.

Pennington, J. T. 1985. The ecology of fertilization of echinoid eggs: the consequences of sperm dilution, adult aggregation, and synchronous spawning. *Biological Bulletin* 169:417–430.

Perpeet, C., and M. Jangoux. 1973. Contribution à l'étude des pieds et des ampoules ambulacraires d'*Asterias rubens* (Echinodermata, Asteroides). *Forma et Functio* 6:191–209.

Perrier, E. 1884. Mémoire sur les étoiles de mer recueillis dans la Mer des Antilles et le Golfe de Mexique. *Nouvelles Archives du Muséum d'Histoire Naturelle de Paris* 6(2):127–276.

Perrier, E. 1894. Stéllerides. *Expédition Scientifique du Travailleur-Talisman* 3:1–143.

Perrin, C., S. R. Wing, and M. S. Roy. 2004. Effects of hydrographic barriers on population genetic structure of the sea star *Coscinasterias muricata* (Echinodermata, Asteroidea) in the New Zealand fiords. *Molecular Ecology* 13:2183–2195.

Peters, K. J., C. D. Amsler, J. B. McClintock, R. W. M. van Soest, and B. J. Baker. 2009. Palatability and chemical defenses of sponges from the western Antarctic Peninsula. *Marine Ecology Progress Series* 385:77–85.

Petes, L., M. Mouchka, R. Milston-Clements, T. Momoda, and B. Menge. 2008. Effects of environmental stress on intertidal mussels and their sea star predators. *Oecologia* 156:671–680.

Piatt, J. 1935. An important parasite of starfish. *Fishery Service Bulletin United States Department of Commerce* 247:3–4.

Picard, A., G. Peaucellier, F. le Bouffant, and M. Dorée. 1985. One millimeter large oocytes as a tool to study hormonal control of meiotic maturation in starfish: role of the nucleus in hormone-stimulated phosphorylation of cytoplasmic proteins. *Development Growth and Differentiation* 27:251–262.

Pickard, G. L. and B. R. Stranton. 1979. Pacific fjords: a review of their water characteristics. Pp. 1–52. In *Fjord Oceanography* (H. J. Freeland, D. M. Farmer and C. D. Levings, eds.). Plenum Press, New York.

Pincebourde, S., E. Sanford, and B. Helmuth. 2008. Body temperature during low tide alters the feeding performance of a top intertidal predator. *Limnology and Oceanography* 53:1562–1573.

Pincebourde, S., E. Sanford, and B. Helmuth. 2009. An intertidal sea star adjusts thermal inertia to avoid extreme body temperatures. *American Naturalist* 174:890–897.

Pincebourde, S., E. Sanford, J. Casas, and B. Helmuth. 2012. Temporal coincidence of environmental stress events modulates predation rates. *Ecology Letters*. doi: 10.1111/j.1461-0248.2012.01785.x.

Pires, A. M. S. 1995. The janirid isopods (Crustacea, Isopoda, Asellota) living on the sea star *Echinaster brasiliensis* Müller & Troschel at São Sebastião Channel, southeastern Brazilian coast, with description of a new species. *Revista Brasileira de Zoologia* 12:303–312.

Pomory, C. M., and M. T. Lares. 2000. Rate of regeneration of two arms in the field and its effect on body components in *Luidia clathrata* (Echinodermata: Asteroidea). *Journal of Experimental Marine Biology and Ecology* 254:211–220.

Poorbagher, H., M. D. Lamare, and M. F. Barker. 2010. The relative importance of parental nutrition and population versus larval diet on development and phenotypic plasticity of *Sclerasterias mollis* larvae. *Journal of the Marine Biological Association of the United Kingdom* 90:527–536.

Power, M. E., D. Tilman, J. A. Estes, B. A. Menge, W. J. Bond, L. S. Mills, G. Daily, J. C Castilla, J. Lubchenco, and R. T. Paine. 1996. Challenges in the quest for keystones. *BioScience* 46:609–620.

Prado, L., and J. C. Castilla. 2006. The bioengineer *Perumytilus purpuratus* (Mollusca: Bivalvia) in central Chile: biodiversity, habitat structural complexity and environmental heterogeneity. *Journal of the Marine Biological Association of the United Kingdom* 86:417–421.

Pratchett, M. S. 2005. Dynamics of an outbreak population of *Acanthaster planci* at Lizard Island, northern Great Barrier Reef (1995–1999). *Coral Reefs* 24:453–462.

Pratchett, M. S. 2010. Changes in coral assemblages during an outbreak of *Acanthaster planci* at Lizard Island, northern Great Barrier Reef (1995–1999). *Coral Reefs* 29:717–725.

Pratchett, M. S., T. J. Schenk, M. Baine, C. Syms, and A. H. Baird. 2009. Selective coral mortality associated with outbreaks of *Acanthaster planci* L. in Bootless Bay, Papua New Guinea. *Marine Environmental Research* 67:230–236.

Pratchett, M., M. Trapon, M. Berumen, and K. Chong-Seng. 2011. Recent disturbances augment community shifts in coral assemblages in Moorea, French Polynesia. *Coral Reefs* 30:183–193.

Preyer, W. 1886. Uber die Bewegungen der Seesterne. Eine vergleichend physiologisch-psychologische Untersuchung. *Mittheilungen aus der Zoologischen Station zu Neapel* 7:27–127.

Prowse, T. A. A., M. A. Sewell, and M. Byrne. 2008. Fuels for development: evolution of maternal provisioning in asterinid sea stars. *Marine Biology* 153:337–349.

Prowse, T. A. A., I., Falkner, M. A. Sewell, and M. Byrne. 2009. Long-term storage lipids and developmental evolution in echinoderms. *Evolutionary Ecology Research* 11:1069–1083.

Prusch, D. R. 1977. Solute secretion by the tube foot epithelium by the starfish *Asterias forbesi*. *Journal of Experimental Biology* 68:35–43.

Prusch, D. R., and F. Whoriskey. 1976. Maintenance of fluid volume in the starfish water vascular system. *Nature* 262:577–578.

Pryor, M., G. J. Parsons, and C. Couturier. 1999. Temporal patterns of larval and post-set distributions of the blue mussel (*Mytilus edulis/M. trossulus*) and the starfish (*Asterias vulgaris*) on Newfoundland mussel culture sites. *National Shellfisheries Association 1999 Annual Meeting* 311.

Puritz, J. B., et al. 2012. Extraordinarily rapid life-history divergence between *Cryptasteara* sea star species. *Proceedings of the Royal Society B* 279:3914–3922.

Raff, R. A., and M. Byrne. 2006. The active evolutionary lives of echinoderm larvae. *Heredity* 97: 244–252.

Raimondi, P. T., et al. 2007. Consistent frequency of color morphs in the sea star *Pisaster ochraceus* (Echinodermata: Asteriidae) across open coast habitats in the northeast Pacific. *Pacific Science* 61:201–210.

Ramsay, K., M. J. Kaiser, M.J., P. G. Moore, and R. N. Hughes. 1997. Consumption of fisheries discards by benthic scavengers: utilization of energy subsides in different marine habitats. *Journal of Animal Ecology* 66:884–896.

Ramsay, K., M. J. Kaiser, and R. N. Hughes. 1998. Responses of benthic scavengers to fishing disturbance by towed gears in different habitats. *Journal of Experimental Marine Biology and Ecology* 224:73–89.

Ramsay, K., M. J. Kaiser, and C. A. Richardson. 2001. Invest in arms: behavioural and energetic implications of multiple autotomy in starfish (*Asterias rubens*). *Behavioral Ecology and Sociobiology* 50:360–365.

Rasmussen, H.W. 1972. Lower Tertiary Crinoidea, Asteroidea and Ophiuridea from Northern Europe and Greenland. *Biologiske Skrifter* 19:1–83.

Raymond, J.-F., J. H. Himmelman, and H. Guderley. 2004. Sex differences in biochemical composition, energy content and allocation to reproductive effort in the brooding sea star *Leptasterias polaris*. *Marine Ecology Progress Series* 283:179–190.

Raymond, J.-F., J. H. Himmelman, and H. E. Guderley. 2007. Biochemical content, energy composition and reproductive effort in the broadcasting sea star *Asterias vulgaris* over the spawning period. *Journal of Experimental Marine Biology and Ecology* 341:32–44.

Regalado, E. L., D. Tasdemir, M. Kaiser, N. Cachet, P. Amade, and O. P. Thomas. 2010. Antiprotozoal steroidal saponins from the marine sponge *Pandaros acanthifolium*. *Journal of Natural Products* 73:1404–1410.

Reidman, M. L., and J. A. Estes. 1990. *The Sea Otter (*Enhydra lutris*): Behavior, Ecology, and Natural History*. U.S. Fish and Wildlife Service Biological Report 90, U.S. Fish and Wildlife Service, Washington, D.C.

Reitzel, A. M., J. Webb, and S. Arellano. 2004. Growth, development and condition of *Dendraster excentricus* (Eschscholtz) larvae reared on natural and laboratory diets. *Journal of Plankton Research* 26:901–908.

Renouf, L. P. W. 1937. On the life cycle of *Luidia ciliaris* (Phiilippi). Pp. 54–55. In *Festschrift Prof. Dr. Embrik Strand 3*, Riga, Latvia.

Rho, J. R., and Y. H. Kim. 2005. Isolation and structure determination of three new ceramides from the starfish *Distolasterias nipon*. *Bulletin of the Korean Chemical Society* 26:1457–1460.

Riccio, R., O. S. Greco, L. Minale, D. Laurent, and D. Duhet. 1986. Highly hydroxylated marine steroids from the starfish *Archaster typicus*. *Journal of the Chemical Society, Perkin Transactions 1* 665–670.

Rice, A. L., M. H. Thurston, and B. J. Bett. 1994. The IOSDL DEEPSEAS programme: introduction and photographic evidence for the presence and absence of a seasonal input of phytodetritus at contrasting abyssal sites in the northeastern Atlantic. *Deep Sea Research (Part I, Oceanographic Research Papers)* 41:1305–1320.

Rieger, R. M., and J. Lombardi. 1987. Ultrastructure of coelomic lining in echinoderm podia: significance for concepts in the evolution of muscle and peritoneal cells. *Zoomorphology* 107:191–208.

Rilov, G., and D. R. Schiel. 2006. Seascape-dependent subtidal-intertidal trophic linkages. *Ecology* 87:731–744.

Rivkin, R. B., I. Bosch, J. S. Pearse, and E. J. Lessard. 1986. Bacteriovery: a novel feeding mode for asteroid larvae. *Science* 233:1311–1314.

Robles, C. 1997. Changing recruitment in constant species assemblages: implications for predation theory in intertidal communities. *Ecology* 78:1400–1414.

Robles, C., and R. Desharnais. 2002. History and current development of a paradigm of predation in rocky intertidal communities. *Ecology* 83:1521–1536.

Robles, C., et al. 1989. Diel variation in intertidal foraging of *Cancer productus* L. in British Columbia. *Journal of Natural History*. 23:1041–1049.

Robles, C., R. Sherwood-Stephens, and M. Alvarado. 1995. Responses of a key intertidal predator to varying recruitment of its prey. *Ecology* 76:565–579.

Robles, C., M. Alvarado, and R. Desharnais. 2001. The shifting balance of littoral predator-prey interaction in regimes of hydrodynamic stress. *Oecologia* 128:142–152.

Robles, C. D., R. A. Desharnais, C. Garza, M. J. Donahue, and C. A. Martinez. 2009. Complex equilibria in the maintenance of boundaries: experiments with mussel beds. *Ecology* 90:985–995.

Robles, C. D., C. Garza, R. A. Desharnais, and M. J. Donahue. 2010. Landscape patterns in boundary intensity: a case study of mussel beds. *Landscape Ecology* 25:745–759.

Rochette. R., and J. H. Himmelman. 1996. Does vulnerability influence trade-offs made by whelks between predation risk and feeding opportunities? *Animal Behavior* 52:783–794.

Rochette, R., J.-F. Hamel, and J. H. Himmelman. 1994. Foraging strategy of the asteroid *Leptasterias polaris*: role of prey odors, current and feeding status. *Marine Ecology Progress Series* 106:93–100.

Rochette, R., S. Morissette, and J. H. Himmelman. 1995. A flexible response to a major predator provides the whelk *Buccinum undatum* L. with nutritional gains. *Journal of Experimental Marine Biology and Ecology* 185:167–180.

Rochette, R., J. N. McNeil, and J. H. Himmelman. 1996. Inter and intra-population variations in the response of the whelk, *Buccinum undatum* L, to the predatory asteroid *Leptasterias polaris* (Müller and Troschel). *Marine Ecology Progress Series* 142:193–201.

Rochette, R., D. J. Arsenault, B. Justome, and J. H. Himmelman. 1998. Chemically-mediated predator-recognition learning in a marine gastropod. *Ecoscience* 5:353–360.

Romanes, G. J. 1885. *Jelly-fish, star-fish and sea-urchins being a research on primitive nervous systems.* Kegan Paul, Trench & Co, London.

Romanes, G. J., and J. C. Ewart. 1881. Observations on the locomotor system of Echinodermata. *Philosophical Transactions of the Royal Society* 3:829–885.

Ross, D. J., C. R. Johnson, and C. L. Hewitt. 2002. Impact of introduced asteroids *Asterias amurensis* on survivorship of juvenile commercial bivalves *Fulvia tenuicostata*. *Marine Ecology Progress Series* 241:99–112.

Ross, D. J., C. R. Johnson, and C. L. Hewitt. 2003a. Assessing the ecological impacts of an introduced seasatar: the importance of multiple methods. *Biological Invasions* 5:3–21.

Ross, D. J., C. R. Johnson, and C. L. Hewitt. 2003b. Variability in the impact of an introduced predator (*Asterias amurensis*: Asteroidea) on soft-sediment assemblages. *Journal of Experimental Marine Biology and Ecology* 288:257–278.

Ross, D. J., C. R. Johnson, C. L. Hewitt, and G. M. Ruiz. 2004. Interaction and impacts of two introduced species on a soft-sediment marine assemblage in SE Tasmania. *Marine Biology* 144:747–756.

Rotjan, R. D., and S. M. Lewis. 2008. The impact of coral predators on tropical reefs. *Marine Ecology Progress Series* 367:73–91.

Rowe, F. W. E., and E. L. Albertson. 1987. A new species in the echinasterid genus *Echinaster* Müller and Troschel, 1840 (Echinodermata: Asteroidea) from southeastern Australia and Norfolk Island. *Proceedings of the Linnean Society of New South Wales* 109:195–202.

Rowe, F. W. E., and A. M. Clark. 1971. *Monograph of the Shallow-Water Indo-West Pacific Echinoderms.* Trustees of the British Museum (Natural History), London.

Rowe, F. W. E., and J. Gates. 1995. Echinodermata. In *Zoologicala Catalogue of Australia* (A Wells, ed.), vol. 33. Melborne: CSIRO Australia.

Rowe, F. W. E., and J. Gates. 1995. Echinodermata. *Zoological Catalogue of Australia* 33:1–510.

Rowe, F. W. E., and M. D. Richmond. 2004. A preliminary account of the shallow-water echinoderms of Rodrigues, Mauritius, western Indian Ocean. *Journal of Natural History* 38:3273–3314.

Rowe, F. W. E., and L. L. Vail. 1982. The distribution of Tasmanian echinoderms in relation to southern Australian biogeographic provinces. Pp. 219–225. In *Echinoderms: Proceedings of the International Conference, Tampa Bay* (J. M. Lawrence, ed.). Balkema, Rotterdam, the Netherlands.

Rubilar, T., C. T. Pastor de Ward, and M. E. Díaz de Vivar. 2005. Sexual and asexual reproduction of *Allostichaster capensis* (Echinodermata: Asteroidea) in Golfo Nuevo. *Marine Biology* 146:1083–1090.

Rumrill, S. S. 1988. Temporal and spatial variability in the intensity of recruitment of a sea star: frequent recruitment and demise. *American Zoologist* 28:123A.

Rumrill, S. S. 1989. Population size-structure, juvenile growth, and breeding periodicity of the sea star *Asterina miniata* in Barkley Sound, British Columbia. *Marine Ecology Progress Series* 56:37–47.

Rumrill, S. S. 1990. Natural mortality of marine larvae. *Ophelia* 32:163–198.

Run, J. Q., C. P. Chen, K. H. Chang, and F.-S. Chia. 1988. Mating behavior and reproductive cycle of *Archaster typicus* (Echinodermata: Asteroidea). *Marine Biology* 99:247–254.

Sadler, K. C., and J. V. Ruderman. 1998. Components of the signaling pathway linking the 1-methyladenine receptor to MPF activation and maturation in starfish oocytes. *Developmental Biology* 197:25–38.

Sagara, J. I., and T. Ino. 1954. The optimum temperature and specific gravity for bipinnaria and young Japanese starfish, *Asterias amurensis* Lutken. *Bulletin of the Japanese Society of Scientific Fisheries* 20:689–693.

Saha, A. K., M. Tamori, M. Inoue, Y. Nakajiman, and T. Motokawa. 2006. NGIW Yamide-induced contraction of tube feet and distribution of NGIW Yamide-like immuno-reactovity in nerves of the starfish *Asterina pectinifera*. *Zoological Science* 23:627–632.

Sameoto, J. A., and A. Metaxas. 2008a. Interactive effects of haloclines and food patches on the vertical distribution of 3 species of temperate invertebrate larvae. *Journal of Experimental Marine Biology and Ecology* 367:131–141.

Sameoto, J. A., and A. Metaxas. 2008b. Can salinity-induced mortality explain larval vertical distribution with respect to a halocline? *Biological Bulletin* 214:329–338.

Sánchez, P. 2000. The sequence of origin of the postmeta-morphic rays in *Heliaster* and *Labidiaster* (Echinodermata: Asteroidea). *Revista Chilena de Historia Natural* 73:573–578.

Sanford, E. 1999. Regulation of keystone predation by small changes in ocean temperature. *Science* 283:2095–2097.

Sanford, E. 2002. Water temperature, predation, and the neglected role of physiological rate effects in rocky intertidal communities. *Integrative and Comparative Biology* 42:881–891.

Sanford, E., and M. D. Bertness. 2009. Latitudinal gradients in species interactions. Pp. 357–391. In *Marine Macroecology* (J. D. Witman and K. Roy, eds.). University of Chicago Press, Chicago, Illinois.

Sanford, E. and B. A. Menge. 2007. Reproductive output and consistency of source populations in the sea star *Pisaster ochraceus*. *Marine Ecology Progress Series* 349:1–12.

San-Martín, A., J. Rovirosa, K. Gaete, A. Olea, and J. Ampuero. 2009. Mantle defensive response of marine pulmonate *Trimusculus peruvianus. Journal of Experimental Marine Biology and Ecology* 376:43–47.

Santos, R., D. Haesaerts, M. Jangoux, and P. Flammang. 2005a. The tube feet of sea urchins and sea stars contain functionally different mutable collagenous tissues. *Journal of Experimental Biology* 208:2277–2288.

Santos, R., D. Haesaerts, M. Jangoux, and P. Flammang. 2005b. Comparative histological and immunohisto-chemical study of sea star tube feet (Echinodermata, Asteroidea). *Journal of Morphology* 263:259–269.

Santos, R., S. Gorb, V. Jamar, and P. Flammang. 2005c. Adhesion of echinoderm tube feet to rough surfaces. *Journal of Experimental Biology* 208:2555–2567.

Santos, R., E. Hennebert, A. Varela Coelho, and P. Flammang. 2009. The echinoderm tube foot and its role in temporary underwater adhesion. Pp. 9–42. In *Functional Surfaces in Biology,* vol. 2 (S. Gorb, ed.). Springer, Heidelberg, Germany.

Saranchova, O. L., and L. P. Flyachinskaya. 2001. The influence of salinity on early ontogeny of the mussel *Mytilus edulis* and the starfish *Asterias rubens* from the White Sea. *Russian Journal of Marine Biology* 27:87–93.

Scandol, J. P. 1999. CotSim: an interactive *Acanthaster planci* metapopulation model for the central Great Barrier Reef. *Marine Models Online* 1:39–81.

Scandol, J. P., and M. K. James. 1992. Hydrodynamics and larval dispersal: a population model of *Acanthaster planci* on the Great Barrier Reef. *Australian Journal of Marine and Freshwater Research* 43:583–596.

Schapiro, J. 1914. Über die Regenerationserscheinungen verschiedener Seesternarten. *Archiv für Entwicklungs-mechanik der Organismen* 38:210–251.

Scheibling, R. E. 1979. *The ecology of* Oreaster reticulatus. Ph.D. Thesis, McGill University, Montreal, Québec.

Scheibling, R. E. 1980a. Abundance, spatial distribution and size structure of populations of *Oreaster reticulatus* (L.) (Echinodermata: Asteroidea) in seagrass beds. *Marine Biology* 57:95–105.

Scheibling, R. E. 1980b. Abundance, spatial distribution and size structure of populations of *Oreaster reticulatus* (L.) (Echinodermata: Asteroidea) on sand bottoms. *Marine Biology* 57:107–119.

Scheibling, R. E. 1980c. Dynamics and feeding activity of high density aggregations of *Oreaster reticulatus* (L.) (Echinodermata: Asteroidea) in a sand patch habitat. *Marine Ecology Progress Series* 2:321–327.

Scheibling, R. E. 1980d. The microphagous feeding behaviour of *Oreaster reticulatus* (L.) (Echinodermata: Asteroidea). *Marine Behavior and Physiology* 7:225–232.

Scheibling, R. E. 1980e. Carbohydrases of the pyloric caeca of *Oreaster reticulatus* (L.) (Echinodermata: Asteroidea). *Comparative Biochemistry and Physiology B* 67:297–300.

Scheibling, R. E. 1980f. Homing movements of *Oreaster reticulatus* (L.) (Echinodermata: Asteroidea) when experimentally translocated from a sand patch habitat. *Marine Behavior and Physiology* 7:213–223.

Scheibling, R. E. 1981a. Optimal foraging movements of *Oreaster reticulatus* (L.) (Echinodermata: Asteroidea). *Journal of Experimental Marine Biology and Ecology* 51:173–185.

Scheibling, R. E. 1981b. The annual reproductive cycle of *Oreaster reticulatus* (L.) (Echinodermata: Asteroidea) and interpopulation comparisons of reproductive capacity. *Journal of Experimental Marine Biology and Ecology* 54:39–54.

Scheibling, R. E. 1981c. Growth and respiration of juvenile *Oreaster reticulatus* (L.) (Echinodermata: Asteroidea) on plant and animal diets. *Comparative Biochemistry and Physiology A* 69:175–176.

Scheibling, R. E. 1982a. Feeding habits of *Oreaster reticulatus* (Echinodermata: Asteroidea). *Bulletin of Marine Science* 32: 504–510.

Scheibling, R. E. 1982b. Habitat utilization and bioturbation by *Oreaster reticulatus* (Asteroidea) and *Meoma ventricosa* (Echinoidea: Spantagoidea) in a subtidal sand patch habitat. *Bulletin of Marine Science* 32:624–629.

Scheibling, R. E. 1982c. Differences in body size and growth rate between morphs of *Echinaster* (Echinodermata: Asteroidea) from the eastern Gulf of Mexico. Pp. 291–298. In *Echinoderms: Proceedings of the International Conference, Tampa Bay* (J. M. Lawrence, ed.). Balkema, Rotterdam, the Netherlands.

Scheibling, R. E. 1985. Directional movement in a sea star (*Oreaster reticulatus*): Adaptive significance and ecological consequences. Pp. 244–256. In *Animal Migration: Mechanisms and Adaptive Significance* Contributions in Marine Science Supplement, vol. 27 (M. A. Rankin, ed.). Marine Science Institute, University of Texas, Port Aransas, Texas.

Scheibling, R. E. 2001. Of size and space: an evolutionary trade-off in fertilization strategy among oreasterid sea stars. P. 207. In *Echinoderms 2000* (M. Barker, ed.). Swets & Zeitlinger, Lisse, the Netherlands.

Scheibling, R. E., and J. M. Lawrence. 1982. Differences in reproductive strategies of morphs of the genus *Echinaster* (Echinodermata: Asteroidea) from the eastern Gulf of Mexico. *Marine Biology* 70:51–62.

Scheibling, R. E., and A. Metaxas. 2001. Population characteristics of the sea star *Oreaster reticulatus* in the Bahamas and across the Caribbean. Pp. 209–214. In *Echinoderms 2000* (M. Barker, ed.). Balkema, Lisse, the Netherlands.

Scheibling, R. E., and A. Metaxas. 2008. Abundance, spatial distribution, and size structure of the sea star *Protoreaster nodosus* in Palau, with notes on feeding and reproduction. *Bulletin of Marine Science* 82:221–235.

Scheibling, R. E., and A. Metaxas. 2010. Mangroves and fringing reefs as nursery habitats for the endangered Caribbean sea star *Oreaster reticulatus. Bulletin of Marine Science* 86:133–148.

Scheltema, R. S. 1986. Long-distance dispersal by planktonic larvae of shoal-water benthic invertebrates among central Pacific islands. *Bulletin of Marine Science* 39:241–256.

Schiel, D. R. 2011. Biogeographic patterns and long-term changes on New Zealand coastal reefs: non-trophic cascades from diffuse and local impacts. *Journal of Experimental Marine Biology and Ecology* 400:33–51.

Schmid, P. H. 1981. Ossicle morphology of four sea star species of the genus *Astropecten* (Asteroidea; Echinodermata) and its bearing on the investigation of a predator's diet. *P.S.Z.N. I: Marine Ecology* 2:199–206.

Schmid, P. H., and R. Schaerer. 1981. Predator-prey interaction between two competing sea star species of the genus *Astropecten*. *P.S.Z.N. I: Marine Ecology* 2:207–214.

Schmitz, F. J., M. B. Ksebati, S. P. Gunasekera, and S. Agarwal. 1988. Sarasinoside A: a saponin containing amino sugars isolated from a sponge. *Journal of Organic Chemistry* 53:5941–5947.

Schoener, T. W. 1983. Field experiments on interspecific competition. *American Naturalist* 122:240–285.

Schoenmakers, H. J. N. 1977. Steroid synthesis of *Asterias rubens*. *Proceedings of the International Union of Physiological Sciences* 13:673.

Schoenmakers, H. J. N. 1979a. *In vitro* biosynthesis of steroids from cholesterol by the ovaries and pyloric caeca of the starfish *Asterias vulgaris*. *Comparative Biochemistry and Physiology* 63B:179–187.

Schoenmakers, H. J. N. 1979b. *Steroids and reproduction of the female Asterias rubens L.* Ph.D. dissertation, University of Utrecht, Utrecht.

Schoenmakers, H. J. N. 1981. The possible role of steroids in vitellogenesis in the starfish *Asterias rubens*. Pp. 127–150. In *Advances in Invertebrate Reproduction* (W. H. Clark Jr. and T. S. Adams, eds.). Elsevier/North-Holland, New York.

Schoenmakers, H. J. N., and S. J. Dieleman. 1981. Progesterone and estrone levels in the ovaries, pyloric ceca, and perivisceral fluid during the annual reproductive cycle of starfish, *Asterias rubens*. *General and Comparative Endocrinology* 43:63–70.

Schoenmarkers, H. J. N., and P. A. Voogt. 1980. *In vitro* biosynthesis of steroids from progesterone by the ovaries and pyloric ceca of the starfish *Asterias rubens*. *General and Comparative Endocrinology* 41(3):408–416.

Schoenmakers, H. J. N., and P. A. Voogt. 1981. *In vitro* biosynthesis of steroids from androstenedione by the ovaries and pyloric caeca of the starfish *Asterias rubens*. *General and Comparative Endocrinology* 45(2):242–248.

Schoenmakers, H. J. N., D. Lambert, and P. A. Voogt. 1976. The steroid synthesizing capacity of the gonads of *Asterias rubens*. *General and Comparative Endocrinology* 29:256.

Schoenmakers, H. J. N., P. H. Colenbrander, and J. Peute. 1977. Ultrastructural evidence for the existence of steroid synthesizing cells in the ovary of the starfish *Asterias rubens* (Echinodermata). *Cell and Tissue Research* 182(2):275–279.

Schoenmakers, H. J. N., C. G. V. Bohemen, and S. J. Dieleman. 1981. Effects of oestradiol-17ß on the ovaries of the starfish *Asterias rubens*. *Development Growth & Differentiation* 23:125–135.

Schöndorf, F. 1909. Paläozoische Seesterne Deutschlands. I Teil. Die echten Asteriden der rhenischen Grauwacke. *Paläontographica* 56:37–112, plates 7–11.

Schram, J., J. B. McClintock, R. A. Angus, and Lawrence, J. M. 2011. Regenerative capacity, biochemical composition and production, and behavior of the sea star *Luidia clathrata* (Say) (Echinodermata: Asteroidea) under conditions of near-future ocean acidification. *Journal of Experimental Marine Biology and Ecology.* doi:10.1016/j.jembe.2011.06.024.

Schuetz, A. W. 2000. Extrafollicular mediation of oocyte maturation by radial nerve factor in starfish *Pisaster ochraceus*. *Zygote* 8:359–368.

Schuetz, A. W., and J. D. Biggers. 1968. Effect of calcium on the structure and functional response of the starfish ovary to radial nerve factor. *Journal of Experimental Zoology* 168:1–10.

Schwartz, F. J., and H. J. Porter. 1977. Fishes, macroinvertebrates, and their ecological interrelationships with a calico scallop bed off North Carolina. *Fishery Bulletin* 75:427–446.

Sebens, K. P. 1987. The ecology of indeterminate growth in animals. *Annual Review of Ecology and Systematics* 18:371–408.

Seldes, A. M., and E. G. Gros. 1985. Main sterols from the starfish *Comasterias lurida*. *Comparative Biochemistry and Physiology* 80B(2): 337–339.

Seto, Y., Y. Moriyama, D. Fujita, and M. Komatsu. 2000. Sexual and asexual reproduction in two populations of the fissiparous asteroid *Coscinasterias acutispina* in Toyama Bay, Japan. *Benthos Research* 55:85–93.

Seto, Y., D. Fujita, M. Komatsu, Y. Yamazaki, and A. Kijima. 2002. Population structure of a fissiparous seastar *Coscinasterias acutispina* in Toyama Bay. *Zoological Science* 19:1499.

Sewell, M. A. 2005. Examination of the meroplankton community in the south-western Ross Sea, Antarctica, using a collapsible plankton net. *Polar Biology* 28:119–131.

Sewell, M. A., and J. A. Jury. In press. Seasonal patterns in diversity and abundance of the high Antarctic meroplankton: plankton sampling using a Ross Sea desalination plant. *Limnology and Oceanography.*

Sewell, M. A., and D. R. Levitan. 1992. Fertilization success during a natural spawning of the dendrochirote sea cucumber *Cucumaria miniata*. *Bulletin of Marine Science* 51:161–166.

Sewell, M. A., and J. C. Watson. 1993. A "source" for asteroid larvae? Recruitment of *Pisaster ochraceus*, *Pycnopodia helianthoides* and *Dermasterias imbricata* in Nootka Sound, British Columbia. *Marine Biology* 117:387–398.

Seymour, R. M., and R. H. Bradbury. 1999. Lengthening reef recovery times from crown-of-thorns outbreaks signal systemic degradation of the Great Barrier Reef. *Marine Ecology Progress Series* 176:1–10.

Shackleton, J. D. 2005. Skeletal homologies, phylogeny and classification of the earliest asterozoan echinoderms. *Systematic Palaeontology* 3:29–114.

Sheikh, Y. M., M. Kaisin, and C. Djerassi. 1973. Steroids from starfish. *Steroids* 22(6): 835–850.

Shepherd, S. A. 1968. The shallow water echinoderm fauna of South Australia. Part I: the asteroids. *Records of the South Australia Museum* 15:729–756.

Sherman, C. D. H., A., Hunt, and D. J. Ayre. 2008. Is life history a barrier to dispersal? Contrasting patterns of genetic differentiation along an oceanographically complex coast. *Biological Journal of the Linnean Society* 95:106–111.

Sherman, K. 1991. The large marine ecosystem concept: research and management strategy for living marine resources. *Ecological Applications* 1:349–360.

Shibata, D., Y. Moriyama, M. Komatsu, and Y. Hirano. 2010. Development of the fissiparous and multiarmed seastar, *Coscinasterias acutispina* (Stimpson). Pp. 479–486. In *Echinoderms: Durham*. (L.G. Harris, S. A. Böttger, C. W. Walker, and M. P. Lesser, eds.). Taylor and Francis, London.

Shick, J. M. 1983. Respiratory gas exchange in echinoderms. Pp. 67–110. In *Echinoderm Studies*, vol. 1 (M. Jangoux and J. M. Lawrence, eds.). Balkema, Rotterdam, the Netherlands.

Shilling, F. M., and I. Bosch. 1994. "Pre-feeding" embryos of Antarctic and temperate echinoderms use dissolved organic material for growth and metabolic needs. *Marine Ecology Progress Series* 109:173–181.

Shilling, F. M., and D. T. Manahan. 1994. Energy metabolism and amino acid transport during early development of Antarctic and temperate echinoderms. *Biological Bulletin* 187:398–407.

Shirai, H., and C. W. Walker. 1988. Chemical control of asexual and sexual reproduction in echinoderms. Pp. 453–476. In *Invertebrate endocrinology*, vol. 2, *Endocrinology of selected invertebrate types* (H. Laufer, and R. G. H. Downer, eds.). Liss, New York.

Sible, J. C., A. G. Marsh, and C. W. Walker. 1991. Effect of extrinsic polyamines on post-spawning testes of the sea star *Asterias vulgaris* (Echinodermata): implications for the seasonal regulation of spermatogenesis. *Invertebrate Reproduction & Development* 19:257–264.

Siddall, S. E. 1979. Development of ossicles in juveniles of the sea star *Echinaster sentus*. *Bulletin of Marine Science* 29:278–282.

Siddon, C. E., and J. D. Witman. 2003. Influence of chronic, low-level hydrodynamic forces on subtidal community structure. *Marine Ecology Progress Series* 261:91–110.

Sköld, M., M. F. Barker, and P. V. Mladenov. 2002. Spatial variability in sexual and asexual reproduction of the fissiparous seastar *Coscinasterias muricata*: The role of food and fluctuating temperature. *Marine Ecology Progress Series* 233:143–155.

Sköld, M., S. R. Wing, and P. V. Mladenov. 2003. Genetic subdivision of a sea star with high dispersal capability in relation to physical barriers in a fjordic seacape. *Marine Ecology Progress Series* 250:163–174.

Sladen, W. P. 1889. The Asteroidea: report on the scientific results of the voyage of H. M. S. *Challenger* during the years 1873–76. *Zoology* 30:1–935.

Slattery, M. 2010. Bioactive compounds from echinoderms: ecological and evolutionary perspectives. Pp. 591–600. In *Echinoderms: Durham* (L. H. Harris, S. A. Boettger, C. W. Walker, and M. P. Lesser, eds.). CRC Press, Boca Raton, Florida.

Slattery, M., and I. Bosch. 1993. Mating behavior of a brooding Antarctic asteroid, *Neosmilaster georgianus*. *Invertebrate Reproduction & Development* 24:97–102.

Sloan, N. A. 1980a. Aspects of the feeding biology of asteroids. *Oceanography and Marine Biology, an Annual Review* 18:57–124.

Sloan, N. A. 1980b. The arm curling and terminal tube-foot responses of the asteroid *Crossaster papposus* (L.). *Journal of Natural History* 14:469–482.

Sloan, N. A. 1984. Interference and aggregation close encounters of the starfish kind. *Ophelia* 23:23–32.

Sloan, N. A., and Aldridge, T. H. 1981. Observations on an aggregation of the starfish *Asterias rubens* L. in Morecambe Bay, Lancashire, England. *Journal of Natural History* 15:407–418.

Sloan, N. A., and A. C. Campbell. 1982. Perception of food. Pp. 3–23. In *Echinoderm Nutrition* (M. Jangoux and J. M. Lawrence, eds.). Balkema, Rotterdam, the Netherlands.

Sloan, N. A., and S. M. C. Robinson. 1983. Winter feeding by asteroids on a subtidal sandbed in British Columbia. *Ophelia*. 22:125–140.

Smiley, S. 1990. A review of echinoderm oogenesis. *Journal of Electron Microscopy Technique* 16:93–114.

Smith, A. B. 1997. Echinoid larvae and phylogeny. *Annual Review of Ecology and Systematics* 28:219–241.

Smith, A. B., and T. H. Tranter. 1985. *Protremaster*, a new Lower Jurassic genus of asteroid from Antarctica. *Geological Magazine* 122:351–359.

Smith, A. G., and L. J. Goad. 1971a. Sterol biosynthesis in the starfish *Asterias rubens* and *Henricia sanguinolenta*. *Biochemistical Journal* 123(4): 671–673.

Smith, A. G., and L. J. Goad. 1971b. The metabolism of cholesterol by the echinoderms *Asterias rubens* and *Solaster papposus*. *FEBS Letters* 12(4): 233–235.

Smith, A. G., and L. J. Goad. 1975. Sterol biosynthesis in the echinoderm *Asterias rubens*. *Biochemistrical Journal* 146(1): 25–33.

Smith, A. G., R. Goodfellow, and L. J. Goad. 1972. The intermediacy of 3-oxo steroids in the conversion of cholest-5-en-3-ol into 5-cholestan-3-ol by the starfish *Asterias rubens* and porania pulvillus. *Biochemistrical Journal* 128(5):1371–1372.

Smith, A. G., I. Rubinstein, and L. J. Goad. 1973. The sterols of the echinoderm *Asterias rubens*. *Biochemistrical Journal* 135(3): 443–455.

Smith, J. E. 1937. The structure and function of the tube feet in certain echinoderms. *Journal of the Marine Biological Association of the United Kingdom* 22:345–357.

Smith, J. E. 1945. The role of the nervous system in some activities of starfishes. *Biological Reviews* 20:29–43.

Smith, J. E. 1947. The activities of the tube feet of *Asterias rubens* L. I. The mechanics of movement and of posture. *Quarterly Journal of Microscopic Science*. 88:1–14.

Smith, L. A., and J. M. Lawrence. 1987. Glycolytic activity in the pyloric caeca of *Luidia clathrata* (Say) (Echinodermata: Asteroidea). *Comparative Biochemistry and Physiology* 86B:693–696.

Smith, R. H. 1971. Reproductive biology of a brooding sea-star, *Leptasterias pusilla* (Fisher), in the Monterey Bay region. Ph.D. Thesis, Stanford University, Stanford, California.

Smithsonian Institution. 2011. Antarctic invertebrates. Available online at http://invertebrates.si.edu/antiz.

Soong, K., D. Chang, and S. M. Chao. 2005. Presence of spawn-inducing pheromones in two brittle stars (Echinodermata: Ophiuroidea). *Marine Ecology Progress Series* 292:195–201.

Soota, T. D., and D. R. K. Sastry. 1977. A note on two species of *Echinaster* Muller and Troschel (Echinodermata: Asteroidea) from Indian Ocean. *Newsletter of the Zoological Survey of India* 3:168–169.

Soto, R. 1996. Estructura gremial de un ensamble de depredadores de la zona intermareal rocosa en Chile central. *Investigaciones Marinas* 24:97–105.

Soto, R., J. C. Castilla, and F. Bozinovic. 2005. The impact of physiological demands on foraging decisions under predation risk: a test with whelk *Acanthina monodon*. *Ethology* 111:1044–1049.

Souza Santos, H., and W. Silva Sasso. 1968. Morphological and histochemical studies on the secretory glands of starfish tube feet. *Acta Anatomica* 69:41–51.

Souza Santos, H., and W. Silva Sasso. 1974. Ultrastructural and histochemical observations of the external epithelium of echinoderm tube feet. *Acta Anatomica* 88:22–33.

Spencer, W. K., and C. W. Wright. 1966. Asterozoans, Pp. U4–107. In *Treatise on Invertebrate Paleontology*, vol. 3, *Echinodermata* (R. C. Moore, ed.). Geological Society of America: University of Kansas Press, Lawrence, Kansas.

Stachowicz, J.J., H. Fried, R.B. Whitlatch, and R.W. Osman. 2002. Biodiversity, invasion resistance and marine ecosystem function: reconciling pattern and process. *Ecology* 83:2575–2590.

Staehli, A., R. Schaerer, K. Hoelzle, and G. Ribi. 2008. Temperature induced disease in the starfish *Astropectin johnstoni*. *JMBA2 Biodiversity Records*. Available at www.mba.ac.uk/jmba/pdf/5846.pdf.

Stanwell-Smith, D., and A. Clarke. 1998. Seasonality of reproduction in the cushion star *Odontaster validus* at Signy Island, Antarctic. *Marine Biology* 131:479–487.

Stanwell-Smith, D., and L. S. Peck. 1998. Temperature and embryonic development in relation to spawning and field occurrence of larvae of three Antarctic echinoderms. *Biological Bulletin* 194:44–52.

Stanwell-Smith, D., L. S. Peck, A. Clarke, A. W. A. Murray, and C. D. Todd. 1999. The distribution, abundance and seasonality of pelagic marine invertebrate larvae in the maritime Antarctic. *Philosophical Transactions of the Royal Society of London B* 354:471–484.

Stearns, S. C. 1992. *The Evolution of Life Histories*. Oxford University Press, Oxford.

Stebbins, T. D. 1988a. Observations and experiments on the natural history and behavior of the commensal isopod *Colidotea rostrata* (Benedict, 1898) (Isopoda: Idioteidae). *Journal of Crustacean Biology* 8:539–547.

Stebbins, T. D. 1988b. The role of sea urchins in mediating fish predation on a commensal isopod (Crustacea: Isopoda). *Journal of Experimental Biology and Ecology* 124:97–113.

Stebbins, T. D. 1989. Population dynamics and reproductive biology of the commensal isopod *Colidotea rostrata* (Crustacea: Isopoda: Idioteidae). *Marine Biology* 101:329–337.

Stevens, M. 1970. Procedures for induction of spawning and meiotic maturation of starfish oocytes by treatment with 1-methyladenine. *Experimental Cell Research* 59:482–484.

Stickle, W. B. 1988. Patterns of nitrogen excretion in the phylum Echinodermata. *Comparative Biochemistry and Physiology* 91:317–321.

Stickle, W. B., and R. Ahokas. 1974. The effects of tidal fluctuation of salinity on the perivisceral fluid composition of several echinoderms. *Comparative Physiology and Biochemistry A* 47:469–476.

Stickle, W. B., and W. J. Diehl. 1987. Effects of salinity on echinoderms. Pp. 235–285. In *Echinoderm Studies*, vol. 2 (M. Jangoux and J. M. Lawrence, eds.). Balkema, Rotterdam, the Netherlands.

Stickle, W. B., and E. N. Kozloff. 2008. Association and distribution of the ciliate *Orchitophrya stellarum* with asteriid sea stars on the west coast of North America. *Diseases of Aquatic Organisms* 80:37–43.

Stickle, W. B., W. Foltz, M. Katoh, and H. L. Nguyen. 1992. Genetic structure and mode of reproduction in five species of sea stars (Echinodermata: Asteroidea) from the Alaskan coast. *Canadian Journal of Zoology* 70:1723–1728.

Stickle W. B., E. H. Weidner, and E. N. Kozloff. 2001a. Parasitism of *Leptasterias spp.* by the ciliate *Orchitophrya stellarum* Cépède. *Invertebrate Biology* 120:88–95.

Stickle W. B., C. Rathbone, and S. Story 2001b. Parasitism of sea stars from Puget Sound, Washington by *Orchitophrya stellarum*. Pp. 221–226. In *Echinoderms 2000* (M. Barker, ed.). Balkema, Lisse, the Netherlands.

Stickle, W. B., E. N. Kozloff, and M. C. Henk. 2007. The ciliate *Orchitophrya stellarum* viewed as a facultative parasite of asteriid sea stars. *Cahiers de Biologie Marine* 48:9–16.

Stonik, V. A., N. V. Ivanchina, and A. A. Kicha. 2008. New polar steroids from starfish. *Natural Product Communications* 3:1587–1610.

Strathmann, M. 1987. *Reproduction and Development of Marine Invertebrates of the Northern Pacific Coast*. University of Washington Press, Seattle, Washington.

Strathmann, R. R. 1971. The feeding behavior of planktotrophic echinoderm larvae: mechanisms, regulation, and rates of suspension feeding. *Journal of Experimental Marine Biology and Ecology* 6:109–160.

Strathmann, R. 1978. Length of pelagic period in echinoderms with feeding larvae from the northeast Pacific. *Biological Bulletin* 34:23–27.

Strathmann, R. R. 1985. Feeding and nonfeeding larval development and life history evolution in marine invertebrates. *Annual Review of Ecology and Systematics* 16:339–361.

Strathmann, R. R. 1989. Existence and functions of a gel filled primary body cavity in development of echinoderms and hemichordates. *Biological Bulletin* 176:25–31.

Strathmann, R. R., and M. F. Strathmann. 1982. The relationship between adult size and brooding in marine invertebrates. *American Naturalist* 119:91–101.

Strathmann, R. R., and M. F. Strathmann. 1995. Oxygen supply and limits on aggregation of embryos. *Journal of the Marine Biological Association of the United Kingdom* 75:413–428.

Sugiyama, S., M. Honda, and T. Komori. 1990. Stereochemistry of the four diastereomers of phyosphingosine. *Liebigs Annalen der Chemie* 1069–1078.

Sukarno, R., and M. Jangoux. 1977. Révision du genre *Archastaer* Müller et Troschel (Echinodermata, Asteroidea: Archasteridae). *Revue de Zoologie Africaine* 91:817–844.

Sumida, P. Y. G., P. A. Tyler, and D. S. M. Billett. 2001. Early juvenile development of deep-sea asteroids of the NE Atlantic Ocean, with notes on juvenile bathymetric distributions. *Acta Zoologica* 82:11–40.

Sunday, J., L. Raeburn, and M. W. Hart. 2008. Emerging infectious disease in sea stars: castrating ciliate parasites in *Patiria miniata. Invertebrate Biology* 128:276–282.

Sweatman, H. P. A. 1995. A field study of fish predation on juvenile crown-of-thorns starfish. *Coral Reefs* 14:47–53.

Sweatman, H. 2008. No-take reserves protect coral reefs from predatory starfish. *Current Biology* 18:R598–R599.

Sweatman, H., A. Cheal, G. Coleman, S. Delean, B. Fitzpatrick, I. Miller, R. Ninio, K. Osborne, C. Page, and A. Thompson. 2001. *Long-term Monitoring of the Great Barrier Reef, Status Report Number 5.* Australian Institute of Marine Science, Townsville, Australia.

Sweatman, H., A. Cheal, N. Coleman, M. Emslie, K. Johns, M. Jonker, I. Miller, and K. Osborne. 2008. *Long-term Monitoring of the Great Barrier Reef, Status Report Number 8.* Australian Institute of Marine Science, Townsville, Australia.

Szulgit, G. K., and R. E. Shadwick. 2000. Dynamic mechanical characterization of a mutable collagenous tissue: response of sea cucumber dermis to cell lysis and dermal extracts. *Journal of Experimental Biology* 203:1539–1550.

Tadenuma, H., K. Takahashi, K. Chiba, M. Hoshi, and T. Katada. 1992. Properties of 1-methyladenine receptors in starfish oocytes membranes: involvement of pertussis toxin-sensitive GTP-binding protein in the receptor-mediated signal transduction. *Biochemical and Biophysical Research Communications* 186:114–121.

Takada, K., Y. Nakao, S. Matsunaga, R. W. M. van Soest, and N. Fusetani. 2002. Nobiloside, a new neuraminidase inhibitory triterpenoidal saponin from the marine sponge *Erylus nobilis. Journal of Natural Products* 65:411–413.

Takahashi, H. and H. Kanatani. 1981. Effect of 17β-estradiol on growth of oocytes in cultured ovarian fragments of the starfish *Asterina pectinifera. Development Growth and Differentiation* 23:565–569.

Takahashi, N. 1982a. Effect of injection of steroids on the starfish testes *Asterina pectinifera. Bulletin of the Japanese Society of Scientific Fisheries* 48:513–515.

Takahashi, N. 1982b. The relation between injection of steroids and ovarian protein amounts in the starfish *Asterina pectinifera. Bulletin of the Japanese Society for the Science of Fish* 48:509–511.

Takashi, I., J. Sagarra, S. Hamada, and M. Tamakawa. 1955. On the spawning season of the starfish *Asterias amurensis*, in Tokyo Bay. *Bulletin Japanese Society Science Fisheries* 21:32–36.

Tang, H., G. Cheng, J. Wu, X. Chen, S. Zhang, A. Wen, and H. Lin. 2009. Cytotoxic asterosaponins capable of promoting polymerization of tubulin from the starfish *Culcita novaeguineae. Journal of Natural Products* 72:284–289.

Tararam, A. S., Y. Wakabara, and M. B. Equi. 1993. Hábitos alimentares de onze espécies de megafauna bêntica da plataform continental de Ubatub, SP. *Publicaçao Especial Instituto Oceanográfico, São Paulo, Brasil* 10:159–167.

Temara, A., I. Gulec, and D. A. Holdway. 1999. Oil-induced disruption of foraging behaviour of the asteroid keystone predator, *Coscinasterias muricata* (Echinodermata). *Marine Biology* 133:501–507.

Teshima, S. I., and A. Kanazawa. 1975. Biosynthesis of sterols in a starfish, *Laiaster leachii. Comparative Biochemistry and Physiology* 52B(3):437–441.

Thayer, C.W 1975. Morphologic adaptations of benthic invertebrates to soft substrata. *Journal of Marine Research* 33:177–189.

Thomas, L. A., and C. O. Hermans. 1985. Adhesive interactions between the tube feet of a starfish, *Leptasterias hexactis*, and substrata. *Biological Bulletin* 169:675–688.

Thomas, M. L. H., and J. H. Himmelman. 1988. Influence of predation on shell morphology of *Buccinum undatum* L. on Atlantic coast of Canada. *Journal of Experimental Marine Biology and Ecology* 115:221–236.

Thompson, G. B. 1982. Some echinoderms collected from coral habitats in Tolo Harbour, Hong Kong. Pp. 651–654. In *The Marine Flora and Fauna of Hong Kong and Southern China* (B. S. Morton and C. K. Tseng, eds.). Hong Kong University Press, Hong Kong.

Thompson, M., D. Drolet, and J. H. Himmelman. 2005. Localization of infaunal prey by the sea star *Leptasterias polaris. Marine Biology* 146:887–894.

Thomson, R. E. 1981. *Oceanography of the British Columbia Coast,* Canadian Special Publication of Fisheries and Aquatic Sciences edition. Department of Fisheries and Oceans, Ottawa.

Thrush, S. F. 1986. Community structure on the floor of a sea-lough: are large epibenthic predators important? *Journal of Experimental Marine Biology and Ecology.* 104:171–183.

Thrush, S. F., and V. J. Cummings. 2011. Massive icebergs, alteration in primary food resources and change in benthic communities at Cape Evans, Antarctica. *Marine Ecology.* doi:10.1111/j.1439-0485.2011.00462.x.

Thrush, S. F., and C. R. Townsend. 1986. The sublittoral macrobenthic community composition of Lough Hyne, Ireland. *Estuarine, Coastal and Shelf Science* 23:551–574.

Timmers M.A., C. E. Bird, D. J. Skillings, P. E. Smouse, and R. J. Toonen. 2012. There's no place like home: crown-of-thorns outbreaks in the Central Pacific are regionally derived and independent events. *PLoS ONE* 7(2):e31159. doi:10.1371/journal.pone.0031159

Tokeshi, M. 1989. Development of a foraging model for a field population of the South American sun-star *Heliaster helianthus. Journal of Animal Ecology* 58:189–206.

Tokeshi, M., and L. Romero. 1995. Quantitative analysis of foraging behaviour in a field population of the South American sun-star *Heliaster helianthus. Marine Biology* 122:297–303.

Tokeshi, M., C. Estrella, and C. Paredes. 1989. Feeding ecology of a size-structured predator population, the South American sun-star *Heliaster helianthus*. *Marine Biology* 100:495–505.

Tominaga, H., M. Komatsu, and C. Oguro. 1994. Aggregation for spawning in the breeding season of the sea-star, *Asterina minor* Hayashi. Pp. 369–373. In *Echinoderms through Time* (B. David, A. Guille, J.-P. Féral, and M. Roux, eds.). Balkema, Rotterdam, the Netherlands.

Tortonese, E. 1950. Differenziazione geografica ed ecologica negli Asteroidi. *Bollettino di Zoologia, Supplemento* 17:339–354.

Tortonese, E. 1965. *Fauna d'Italia*, vol. 6, *Echinodermata*. Edizioni Calderini, Bologna.

Tortonese, E., and M. E. Downey. 1977. On the genera *Echinaster* Mueller and Troschel and *Othilia* Gray, and the validity of *Verrillaster* Downey (Echinodermata: Asteroidea). *Proceedings of the Biological Society of Washington* 90:829–830.

Trotter, J. A., and T. J. Koob. 1995. Evidence that calcium-dependent cellular processes are involved in the stiffening response of holothurian dermis and that dermal cells contain an organic stiffening factor. *Journal of Experimental Biology* 198:1951–1961.

Trotter, J. A., J. Tipper, G. Lyons-Levy, K. Chino, A. H. Heuer, Z. Liu, M. Mrksich, C. Hodneland, W. S. Dillmore, T. J. Koob, M. M. Koob-Emunds, K. Kadler, and D. Holmes. 2000. Towards a fibrous composite with dynamically controlled stiffness: lessons from echinoderms. *Biochemical Society Transactions* 28:357–362.

Trussell G. C., P. J. Ewanchuk, and M. D. Bertness. 2003. Trait-mediated effects in rochy intertidal food chains: predator risk cues alter prey feeding rates. *Ecology* 84:629–640.

Turner, E. 1992. A northern Pacific seastar, *Asterias amurensis*, in Tasmania. *Bulletin Australian Marine Science Association* 120:18–19.

Turner, E., R. Klevit, L. J. Hager, and B. M. Shapiro. 1987. Ovothiols, a family of redox-active mercaptohistidine compounds from marine invertebrate eggs. *Biochemistry* 26:4028–4036.

Turner, R. L., and J. M. Lawrence. 1979. Volume and composition of echinoderm eggs: implications for the use of egg size in life-history models. Pp. 25–40. In *Reproductive Ecology of Marine Invertebrates* (S. E. Stancyk, ed.). University of South Carolina Press, Columbia, South Carolina.

Tuttle, R. D., and R. Lindahl. 1980. Genetic variability in 3 co-occurring forms of the starfish genus *Othilia* (= *Echinaster*). *Experientia* 36:923–925.

Tyler, P. A., D. S. M. Billett, and J. D. Gage. 1990. Seasonal reproduction in the sea star *Dytaster grandis* from 4000 m in the northeast Atlantic Ocean. *Journal of the Marine Biological Association of the United Kingdom* 70:173–180.

Tyler, P. A., J. D. Gage, G. J. L. Paterson, and A. L. Rice. 1993. Dietary constraints on reproductive periodicity in two sympatric deep-sea astropectinid seastars. *Marine Biology* 115:267–277.

Ummels, F. 1963. Asteroids from the Netherlands Antilles and Caribbean localities. *Studies on the Fauna of Curaçao* 15:72–101.

Underwood, A. J, E. J. Denley, and M. J. Moran. 1983. Experimental analyses of the structure and dynamics of mid-shore rocky intertidal communities in New South Wales. *Oecologia* 56:202–219.

Unger, H. 1962. Experimented und histologische Untersuchungen uber Wirkfaktoren aus dem Nervensystem von *Asterias glacialis*. *Zoologische Jahrbucher Abteilung fuer Allgemeine Zoologie* 69:481–536.

Urriago, J. D., J. H. Himmelman and C. F. Gaymer. 2011. Responses of the black sea urchin *Tetrapygus niger* to its sea-star predators *Heliaster helianthus* and *Meyenaster gelatinosus* under field conditions. *Journal of Experimental Marine Biology and Ecology* 399:17–24.

Ursin, E. 1960. A quantitative investigation of the echinoderm fauna of the central North Sea. *Meddelelser fra Danmarks Fiskeri-og Havundersøgelser Ny Serie* 2(2):1–204.

Uter, A. 1967. Physiological location of shedding substance in radial nerve complex of starfish (*Asterias forbesi*). M.S. Thesis, American University, Washington, D.C.

Uthicke, S, B. Schaffelke, and M. Byrne. 2009. A boom and bust phylum? Ecological and evolutionary consequences of large population density variations in echinoderms. *Ecological Monographs* 79:3–24.

Valentinčič, T. 1983. Innate and learned responses to external stimuli in asteroids. Pp. 111–137. In *Echinoderm Studies*, vol. 1 (M. Jangoux and J. M. Lawrence, eds.). Balkema, Rotterdam, the Netherlands.

Valentine, J. W., and D. Jablonski. 1986. Mass extinctions: sensitivity of marine larval types. *Proceedings National Academy of Science* 83:6912–6914.

VanBlaricom, G. R. 1987. Effects of foraging by sea otters on mussel-dominated intertidal communities. Pp. 48–92. In *The Community Ecology of Sea Otters* (G. R. VanBlaricom and J. A. Estes, eds.). Springer-Verlag, Berlin.

Vance, D. E. and H. van den Bosch. 2000. Cholesterol in the year 2000. *Biochimica et Biophysica Acta* 1529(1–3): 1–8.

VandenSpiegel, D., D. J. W. Lane, S. Stampanato, and M. Jangoux. 1998. The asteroid fauna (Echinodermata) of Singapore, with a distribution table and an illustrated identification to the species. *Raffles Bulletin of Zoology* 46:431–470.

Van der Plas, A. J., and R. C. H. M. Oudejans. 1982. Changes in the activities of selected enzymes of intermediary metabolism in the pyloric ceca and ovaries of *Asterias rubens* during the annual reproductive cycle. *Comparative Biochemistry & Physiology B* 71:379–386.

Van der Plas, A. J., A. H. Koenderman, G. J. Deibel-van Schijndel, and P. A. Voogt. 1982. Effects of oestradiol-17 beta on the synthesis of RNA, proteins and lipids in the pyloric caeca of the female starfish *Asterias rubens*. *Comparative Biochemistry and Physiology B* 73(4):965–970.

Vasserot, J. 1961. Caractère hautement spécialisé du régime alimentaire chez les Astérides *Echinaster sepositus* et *Henricia sanguinolenta*, prédateurs de Spongiaires. *Bulletin de la Société Zoologique de France* 86:796–809.

Vázquez, E., J. Ameneiro, S. Putzeys., C. Gordo, and P. Sangrà. 2007. Distribution of meroplankton communities in the Bransfield Strait, Antarctica. *Marine Ecology Progress Series* 338:119–129.

Ventura, C. R. R. 1999. Gametogenesis and growth of two *Astropecten* species from a midshelf upwelling region, Cabo Frio, Brazil. P. 167. In *Echinoderm Research 1998* (C. Carnevali, M. D. Bonasuro, and F. Bonasuro, eds.). Balkema, Rotterdam, the Netherlands.

Ventura, C. R. R., and F. C. Fernandes. 1995. Bathymetric distribution and population size structure of paxilosid seastars (Echinodermata) in the Cabo Frio upwelling ecosystem of Brazil. *Bulletin of Marine Science* 56:268–282.

Ventura, C. R. R, A. O. R. Junqueira, and F. C. Fernandes. 1994. The relation between body size and number of prey in starfish (Echinodermata: Asteroidea). Pp. 375–380. In *Echinoderms through Time* (B. David, A. Guille, A., J.-P. Féral, and M. Roux, eds.). Balkema, Rotterdam, the Netherlands.

Ventura, C. R. R, A. P. C. Falcão, J. S. Santos, and C. S. Fiori. 1997. Reproductive cycle and feeding periodicity in the starfish *Astropecten brasiliensis* in the Cabo Frio upwelling ecosystem (Brazil). *Invertebrate Reproduction and Development* 31:135–141.

Ventura, C. R. R, J. S. Santos, A. P. C. Falcão, and C. S. Fiori. 1998. Reproduction and food intake in *Astropecten cingulatus* (Asteroidea: Paxillosida) in the upwelling environment of Cabo Frio (Brazil). Pp. 313–318. In *Echinoderms: San Francisco* (R. Mooi, and M. Telford, eds.). Balkema, Rotterdam, the Netherlands.

Ventura, C. R. R., M. C. G. Grillo, and F. Fernandes. 2001. Feeding niche breadth and feeding niche overlap of paxillosid starfishes (Echinodermata: Asteroidea) from a midshelf upwelling region, Cabo Frio, Brazil. Pp. 227–233. In *Echinoderms, 2000* (M. Barker, ed.). Swets & Zeitlinger, Lisse, the Netherlands.

Ventura, C. R. R., S. L. S. Alves, C. E. P. Maurício, and E. P. Silva. 2004. Reproduction and population genetics of *Coscinasterias tenuispina* (Asteroidea: Asteriidae) on the Brazilian coast. Pp. 73–77. In *Echinoderms: München* (T. Heinzeller and J.H. Nebelsick eds.). Balkema, Leiden, the Netherlands.

Verling, E., A. C. Crook, D. K. A. Barnes, and S. S. C. Harrison. 2003. Structural dynamics of a sea-star (*Marthasterias glacialis*) population. *Journal of the Marine Biological Association of the United Kingdom* 83:583–592.

Verrill, A. E. 1914. Monograph of the shallow-water starfishes of the North Pacific Coast from the Arctic Ocean to California. *Smithsonian Institution, Harriman Alaska Series* 14:1–428.

Vevers H. G. 1951. The biology of *Asterias rubens* L. II. Parasitization of the gonads by the ciliate *Orchitophrya stellarum* Cépède. *Journal of the Marine Biological Association of the United Kingdom* 29:169–624.

Vickery, M. S., and J. B. McClintock. 1998. Regeneration in metazoan larvae. *Nature* 394:140.

Vickery, M. S., and J. B. McClintock. 2000a. Comparative morphology of tube feet among the Asteroidea: phylogenetic implications. *American Zoologist* 40:355–364.

Vickery, M. S., and J. B. McClintock. 2000b. Effects of food concentration and availability on the incidence of cloning in planktotrophic larvae of the sea star *Pisaster ochraceus*. *Biological Bulletin* 199:298–304.

Viguier, C. 1879. Anatomie comparée du squelette des stéllérides. *Archives de Zoologie Expérimentale et Genérale, Paris* 7:33–250.

Villamor, A., and M. A. Becerro. 2010. Matching spatial distributions of the sea star *Echinaster sepositus* and crustose coralline algae in shallow rocky Mediterranean communities. *Marine Biology* 157:2241–2251.

Villier, L., D. B. Blake, J. W. M. Jagt, and M. Kutscher. 2004. A preliminary phylogeny of the Pterasteridae (Echinodermata, Asteroidea) and the first fossil record: Late Cretaceous of Belgium and Germany. *Paläontologische Zeitschrift* 78:281–299.

Villinski, J. T., J. L. Villinski, M. Byrne, and R. R. Raff. 2002. Convergent maternal provisioning and life history evolution in echinoderms. *Evolution* 56:1764–1775.

Vincent, J. 1990. *Structural Biomaterials*. Princeton University Press, Princeton, New Jersey.

Viviani, C. A. 1978. *Predación interespecíca, canibalismo y autotomía como mecanismo de escape en las especies de Asteroídea (Echinodermata) en el litoral del desierto del Norte Grande de Chile*. Laboratorio de Ecología Marina, Universidad del Norte, Iquique, Chile.

Viviani, C. A. 1979. Ecogeografía del litoral chileno. *Studies on Neotropical Fauna and Environment* 14:65–123.

Vogel, S. 2003. *Comparative Biomechanics: Life's Physical World*. Princeton University Press, Princeton, New Jersey.

Vogler C, J. A. H. Benzie, H. Lessios, P. Barber, and G. Wörheide. 2008. A threat to coral reefs multiplied? Four species of crown-of-thorns starfish. *Biology Letters* 4:696–699.

Voogt, P. A. 1973. On the biosynthesis and composition of 3β-sterols in some representatives of the asteroidea. *International Journal of Biochemistry* 4: 42–50.

Voogt, P. A. 1982. Steroid metabolism. Pp. 417–436. In *Echinoderm nutrition* (M. Jangoux, and J. M. Lawrence, eds.). Balkema, Rotterdam, the Netherlands.

Voogt, P. A., and S. J. Dieleman. 1984. Progesterone and oestrone levels in the gonads and pyloric caeca of the male sea star *Asterias rubens*: a comparison with the corresponding levels in the female sea star. *Comparative Biochemistry and Physiology A* 79(4):635–639.

Voogt, P. A., and R. Huiskamp. 1979. Sex-dependence and seasonal variation of saponins in the gonads of the starfish *Asterias rubens*: their relation to reproduction. *Comparative Biochemistry & Physiology A* 62:1049–1056.

Voogt, P. A. and H. J. N. Schoenmakers. 1973. Some aspects of the sterol metabolism in the echinoderm *Asterias rubens*. *Comparative Biochemistry and Physiology B* 45(313):509–514.

Voogt, P. A., and H. J. N. Schoenmakers. 1980. On the possible physiological functions of steroids in echinoderms. Pp. 349–358. In *Echinoderms: Past and Present* (M. Jangoux, ed.). Balkema, Rotterdam, the Netherlands.

Voogt, P. A., and J. W. van Rheenen. 1976a. On the fate of dietary sterols in the sea star *Asterias rubens*. *Comparative Biochemistry and Physiology B* 54(4):473–477.

Voogt, P. A. and J. W. van Rheenen. 1976b. On the origin of the sterols in the sea star *Asterias rubens*. *Comparative Biochemistry and Physiology B* 54(4):479–482.

Voogt, P. A., and J. W. van Rheenen. 1986. Androstenedione metabolism in the sea star *Asterias rubens* L. studied in homogenates and intact tissue: biosynthesis of the novel steroid fatty-acyl testosterone. *Comparative Biochemistry and Physiology B* 85(2):497–501.

Voogt, P. A., R. C. H. M. Oudejans, and J. J. S. Broertjes. 1984. Steroids and reproduction in starfish. Pp. 151–161. In *Advances in Invertebrate Reproduction* (W. Engels, ed.). Elsevier Science, Amsterdam.

Voogt, P. A., J. W. A. van Rheenen, J. G. D. Lambert, B. F. De Groot, and C. Mollema. 1986. Effects of different experimental conditions on progesterone metabolism in the sea star *Asterias rubens* L. *Comparative Biochemistry and Physiology B* 84:397–402.

Voogt, P. A., P. J. den Besten, and M. Jansen. 1990. The D5 pathway in steroid metabolism in the sea star *Asterias rubens*. *Comparative Biochemistry and Physiology B* 97:555–562.

Voogt, P. A., S. van Ieperen, M. B. J. C. B. van Rooyen, H. Wynne, and M. Jansen. 1991. Effect of photoperiod on steroid metabolism in the sea star *Asterias rubens* L. *Comparative Biochemistry and Physiology B* 100(1):37–43.

Voogt, P. A., J. G. D. Lambert, J. C. M. Granneman, and M. Jansen. 1992. Confirmation of the presence of oestradiol-17β in sea star *Asterias rubens* by GC-MS. *Comparative Biochemistry and Physiology* 101B(1–2):13–16.

Wada, H., M. Komatsu, and N. Satoh, N. 1996. Mitochondrial rDNA phylogeny of the Asteroidea suggests the primitiveness of the Paxillosida. *MolecularPhylogenetics and Evolution* 6:97–106.

Waddell, B., and J. R. Pawlik. 2000. Defenses of Caribbean sponges against invertebrate predators. II. Assays with sea stars. *Marine Ecology Progress Series* 195:133–144.

Wakabayashi, K., M. Komatsu, M. Murakami, I. Hori, and T. Takegami. 2008. Morphology and gene analysis of hybrids between two congeneric sea stars with different modes of development. *Biological Bulletin*, 215:89–97.

Walker, C. W. 1982. Nutrition of gametes. Pp. 449–468. In *Echinoderm Nutrition* (M. Jangoux and J. M. Lawrence, eds.). Balkema, Rotterdam, the Netherlands.

Walton, M. J., and J. F. Pennock. 1972. Some studies on the biosynthesis of ubiquinone, isoprenoid alcohols, squalene and sterols by marine invertebrates. *Biochemical Journal* 127(3):471–479.

Ward, R. D., and J. Andrew. 1995. Population genetics of the northern Pacific seastar *Asterias amurensis* (Echinodermata: Asteriidae): allozyme differentiation among Japanese, Russian and recently introduced Tasmanian populations. *Marine Biology* 124:99–109.

Wares, J. P. 2001. Biogeography of *Asterias*: North Atlantic climate change and speciation. *Biological Bulletin* 201:93–103.

Warner, G. F. 1979. Aggregation in echinoderms. Pp. 375–396. In *Biology and Systematics of Colonial Organisms* (G Larwood and B.R. Rosen, eds.). Academic Press, London.

Wasson, K. M., and T. S. Klinger. 1994. Changes in nucleic acid levels of the pyloric caeca of *Asterias forbesi* (Desor) (Asteroidea) and the gut of *Strongylocentrotus droebachiensis* (Muller) (Echinoidea) in conjunction with the annual reproductive cycle. Pp. 149–153. In *Echinoderms through Time* (B. David, A. Guille, J.-P. Féral, and M. Roux, eds.). Balkema, Rotterdam, the Netherlands.

Wasson, K. M. and S. A. Watts. 2007. Endocrine regulation of sea urchin reproduction. Pp. 55–70. In *Edible Sea Urchins: Biology and Ecology,* 2nd ed. (J. M. Lawrence, ed.). Elsevier, New York.

Waters, J. M. 2008. Marine biogeographical disjunction in temperate Australia: historical land bridge, contemporary currents, or both? *Diversity and Distributions* 14:692–700.

Waters, J. M., and M. S. Roy. 2003a. Global phylogeography of the fissiparous sea-star genus *Coscinasterias*. *Marine Biology* 142:185–191.

Waters, J. M., and M. S. Roy. 2003b. Marine biogeography of southern Australia: phylogeographical structure in a temperate sea-star. *Journal of Biogeography*. 30:1787–1796.

Waters, J. M., and M. S. Roy. 2004. Out of Africa: the slow train to Australasia. *Systematic Biology* 53:18–24.

Waters, J. M., P. M. O'Loughlin, and M. S. Roy. 2004. Cladogenesis in a starfish species complex from southern Australia: evidence for vicariant speciation? *Molecular Phylogenetics and Evolution* 32:236–245.

Watts, S. A., and J. M. Lawrence. 1986. Seasonal effects of temperature and salinity on the organismal activity of the seastar *Luidia clathrata* (Say) (Echinodermata: Asteroidea). *Marine Behavior and Physiology* 12:161–169.

Watts, S. A., and J. M. Lawrence. 1987. The effects of 17 β-estradiol and estrone on intermediary metabolism of the pyloric caeca of the asteroid *Luidia clathrata* (Say) maintained under different nutritional regimes. *Development, Growth and Differentiation*. 29:153–160.

Watts, S. A., and J. M. Lawrence. 1990a.The effect of temperature and salinity interactions on righting, feeding and growth in the sea star *Luidia clathrata* (Say). *Marine Behavior and Physiology* 17:159–165.

Watts, S. A., and J. M. Lawrence.1990b. The effect of reproductive state, temperature, and salinity on DNA and RNA levels and activities of metabolic enzymes of the pyloric caeca of the sea star *Luidia clathrata* (Say). *Physiological Zoology* 63:1196–1215.

Watts, S. A., R. E. Scheibling, A. G. Marsh, and J. B McClintock. 1982. Effect of temperature and salinity on larval development of sibling species of *Echinaster* (Echinodermata: Asteroidea) and their hybrids. *Biological Bulletin* 163:348–354.

Watts, S. A., R. E. Scheibling, A. G. Marsh, and J. B. McClintock. 1983. Induction of aberrant ray numbers in *Echinaster* sp. (Echinodermata: Asteroidea) by high salinity. *Florida Scientist* 46:125–127.

Watts, S. A., G. Hines, K. Lee, D. Jaffurs, J. Roy, F. F. Smith, and C. W. Walker. 1990a. Seasonal patterns of ornithine decarboxylase activity and levels of polyamines in relation to the cytology of germinal cells during spermatogenesis in the sea star, *Asterias vulgaris*. *Tissue & Cell* 22:435–448.

Watts, S. A., J. Roy, and C. W. Walker. 1990b. Ornithine decarboxylase exhibits negative thermal modulation in the sea star *Asterias vulgaris*: potential regulatory role during temperature-dependent testicular growth. *Biological Bulletin* 179:159–162.

Wells, H. W, M. J. Wells, and I. E. Gray. 1961. Food of the sea star *Astropecten articulatus*. *Biological Bulletin* 120:265–271.

Wiens, J. A. 1989. Spatial scaling in ecology. *Functional Ecology* 3:385–397.

Wieters, E. 2005. Upwelling control of positive interactions over mesoscales: a new link between bottom-up and top-down processes on rocky shores. *Marine Ecology Progress Series* 301:43–54.

Wieters, E. A., S. D. Gaines, S. A. Navarrete, C. A. Blanchette, and B. A. Menge. 2008. Scales of dispersal and the biogeography of marine predator-prey interactions. *American Naturalist* 171:405–417.

Wilkie, I. C. 1996 Mutable collagenous tissues: extracellular matrix as mechano-effector. Pp. 61–102. In *Echinoderm Studies,* vol. 5 (M. Jangoux and J. M. Lawrence, eds.). Balkema, Rotterdam, the Netherlands.

Wilkie, I.C., 2001. Autotomy as a prelude to regeneration in echinoderms. *Microscopy Research and Technique* 55:369–396.

Wilkie, I. C. 2002. Is muscle involved in the mechanical adaptability of echinoderm mutable collagenous tissue? *Journal of Experimental Biology* 205:159–165.

Wilkie, I. C. 2005. Mutable collagenous tissue: overview and biotechnological perspectives. Pp. 221–250. In *Marine Molecular Biotechnology,* vol. 39, *Progress in Molecular and Subcellular Biology* (V. Matranga, ed.). Springer-Verlag, Berlin.

Wilkie, I. C., G. V. R. Griffiths, and S. F. Glennie. 1990. Morphological and physiological aspects of the autotomy plane in the aboral integument of *Asterias rubens* L. (Echinodermata). Pp. 301–313. In *Echinoderm Research 1990* (C. De Ridder, P. Dubois, M.-C. Lahaye and M. Jangoux, eds.). Balkema, Rotterdam, the Netherlands.

Wilkie, I. C., R. H. Emson, and P. V. Mladenov. 1995. Autotomy mechanism and its control in the starfish *Pycnopodia helianthoides.* Pp. 137–146. In *Echinoderm Research 1995* (R. H. Emson, A. B. Smith, and A. C. Campbell, eds.). Balkema, Rotterdam the Netherlands.

Williams, S. M., and J. García-Sais. 2010. Temporal and spatial distribution patterns of echinoderm larvae in La Parguera, Puerto Rico. *Revista de Biologia Tropical* 58:81–88.

Wilson, D. P. 1978. Some observations on bipinnaria and juveniles of the starfish genus *Luidia. Journal of the Marine Biological Association of the United Kingdom* 58:467–478.

Wing, S. R., J. L. Largier, L. W. Botsford, and J. F. Quinn. 1995. Settlement and transport of benthic invertebrates in an intermittent upwelling region. *Limnology and Oceanography* 40:316–329.

Witman, J. D., and K. R. Grange. 1998. Links between rain, salinity, and predation in a rocky subtidal community *Ecology* 79:2429–2447.

Witman, J. D., S. J. Genovese, J. F. Bruno, J. W. McLaughlin, and B. I. Pavlin. 2003. Massive prey recruitment and the control of rocky subtidal communities on large spatial scales. *Ecological Monographs* 73:441–462.

Wong, M. C., and M. A. Barbeau. 2005. Prey selection and the functional response of sea stars (*Asterias vulgaris* Verrill) and to rock crabs (*Cancer irroratus* Say) preying on juvenile sea scallops (*Placopecten magellanicus* (Gmelin)) and blue mussels (*Mytilus edulis* Linnaeus). *Journal of Experimental Marine Biology and Ecology* 327:1–21.

Wood, R. L., and M. J. Cavey. 1981. Ultrastructure of the coelomic lining in the podium of the starfish *Stylasterias forreri*. *Cell and Tissue Research* 218:449–473.

World Asteroidea Database. 2012. www.marinespecies.org/asteroidea.

Worley, E. K., D. R. Franz, and G. Hendler. 1977. Seasonal patterns of gametogenesis in a North Atlantic brooding asteroid *Leptasterias tenera. Biological Bulletin* 153:237–253.

Wray, G. A. 1996. Parallel evolution of nonfeeding larvae in echinoderms. *Systematic Biology* 45:308–322.

Wulff, J. L. 1995. Sponge-feeding by the Caribbean starfish *Oreaster reticulatus. Marine Biology* 123:313–325.

Wulff, J. L. 2000. Sponge predators may determine differences in sponge fauna between two sets of mangrove cays, Belize Barrier Reef. *Atoll Research Bulletin* 477:251–263.

Wulff, J. L. 2006. Sponge systematics by starfish: predators distinguish cryptic sympatric species of Caribbean fire sponges, *Tedania ignis* and *Tedania klausi* n. sp. (Demospongiae, Poecilosclerida). *Biological Bulletin* 211:83–94.

Wulff, J. L. 2008. Collaboration among sponge species increases sponge diversity and abundance in a seagrass meadow. *Marine Ecology* 29:193–204.

Xu, R. A., and M. F. Barker. 1990a. Effect of diet on steroid levels and reproduction in the starfish *Sclerasterias mollis*. *Comparative Biochemistry & Physiology A* 96:33–40.

Xu, R. A., and M. F. Barker. 1990b. Photoperiodic regulation of oogenesis in the starfish *Sclerasterias mollis* (Hutton 1872) (Echinodermata: Asteroidea). *Journal of Experimental Marine Biology and Ecology* 141:159–168.

Xu, R. A., and M. F. Barker, 1990c. Annual changes in the steroid levels in the ovaries and the pyloric caeca of *Sclerasterias mollis* (Echinodermata: Asteroidea) during the reproductive cycle. *Comparative Biochemistry and Physiology A* 95(1):127–133.

Xu, R. A., and M. F. Barker. 1993. Effects of estrogens on gametogenesis and steroid levels in the ovaries and pyloric caeca of *Sclerasterias mollis* (Echinodermata: Asteroidea). *Invertebrate Reproduction and Development* 24:53–58.

Yakolev,Y. M. 1998. The temperature tolerance of adult sea-stars (*Asterias amurensis*) in the laboratory. P. 319. In *Echinoderms: San Francisco* (R. Mooi and M. Telford, eds.). Balkema, Rotterdam, the Netherlands.

Yamaguchi, M. 1973. Early life histories of coral reef asteroids, with special reference to *Acanthaster planci* (L.). Pp. 369–387. In *Biology and Geology of Coral Reefs: Biology 1* (O. A. Jones and R. Endean, eds.). Academic Press, New York.

Yamaguchi, M. 1974. Growth of juvenile *Acanthaster planci* (L.) in the laboratory. *Pacific Science* 28:123–38.

Yamaguchi, M. 1975. Coral-reef asteroids of Guam. *Biotropica* 7:12–23.

Yamaguchi, M., and J. S. Lucas. 1984. Natural parthenogenesis, larval and juvenile development and geographical distribution of the coral reef asteroid *Ophidiaster granifer*. *Marine Biology* 83:33–42.

Yamazi, I. 1950. Autonomy and regeneration in Japanese sea-stars and ophiurans. I. Observations on a sea-star, *Coscinasterias acutispina* Stimpson and four species of ophiurans. *Annotationes Zoologicae Japonenses* 23:175–186.

Yang, X. W., X. Q. Chen, G. A. Dong, X. F. Zhou, X. Y. Chai, Y. Q. Li, B. Yang, W. D. Zhang, and Y. H. Liu. 2011. Isolation and structural characterization of five new and 14 known metabolites from the commercial starfish *Archaster typicus*. *Food Chemistry* 124:1634–1638.

Yasuda, N., S. Nagai, M. Hamaguchi, K. Okaji, K. Gerard, and K. Nadaoka. 2009. Gene flow of *Acanthaster planci* (L.) in relation to ocean currents revealed by microsatellite analysis. *Molecular Ecology* 18:1574–590.

Yasumoto, T., and Y. Hashimoto. 1965. Properties and sugar components of asterosaponin A isolated from starfish. *Agricultural and Biological Chemistry* 29:804–808.

Yasumoto, T., and Y. Hashimoto. 1967. Properties of asterosaponin B isolated from a starfish,

Yasumoto, T., T. Watanabe, and Y. Hashimoto. 1964. Physiological activities of starfish saponin. *Bulletin of the Japanese Society for the Science of Fisheries* 30:357–364.

Yasumoto, T., M. Tanaka, and Y. Hashimoto. 1966. Distribution of saponins in echinoderms. *Bulletin of the Japanese Society of Scientific Fisheries* 32:673–976.

Yokochi, H., and M. Ogura. 1987. Spawning period and discovery of juvenile *Acanthaster planci* (L.) (Echinodermata: Asteroidea) at northwestern Iriomote-Jima, Ryukyu Islands. *Bulletin of Marine Science* 41:611–616.

Young, C. M. 1995. Behavior and locomotion during the dispersal phase of larval life. Pp. 249–277. In *Ecology of Marine Invertebrate Larvae* (L. R. McEdward, ed.). CRC Press, Boca Raton, Florida.

Young, C. M., and F.-S. Chia. 1987. Abundance and distribution of pelagic larvae as influenced by predation, behavior, and hydrographic factors. Pp. 385–463. In *Reproduction of Marine Invertebrates*, vol. 9, *General Aspects: Seeking Unity in Diversity* (A. C. Gease, J. S. Pearse and V. B. Pearse, eds.). Blackwell, Palo Alto, California.

Young, C. M., P. A. Tyler, J. L. Cameron, and S. G. Rumrill. 1992. Seasonal breeding aggregations in low-density populations of the bathyal echinoid *Stylocidaris lineata*. *Marine Biology* 113:603–612.

Young, C. M., and P. A. Tyler, and J. D. Gage. 1996. Vertical distribution correlates with pressure tolerances of early embyros in the deep-sea asteroid *Plutonaster bifrons*. *Journal of the Marine Biological Association of the United Kingdom* 76:749–757.

Zaixso, H. E., P. Stoyanoff, and D. G. Gil. 2009. Detrimental effects of the isopod, *Edotia doellojuradoi*, on gill morphology and host condition of the mussel, *Mytilus edulis platensis*. *Marine Biology* 156:2369–2378.

Zamorano, J. H., W. E. Duarte, and C. A. Moreno. 1986. Predation upon *Laternula elliptica* (Bivalvia, Anatinidae): a field manipulation in South Bay, Antarctica. *Polar Biology* 6:139–143.

Zann, L., and K. Weaver. 1988. An evaluation of crown of thorns starfish control programs undertaken on the Great Barrier Reef. Pp. 183–88. In *Sixth International Coral Reef Symposium* (J. H. Choat, D, Barnes, M. A. Borowitzka, J. C. Coll, P. J. Davies, P. Flood, and B. G. Hatcher, eds.). Organizing Committee of the Sixth International Coral Reef Symposium, Townsville, Australia.

Zann, L., J. Brodie, C. Berryman, and M. Naqasima. 1987. Recruitment, ecology, growth and behavior of juvenile *Acanthaster planci* (L.) (Echinodermata: Asteroidea). *Bulletin of Marine Science* 41:561–575.

Zann, L., J. Brodie, and V. Vuki. 1990. History and dynamics of the crown-of-thorns starfish *Acanthaster planci* (L.) in the Suva area, Fiji. *Coral Reefs* 9:135–44.

Zeidler, W. 1992. Introduced starfish pose threat to scallops. *Australian Fisheries* 51:28–29.

Zollo, F., E. Finamore, and L. Minale. 1985. Starfish saponins XXIV. Two novel steroidal glycoside sulphates from the starfish *Echinaster sepositus*. *Gazzetta Chimica Italiana* 115:303–306.

Zollo, F., E. Finamore, R. Riccio, and L. Minale. 1989. Starfish saponins. XXXVII. Steroidal glycoside sulfates from starfishes of the genus *Pisaster*. *Journal of Natural Products* 52:693–700.

Zolotarev, P. 2002. Population density and size structure of sea stars on beds of Iceland scallop, *Chlamys islandica*, in the southeastern Barents Sea. *Sarsia* 87:91–96.

Zulliger, D. 2009. Phylogeography, evolutionary history and genetic diversity of sea stars of the genus Astropecten and genetic structure within the Atlanto-Mediterranean species A. aranciacus. Ph.D. Thesis, University of Zürich, Zürich, Switzerland.

Zulliger, D. E., and H. A. Lessios. 2010. Phylogenetic relationships in the genus *Astropecten* (Paxillosida: Astropectinidae) on a global scale: molecular evidence for morphological convergence, species-complexes and possible cryptic speciation. *Zootaxa* 2504:1–19.

Index